AF293968

Birkhäuser

Frontiers in Mathematics

Advisory Editors

Laurent Saloff-Coste, Cornell University, Ithaca, NY, USA

Igor Shparlinski, The University of New South Wales, Sydney, NSW, Australia

Wolfgang Sprößig, TU Bergakademie Freiberg, Freiberg, Germany

This series is designed to be a repository for up-to-date research results which have been prepared for a wider audience. Graduates and postgraduates as well as scientists will benefit from the latest developments at the research frontiers in mathematics and at the "frontiers" between mathematics and other fields like computer science, physics, biology, economics, finance, etc. All volumes are online available at SpringerLink.

Anatolij Dvurečenskij · Omid Zahiri ·
Mona Aaly Kologani · Rajab Ali Borzooei

Hoop Algebras

An Introduction

Birkhäuser

Anatolij Dvurečenskij
Mathematical Institute
Slovak Academy of Sciences
Bratislava, Slovakia

Omid Zahiri
Mathematical Institute
Slovak Academy of Sciences
Bratislava, Slovakia

Mona Aaly Kologani
Hatef Higher Education Institute
Zahedan, Iran

Rajab Ali Borzooei
Department of Mathematics
Shahid Beheshti University
Teheran, Iran

ISSN 1660-8046 ISSN 1660-8054 (electronic)
Frontiers in Mathematics
ISBN 978-3-032-11735-9 ISBN 978-3-032-11736-6 (eBook)
https://doi.org/10.1007/978-3-032-11736-6

Mathematics Subject Classification: 06C15, 06D35

© The Editor(s) (if applicable) and The Author(s), under exclusive license to Springer Nature Switzerland AG
2026

This work is subject to copyright. All rights are solely and exclusively licensed by the Publisher, whether the whole or part of the material is concerned, specifically the rights of translation, reprinting, reuse of illustrations, recitation, broadcasting, reproduction on microfilms or in any other physical way, and transmission or information storage and retrieval, electronic adaptation, computer software, or by similar or dissimilar methodology now known or hereafter developed.
The use of general descriptive names, registered names, trademarks, service marks, etc. in this publication does not imply, even in the absence of a specific statement, that such names are exempt from the relevant protective laws and regulations and therefore free for general use.
The publisher, the authors and the editors are safe to assume that the advice and information in this book are believed to be true and accurate at the date of publication. Neither the publisher nor the authors or the editors give a warranty, expressed or implied, with respect to the material contained herein or for any errors or omissions that may have been made. The publisher remains neutral with regard to jurisdictional claims in published maps and institutional affiliations.

This book is published under the imprint Birkhäuser, www.birkhauser-science.com by the registered company Springer Nature Switzerland AG
The registered company address is: Gewerbestrasse 11, 6330 Cham, Switzerland

If disposing of this product, please recycle the paper.

Preface

The study of algebraic structures has been a cornerstone of modern mathematics, providing a powerful framework for understanding logical systems, ordered structures, and their applications across diverse fields, including computer science, theoretical physics, and artificial intelligence. Among these structures, hoops stand out as a fascinating and rich subject of study, bridging the worlds of residuated lattices, BL-algebras, MV-algebras, and related systems.

Hoop algebras, initially introduced by B. Bosbach [48, 49] under the name "complemented semigroups" were further explored by J. R. Büchi and T. M. Owens in [59], in their unpublished work. These algebraic structures hold significant importance in both universal algebra and algebraic logic. Hoops find diverse applications, particularly within the realm of fuzzy logic and non-classical logic. F. Esteva et al. [138] provided axiomatizations for logics corresponding to hoops, which are now known as hoop logics. Since their introduction, hoops have been extensively studied by numerous researchers, leading to a rich and multifaceted theory.

This book, *Hoop Algebras: An Introduction*, provides a systematic and comprehensive exploration of hoops and their theoretical underpinnings. This book is designed for self-study and serves as both a foundational reference for newcomers to the field and a detailed guide for researchers seeking to deepen their understanding of ordered algebraic systems.

The journey begins with the fundamental concepts of algebraic and lattice-theoretical structures, laying the necessary groundwork for understanding the more advanced topics presented in later chapters. As the book progresses, we delve into the intricate structure of hoops, their filters, congruences, subvarieties, and their relationships with other important algebraic structures. We also devote significant attention to the lattice-theoretical properties of hoops, investigating their lattice-theoretic structure, including concepts like simple hoops, semisimple hoops, and subdirectly irreducible hoops. Furthermore, we explore the emerging field of state hoops, emphasizing the interplay between algebraic and logical perspectives.

This book presents a comprehensive overview of hoop theory, drawing upon years of research and reflection by prominent scholars in the field. While preparing this manuscript, we have aimed to provide a balanced combination of theory, examples, and exercises to ensure accessibility while maintaining mathematical rigor. At the end of each chapter, additional and more advanced resources are suggested for readers seeking further exploration. The book is organized as follows:

Chapter 1 establishes a foundation by introducing fundamental concepts in lattices and partially ordered sets, algebraic structures (including varieties and their related theorems), and ℓ-groups, and MV-algebras, along with their respective notations. Finally, we gather some fundamental results concerning equational classes and identities, including Birkhoff's theorem, which elucidates the profound connection between equational classes and varieties. A thorough understanding of these foundational concepts, particularly ℓ-groups, MV-algebras, and BCK-algebras, is essential for grasping the material covered in the subsequent chapters.

Chapter 2 commences by establishing the formal definition of pocrims and exploring a range of illustrative examples, setting the stage for a deeper investigation. The second section of the chapter is dedicated to unraveling the main properties of pocrims, including boundedness, involutivity, and the double negation property (DNP), providing essential tools for their analysis. A significant contribution is the introduction of the ordinal sum of pocrims, a powerful technique for understanding their structural composition. Furthermore, we examine the relationships between filters and congruence relations, crucial for algebraic investigations. Finally, we broaden our scope to encompass BCK-algebras, demonstrating that the quasivariety of pocrims is termwise equivalent to a specific subclass of BCK-algebras with the product property, thus highlighting the connections between these algebraic structures. For those acquainted with BCK-algebras, this equivalence offers a more insightful understanding of pocrims and their applications and properties.

Chapter 3 introduces hoops, a class of algebras derived from pocrims, defining them as naturally ordered pocrims forming a variety, whereas the class of pocrims does not. It establishes that hoops are meet-semilattices and explores their fundamental properties and the relationships between their operations. The chapter discusses when hoops are lattices, introduces $\vee$-hoops and $\sqcup$-hoops, and examines bounded hoops derived from the elements of their join-centers and cancellative-centers. It then details various significant subclasses like cancellative, Wajsberg, and basic hoops, including their properties, interrelationships, and the categorical equivalence of cancellative hoops with Abelian ℓ-groups. Every cancellative hoop is a Wajsberg hoop, both product hoops and Wajsberg hoops are basic hoops, and every basic hoop is a $\vee$-hoop. Finally, it provides a range of examples to illustrate hoop theory.

Chapter 4 delves into the intricate relationship between congruence relations and filters within the context of hoops. The chapter meticulously examines the properties of prime, maximal, implicative, and positive implicative filters, along with their interconnections.

The study of these filters is important, as each corresponds to a specific type of congruence. While the existence of maximal filters is not guaranteed in hoops, every bounded hoop possesses at least one maximal filter. We define two distinct types of prime filters, which are equivalent in basic hoops. Furthermore, every proper filter of a basic hoop is the intersection of all prime filters containing it. These investigations provide valuable insights into the underlying structure of hoops and the quotient hoops corresponding to each type of filter. For instance, quotient hoops associated with implicative filters are idempotent hoops, while quotient hoops associated with fantastic filters are Wajsberg hoops.

Chapter 5 explores the intricate connections between hoops and various other well-established algebraic structures, including BL-algebras, MV-algebras, BCK-algebras, Wajsberg algebras, Heyting algebras, and L-algebras. Specifically, hoops are termwise equivalent to BCK-algebras with special conditions, MV-algebras are termwise equivalent to bounded Wajsberg hoops, and BL-algebras are termwise equivalent to bounded basic hoops. These equivalences provide powerful tools for understanding hoops by leveraging the established theories of these related algebras. These relationships can be beneficial for individuals interested in any of the aforementioned algebraic structures.

Chapter 6 explores more advanced topics in the algebraic and lattice structures of hoops. The chapter delves into the structural analysis of hoops, focusing on specific classes and their representations in seven sections. We begin by exploring ordinal sums and $\oplus$-irreducible hoops, characterizing linear $\oplus$-irreducible hoops as Wajsberg hoops and demonstrating that linear Wajsberg hoops are either cancellative or bounded. A key theorem establishes that any linearly ordered hoop is an ordinal sum of Wajsberg hoops, leading to representations for BL-algebras and cancellative hoops. We continue with a study of simple, semisimple, perfect, and local hoop algebras. Simple hoops are linearly ordered Wajsberg hoops, and semisimple hoops are Wajsberg hoops, showing that every finite Wajsberg hoop is semisimple. Local and perfect hoops are studied in the third section of this chapter, gathering their basic properties and examples. Representable hoops, explored in the fourth section, are shown to be equivalent to basic hoops, leading to the conclusion that varieties of cancellative, Wajsberg, or product hoops are generated by their linearly ordered elements. A representation for Wajsberg hoops based on ultrafilters of bounded Wajsberg hoops is studied in the sixth section. Moreover, we demonstrate that every Wajsberg hoop is either bounded or can be regarded as an ultrafilter of a bounded Wajsberg hoop, which is equivalent to an MV-algebra. Finally, we aim to provide a brief study of the variety of hoops. We begin by examining some fundamental properties, such as the congruence extension property (CEP) and arithmeticity. We then demonstrate that the lattice of varieties of hoops possesses exactly two atoms, namely $V(\mathbf{L}_1)$ and $V(\mathbf{L}_\infty)$. Following an investigation of basic results concerning the variety, we establish that the variety of all hoops has the finite embeddability property (FEP).

Chapter 7 discusses the state hoops and state-morphism hoops. The notions of state operator, strong state operator, state-morphism operator, and weak state-morphism operator are introduced, and their properties are investigated. We will see that every strong state hoop is a state hoop and any state operator on an idempotent hoop is a weak state-morphism operator. Glivenko's property is defined, and it is proved that for an idempotent hoop $\mathbf{H}$ having these properties a state operator on $\mathrm{Reg}(\mathbf{H})$ can be extended to a state operator on $\mathbf{H}$. Every perfect hoop admits a non-trivial state operator.

The book contains two appendices. In Appendix A, we concentrate on pseudo MV-algebras, showing their considerable complexity that can also be studied as pseudo hoops or pseudo BL-algebras. It presents an advanced setup for some non-commutative algebraic structures with possible applications in non-commutative logic and reasoning. This chapter introduces pseudo MV-algebras as a non-commutative generalization of MV-algebras. A representation theorem is then established for this structure, expressing it as $\mathbf{0}(\mathbf{G}, u)$, where $(\mathbf{G}, u)$ is a unital ℓ-group. Subsequently, the chapter demonstrates the categorical equivalence between the category of pseudo MV-algebras and the category of unital ℓ-groups. Further exploration delves into perfect and n-perfect pseudo MV-algebras, with a particular focus on symmetric and non-symmetric perfect MV-algebras. Building upon this foundation, the kite construction is employed to introduce the concepts of pseudo BL-algebras and pseudo hoops. The chapter then culminates with an examination of the Loomis-Sikorski Theorem for σ-complete MV-algebras. This theorem asserts that for every σ-complete MV-algebra $\mathbf{M}$, there exist a tribe $\mathcal{T} = \mathcal{T}(\Omega)$ of fuzzy sets on a compact Hausdorff topological space Ω and an MV-σ-homomorphism h from $\mathcal{T}$ onto $\mathbf{M}$. Finally, the chapter concludes with an investigation of pseudo EMV-algebras, a generalization encompassing both pseudo MV-algebras and generalized Boolean algebras.

Appendix B presents a comprehensive compilation of all pocrims of order five or smaller. This resource serves as a valuable reference for readers, aiding them in constructing finite pocrims by utilizing direct products and ordinal sums.

We are deeply grateful to our colleagues, mentors, and students who have supported us in this endeavor, offering their valuable feedback and insights. Their encouragement and collaboration have been instrumental in shaping this work.

We hope that this book will inspire further research and exploration in the field of hoop algebra and contribute to its ongoing development. Whether you are a student encountering these ideas for the first time or a seasoned researcher delving into advanced topics, we trust you will find this book both informative and engaging.

While following the chapters sequentially is the primary recommendation for reading this book, readers who prefer a different approach can utilize the following diagram as an alternative reading path:

Thank you for joining us on this journey into the world of hoop algebras and related structures. We gratefully acknowledge partial support from the following organizers:

- Mathematical Institute of the Slovak Academy of Sciences by the grant of the Slovak Research and Development Agency under contract APVV-20-0069 and the grant VEGA No. 2/0128/24 SAV, Anatolij Dvurečenskij—Bratislava, Slovakia.

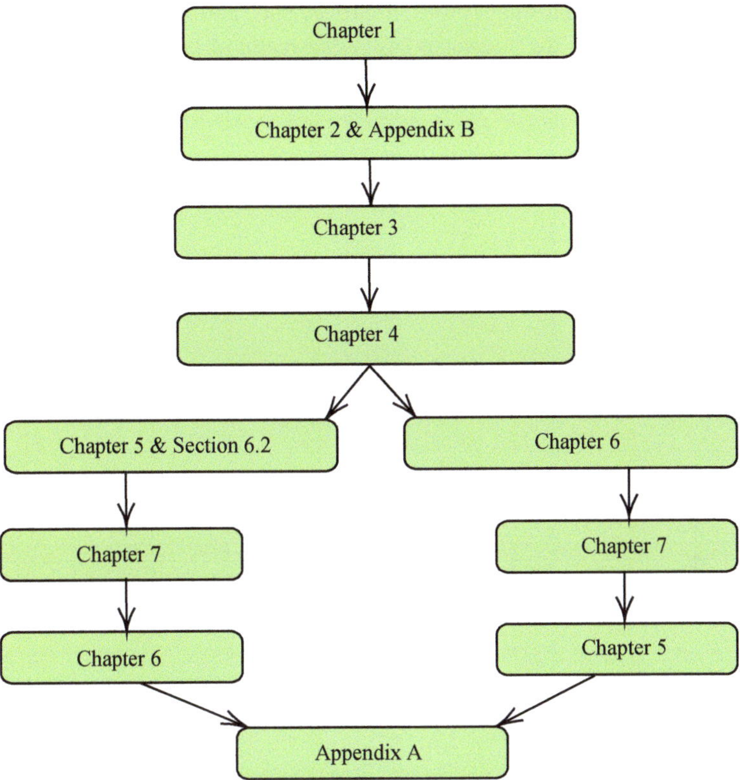

Fig. 1 Alternative approaches to studying this book

- Mathematical Institute of the Slovak Academy of Sciences and the European Union's Horizon 2020 Research and Innovation Programme based on the Grant Agreement under the Marie Skłodowska-Curie funding scheme No. 945478—SASPRO2, Omid Zahiri—Bratislava, Slovakia.
- Soft Computing Center, Department of Mathematics, Faculty of Mathematical Sciences, Shahid Beheshti University, Mona Aaly Kologani and Rajab Ali Borzooei—Tehran, Iran.

We are also thankful to our colleagues at the Mathematical Institute of the Slovak Academy of Sciences for their support, and to the members of the Soft Computing Center of Shahid Beheshti University, who read and reviewed the early versions of this book and provided valuable suggestions, improvements, and corrections that enhanced its readability and deepened the understanding of the realm of hoops.

Bratislava, Slovakia   Anatolij Dvurečenskij
September 2025   Omid Zahiri

Competing Interests Anatolij Dvurečenskij has a received research grant from Mathematical Institute of the Slovak Academy of Sciences by the grant of the Slovak Research and Development Agency under contract APVV-20-0069 and the grant VEGA No. 2/0128/24 SAV.

Omid Zahiri* has a received research grant from Mathematical Institute of the Slovak Academy of Sciences and the European Union's Horizon 2020 Research and Innovation Programme based on the Grant Agreement under the Marie Skłodowska-Curie funding scheme No. 945478—SASPRO2.

Mona Aaly Kologani and Rajab Ali Borzooei are member of Soft Computing Center, Department of Mathematics, Faculty of Mathematical Sciences, Shahid Beheshti University—Tehran, Iran.

Ethics Approval Does not apply to this manuscript.

Contents

Basic Notions **1**

This chapter provides a concise overview of fundamental concepts, designed for readers with a strong background in the subject. We assume familiarity with posets, lattices, and universal algebra. We begin by establishing the foundational definitions of partially ordered sets and algebraic structures, laying the groundwork for exploring the interaction between order and algebra. We then proceed to examine semilattices and lattices, essential tools for analyzing more intricate algebraic systems. Subsequently, we explore MV-algebras and ℓ-groups, emphasizing their relevance to many-valued logic. The study of congruence relations, classes of algebras, and varieties enables us to classify and analyze algebraic structures. Finally, we gather some fundamental results concerning equational classes and identities, including Birkhoff's theorem, which elucidates the profound connection between equational classes and varieties. Altogether, this chapter lays a solid foundation for the developments in the subsequent chapters.

1.1 Partially Ordered Sets

A relation $\leq$ on a set P is said to be a **partially ordered relation** or an **order relation** if it is reflexive, anti-symmetric, and transitive. A **partially ordered set**, or **poset** for short, is a pair $(P, \leq)$ where $\leq$ is an order relation on P. If there is no risk of ambiguity, we say that P is a poset.

Given a set X, the powerset $\mathcal{P}(X)$, consisting of all subsets of X, is ordered by set theoretical inclusion $\subseteq$, i.e., $(\mathcal{P}(X), \subseteq)$ is a partially ordered set. On the other hand, the relation $x \leq y$ if and only if $x = y$, is a partially ordered relation on X which is called a **trivial order relation** or **discrete order relation**.

In a poset $(P, \leq)$, if $x, y \in P$ such that $x \leq y$ and $x \neq y$, then we write $x < y$. An element $y \in P$ is called a **cover** of an element $x \in P$ if (1) $x < y$, and (2) for every $z \in P$,

© The Author(s), under exclusive license to Springer Nature Switzerland AG 2026

A. Dvurečenskij et al., *Hoop Algebras*, Frontiers in Mathematics,
https://doi.org/10.1007/978-3-032-11736-6_1

if $x \leq z < y$, then $x = z$. In other words, there is no element strictly between x and y. If $x \leq y$ we can also write $y \geq x$, similarly, if $x < y$, we may also write $y > x$.

For each poset $(P, \leq)$, the relation R is defined by $(x, y) \in R$ if and only if $y \leq x$ is a partial order relation on P which is denoted by $\leq_d$. It is called a **dual relation** of $\leq$, and the poset $(P, \leq_d)$ is said to be a dual poset of $(P, \leq)$. Trivially, if $(P, \leq)$ is a poset and $Q \subseteq P$, then the intersection of $\leq$ with $Q \times Q$, i.e., $(Q \times Q) \cap \leq$, is a partially ordered relation on Q which is denoted by $\leq_Q$ or simply $\leq$ on Q.

Definition 1.1.1 Assume that $(P, \leq)$ is a partially ordered set and $Q \subseteq P$.

(i) The element $a \in Q$ is said to be the **greatest** (or **top**) element of Q and denoted by $\max Q = a$, if for any $x \in Q, x \leq a$.

(ii) The element $b \in Q$ is said to be the **least** element of Q and is denoted by $\min Q = b$, if for any $x \in Q, b \leq x$.

(iii) The element $a \in P$ is called an **upper bound** of Q in P if for any $x \in Q, x \leq a$ and the element $b \in P$ is called a **lower bound** of Q in P if for any $x \in Q, b \leq x$. The set of all upper bounds of Q in P is denoted by $\mathcal{U}(Q)$ and the set of all lower bounds of Q in P is denoted by $\mathcal{L}(Q)$.

(iv) If $\min \mathcal{U}(Q)$ exists, then it is called a **supremum of the set** Q in P and is denoted by $\sup Q$ or $\bigvee Q$. In a similar way, if $\max \mathcal{L}(Q)$ exists, then it is called an **infimum of the set** Q in P, denoted by $\inf Q$ or $\bigwedge Q$. Clearly, $\sup Q$ and $\inf Q$ are the least upper bound and the greatest lower bound of the set Q in P, respectively.

(v) The element $a \in P$ is called a **maximal** element if there is no element $x \in P$ such that $a < x$.

(vi) The element $a \in P$ is called a **minimal** element if there is no element $x \in P$ such that $x < a$.

A poset $(P, \leq)$ is called **bounded** if $\min P$ and $\max P$ exist. Usually, $\max P$ is denoted by 1 and $\min P$ is denoted by 0.

Definition 1.1.2 Let $(P, \leq)$ be a bounded poset with the least element 0 and top element 1. A mapping $n : P \to P$ is called a **weak negation** on P if

(i) $x \leq y$ implies $n(y) \leq n(x)$.

(ii) $x \leq n(n(x))$.

(iii) $n(0) = 1$.

If $n(n(x)) = x$ for all $x \in P$, then n is a **strong negation**, or **double negation** and is also referred to as an **involution**.

In a poset $(P, \leq)$, for any subset $X \subseteq P$, we define

$$\downarrow X := \{y \in P : \exists x \in X \text{ s.t } y \leq x\}, \quad \uparrow X := \{y \in P : \exists x \in X \text{ s.t } x \leq y\}.$$

Clearly, $X \subseteq \downarrow X$ and $X \subseteq \uparrow X$. The subset X is said to be a **lower (upper) set** if $\downarrow X = X$ ($\uparrow X = X$). A lower set is also called a **down-set**, and an upper set is called an **up-set**. In addition, if $x \leq y$ are elements of $(P, \leq)$, then we define the following intervals on $(P, \leq)$:

$$[x, y] = \{a \in P : x \leq a \leq y\}, \quad [x, y) = [x, y] \setminus \{y\},$$
$$(x, y] = [x, y] \setminus \{x\}, \quad (x, y) = [x, y] \setminus \{x, y\}.$$

There is an interesting point in the definitions given above. The concepts of maximum and minimum (supremum and infimum, maximal and minimal) are dual to each other. That means $\max Q$ ($\sup Q$) in a partially ordered set $(P, \leq)$ is in fact, $\min Q$ ($\inf Q$) in the poset $(P, \leq_d)$.

There is a visual method for displaying partially ordered sets called Hasse diagrams. We will mention it briefly below.

To draw a Hasse diagram, we do the following:

(1) Every element $x \in P$ is shown by a point.
(2) If y is a cover of x, then the point x is lower of y and we draw a line from x to y.
(3) If there is a bottom-up path from x to y, then $x \leq y$. Otherwise, x and y are incomparable.

For example, consider the partially ordered relation $\leq$ on the set $X = \{a, b, c, d\}$.

$$\leq := \{(x, x) : x \in X\} \cup \{(a, b), (b, d), (a, c), (c, d), (a, d)\}.$$

The Hasse diagram of the above poset $(X, \leq)$ is shown in Fig. 1.1:

Definition 1.1.3 Suppose $(P, \leq)$ and $(Q, \leq')$ are two posets. Then the map $f : P \to Q$ is called a **homomorphism of partially ordered sets** (or **monotone map**) or an **order-preserving map** if for any $x, y \in P$, $x \leq y$ implies $f(x) \leq' f(y)$.

Fig. 1.1 Hasse diagram of the partially ordered set $(X, \leq)$

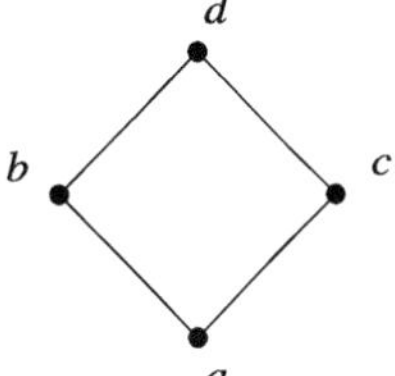

Clearly, for each poset $(P, \leq)$ the identity map $\mathrm{Id}_P : P \to P$ is a homomorphism of partially ordered sets.

A homomorphism $f : P \to Q$ is called an **isomorphism** if there exists a homomorphism $g : Q \to P$ such that $f \circ g = \mathrm{Id}_Q$ and $g \circ f = \mathrm{Id}_P$. Easily, we can see that, f is an isomorphism of partially ordered sets if f is a bijection function and for any $x, y \in P$, $f(x) \leq' f(y)$ if and only if $x \leq y$.

A **linearly ordered set** or **totally ordered set**, or **chain** is a poset $(P, \leq)$ such that for each $x, y \in P$ we have $x \leq y$ or $y \leq x$, that is every pair of elements of P is comparable with respect to $\leq$. The set of real numbers $\mathbb{R}$, rational numbers $\mathbb{Q}$, integer numbers $\mathbb{Z}$, natural numbers $\mathbb{N}$ as well as non-negative integers $\mathbb{N}_0$, with the natural ordering, are examples of linearly ordered sets.

Lemma 1.1.4 (Zorn's lemma) *If $(P, \leq)$ is a non-empty poset such that every chain in P has an upper bound in P, then P contains a maximal element.*

1.2   Algebraic Structures

Definition 1.2.1 Let A be a set and n be a non-negative integer. Each function $\lambda : A^n \to A$ is called an n-**ary operation** and n is the arity or rank of λ. The set of all n-ary operations on A is denoted by $F^n(A)$ and every element of the set $F = \bigcup\{F^n : n \in \mathbb{N}_0\}$ is said to be a **finitary operation** on A. If λ is an n-ary operation on A, then we write $\mathrm{arity}(\lambda) = n$.

An operation f on A is called a **nullary** operation or **constant** if its arity is zero. Given a set A, clearly, $A^0 = \{\emptyset\}$. Now, if $\lambda \in F^0(A)$, then this function is identified exactly by its image on the member $\emptyset$, which means $\lambda(\emptyset) \in A$. For simplicity, we usually combine this function with its image. Therefore, each nullary operation defines a member of the set A, and from now on, if $a \in A$, then when we write a 0-ary operation a, we mean that it is a function where $\lambda(\emptyset) = a$. So, we conclude that $F^0(A) = A$. An operation λ on A is **unary**, or **binary** if its arity is 1 or 2, respectively.

Definition 1.2.2 A set $\mathcal{F}$ of function symbols such that a non-negative integer n is assigned to each member λ of $\mathcal{F}$ is called a **type of algebra** and the assigned integer is called the **arity** or **rank** of λ; and λ is said to be an n-ary function symbol. This defines a function

$$\mathrm{arity} : \mathcal{F} \to \mathbb{N}_0, \quad f \mapsto \mathrm{arity}(f), \quad \forall f \in \mathcal{F}.$$

For any non-negative integer number n, the set of all symbols of n-ary functions in $\mathcal{F}$ is denoted by $\mathcal{F}^n$. Thus, $\mathcal{F} = \bigcup\{\mathcal{F}^n : n \in \mathbb{N}_0\}$.

Definition 1.2.3 Suppose $\mathcal{F}$ is a type of algebra, A is a set, and F is a family of finitary operations on A indexed by $\mathcal{F}$ such that, corresponding to any n-ary function $f \in \mathcal{F}$, there exists an n-ary operation $f^A \in F$. Then the pair $\mathbf{A} = (A, F)$ is called an **algebraic structure of type** $\mathcal{F}$ or simply an **algebra of type** $\mathcal{F}$. The set A, is said to be the **universe** or **underlying set** of (A, F), so the algebraic structure (A, F) is shown by $\mathbf{A}$. If $\mathcal{F}$ is finite, for instance $\mathcal{F} = \{\lambda_1^A, \lambda_2^A, \ldots, \lambda_n^A\}$, then the algebraic structure (A, F) is shown by $(A; \lambda_1^A, \lambda_2^A, \ldots, \lambda_n^A)$ such that

$$\text{arity}(\lambda_1^A) \geq \text{arity}(\lambda_2^A) \geq \cdots \geq \text{arity}(\lambda_n^A).$$

In such a case, $(A; \lambda_1^A, \lambda_2^A, \ldots, \lambda_n^A)$ is an algebraic structure of type

$$\left(\text{arity}(\lambda_1^A), \text{arity}(\lambda_2^A), \ldots, \text{arity}(\lambda_n^A)\right).$$

An algebraic structure (A, F) is **trivial** if $|A| = 1$.

As an example, suppose A is a non-empty set and $\mathcal{F} = \{+, \oplus, \vee, ', a_1, a_2, a_3\}$ such that $+, \oplus$ and $\vee$ are binary operations, $'$ is unary, a_1, a_2 and a_3 are nullary operations on A, and the algebraic structure (A, F) is denoted by $(A; +, \oplus, \vee, ', a_1, a_2, a_3)$ which is an algebra of type $(2, 2, 2, 1, 0, 0, 0)$.

Definition 1.2.4 Let $\mathbf{A} = (A, F^A)$ and $\mathbf{B} = (B, F^B)$ be algebras of the same type $\mathcal{F}$. Then $\mathbf{B}$ is a **subalgebra of A** and is written $\mathbf{B} \leq \mathbf{A}$ if every operation of $\mathbf{B}$ is the restriction of the corresponding operation of $\mathbf{A}$, that is, for each function symbol $\lambda \in \mathcal{F}$, λ^B is λ^A restricted to B, i.e.,

$$\forall n \in \mathbb{N}, \ \forall \lambda \in \mathcal{F}^n, \quad \lambda^B = \lambda^A \big|_{B^n}.$$

A **subuniverse** of $\mathbf{A}$ is a subset B of A which is closed under the operations of $\mathbf{A}$, that is

$$\forall n \in \mathbb{N}_0 \ \forall \lambda^A \in \mathcal{F}^n \ \forall (b_1, \ldots, b_n) \in B^n \ \lambda^A(b_1, b_2, \ldots, b_n) \in B.$$

The set of all subalgebras of $\mathbf{A}$ is denoted by $\text{Sub}(\mathbf{A})$.

According to Definition 1.2.4, given every subalgebra $\mathbf{B}$ of an algebra $\mathbf{A}$ the set B is a subuniverse of $\mathbf{A}$.

Example 1.2.5 ([175]) (i) A **semigroup** is an algebraic structure $(A; *)$ of type (2) such that $*$ is an associative binary operation, that is $x * (y * z) = (x * y) * z$ for all $x, y, z \in A$. A semigroup $(A; *)$ is called **commutative** if $*$ is a commutative binary operation, i.e., $x * y = y * x$ for all $x, y \in A$.

(ii) A **monoid** is an algebraic structure $(A; *, e)$ of type $(2, 0)$ such that $(A; *)$ is a semigroup and $e * x = x * e = x$ for all $x \in A$. A monoid $(A; *, e)$ is commutative if the semigroup $(A; *)$ is commutative. The element e is called the **identity** or **neutral element** of the monoid $(A; *, e)$. Typically, when the binary operation of a monoid is represented using the additive symbol, the identity element is denoted by 0. Also, when the binary operation is represented using the multiplicative symbol, the identity element is denoted by 1.

(iii) A **group** is an algebraic structure $(G; *, ^{-1}, e)$ of type $(2, 1, 0)$ such that $(G; *, e)$ is a monoid and $x * x^{-1} = e = x^{-1} * x$ for all $x \in G$. A group $(G; *, ^{-1}, e)$ is said to be **Abelian** if $(G; *)$ is a commutative semigroup.

Example 1.2.6 ([175]) A **ring** is an algebraic structure $(R; +, \cdot, -, 0)$ of type $(2, 2, 1, 0)$ such that $(R; +, -, 0)$ is an Abelian group and $(R; \cdot)$ is a semigroup satisfying the condition $x \cdot (y + z) = (x \cdot y) + (x \cdot z)$ and $(y + z) \cdot x = (y \cdot x) + (z \cdot x)$ for all $x, y, z \in R$. A ring is said to be **commutative** if $x \cdot y = y \cdot x$ for all $x, y \in R$. Furthermore, if a ring $(R; +, \cdot, -, 0)$ contains an element 1 such that $1 \cdot x = x \cdot 1 = x$ for all $x \in R$, then R is a ring with identity. A **commutative integral domain** is an algebraic structure $(R; +, \cdot, -, 0, 1)$ of type $(2, 2, 1, 0, 0)$ where $(R; +, \cdot, -, 0)$ is a ring with the identity 1 such that $x \cdot y = 0$ implies that $x = 0$ or $y = 0$ for all $x, y \in R$.

Remark 1.2.7 Suppose $\mathbf{A} = (A, F)$ is an algebra of type $\mathcal{F}$ and $B \subseteq A$.

(i) By definition, it is obvious that for each nullary operation $a \in \mathcal{F}^n$, B is closed under a if and only if $a \in B$.

(ii) If B is a subuniverse of $\mathbf{A}$, then for any $n \in \mathbb{N}_0$ and every n-ary operation $\lambda^A \in \mathcal{F}^n$, the restriction of the operation λ^A to the set B^n, means $\lambda^B := \lambda^A|_{B^n}$ is an n-ary operation on B. Set

$$F' := \bigcup_{n \in \mathbb{N}_0} \{\lambda^A|_{B^n} : \lambda \in \mathcal{F}^n\}.$$

Then (B, F') is an algebra of type $\mathcal{F}$ and also a subalgebra of $\mathbf{A}$. Clearly, for any $n \in \mathbb{N}_0$, the n-ary operations corresponding to $\lambda \in \mathcal{F}^n$ in (A, F) and (B, F') are λ^A and $\lambda^B = \lambda^A|_{B^n}$, respectively.

To simplify notation and avoid cluttering proofs with numerous symbols, we will denote F' by F. Additionally, we will use the same symbol for the n-ary operations λ^A and λ^B. However, when working with these operations, we will always be mindful of the underlying set, whether it is A or B). Furthermore, if there is no risk of ambiguity, we write that B is a subalgebra of $\mathbf{A}$ while actually referring to (B, F) as a subalgebra of $\mathbf{A}$.

In a similar way, if $\mathbf{C} = (C, F^C)$ is a subalgebra of $\mathbf{A}$, then C is a subuniverse of $\mathbf{A}$. For simplicity, we prefer to use the same symbol F for the operator symbol of $\mathbf{C}$ instead of F^C.

(iii) Suppose (A, F) is an algebra of type $\mathcal{F}$. If $\mathcal{F}^0 = \emptyset$, then clearly, $\emptyset$ is a subalgebra of (A, F). Now, if $\mathcal{F}^0 \neq \emptyset$, then F contains at least one 0-ary operation such as $a \in A$. Since $a \notin \emptyset$, the empty set is not a subalgebra of (A, F).

(iv) Every set is an algebra of type $\mathcal{F} = \emptyset$, i.e., there is not any operation.

(v) Let (A, F^A) be an algebra of type $\mathcal{F}$. Then the intersection of any family of subalgebras of A is a subalgebra of A.

Definition 1.2.8 An algebra (A, F) of type $\mathcal{F}$ is a **reduct** of algebra (B, F') of type $\mathcal{F}'$ if $\mathcal{F} \subseteq \mathcal{F}'$ and F is restriction of F' to $\mathcal{F}$.

Indeed, a reduct of an algebraic structure is obtained by omitting some of the operations of that structure. For example, the monoid $(\mathbb{Z}; +, 0)$ is a reduct of the group $(\mathbb{Z}; +, -, 0)$.

Definition 1.2.9 Assume that $\mathbf{A} = (A, F)$ is an algebra of type $\mathcal{F}$ and $X \subseteq A$. The smallest subalgebra of $\mathbf{A}$ containing X is called the **generated subuniverse by** X or the subuniverse of $\mathbf{A}$ **generated by** X and is denoted by $\mathrm{Alg}(X)$. By Remark 1.2.7, we have

$$\mathrm{Alg}(X) = \bigcap \{B \leq A \colon X \subseteq B\}.$$

Considering the notations in Definition 1.2.9 and Remark 1.2.7(ii), $\mathbf{Alg}(X) = (\mathrm{Alg}(X), F)$ is a subalgebra of $\mathbf{A}$.

Definition 1.2.10 Suppose $\mathbf{A} = (A, F^A)$ and $\mathbf{B} = (B, F^B)$ are two algebras of type $\mathcal{F}$. A map $f : A \to B$ is called a **homomorphism** if f preserves the operations of F^A, it means for any $n \in \mathbb{N}_0, \lambda \in \mathcal{F}^n$ we have

$$f(\lambda^A(a_1, \ldots, a_n)) = \lambda^B(f(a_1), \ldots, f(a_n)), \qquad \forall a_1, \ldots, a_n \in A,$$

and we write $f : \mathbf{A} \to \mathbf{B}$ is a homomorphism. The homomorphism f is called an **isomorphism** if there exists a homomorphism $g : \mathbf{B} \to \mathbf{A}$ such that $f \circ g = \mathrm{Id}_B$ and $g \circ f = \mathrm{Id}_A$, and algebraic structures (A, F^A) and (B, F^B) are **isomorphic** if there is an isomorphism between them. Then we write $(A, F^A) \cong (B, F^B)$ or $\mathbf{A} \cong \mathbf{B}$, for short. Every one-to-one (surjective) homomorphism is called a **monomorphism** or an **embedding (an epimorphism)**. We say that the algebraic structure (A, F^A) can be **embedded** into the algebraic structure (B, F^B) if there is a monomorphism $f : (A, F^A) \to (B, F^B)$.

Suppose $\mathbf{A} = (A, F)$ is an algebra of type $\mathcal{F}$. If $\mathbf{B} \leq \mathbf{A}$, then the inclusion map $i : B \to A$, which attributes each member to itself, is a monomorphism.

Theorem 1.2.11 *Let $\mathbf{A} = (A, F^A)$ and $\mathbf{B} = (B, F^B)$ be two algebras of type $\mathcal{F}$. If $f : \mathbf{A} \to \mathbf{B}$ is a homomorphism, then $\mathrm{Im}(f) := f(A) = \{f(a) \colon a \in A\}$ is a subuniverse of the algebra (B, F^B). Considering the notations in* Remark 1.2.7(ii) $f(\mathbf{A}) := (f(A), F^B)$ *is a subalgebra of (B, F^B)*

Definition 1.2.12 Suppose $\mathbf{A} = (A, F)$ is an algebra of type $\mathcal{F}$ and θ is an equivalence relation on A. Then θ is called a **congruence relation** on (A, F) if for any $n \in \mathbb{N}_0$ and every operation $\lambda^A \in \mathcal{F}^n$ and $(x_1, y_1), \ldots, (x_n, y_n) \in \theta$ imply

$$(\lambda^A(x_1, \ldots, x_n), \lambda^A(y_1, \ldots, y_n)) \in \theta.$$

The set of all congruence relations on (A, F) is denoted by $\text{Con}(A, F)$ or $\text{Con}(\mathbf{A})$, for short.

As we noted, every set X is an algebra of type $\mathcal{F} = \emptyset$. Clearly, in this algebra $\text{Con}(A, \emptyset) = \text{Eq}(A)$, where $\text{Eq}(A)$ is the set of all equivalence relations of A.

Theorem 1.2.13 *Let* $\mathbf{A} = (A, F^A)$ *and* $\mathbf{B} = (B, F^B)$ *be two algebras of type* $\mathcal{F}$ *and* $f : \mathbf{A} \to \mathbf{B}$ *be a homomorphism. Then the relation* $\text{Ker}(f) := \{(x, y) \in A \times A : f(x) = f(y)\}$ *is a congruence relation on* $\mathbf{A}$.

Remark 1.2.14 Assume that θ is a congruence relation on an algebra $\mathbf{A} = (A, F)$ of type $\mathcal{F}$. Set $[x]_\theta = x/\theta := \{y \in A : (x, y) \in \theta\}$, and $A/\theta := \{x/\theta : x \in A\}$. Thus, for any $n \in \mathbb{N}_0$ and any n-ary operation $\lambda \in \mathcal{F}^n$, we define an n-ary operation $\lambda^{\frac{A}{\theta}} : (A/\theta)^n \to A/\theta$ where

$$\lambda^{\frac{A}{\theta}}(\frac{a_1}{\theta}, \ldots, \frac{a_n}{\theta}) := \frac{\lambda^A(a_1, \ldots, a_n)}{\theta}, \qquad \forall a_1, \ldots, a_n \in A.$$

Obviously, $\lambda^{\frac{A}{\theta}}$ is well-defined and we call it an **operation induced by** λ^A **on** A/θ. Then $\mathbf{A}/\theta = (A/\theta, F^{\frac{A}{\theta}})$ is an algebraic structure of type $\mathcal{F}$ such that $F^{\frac{A}{\theta}} = \{\lambda^{\frac{A}{\theta}} : \lambda \in \mathcal{F}\}$. This new algebraic structure is called a **quotient structure** of $\mathbf{A}$ on the congruence relation θ.

Theorem 1.2.15 *Suppose* θ *is a congruence relation on* (A, F) *of type* $\mathcal{F}$. *Then the map* $\pi_\theta : A \to A/\theta$, *where* $\pi_\theta(a) = a/\theta$ *for any* $a \in A$, *is an epimorphism, and is called the* **natural epimorphism**.

Assume $\mathbf{A} = (A, F^A)$ and $\mathbf{B} = (B, F^B)$ are algebras of type $\mathcal{F}$ and $f : \mathbf{A} \to \mathbf{B}$ is a homomorphism. By Theorem 1.2.11, $\text{Im}(f)$ is a subuniverse of (B, F^B). Thus, $\text{Im}(f)$ with the operations of B is an algebra of type $\mathcal{F}$. Also, $\text{Ker}(f)$ is a congruence relation on (A, F^A) and so $A/\text{Ker}(f)$ with the operations induced by F^A is an algebra of type $\mathcal{F}$.

Building on this concept, we present the first isomorphism theorem for algebras:

Theorem 1.2.16 (The first theorem of isomorphism) *Let* $\mathbf{A} = (A, F^A)$ *and* $\mathbf{B} = (B, F^B)$ *be two algebras of type* $\mathcal{F}$ *and* $f : \mathbf{A} \to \mathbf{B}$ *be a homomorphism. Then* $\bar{f} : \mathbf{A}/\text{Ker}(f) \to f(\mathbf{A})$ *defined by* $\bar{f}(x/\text{Ker}(f)) = f(x)$ *for all* $x \in A$ *is an isomorphism and* $\mathbf{A}/\text{Ker}(f) \cong f(\mathbf{A})$.

Remark 1.2.17 Let $\{X_i : i \in I\}$ be a family of sets. The direct product of this family, denoted by $\prod_{i \in I} X_i$ is the set

$$\prod_{i \in I} X_i := \{a : I \to \bigcup_{i \in I} X_i : a \text{ is a map such that } a(i) \in X_i, \forall i \in I\}.$$

We usually denote the map a by its image $(a_i)_{i \in I}$ meaning $a(i) = a_i$ for all $i \in I$. In other words,

$$\prod_{i \in I} X_i := \{(a_i)_{i \in I} : a_i \in X_i, \ \forall i \in I\}.$$

Suppose $\{\mathbf{A}_i = (A_i, F^{A_i})\}_{i \in I}$ is a family of algebras of type $\mathcal{F}$. If $A := \prod_{i \in I} A_i$, then for any $n \in \mathbb{N}_0$ and every $\lambda \in \mathcal{F}^n$, the map $\lambda^A : A^n \to A$ for any $j \in \{1, 2, \ldots, n\}$ and every $(a_i^j)_{i \in I} \in A$, is defined as follows:

$$\lambda^A((a_i^1)_{i \in I}, \ldots, (a_i^n)_{i \in I}) = (\lambda^{A_i}(a_i^1, \ldots, a_i^n))_{i \in I}.$$

Clearly, λ^A is an n-ary operation on A. If $F^A = \bigcup_{n \in \mathbb{N}_0}\{\lambda^A : \lambda \in \mathcal{F}^n\}$, then $\mathbf{A} = (A, F^A)$ is an algebra of type $\mathcal{F}$ which is called the **direct product** of the family of algebras $\{(A_i, F^{A_i})\}_{i \in I}$ and denoted by $\prod_{i \in I} \mathbf{A}_i$. By definition of $\prod_{i \in I} \mathbf{A}_i$, its operations act component-wise[1]. So, these operations are called **component-wise** operations. In some literature, the operation λ^A which defined above is denoted by $\prod_{i \in I} \lambda^{A_i}$ or $(\lambda_i)_{i \in I}$. In addition, for any $j \in I$, the map $\pi_j : \prod_{i \in I} \mathbf{A}_i \to \mathbf{A}_j$ defined by $\pi_j((a_i)_{i \in I}) = a_j$ is an epimorphism which is called a j-**th natural projection map**.

1.3 Semilattices and Lattices

Lattices, versatile structures foundational to various mathematical domains, can be defined equivalently from either an algebraic or order-theoretic perspective. We commence by presenting the algebraic definition.

Definition 1.3.1 ([165]) An algebra $(L; \circ)$ of type (2) is said to be a **semilattice** if for any $x, y, z \in L$, the following conditions hold:

  (i) (Commutative law) $x \circ y = y \circ x$.
 (ii) (Associative law) $x \circ (y \circ z) = (x \circ y) \circ z$.
(iii) (Idempotency law) $x \circ x = x$.

Remark 1.3.2 On each semilattice $(L; \circ)$ the relations $x \leq y$ if and only if $x \circ y = x$ and $x \leq_d y$ if and only if $x \circ y = y$ are partially ordered relations on L.

---

[1] It is also called coordinate-wise.

Definition 1.3.3 ([92, 165]) An algebra $(L; \vee, \wedge)$ of type $(2, 2)$ is called a **lattice** if for any $x, y \in L$, the following conditions hold:

(i) $(L; \vee)$ and $(L; \wedge)$ are semilattices.
(ii) (Absorption laws) $x \vee (x \wedge y) = x$ and $x \wedge (x \vee y) = x$.

    Trivially, if $\mathbf{L} = (L; \vee, \wedge)$ is a lattice, then $(L; \wedge, \vee)$ is a lattice, which is denoted by $\mathbf{L}_d$.
    In each lattice $(L; \vee, \wedge)$, the algebraic structure $(L; \vee)$ is called a **join-semilattice** or $\vee$**-semilattice** and $(L; \wedge)$ is called a **meet-semilattice** or $\wedge$**-semilattice**.
    Like every algebraic structure, considering the agreement established in Remark 1.2.7(ii), a non-empty subset S of a lattice $(L; \wedge, \vee)$ which is closed under $\vee$ and $\wedge$ is a **sublattice**. Consequently, $(S; \vee, \wedge)$ is a lattice, too.

Example 1.3.4 (i) The most familiar example of a lattice is the set of all subsets of a set X, which means $\mathcal{P}(X)$ with the operations $\cup$ and $\cap$.
    (ii) Let $(P, \leq)$ be a partially ordered set such that for any $x, y \in P$, there exists $\sup\{x, y\}$ and $\inf\{x, y\}$. Then for any $x, y \in P$ define 2-ary operations $\vee$ and $\wedge$ on P as follows:

$$x \vee y = \sup\{x, y\} \qquad \& \qquad x \wedge y = \inf\{x, y\}.$$

Obviously, $(P; \vee, \wedge)$ is a lattice.
    (iii) By (ii), every totally ordered set $(P, \leq)$ is a lattice.

Theorem 1.3.5 [165] *Suppose $(L; \vee, \wedge)$ is a lattice such that for any $x, y \in L$, the relation $\leq$ on L is defined as $x \leq y$ if and only if $x \vee y = y$. Then the following statements hold:*

(i) *$(L, \leq)$ is a partially ordered set.*
(ii) *For any $x, y \in L$, $x \vee y = y$ if and only if $x \wedge y = x$.*
(iii) *In the poset $(L, \leq)$, for any $x, y \in L$, there exist $\sup\{x, y\}$ and $\inf\{x, y\}$ such that $\sup\{x, y\} = x \vee y$ and $\inf\{x, y\} = x \wedge y$.*

Theorem 1.3.6 [92, 165] *There exists a one-to-one correspondence between the family of all lattices and the family of all partially ordered sets such that* sup *and* inf *exist for each pair of its elements.*

    By Theorem 1.3.6, an equivalent characterization of a lattice can be given in the framework of partially ordered sets, a formulation frequently encountered in the literature.

Definition 1.3.7 A **lattice** is a partially ordered set in which every pair of elements possesses both a least upper bound (join) and a greatest lower bound (meet).

In this book, we use both definitions, when we say $(P, \leq)$ is a lattice, we mean $\vee = \sup$ and $\wedge = \inf$. Also, in a lattice $\mathbf{L} = (L; \vee, \wedge)$, the partially ordered relation is the relation defined according to Theorem 1.3.5. In addition, if $(L, \leq)$ is a lattice, then the dual lattice $\mathbf{L}_d$ is $(L, \leq_d)$.

Definition 1.3.8 A lattice $(L; \vee, \wedge)$ is called **distributive** if for any $x, y, z \in L$, the following conditions hold:

(D1) $x \vee (y \wedge z) = (x \vee y) \wedge (x \vee z)$.
(D2) $x \wedge (y \vee z) = (x \wedge y) \vee (x \wedge z)$.

Theorem 1.3.9 *A lattice $(L; \vee, \wedge)$ is distributive if and only if one of the conditions* (D1) *or* (D2) *holds.*

Example 1.3.10 For any set X the lattice $(\mathcal{P}(X); \cup, \cap)$ is a distributive lattice.

Theorem 1.3.11 ([28]) *A lattice $(L; \vee, \wedge)$ is distributive if and only if there is no one-to-one homomorphism from $\mathbf{M}_5$ or $\mathbf{N}_5$ to L (see Fig. 1.2).*

A lattice $(L; \vee, \wedge)$ is **bounded** if the partially ordered set $(L, \leq)$ is bounded. Then the greatest and smallest elements of L are denoted by 1 and 0, respectively. A bounded lattice $(L; \vee, \wedge)$ is shown by $(L; \vee, \wedge, 0, 1)$, which means that $(L; \vee, \wedge, 0, 1)$ is an algebra of type $(2, 2, 0, 0)$ where $(L; \vee, \wedge)$ is a lattice and for any $x \in L, 0 \leq x \leq 1$.

Remark 1.3.12 Given each lattice $(L; \vee, \wedge)$ and each element $a, b \in L$ with $a \leq b$, the interval $([a, b]; \vee, \wedge, a, b)$ is a bounded lattice, simply denoted by $[a, b]$.

Definition 1.3.13 ([35]) An element a of a lattice $\mathbf{L} = (L; \vee, \wedge)$ is said to be **join-irreducible** if

- $a \neq 0$ (in case $\mathbf{L}$ has a zero);
- $a = x \vee y$ entails $a = x$ or $a = y$ for all $x, y \in L$.

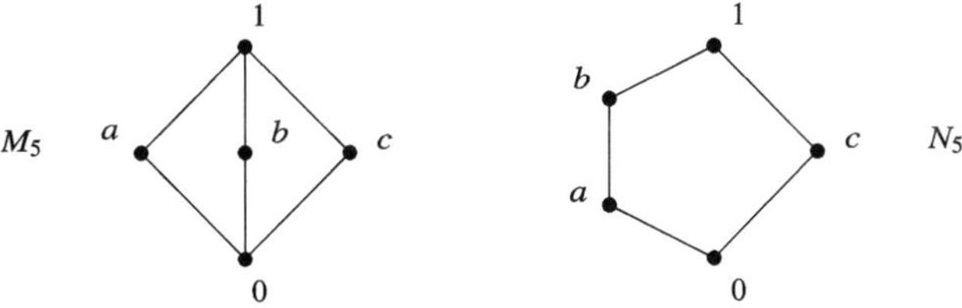

Fig. 1.2 Hasse diagrams of M_5 and N_5

A **meet-irreducible** element of $\mathbf{L}$ is defined in a dual way, that is $b \in L$ is meet-irreducible if it is a join-irreducible element in the lattice $\mathbf{L}_d$, that is $b \neq 1$ (in case $\mathbf{L}$ has a 1) and $b = x \wedge y$ implies that $b = x$ or $b = y$ for all $x, y \in L$.

A partially ordered set $(P, \leq)$ is called **complete** if $\sup X$ and $\inf X$ exist for each subset X of P. Then $\sup X$ and $\inf X$ is shown by $\bigvee X$ and $\bigwedge X$, respectively. A lattice $(L; \vee, \wedge)$ is called a **complete lattice** if the partially ordered set $(L, \leq)$ is complete.

Therefore, there is no difference between the concepts of a complete partially ordered set or a complete lattice.

Remark 1.3.14 Let $(L, \leq)$ be a complete lattice. Then $\bigwedge L$ and $\bigvee L$ exist. By definition of inf and sup, $\bigwedge L$ is the smallest element of $(L, \leq)$ and $\bigvee L$ is the greatest element $(L, \leq)$. Hence, $(L, \leq)$ is bounded. Note that $\bigvee \emptyset = \bigwedge L$ and $\bigwedge \emptyset = \bigvee L$.

Theorem 1.3.15 ([92, Theorem 2.31]) *Let $(P, \leq)$ be a partially ordered set. The following statements are equivalent:*

(i) *$(P, \leq)$ is a complete lattice.*
(ii) *For any subset S of P, $\bigwedge S$ exists.*
(iii) *For any subset S of P, $\bigvee S$ exists.*

Now, suppose $\mathbf{A} = (A, F^A)$ is an algebraic structure of type $\mathcal{F}$. Then $(\mathrm{Con}(\mathbf{A}), \subseteq)$ is a complete lattice. Since obviously, the intersection of a family of congruence relations on $\mathbf{A}$ is a congruence relation on (A, F), by Theorem 1.3.15, for any family S of elements of $\mathrm{Con}(\mathbf{A})$, we have

$$\bigwedge S = \bigcap_{R \in S} R \quad \& \quad \bigvee S = \bigcap \{R \in \mathrm{Con}(\mathbf{A}) : T \subseteq R, \ \forall T \in S\}.$$

By a similar argument, $(\mathrm{Sub}(\mathbf{A}), \subseteq)$ is a complete lattice, where $\mathrm{Sub}(\mathbf{A})$ is the set of all subalgebras of $\mathbf{A}$.

Furthermore, let $\mathrm{Eq}(Y)$ be the set of all equivalence relations on a set Y. Given $X \subseteq \mathrm{Eq}(Y)$, we can easily verify that $\bigwedge X = \bigcap \{\theta : \theta \in X\}$ is the greatest lower bound of X and

$$\bigvee X = \bigcap \{E \in \mathrm{Eq}(Y) : \theta \subseteq E, \ \forall \theta \in X\},$$

is the smallest upper bound of X. Hence, $(\mathrm{Eq}(Y), \subseteq)$ is a complete lattice.

Theorem 1.3.16 ([92, Theorem 2.35]) (The Knaster-Tarski fixpoint theorem) *Suppose $(L, \leq)$ is a complete lattice and a map $f : L \rightarrow L$ is an order-preserving map. Then f has at least one fixed point, which means there exists $p \in L$ such that $f(p) = p$.*

1.3.1 Ideals and Filters of Lattices

In this section, we continue our exploration of specific substructures of lattices, focusing on ideals and their significance in the development of these algebraic structures.

Definition 1.3.17 ([241]) A non-empty subset A of a lattice $\mathbf{L} = (L; \wedge, \vee)$ is called:

(i) An **ideal** of L if A is a lower set which is closed under $\vee$. The set of ideals of the lattice L is denoted by $I(\mathbf{L})$.

(ii) A **filter** of L if A is an upper set which is closed under $\wedge$. The set of filters of the lattice L is denoted by $F(\mathbf{L})$.

If $\mathbf{L}$ is a bounded lattice, then $\{0\}$ is an ideal of L which is called a **zero ideal**.

From the definition, it follows that every filter (or ideal) forms a sublattice. Clearly, the concept of an ideal is a dual concept of a filter, i.e., A is an ideal of $(L, \leq)$ if and only if A is a filter of $L_d = (L, \leq_d)$.

It is worth noting that in certain literature, the empty set, $\emptyset$, is considered a filter (ideal) of a lattice, as it trivially satisfies the conditions of being an upper set and closed under $\wedge$ (a lower set and closed under $\vee$).

Example 1.3.18 Consider the lattice with the Hasse diagram in Fig. 1.3:

Clearly, $I = \{0, a\}$ and $J = \{0, b\}$ are ideals and $F = \{a, 1\}$ and $G = \{b, 1\}$ are non-trivial filters of L.

Remark 1.3.19 Suppose $(L, \leq)$ is a lattice and $x \in L$. The set $\downarrow x = \{a \in L : a \leq x\}$ is an ideal of L which is called a **principal ideal generated** by x. Similarly, $\uparrow x$ is a filter of $(L, \leq)$ which is called a **principal filter generated** by x.

Proposition 1.3.20 *The intersection of each family $\{A_i : i \in I\}$ of filters (ideals) of a lattice $(L; \vee, \wedge)$ is a filter (ideal) of L. But, this statement does not hold for the union of filters (ideals) of L.*

Fig. 1.3 Hasse diagram of Example 1.3.18

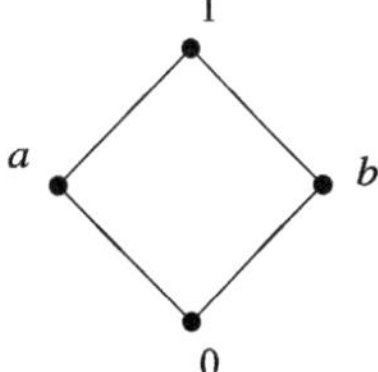

Example 1.3.21 Suppose **L** is the lattice in Example 1.3.18. Clearly, $I_1 = \{0, a\}$ and $I_2 = \{0, b\}$ are two ideals of L. However, $I = I_1 \cup I_2 = \{0, a, b\}$ is not an ideal of **L**, because $a, b \in I$ and $1 = a \vee b \notin I$.

Proposition 1.3.22 *The union of two ideals of a lattice* $(L; \wedge, \vee)$ *is an ideal if and only if one of them includes the other.*

Theorem 1.3.23 ([165]) *For any lattice* $\mathbf{L} = (L; \vee, \wedge)$, $(I(\mathbf{L}), \subseteq)$ *is a lattice which is called a* **lattice of ideals** *of* L. *If* $\{A_i : i \in I\}$ *is a family of ideals of* $\mathbf{L}$, *then*

$$\bigwedge_{i \in I} A_i = \bigcap_{i \in I} A_i, \quad \bigvee_{i \in I} A_i = \bigcap \{B \in I(\mathbf{L}) : \bigcup_{i \in I} A_i \subseteq B\}.$$

Proposition 1.3.24 ([92]) *For each element* x *and* y *of a lattice* $(L, \leq)$, *the following statements hold:*

(i) $\downarrow x \wedge \downarrow y = \downarrow(x \wedge y)$.
(ii) $\downarrow x \vee \downarrow y = \downarrow(x \vee y)$.
(iii) $x \leq y$ *if and only if* $\downarrow x \subseteq \downarrow y$.

Theorem 1.3.25 ([92]) *A lattice* $\mathbf{L} = (L; \wedge, \vee)$ *is distributive if and only if* $(I(\mathbf{L}), \subseteq)$ *is a distributive lattice.*

Similarly, as presented in Definition 1.2.10, lattice homomorphisms are defined as follows:

Definition 1.3.26 Suppose $\mathbf{L}_1 = (L_1; \wedge, \vee)$ and $\mathbf{L}_2 = (L_2; \wedge, \vee)$ are two lattices. The function $\varphi : L_1 \rightarrow L_2$ where for any $x, y \in L_1$, $\varphi(x \vee y) = \varphi(x) \vee \varphi(y)$ is called a $\vee$-**semilattice homomorphism**. Similarly, if for any $x, y \in L_1, \varphi(x \wedge y) = \varphi(x) \wedge \varphi(y)$, then φ is called a $\wedge$-**semilattice homomorphism**. The map $\varphi : L_1 \rightarrow L_2$ is called a **lattice homomorphism** if φ is both a $\vee$-semilattice homomorphism and a $\wedge$-semilattice homomorphism. A lattice homomorphism φ is called (i) a **monomorphism** if it is one-to-one; (ii) an **epimorphism** if φ is surjective; and (iii) an **isomorphism** if φ is both a monomorphism and an epimorphism.

If $\mathbf{L}_1$ and $\mathbf{L}_2$ are two bounded lattices and $f : \mathbf{L}_1 \rightarrow \mathbf{L}_2$ is a lattice epimorphism, then we can prove $f(0_{L_1}) = 0_{L_2}$ and $f(1_{L_1}) = 1_{L_2}$.

Proposition 1.3.27 *Let* $f : \mathbf{L}_1 \rightarrow \mathbf{L}_2$ *be a lattice homomorphism.*

(i) f *is an order-preserving map.*
(ii) *If* $\mathbf{L}_1$ *has the least element and* f *is an epimorphism, then* $\mathbf{L}_2$ *has the least element.*

Theorem 1.3.28 *For each lattice $(L, \leq)$, the mapping $f : \mathbf{L} \to I(\mathbf{L})$ defined by $f(x) = \downarrow x$, is a lattice monomorphism. Thus, every lattice $(L, \leq)$ can be embedded into the lattice $(I(\mathbf{L}), \subseteq)$.*

We now recall the concept of a congruence relation on a lattice, through which we introduce the corresponding quotient structure and show that this new structure is itself a lattice, and in certain cases, may even possess additional properties.

Definition 1.3.29 ([92]) An equivalence relation $\sim$ on a lattice $\mathbf{L} = (L; \wedge, \vee)$ is called a **congruence relation on L** if for any $x_1, x_2, y_1, y_2 \in L$, $x_1 \sim y_1$ and $x_2 \sim y_2$ imply $(x_1 \wedge x_2) \sim (y_1 \wedge y_2)$ and $(x_1 \vee x_2) \sim (y_1 \vee y_2)$. Then, by using this congruence relation $\sim$, we make a quotient structure $L/\sim$ containing all equivalence classes that are made by $\sim$ on L and define two new operations on $L/\sim$ as follows:

$$[a] \vee [b] = [a \vee b] \quad \& \quad [a] \wedge [b] = [a \wedge b], \qquad \forall [a], [b] \in L/\sim .$$

Easily, we can see that all these operations are well-defined on $L/\sim$. Hence, $\mathbf{L}/\sim = (L/\sim; \wedge, \vee)$ is a lattice which is called a **quotient lattice** of $\mathbf{L}$ **modulo** $\sim$.

Next, we introduce one of the most significant types of lattices: a Boolean lattice, also known as a Boolean algebra, and examine its properties.

Definition 1.3.30 An element y of a bounded lattice $(L; \vee, \wedge, 0, 1)$ is called a **complement** of x if $x \vee y = 1$ and $x \wedge y = 0$. In this case, we say that x is a complemented element of the lattice $(L; \vee, \wedge, 0, 1)$. Clearly, every complement of a complemented element is itself complemented. A bounded lattice $(L; \vee, \wedge, 0, 1)$ is **complemented** if every element of L is complemented.

Theorem 1.3.31 ([35, Theorem 6.2]) *In each distributive lattice, all complements that exist are unique.*

By a Boolean lattice, we mean a complemented distributive lattice. According to Theorem 1.3.31, if $(L; \vee, \wedge, 0, 1)$ is a Boolean lattice, then the complement of x is unique for each $x \in L$, so the complementation is a unary operation on L denoted by $'$, that is x' is the unique element of L such that $x \vee x' = 1$ and $x \wedge x' = 0$. There exists an equivalent definition for Boolean lattices as follows:

Definition 1.3.32 An algebraic structure $(B; \vee, \wedge, ', 0, 1)$ of type $(2, 2, 1, 0, 0)$ is called a **Boolean lattice** or **Boolean algebra** if

 (i) $(B; \vee, \wedge, 0, 1)$ is a bounded distributive lattice.
 (ii) $x \vee x' = 1$ and $x \wedge x' = 0$ for all $x \in B$.

A **generalized Boolean algebra** is a distributive lattice $\mathbf{L} = (L; \vee, \wedge)$ with the least element 0 such that each interval $[a, b]$ of $\mathbf{L}$ is complemented (equivalently, $[a, b]$ is a Boolean algebra). For each elements $a \leq b$ of $\mathbf{L}$ and each $x \in [a, b]$, the complement of x in $[a, b]$ is said to be a **relative complement** of x with respect to a and b.

Example 1.3.33 (i) Suppose X is a set. The algebraic structure $(\mathcal{P}(X); \cup, \cap)$ is a distributive lattice. Clearly, X is the greatest, and $\emptyset$ is the least element of this lattice. For any $S \in \mathcal{P}(X)$, the set $S' = X \setminus S$ is complement of S. Hence, $(\mathcal{P}(X); \cup, \cap, ', \emptyset, X)$ is a Boolean algebra.
(ii) The set with two elements $\{0, 1\}$, where $0 \leq 1$, is a bounded distributive lattice (by corollary of Theorem 1.3.11). Define $0' = 1$ and $1' = 0$. Then $(\{0, 1\}; \vee, \wedge, ', 0, 1)$ is a Boolean algebra.
(iii) For each bounded distributive lattice $\mathbf{L} = (L; \vee, \wedge, 0, 1)$ the set $\mathrm{B}(\mathbf{L})$ of all complemented elements of $\mathbf{L}$ is closed under $\vee$ and $\wedge$ and $\mathbf{B}(\mathbf{L}) = (\mathrm{B}(\mathbf{L}); \vee, \wedge, ', 0, 1)$ is a Boolean algebra. Indeed, $0, 1 \in \mathrm{B}(\mathbf{L})$, clearly. In addition, for each $x, y \in \mathrm{B}(\mathbf{L})$ we have

$$(x \vee y) \wedge (x' \wedge y') = \big(x \wedge (x' \wedge y')\big) \vee \big(y \wedge (x' \wedge y')\big) = 0, \text{ since } \mathbf{L} \text{ is distributive}$$
$$(x \vee y) \vee (x' \wedge y') = \big((x \vee y) \vee x'\big) \wedge \big((x \vee y) \vee y'\big) = 1, \text{ since } \mathbf{L} \text{ is distributive}.$$

Thus, $x \vee y$ is complemented, i.e., $x \vee y \in \mathrm{B}(\mathbf{L})$. In a similar way, we can show that, $(x \wedge y) \wedge (x' \vee y') = 0$ and $(x \wedge y) \vee (x' \vee y') = 1$, that is $x \wedge y \in \mathrm{B}(\mathbf{L})$. Therefore, $(\mathrm{B}(\mathbf{L}); \vee, \wedge, 0, 1)$ is a complemented distributive lattice, consequently, $\mathbf{B}(\mathbf{L})$ is a Boolean algebra (see [243, Example 4.2.15]).

Proposition 1.3.34 *Every Boolean algebra* $(B; \vee, \wedge, ', 0, 1)$ *satisfies in the following statements for any* $x, y \in B$:

 (i) $0' = 1,$ *and* $1' = 0.$
 (ii) $x'' = x,$ *where* $x'' := (x')'.$
 (iii) $x \leq y$ *if and only if* $y' \leq x'.$
 (iv) $(x \vee y)' = x' \wedge y'$ *and* $(x \wedge y)' = x' \vee y'.$
 (v) $x \wedge y = (x' \vee y')'$ *and* $x \vee y = (x' \wedge y')'.$
 (vi) $x \vee (x' \wedge y) = x \vee y$ *and* $x \wedge (x' \vee y) = x \wedge y.$
 (vii) $x \wedge y = x$ *if and only if* $x \wedge y' = 0.$

Theorem 1.3.35 *Suppose* $(L; \vee, \wedge, 0, 1)$ *is a bounded distributive lattice and* $'$ *is an unary operation on* L *such that the following conditions hold:*

$$(x \vee y)' = x' \wedge y' \quad and \quad (x \wedge y)' = x' \vee y'.$$

Then the algebraic structure $(L; \vee, \wedge, ', 0, 1)$ *is called a* **De Morgan algebra**.

Definition 1.3.36 ([36, p. 202]) Let $(L; \vee, \wedge)$ be a lattice and $x, y \in L$. The **relative pseudo-complement** of x with respect to y, provided it exists, is the greatest element $z \in L$ such that $x \wedge z \leq y$. The relative pseudo-complement of x with respect to y is denoted by $x \to y$. A lattice $(L; \vee, \wedge)$ is said to be **relatively pseudo-complemented** provided the relative pseudo-complement $x \to y$ of x with respect to y exists for every $x, y \in L$, then $\to$ can be viewed as a binary operation on L. A **Heyting algebra** or **Brouwerian algebra** is a relatively pseudo-complemented bounded lattice.

Easy calculation shows that each Heyting algebra satisfies the following condition:

$$x \leq y \to z \text{ if and only if } x \wedge y \leq z, \qquad \forall x, y, z \in L, \tag{1.1}$$

where $x \to y$ is the relative pseudo-complement of x with respect to y. Hence, there is another equivalent definition for Heyting algebra as follows:

Proposition 1.3.37 *A Heyting algebra is an algebra* $(L; \vee, \wedge, \to, 0, 1)$ *of type* $(2, 2, 2, 0, 0)$ *where* $(L; \vee, \wedge, 0, 1)$ *is a bounded lattice satisfying* (1.1).

For more details on lattice theory, we recommend to study [29, 31, 35, 162, 164, 165, 241].

1.4 MV-Algebras and ℓ-Groups

MV-algebras were introduced by C.C. Chang in [70, 71] as basic algebraic tools for many-valued evaluation to provide an algebraic proof of the completeness of the Łukasiewicz infinite-valued proposition calculus. Another fundamental study on MV-algebras was done in [221], which provides a one-to-one characterization of MV-algebras as intervals in unital Abelian ℓ-groups. In this section, we recall some definitions, properties, and main results related to MV-algebras and ℓ-groups.

1.4.1 MV-Algebras

Definition 1.4.1 ([73]) An **MV-algebra** is an algebraic structure $\mathbf{M} = (M; \oplus, ', 0, 1)$ of type $(2, 1, 0, 0)$, where $(M; \oplus, 0)$ is a commutative monoid with the neutral element 0 and for all $x, y \in M$, we have:

(i) $x'' = x$.

(ii) $x \oplus 1 = 1$.

(iii) $x \oplus (x \oplus y')' = y \oplus (y \oplus x')'$.

(iv) $0' = 1$.

An MV-algebra $(M; \oplus, ', 0, 1)$ is said to be **non-trivial** if $0 \neq 1$. In any MV-algebra M, we define three more binary operations as follows:

$$x \odot y := (x' \oplus y')', \quad x \to y := x' \oplus y, \quad x \ominus y := (x' \oplus y)'. \tag{1.2}$$

In addition, for any integer $n \in \mathbb{N}$ and any $x \in M$, we define $0.x = 0$, and $n.x = (n-1).x \oplus x, n \geq 1$. Given $x \in M$, $\mathrm{ord}(x)$ is the least integer m such that $m.x = 1$. If no such integer m exists then $\mathrm{ord}(x) = \infty$. On each MV-algebra $\mathbf{M}$, the relation $\leq$ is defined by $x \leq y$ if and only if $x \to y = 1$ is a partially ordered relation, and $(M, \leq)$ is a bounded distributive lattice with 0 as the least element and 1 as the greatest element. Indeed for all $x, y, z \in M$

$$x \vee y = (x' \oplus y)' \oplus y \quad \& \quad x \wedge y = (x' \vee y')' = (x' \oplus (x' \oplus y)')'. \tag{1.3}$$

In addition, $x \odot y \leq z$ if and only if $x \leq y \to z$.

An element x of an MV-algebra $\mathbf{M} = (M; \oplus, ', 0, 1)$ is **Boolean** if $x \oplus x = x$. In addition, an element x is Boolean if and only if one of the following conditions holds (see [73, Theorem 1.5.3]):

(i) $x \odot x = x$.

(ii) $x \oplus y = x \vee y$ for all $x, y \in M$.

(iii) $x \odot y = x \wedge y$ for all $x, y \in M$.

(iv) $x \vee x' = 1$.

(v) $x \wedge x' = 0$.

(vi) $x \oplus x = x$.

The set of all Boolean elements of $\mathbf{M}$ is denoted by $B(\mathbf{M})$. Then, $(B(\mathbf{M}); \oplus, ', 0, 1)$ is a subalgebra of $\mathbf{M}$ and also it is a Boolean algebra, that is $\mathbf{B}(\mathbf{M}) = (B(\mathbf{M}); \vee, \wedge, ', 0, 1)$ is a Boolean algebra.

Example 1.4.2 (i) Let [0, 1] be the unit interval of real numbers. Consider the operations $\oplus$ and $'$ defined as follows for all $x, y \in [0, 1]$:

$$x \oplus y = \min\{x + y, 1\} \quad \& \quad x' = 1 - x. \tag{1.4}$$

Then $\mathbf{C} = ([0, 1]; \oplus, ', 0, 1)$ is an MV-algebra. It is called the **standard** MV-algebra of the real unit interval and the binary operation $\oplus$ is referred to as the **truncated sum**. Easy calculations show that $([0, 1] \cap \mathbb{Q}; \oplus, ', 0, 1)$ is an MV-algebra, too.
(ii) Set $L_0 = \{1\}$ and given an integer $n \in \mathbb{N}$, the set $L_n = \{0, 1/n, 2/n, \ldots, (n-1)/n, 1\}$ with the operations defined in (1.4) form an MV-algebra, $(L_n; \oplus, ', 0, 1)$.
(iii) Given a Boolean algebra $(L; \vee, \wedge, ', 0, 1)$, the algebraic structure $(L; \vee, ', 0, 1)$ is an MV-algebra.
(iv) Let $(M; \oplus, ', 0, 1)$ be an MV-algebra and X be an arbitrary set. Then the set of all functions from X to A, A^X, together with the pointwise operations, form an MV-algebra. For all $f, g \in A^X$:

$$(f \oplus g)(x) = f(x) \oplus g(x) \quad \& \quad f'(x) = f(x)', \quad \forall x \in X.$$

(v) If $\mathbf{M} = (M; \oplus, ', 0, 1)$ is an MV-algebra, then the algebraic structure $(M; \odot, ', 1, 0)$ is an MV-algebra, too. It is called the dual MV-algebra of $\mathbf{M}$ and it is denoted by $\mathbf{M}^d$. A straightforward computation shows that the join (meet) operation in $\mathbf{M}$ coincides with the meet (join) operation in $\mathbf{M}^d$.

Definition 1.4.3 Let $(M_1; \oplus, ', 0, 1)$ and $(M_2; \oplus, ', 0, 1)$ be two MV-algebras. A map $f : M_1 \to M_2$ is called an **MV-homomorphism** if (i) $f(x \oplus y) = f(x) \oplus f(y)$, (ii) $f(x') = f(x)'$, and (iii) $f(0) = 0$. Similarly to every algebraic structure, f is an **MV-isomorphism** if there is a homomorphism $g : M_2 \to M_1$ such that $f \circ g = \mathrm{Id}_{M_2}$ and $g \circ f = \mathrm{Id}_{M_1}$.

We can easily see that if f is a homomorphism, then $f(1) = 1$, $f(x \vee y) = f(x) \vee f(y)$, and $f(x \wedge y) = f(x) \wedge f(y)$ for all $x, y \in M$. In addition, a homomorphism f is an isomorphism if f is one-to-one and onto.

A non-empty subset I of an MV-algebra $\mathbf{M}$ is said to be an **ideal** if it is closed under $\oplus$ and $x \leq y \in I$ implies that $x \in I$ for all $x, y \in M$. The set of all ideals of M is denoted by $\mathcal{I}(\mathbf{M})$. An easy calculation shows that the intersection of any family $\{I_j : j \in J\}$ of ideals of $\mathbf{M}$ is an ideal. Hence, by [92, Theorem 2.31], the poset $(\mathcal{I}(\mathbf{M}), \leq)$ is a complete lattice. An ideal $I \in \mathcal{I}(\mathbf{M})$ is called (1) **proper** if $I \neq M$; (2) **Boolean** if $x \vee x' \in I$ for all $x \in M$.

Given an ideal I of an MV-algebra $\mathbf{M}$, we can define a relation θ_I as follows:

$$(x, y) \in \theta_I \text{ if and only if } x \ominus y, y \ominus x \in I, \quad \forall x, y \in M. \tag{1.5}$$

Trivially, θ_I is a congruence relation on M, and x/I and M/I will denote $\{y \in M : (x, y) \in \theta_I\}$ and $\{x/I : x \in M\}$, respectively. Hence, $\mathbf{M}/I := (M/I; \oplus, ', 0/I, 1/I)$ is an MV-algebra with the following operations:

$$\frac{x}{I} \oplus \frac{y}{I} = \frac{x \oplus y}{I} \quad \& \quad \left(\frac{x}{I}\right)' = \frac{x'}{I}, \quad \forall x, y \in M.$$

In addition, if θ is an arbitrary congruence relation on $\mathbf{M}$, then there exists an ideal I of $\mathbf{M}$ such that $\theta = \theta_I$. Indeed, $0/I := \{x \in M : (x, 0) \in \theta\}$ is an ideal of $\mathbf{M}$ and $I = 0/I$. Furthermore, there exists a one-to-one correspondence between, $\mathrm{Con}(\mathbf{M})$, the set of all congruence relations of the MV-algebra $\mathbf{M}$, and $\mathcal{I}(\mathbf{M})$. A map φ defined by

$$\varphi : \mathcal{I}(\mathbf{M}) \to \mathrm{Con}(\mathbf{M}), \qquad \varphi(I) = \theta_I, \qquad \forall I \in \mathcal{I}(\mathbf{M}),$$

is a bijection map that preserves $\subseteq$ and $\varphi^{-1}(\theta) = 0/I$ for all $\theta \in \mathrm{Con}(\mathbf{M})$.

Definition 1.4.4 Let I be a proper ideal of an MV-algebra $\mathbf{M}$. Then I is:

(i) A **prime** ideal if $x \wedge y \in I$ implies that $x \in I$ or $y \in I$ for all $x, y \in M$. The set of all prime ideals of $\mathbf{M}$ is denoted by $\mathrm{Spec}(\mathbf{M})$.
(ii) A **maximal** ideal if I is a maximal element on the poset $(\mathcal{I}(\mathbf{M}), \subseteq)$.

The intersection of all maximal ideals of $\mathbf{M}$ is denoted by $\mathrm{Rad}(\mathbf{M})$.

Definition 1.4.5 ([73, 98]) An MV-algebra is called:

(i) **Local** if it has a unique maximal ideal.
(ii) **Perfect** if for each $x \in M$, $\mathrm{ord}(x) < \infty$ if and only if $\mathrm{ord}(x') = \infty$.
(iii) **Hyperarchimedean** if all its elements are Archimedean, i.e., for each $x \in M$, there is an integer n such that $n.x$ is a Boolean element.
(iv) **Simple** if $\mathcal{I}(\mathbf{M}) = \{\{0\}, M\}$ or equivalently, $\mathrm{Con}(\mathbf{M}) = \{\Delta, \nabla\}$.
(v) **Semisimple** if $\mathrm{Rad}(\mathbf{M}) = \{0\}$.

Proposition 1.4.6 *Let I be an ideal of a non-trivial MV-algebra* $\mathbf{M} = (M; \oplus, ', 0, 1)$. *Then:*

(i) *I is prime if and only if* $\mathbf{M}/I$ *is linearly ordered* (see [73, Sec 1.2]).
(ii) *I is maximal if and only if* $\mathbf{M}/I$ *is simple* (see [73, Sec 3.5]).
(iii) $\mathbf{M}/I$ *is semisimple if and only if I is an intersection of maximal ideals of* $\mathbf{M}$ (see [73, Lemma 3.6.6]).
(iv) $\bigcap \mathrm{Spec}(\mathbf{M}) = \{0\}$ (see [73, Proposition 1.2.13]).
(v) $\mathbf{M}$ *is a subdirect product of a family of linearly ordered MV-algebras* (see [73, Sec 1.3]).

(vi) **M** *is simple if and only if M is isomorphic to a subalgebra of the standard MV-algebra of the real unit interval* [0, 1] (see [73, Proposition 3.5.1]).

(vii) **M** *is a finite simple MV-algebra if and only if M is isomorphic to a* $(L_n; \oplus, ', 0, 1)$ *for some* $n \in \mathbb{N}$ (see [73, Corollary 3.5.4]).

In the following lemma, we will see a generalization of this result.

Theorem 1.4.7 (Chang Completeness Theorem [73, Theorem 2.5.3]) *An MV-identity holds in all MV-algebras if and only if it holds in the standard MV-algebra of the real unit interval* [0, 1].

To delve deeper into the theory of MV-algebras, consider exploring the following recommended monographs [54, 73, 99, 123, 226].

1.4.2 Lattice-Ordered Groups

Now, we are going to define ℓ-groups as an important ordered algebraic structure related to MV-algebras.

Definition 1.4.8 A **partially ordered monoid (group)** is a monoid $\mathbf{G} = (G; +, 0)$ (group $(G; +, -, 0)$) equipped with a partially ordered relation $\leq$ on G such that the binary operation $+$ is order-preserving on both sides, i.e., for each $x, y, z \in G$, $x \leq y$ implies that $z + x \leq z + y$ and $x + z \leq y + z$.

We denote by $\mathbf{G}^+ = \{g \in G : g \geq 0\}$ and $\mathbf{G}^- = \{g \in G : g \leq 0\}$. They are called the **positive** and **negative cones** of $\mathbf{G}$, respectively. It is easily verified that the partial order on the partially ordered group $\mathbf{G}$ is determined by $\mathbf{G}^+$ in the sense that $x \leq y$ if and only if $y - x \in \mathbf{G}^+$.

Definition 1.4.9 ([89]) A **lattice-ordered group** (or ℓ**-group** for short) is an algebraic structure $(G; \vee, \wedge, +, -, 0)$ such that $(G; \vee, \wedge)$ is a lattice, $(G; +, -, 0)$ is a group, and $+$ is an order preserving map.

Clearly, each ℓ-group is a partially ordered group, and each linearly ordered group is an ℓ-group. For each $g \in G$ we define the **positive part** g^+ of g and **negative part** g^- of g as follows

$$g^+ = g \vee 0 \quad \& \quad g^- = -g \vee 0.$$

Furthermore, the **absolute value** $|g|$ of g is $g^+ - g^-$.

An ℓ-group is **Archimedean** if for any element g and h, $ng \leq h$ for all positive integers n implies that $g \leq 0$.

Proposition 1.4.10 ([21]) *Let* $\mathbf{G} = (G; \vee, \wedge, +, -, 0)$ *be an* ℓ*-group. Then the following statements hold for all* $x, y, z \in G$:

   (i) $-(x \vee y) = -x \wedge -y$ *and* $-(x \wedge y) = -x \vee -y$.
  (ii) $z + (x \vee y) = (z + x) \vee (z + y)$ *and* $z + (x \wedge y) = (z + x) \wedge (z + y)$.
 (iii) $(x \vee y) + z = (x + z) \vee (y + z)$ *and* $(x \wedge y) + z = (x + z) \wedge (y + z)$.
 (iv) $x = x^+ - x^- = -x^- + x^+$.
  (v) $x \wedge y = 0$ *implies that* $x + y = x \vee y$.
 (vi) $|x| = x^+ \vee x^-$.
 (vii) $\mathbf{G}$ *is torsion free, that is* $nx = 0$ *implies that* $x = 0$.
(viii) *If* $Y \subseteq G$ *such that* $\bigvee Y$ $(\bigwedge Y)$ *exists, then*

$$x \wedge \bigvee Y = \bigvee_{y \in Y} (x \wedge y) \quad \left(x \vee \bigwedge Y = \bigwedge_{y \in Y} (x \vee y) \right).$$

 (ix) $z \wedge (x + y) \leq (z \wedge x) + (z \wedge y)$.

Definition 1.4.11 ([148]) A **lattice-ordered monoid** or ℓ**-monoid** is an algebra $\mathbf{S} = (S; \vee, \wedge, \cdot, 1)$ of type $(2, 2, 2, 0)$ where $(S; \vee, \wedge)$ is a lattice, $(S; \cdot, 1)$ is a monoid and $\cdot$ distributes over $\vee$ and $\wedge$ of both side, i.e., for all $\circ \in \{\vee, \wedge\}$ and $x, y, z \in S$ we have

$$z \cdot (x \circ y) = (z \cdot x) \circ (z \cdot y) \quad \& \quad (x \circ y) \cdot z = (x \cdot z) \circ (y \cdot z). \tag{1.6}$$

Meet-semilattice (join-semilattice) ordered monoid is defined similarly.

Evidently, every ℓ-group is an ℓ-monoid. In addition, each ℓ-monoid $\mathbf{S}$ is a partially ordered monoid, since $x \leq y$ implies that $x \cdot z = (x \wedge y) \cdot z = (x \cdot z) \wedge (y \cdot z)$, consequently, $x \cdot z \leq y \cdot z$, similar proof shows that $z \cdot x \leq z \cdot y$ for all $z \in S$.

Definition 1.4.12 ([89]) An ℓ**-subgroup** of an ℓ-group $\mathbf{G} = (G; \vee, \wedge, +, -, 0)$ is a non-empty subset $A \subseteq G$ which is closed under the operations $\vee$, $\wedge$, $+$, and $-$. So, $\mathbf{A} = (A; \vee, \wedge, +, -, 0)$ is an ℓ-group. An ℓ-subgroup A is said to be **normal** if A is a normal subgroup of $\mathbf{G}$ and is said to be **convex** if for all $x, y \in A$, $x \leq z \leq y$ implies that $z \in A$. A normal convex ℓ-subgroup is called an ℓ**-ideal**. Denote by $\mathcal{I}(\mathbf{G})$ the set of all ℓ-ideals of the ℓ-group $\mathbf{G}$. An ℓ-ideal I of $\mathbf{G}$ is **proper** if $I \neq G$. A proper ℓ-ideal I is said to be **prime** if $x \wedge y \in I$ implies that $x \in I$ or $y \in I$.

Let $\mathbf{G} = (G; \vee, \wedge, +, -, 0)$ and $\mathbf{H} = (H; \vee, \wedge, +, -, 0)$ be two ℓ-groups. A map $f : G \rightarrow H$ is an ℓ**-homomorphism** if f preserves all of the operations $\{\vee, \wedge, +, -, 0\}$, then we write $f : \mathbf{G} \to \mathbf{H}$ is an ℓ-homomorphism. For each ℓ-homomorphism $f : \mathbf{G} \to \mathbf{H}$, $f^{-1}(0)$ is an ℓ-ideal of $\mathbf{G}$.

For each ℓ-ideal I of an ℓ-group $\mathbf{G}$, the set $G/I = \{x + I : x \in G\}$, considering the following operations from a group with the identity element $0 + I$:

$$(x + I) + (y + I) = (x + y) + I \quad \& \quad -(x + I) = -x + I. \tag{1.7}$$

On the other hand, the relation $x + I \leq y + I$ if and only if $x \leq y + a$ for some $a \in I$ is a partially ordered relation on G/I and G/I is a partially ordered group. Furthermore, $(x + I) \vee (y + I)$ and $(x + I) \wedge (y + I)$ exist for all $x, y \in G$ and

$$(x + I) \vee (y + I) = (x \vee y) + I, \qquad (x + I) \wedge (y + I) = (x \wedge y) + I.$$

So, $(G/I; \vee, \wedge, +, -, 0/I)$ is an ℓ-group which is simply denoted by $\mathbf{G}/I$. Trivially, the natural projection map $\pi_I : \mathbf{G} \to \mathbf{G}/I$ defined by $\pi_I(x) = x + I$ is an ℓ-homomorphism.

An ℓ-group $\mathbf{G}$ is **representable** if $\mathbf{G}$ is a subdirect product of linearly ordered groups, which means there are a family $\{\mathbf{G}_i : i \in I\}$ of ℓ-groups and an injective homomorphism $f : \mathbf{G} \to \prod_{i \in I} \mathbf{G}_i$ such that $\pi_i(f(G)) = G_i$ for all $i \in I$.

Theorem 1.4.13 *Let* $\mathbf{G} = (G; \vee, \wedge, +, -, 0)$ *be an ℓ-group. Then the following statements hold:*

 (i) $\mathbf{G}$ *is representable if and only if* $2(x \wedge y) = 2x \wedge 2y$ *for all* $x, y \in G$ *if and only if the intersection of all prime ℓ-ideals of* $\mathbf{G}$ *is* $\{0\}$ *(see [21, Theorem 4.1.1]).*
 (ii) *An ℓ-ideal I is prime if and only if* $\mathbf{G}/I$ *is linearly ordered if and only if* $a \wedge b \in I$ *implies that* $a \in I$ *or* $b \in I$ *(see [21, Theorem 1.2.10]).*
(iii) *Every Abelian ℓ-group is representable (see [21, Theorem 4.1.2]).*
(iv) *Each Archimedean ℓ-group is Abelian (see [21, Theorem 2.2]).*

For a comprehensive understanding of ℓ-groups, we recommend exploring the monographs [21, 89, 148, 159, 198].

1.4.3 Relation Between Abelian ℓ-Groups and MV-Algebras

An element $u \in \mathbf{G}^+$ is said to be a **strong unit** if given $g \in G$, there is an integer $n \geq 1$ such that $g \leq nu$. A couple $(\mathbf{G}, u)$, where $\mathbf{G}$ is an ℓ-group with a fixed strong unit u, is said to be a **unital ℓ-group**. If $(G; +, -, 0)$ is an Abelian group, then the ℓ-group $(G; \vee, \wedge, +, -, 0)$ is said to be an Abelian ℓ-group.

If $\mathbf{G} = (G; \vee, \wedge, +, -, 0)$ is an Abelian ℓ-group, given $u \in \mathbf{G}^+$, then we define an interval $[0, u] := \{x \in G : 0 \leq x \leq u\}$. The interval $[0, u]$ can be converted into an MV-algebra $\Gamma(\mathbf{G}, u) = ([0, u]; \oplus, ', 0, u)$ as follows:

$$x \oplus y = (x + y) \wedge u, \quad x \odot y = (x - u + y) \vee 0, \quad x' = u - x, \quad \forall x, y \in [0, u].$$

Example 1.4.14 (i) Consider the Abelian group $(\mathbb{R}; +, -, 0)$ which is a linearly ordered group with the natural order relation on $\mathbb{R}$. Then 1 is a strong unit of this ℓ-group. Trivially, $\Gamma(\mathbb{R}, 1)$ is exactly the standard MV-algebra of the real unit interval.
(ii) Similar to part (i), $(\mathbb{Z}; +, -, 0)$ is a linearly ordered group and each positive integer n is a strong unit. If $n = 1$, then $\Gamma(\mathbb{Z}, 1)$ is the Boolean algebra $(\{0, 1\}; \vee, \wedge, ', 0, 1)$. If $n \geq 1$, then $\Gamma(\mathbb{Z}, n)$ is isomorphic to the MV-algebra $(L_n; \oplus, ', 0, 1)$ in Example 1.4.2(ii). Indeed, the map $f : \Gamma(\mathbb{Z}, n) \to (L_n; \oplus, ', 0, 1)$ defined by $f(m) = m/n$ for each $m \in \{0, 1, \ldots, n\}$, is the isomorphism. It is worthy of note that each linearly ordered finite MV-algebra with $n + 1$ elements is isomorphic to $\Gamma(\mathbb{Z}, n)$ (see [73]).

Conversely, due to an important result by D. Mundici [221], every MV-algebra is isomorphic to $\Gamma(\mathbf{G}, u)$ for some unital Abelian ℓ-group $(\mathbf{G}, u)$. The unital Abelian ℓ-group $(\mathbf{G}, u)$ is unique up to isomorphism of unital ℓ-groups. Moreover, the category of unital Abelian ℓ-groups is categorically equivalent to the category of MV-algebras.

Let $\mathcal{U}\mathcal{A}\ell\mathcal{G}$ be the category of unital Abelian ℓ-groups whose objects are unital Abelian ℓ-groups $(\mathbf{G}, u)$ and morphisms between objects are ℓ-homomorphisms preserving fixed strong units. We denote by $\mathcal{MV}$ the category of MV-algebras whose objects are MV-algebras and morphisms are MV-homomorphisms. Consider the functor $\boldsymbol{\Gamma} : \mathcal{U}\mathcal{A}\ell\mathcal{G} \to \mathcal{MV}$ which is defined as follows

$$\boldsymbol{\Gamma}(\mathbf{G}, u) := \Gamma(\mathbf{G}, u),$$

and if $h : (\mathbf{G}, u) \to (\mathbf{H}, v)$ is a morphism, then $\boldsymbol{\Gamma}(h) = h|_{[0,u]}$ is the restriction of h onto the interval $[0, u]$. On the other side, there is a functor from the category of $\mathcal{MV}$ to $\mathcal{U}\mathcal{A}\ell\mathcal{G}$ sending an MV-algebra $\mathbf{M}$ to a unital Abelian ℓ-group $(\mathbf{G}, u)$ such that $\mathbf{M} \cong \Gamma(\mathbf{G}, u)$ which is denoted by $\boldsymbol{\Psi} : \mathcal{MV} \to \mathcal{U}\mathcal{A}\ell\mathcal{G}$. If $\phi : \mathbf{M}_1 \to \mathbf{M}_2$ is a morphism of MV-algebras, then $\boldsymbol{\Psi}(\phi) : \boldsymbol{\Psi}(\mathbf{M}_1) \to \boldsymbol{\Psi}(\mathbf{M}_2)$ is the unique extension of ϕ onto a morphism of unital ℓ-groups. Indeed, for $0 \leq g \in G$ there exists $n \in \mathbb{N}$ such that $g \leq nu$ and we can find $a_1, \ldots, a_n \in M$ such that $g = a_1 + \cdots + a_n$ and $\boldsymbol{\Psi}(\phi)(g) = \phi(a_1) + \cdots + \phi(a_n)$. Also, $\boldsymbol{\Psi}(\phi)(x) = \boldsymbol{\Psi}(\phi)(x^+) + \boldsymbol{\Psi}(\phi)(x^-)$ for all $x \in G$. It is possible to show that if $\phi : \mathbf{M}_1 \to \mathbf{M}_2$ is an injective (surjective) homomorphism, so is $\boldsymbol{\Psi}(\phi)$.

Theorem 1.4.15 ([73, 221, Chap. 7]) *The composite functors $\boldsymbol{\Gamma} \circ \boldsymbol{\Psi}$ and $\boldsymbol{\Psi} \circ \boldsymbol{\Gamma}$ are naturally equivalent to the identity functors of $\mathcal{MV}$ and $\mathcal{U}\mathcal{A}\ell\mathcal{G}$, respectively. Therefore, $\mathcal{MV}$ and $\mathcal{U}\mathcal{A}\ell\mathcal{G}$ are categorically equivalent.*

Let $\mathbf{H}$ and $\mathbf{G}$ be (Abelian) ℓ-groups. We define the lexicographic product $\mathbf{H} \overset{\rightarrow}{\times} \mathbf{G}$ as the group addition on the direct product $H \times G$ endowed with the lexicographic order $(h_1, g_1) \leq (h_2, g_2)$ for $(h_1, g_1), (h_2, g_2) \in H \times G$ if and only if either $h_1 < h_2$ or $h_1 = h_2$ and $g_1 \leq g_2$. Then $\mathbf{H} \overset{\rightarrow}{\times} \mathbf{G}$ is a partially-ordered (Abelian) group that is an (Abelian) ℓ-group if and only if $\mathbf{H}$ is linearly ordered, see [148, (d) p. 26].

Example 1.4.16 (i) Let $\mathbf{G} = (G; \vee, \wedge, +, -, 0)$ be an Abelian ℓ-group and $(\mathbb{R}; +, -, 0)$ be the group of real numbers with natural ordering which is totally ordered. Then $(1, 0)$ is a strong unit of the ℓ-group $\mathbf{H} := \mathbb{R} \overset{\rightarrow}{\times} \mathbf{G}$ and $\mathbf{H}^+ = (\{0\} \times \mathbf{G}^+) \cup ((0, +\infty) \times \mathbf{G}^-)$. In addition, $\Gamma(\mathbf{H}, (1, 0))$ is an MV-algebra.

(ii) Consider the totally ordered group $(\mathbb{Z}; +, -, 0)$. Clearly, $(1, 0)$ is a strong unit of $\mathbb{Z} \overset{\rightarrow}{\times} \mathbb{Z}$ and $\mathbf{M} := \Gamma(\mathbb{Z} \overset{\rightarrow}{\times} \mathbb{Z}, (1, 0))$ is a totally ordered MV-algebra, called **Chang MV-algebra**. It is easy to see that $M = (\{0\} \times \mathbb{Z}^+) \cup (\{1\} \times \mathbb{Z}^-)$, where $\mathbb{Z}^- = \{0, -1, -2, \ldots\}$ and $Z^+ = \{0, 1, 2, \ldots\}$. Moreover, M is a perfect MV-algebra.

Theorem 1.4.17 ([74]) *Let $(\mathbf{G}, u)$ be a unital Abelian ℓ-group. The correspondence*

$$f : \mathcal{I}(\Gamma(\mathbf{G}, u)) \to \mathcal{I}(\mathbf{G}), \quad f(I) = \{g \in G : |g| \leq u\}, \tag{1.8}$$

defines an isomorphism from the poset $(\mathcal{I}(\Gamma(\mathbf{G}, u)), \subseteq)$ onto the poset $\mathcal{I}(\mathbf{G})$. The inverse isomorphism is given by the correspondence $g(J) = J \cap \Gamma(\mathbf{G}, u)$. Furthermore, f transforms prime ideals of the MV-algebra $\Gamma(\mathbf{G}, u)$ to into prime ℓ-ideals of the ℓ-group G (similar statement holds for g).

For more details, we recommend [21, 89, 148, 159] for the theory of ℓ-groups and [73, 123, 226] for MV-algebras.

1.5   Congruences, Class of Algebras, and Varieties

1.5.1   Congruences, Subdirect Products, Subdirectly Irreducible Algebras

Let $\mathbf{A} = (A, F^A)$ be an algebra of type $\mathcal{F}$. It is well known the poset $(\mathrm{Con}(\mathbf{A}), \subseteq)$ is a complete lattice. Indeed, for each family $\{\theta_i : i \in I\}$ of elements $\mathrm{Con}(\mathbf{A})$ we have $\bigcap_{i \in I} \theta_i \in \mathrm{Con}(\mathbf{A})$. In addition,

$$\bigwedge_{i \in I} \theta_i = \bigcap_{i \in I} \theta_i, \quad \bigvee_{i \in I} \theta_i = \bigcap \{\theta \in \mathrm{Con}(\mathbf{A}) : \bigcup_{i \in I} \theta_i \subseteq \theta\}. \tag{1.9}$$

$\Delta := \{(x, x) : x \in A\}$ and $\nabla := A \times A$ are the least and greatest elements of $(\mathrm{Con}(\mathbf{A}), \subseteq)$, respectively. From now on, in this section, we simply denote the complete lattice $(\mathrm{Con}(\mathbf{A}); \vee, \wedge)$ by $\mathrm{Con}(\mathbf{A})$.

Definition 1.5.1 An algebra $\mathbf{A} = (A, F^A)$ is said to be **congruence-distributive** or **congruence-modular** if $\mathrm{Con}(\mathbf{A})$ is a distributive or modular lattice, respectively. In addition, for $\theta_1, \theta_2 \in \mathrm{Con}(\mathbf{A})$ we say that θ_1 and θ_2 permute if $\theta_2 \circ \theta_1 = \theta_1 \circ \theta_2$. The algebra $\mathbf{A}$ is **congruence-permutable** if every pair of congruences on $\mathbf{A}$ permutes.

Theorem 1.5.2 ([61]) *Let* $\mathbf{A} = (A, F^A)$ *be an algebra of type* $\mathcal{F}$ *and* $\theta_1, \theta_2 \in \mathrm{Con}(\mathbf{A})$. *Then the following statements are equivalent:*

(i) $\theta_1 \circ \theta_2 = \theta_2 \circ \theta_1$.
(ii) $\theta_1 \circ \theta_2 = \theta_1 \vee \theta_2$.

An important theorem was proved by Birkhoff [29] showing the relation between congruence-permutable and congruence-modular algebras.

Theorem 1.5.3 ([61, Theorem 5.5.1]) *Every congruence-permutable algebra* (A, F^A) *of type* $\mathcal{F}$ *is congruence-modular.*

Example 1.5.4 ([61, P. 42, Exercise 4, 5]) Both groups and rings, as well as lattices, are congruence-permutable but not necessarily congruence-distributive.

Definition 1.5.5 A congruence θ on an algebra $\mathbf{A} = (A, F^A)$ is a **factor congruence** if there is a congruence $\theta' \in \mathrm{Con}(\mathbf{A})$ such that $\theta \wedge \theta' = \Delta, \theta \vee \theta' = \nabla$, and θ and θ' permute. In addition, θ and θ' are called a **pair of factor congruences** on $\mathbf{A}$.

Theorem 1.5.6 ([61]) *Let* θ *and* θ' *be a pair of factor congruences on* $\mathbf{A} = (A, F^A)$. *Then the map* $f : \mathbf{A} \to \mathbf{A}/\theta \times \mathbf{A}/\theta'$ *defined by* $f(x) = (x/\theta, x/\theta')$ *is an isomorphism.*

An algebra $\mathbf{A} = (A, F^A)$ has the **congruence extension property** (CEP) if for every subalgebra $\mathbf{B}$ of $\mathbf{A}$ and each $\theta \in \mathrm{Con}(\mathbf{B})$ there is a $\phi \in \mathrm{Con}(\mathbf{A})$ such that $\theta = \phi \cap B^2$.

Definition 1.5.7 An algebra $\mathbf{A} = (A, F^A)$ is called **directly indecomposable** if $\mathbf{A}$ is not isomorphic to a direct product of two non-trivial algebras of type $\mathcal{F}$.

By applying Theorem 1.5.6, we obtain the following proposition:

Proposition 1.5.8 *An algebra* $\mathbf{A} = (A, F^A)$ *is directly indecomposable if and only if* Δ *and* ∇ *are the only factor congruences on* $\mathbf{A}$.

Theorem 1.5.9 ([61, Theorem 7.10]) *Every finite algebra is isomorphic to a direct product of directly indecomposable algebras.*

Definition 1.5.10 An algebra $\mathbf{A} = (A, F^A)$ of type $\mathcal{F}$ is a **subdirect product** of a family $\{\mathbf{A}_i = (A_i, F_i)\}_{i \in I}$ of algebras of type $\mathcal{F}$ if $\mathbf{A}$ is a subalgebra of $\prod_{i \in I} \mathbf{A}_i$ such that $\pi_i(A) = A_i$ for all $i \in I$.

Furthermore, a monomorphism (an embedding) $f : \mathbf{A} \to \prod_{i \in I} \mathbf{A}_i$ is said to be a **subdirect embedding** if $f(\mathbf{A})$ is a subdirect product of $\{\mathbf{A}_i : i \in I\}$.

Proposition 1.5.11 *Let* $\mathbf{A} = (A, F^A)$ *be an algebra of type* $\mathcal{F}$ *and* $\{\theta_i : i \in I\} \subseteq \mathrm{Con}(\mathbf{A})$ *such that* $\bigcap_{i \in I} \theta_i = \Delta$. *Then the mapping* $f : \mathbf{A} \to \prod_{i \in I} \mathbf{A}/\theta_i$ *defined by* $f(a) = (a/\theta_i)_{i \in I}$ *is a subdirect embedding.*

Definition 1.5.12 An algebra $\mathbf{A} = (A, F^A)$ of type $\mathcal{F}$ is called:

- **Subdirectly irreducible** if for every subdirect embedding $f : \mathbf{A} \to \prod_{i \in I} \mathbf{A}_i$ there exists $i \in I$ such that $\pi_i \circ f : \mathbf{A} \to \mathbf{A}_i$ is an isomorphism.
- **Simple** if $\mathrm{Con}(\mathbf{A}) = \{\Delta, \nabla\}$. In addition, a congruence $\theta \in \mathrm{Con}(\mathbf{A})$ is **maximal** if $[\theta, \nabla] = \{\theta, \nabla\}$, where $[\theta, \nabla] = \{\beta \in \mathrm{Con}(\mathbf{A}) : \theta \subseteq \beta \subseteq \nabla\}$.

Theorem 1.5.13 *Let* $\theta \in \mathrm{Con}(\mathbf{A})$. *Then* $\mathbf{A}/\theta$ *is simple if and only if* θ *is a maximal congruence of* $\mathbf{A}$ *or* $\theta = \nabla$.

The next theorem provides us with a characterization of subdirectly irreducible algebras.

Theorem 1.5.14 ([61, Theorem 8.4]) *An algebra* $\mathbf{A} = (A, F^A)$ *of type* $\mathcal{F}$ *is subdirectly irreducible if and only if* $\mathbf{A}$ *is trivial or* $\mathrm{Con}(\mathbf{A}) \setminus \{\Delta\}$ *has a minimum element, equivalently,*

$$\bigcap (\mathrm{Con}(\mathbf{A}) \setminus \{\Delta\}) \neq \Delta.$$

Theorem 1.5.15 *Every subdirectly irreducible algebra is directly indecomposable.*

Theorem 1.5.16 ([61, Theorem 8.6], Birkhoff) *Every algebra* $\mathbf{A} = (A, F^A)$ *of type* $\mathcal{F}$ *is isomorphic to a subdirect product of subdirectly irreducible algebras those are homomorphic images of* $\mathbf{A}$.

1.5.2   Operators and Varieties

This section focuses on studying classes of algebras of the same type. We introduce several key operators that map a class of algebras of type $\mathcal{F}$ to another class of algebras of the same type.

Definition 1.5.17 ([61]) Let C be a class of algebras of type $\mathcal{F}$.

- $\mathbf{A} \in \mathrm{I}(\mathsf{C})$ if and only if $\mathbf{A}$ is isomorphic to some elements of C.
- $\mathbf{A} \in \mathrm{H}(\mathsf{C})$ if and only if $\mathbf{A}$ is a homomorphic image of some elements of C.
- $\mathbf{A} \in \mathrm{S}(\mathsf{C})$ if and only if $\mathbf{A}$ is a subalgebra of some elements of C.
- $\mathbf{A} \in \mathrm{P}(\mathsf{C})$ if and only if $\mathbf{A}$ is a direct product of a non-empty family of elements of C.
- $\mathbf{A} \in \mathrm{P_s}(\mathsf{C})$ if and only if $\mathbf{A}$ is a subdirect product of a non-empty family of elements of C.

Each of I, H, S, P and $\mathrm{P_s}$ is called an **operator** on classes of algebras. A class C of algebras of type $\mathcal{F}$ is said to be **closed** under an operator O if $\mathrm{O}(\mathsf{C}) \subseteq \mathsf{C}$.

Definition 1.5.18 Assume that O_1 and O_2 are operators on classes of algebras.

 (i)   We write $\mathrm{O}_1\mathrm{O}_2$ for the composition of the two operators, that is $\mathrm{O}_1\mathrm{O}_2(\mathsf{C}) := \mathrm{O}_1(\mathrm{O}_2(\mathsf{C}))$ for all classes of algebras C.
 (ii) We write $\mathrm{O}_1 \leq \mathrm{O}_2$ if $\mathrm{O}_1(\mathsf{C}) \subseteq \mathrm{O}_2(\mathsf{C})$ for all classes of algebras C.
(iii) An operator O is called **idempotent** if $\mathrm{O}^2 = \mathrm{O}$, where $\mathrm{O}^2 := \mathrm{OO}$.

Clearly, $\mathrm{O}_1 = \mathrm{O}_2$ if and only if $\mathrm{O}_1 \leq \mathrm{O}_2 \leq \mathrm{O}_1$.

Proposition 1.5.19 ([61]) SH $\leq$ HS, PS $\leq$ SP, *and* PH $\leq$ HP. *In addition, the operators,* H, S, *and* IP *are idempotent.*

Definition 1.5.20 ([61]) A non-empty class C of algebras of type $\mathcal{F}$ is called a **variety** if it is closed under H, S, and P.

The intersection of a class of varieties of type $\mathcal{F}$ is a variety with the same type $\mathcal{F}$. Also, all algebras of type $\mathcal{F}$ form a variety, so for every class C of algebras of type $\mathcal{F}$, there is a smallest variety containing C denoted by $\mathrm{V}(\mathsf{C})$. It is called **the variety generated by** C. Therefore, we can consider V as an operator on classes of algebras. If $\mathsf{C} = \{\mathbf{A}\}$ for some algebra $\mathbf{A} = (A, F)$, then we simply denote the variety generated by C by $\mathrm{V}(\mathbf{A})$.

Theorem 1.5.21 ([61, Theorem 2.9.5]) $\mathrm{V}(\mathsf{C}) = \mathrm{HSP}(\mathsf{C})$.

Proposition 1.5.22 ([159, Theorem 10B]) *The variety* A *of Abelian ℓ-groups is the least non-trivial subvariety of ℓ-groups. In addition, $\mathbb{Z}$, the group of integers, generates the variety Abelian ℓ-groups, that is* $\mathsf{A} = \mathrm{V}(\mathbb{Z})$.

Corollary 1.5.23 ([73, Chap. 8]) *If* $\mathbf{C} = ([0, 1]; \oplus, 0, 1)$ *is the standard MV-algebra of the real unit interval, then* $\mathsf{MV} = \mathsf{V}(\mathbf{C})$. *In addition, if* $\mathbf{A}$ *is an infinite subalgebra of* $\mathbf{C}$, *then* $\mathsf{V}(\mathbf{A}) = \mathsf{MV}$.

We note that the class of perfect MV-algebras does not form a variety, as it is not closed under the product operator, (P). Let V_1 be the variety generated by all perfect MV-algebras. Evidently, $\mathsf{V}_1 \subseteq \mathsf{MV}$. According to [100, Theorem 3.8], $\mathsf{V}_1 = \mathsf{V}(\Gamma(\mathbb{Z} \overrightarrow{\times} \mathbb{Z}, (1, 0)))$. In addition, an MV-algebra $\mathbf{M}$ belongs to V_1 if and only if it satisfies the following identity (see [98, Theorem 7.2]):

$$(x \oplus x) \odot (x \oplus x) = (x \odot x) \oplus (x \odot x). \tag{1.10}$$

The following version of Birkhoff's theorem is valuable for analyzing varieties.

Theorem 1.5.24 ([61]) *Let* V *be a variety. Then every member of* V *is isomorphic to a subdirect product of subdirectly irreducible members of* V. *In addition, every variety is generated by its subdirectly irreducible members.*

Let $\{\mathbf{A}_i = (A_i, F_i)\}_{i \in I}$ be an indexed family of algebras of type $\mathcal{F}$ and F be a proper filter of the lattice $(\mathcal{P}(I), \subseteq)$. Define the binary relation θ_F on $\prod_{i \in I} \mathbf{A}_i$ by

$$(a, b) \in \theta_F \Leftrightarrow [[a = b]] \in F, \tag{1.11}$$

where $[[a = b]] = \{i \in I : a(i) = b(i)\}$. We can easily verify that θ_F is a congruence relation on the direct product $\prod_{i \in I} \mathbf{A}_i$ (see [61, Lemma 2.2]).

Recall that a proper filter of a Boolean algebra $\mathbf{B} = (B; \vee, \wedge, ', 0, 1)$ is said to be an **ultrafilter** if for every $x \in B$, either $x \in F$ or $x' \in F$. An easy calculation shows that ultrafilters and maximal filters coincide in Boolean algebras.

Definition 1.5.25 Let $\{\mathbf{A}_i = (A_i, F_i)\}_{i \in I}$ be an indexed family of algebras of type $\mathcal{F}$ and F be a proper filter of the lattice $(\mathcal{P}(I), \subseteq)$. The **reduced product** of the family $\{\mathbf{A}_i\}_{i \in I}$ is the quotient algebra $(\prod_{i \in I} \mathbf{A}_i)/F$. For each n-ary operation $\lambda \in \mathcal{F}$ and each $a_1, \ldots, a_n \in \prod_{i \in I} \mathbf{A}_i$ we have

$$f(a_1/F, \ldots, a_n/F) = f(a_1, \ldots, a_n)/F. \tag{1.12}$$

If K is a non-empty class of algebraic structures of type $\mathcal{F}$, then denote by $\mathsf{P_r}(\mathsf{K})$ the class of all reduced products $(\prod_{i \in I} \mathbf{A}_i)/F$ where $\mathbf{A}_i \in \mathsf{K}$. When F is an ultrafilter, the reduced product of the family $\{\mathbf{A}_i\}_{i \in I}$ is termed an **ultraproduct** and $\mathsf{P_r}(\mathsf{K})$ is denoted by $\mathsf{P_u}(\mathsf{K})$.

Definition 1.5.26 ([163]) A class C of algebras of type F is a **quasivariety** if it is closed under the operators I (isomorphic), S (subalgebra), P (direct product), $\mathsf{P_r}$ (reduced products) and contains a trivial algebra.

Theorem 1.5.27 ([61, Theorem 2.2.25]) C *is a quasivariety if and only if it closed under* $\mathrm{ISPP_u}$ *and contains a trivial algebra.*

1.6 Equational Classes of Algebras

Let X be a set of (distinct) variables and $\mathcal{F}$ be a type of algebras. The set $T(X)$ of terms of type $\mathcal{F}$ over X is the smallest set such that

(T1) $X \cup \mathcal{F}_0 \subseteq T(X)$.

(T2) If $p_1, \ldots, p_n \in T(X)$ and $\lambda \in \mathcal{F}_0$, then $\lambda(p_1, \ldots, p_n) \in T(X)$. If f is a binary operation $*$ we usually prefer $p_1 * p_2$ to $*(p_1, p_2)$.

We remark that $T(X) \neq \emptyset$ if and only if $X \cup \mathcal{F}_0 \neq \emptyset$. We can easily see tha t $(T(X), \mathcal{F}^{T(X)})$ is an algebra of type $\mathcal{F}$. For each $n \in \mathbb{N}$ and $f \in \mathcal{F}_n$ we have

$$f^{T(X)} : T(X)^n \to T(X), \qquad f^{T(X)}(p_1, \ldots, p_n) = f(p_1, \ldots, p_n).$$

Furthermore, $T(\emptyset)$ exists if and only if $\mathcal{F}_0 \neq \emptyset$. Evidently, $(T(X), F^{T(X)})$ is generated by X, i.e., $\mathrm{Alg}(X)$ in $(T(X), F^{T(X)})$ is equal to $T(X)$.

For each element $p \in T(X)$, we often write p as $p(x_1, \ldots, x_n)$ to indicate that the variables occurring in p are among the elements $x_1, \ldots x_n$ of X. In addition, a term p is **n-ary** if the number of variables appearing explicitly in p is equal to or less than n. Every element $p(x_1, \ldots, x_n)$ of the algebra $(T(X), F^{T(X)})$ is said to be a **term of type** $\mathcal{F}$ over X (for more details see [61, Ch 10]).

Definition 1.6.1 ([61, Definition 10.2]) Suppose that (A, F^A) is an algebra of type $\mathcal{F}$ and $p(x_1, \ldots, x_n)$ is a term of type $\mathcal{F}$ over X. We define the mapping $p^A : A^n \to A$ as follows:

(i) If p is a variable x_i (that is, $p = x_i$), then $p^A(a_1, \ldots, a_n) := a_i$, for all $a_1, \ldots, a_n \in A$, i.e., p^A is the ith projection map.

(ii) If p is of the form $f(p_1(x_1, \ldots, x_n), \ldots, p_k(x_1, \ldots, x_n))$ where $f \in \mathcal{F}^k$, then we define

$$p^A(a_1, \ldots, a_n) := f^A\big(p_1^A(a_1, \ldots, a_n), \ldots, p_k^A(a_1, \ldots, a_n)\big).$$

In particular, if $p = f \in \mathcal{F}$, then $p^A = f^A$. We define p^A as the term function on (A, F^A) corresponding to the term p. For brevity, we will often omit the superscript A.

Definition 1.6.2 ([61, Definition 11.1]) An **identity of type** $\mathcal{F}$ over a set X is an expression of the form $p \approx q$, where $p, q \in T(X)$. An algebra (A, F^A) of type $\mathcal{F}$ satisfies an identity $p(x_1, \ldots, x_n) \approx q(x_1, \ldots, x_n)$ if

$$p^A(a_1, \ldots, a_n) = q^A(a_1, \ldots, a_n) \qquad \forall a_1, \ldots, a_n \in A.$$

Then we say that the identity is true in (A, F^A), or holds in (A, F^A), and write

$$(A, F^A) \models p(x_1, \ldots, x_n) \approx q(x_1, \ldots, x_n), \text{ briefly } (A, F^A) \models p \approx q \text{ or } \mathbf{A} \models p \approx q.$$

In addition, if Σ is a set of identities, we say (A, F^A) satisfies Σ, written $(A, F^A) \models \Sigma$ if $(A, F^A) \models p \approx q$ for each identity $p \approx q \in \Sigma$. Furthermore, a class K of algebras satisfies $p \approx q$, written $\mathsf{K} \models p \approx q$, if each member of K satisfies $p \approx q$. Similarly, given a set Σ of identity, we say K satisfies Σ, written $\mathsf{K} \models \Sigma$ if $\mathsf{K} \models p \approx q$ for each identity $p \approx q \in \Sigma$.

Definition 1.6.3 ([61, Definition 11.7]) Let Σ be a set of identities of type $\mathcal{F}$ and define $\mathsf{M}(\Sigma)$ to be the class of algebras (A, F^A) satisfying $(A, F^A) \models \Sigma$. A class K of algebras of type $\mathcal{F}$ is an equational class if there is a set of identities Σ such that $\mathsf{K} = \mathsf{M}(\Sigma)$.

Theorem 1.6.4 ([61, Theorem 11.9]) (Birkhoff). *A class of algebra* K *is an equational class if and only if* K *is a variety.*

Theorem 1.6.5 ([61, Theorems 12.2 and 12.3]) *Let* V *be a variety of type* $\mathcal{F}$.

(i) *The variety* V *is congruence-permutable if and only if there is a term* $p(x, y, z)$ *such that* $\mathsf{V} \models p(x, x, y) \approx y$ *and* $\mathsf{V} \models p(x, y, y) \approx x$.
(ii) *If there exists a term* $q(x, y, z)$ *such that* $\mathsf{V} \models q(x, x, y) \approx q(x, y, x) \approx q(y, x, x)$, *then* V *is congruence-distributive.*

Definition 1.6.6 ([61]) A variety V of type $\mathcal{F}$ is **arithmetical** if it is both congruence-distributive and congruence-permutable.

Theorem 1.6.7 ([61, Theorem 12.5]) (Pixley). *A variety* V *of type* $\mathcal{F}$ *is arithmetical if and only if it satisfies either of the equivalent conditions:*

(i) *There exist terms* $p(x, y, z)$ *and* $q(x, y, z)$ *satisfying the conditions* (i) *and* (ii) *of Theorem 1.6.5, respectively.*
(ii) *There exists a term* $r(x, y, z)$ *such that* $\mathsf{V} \models r(x, y, x) \approx r(x, y, y) \approx r(y, y, x)$.

At the end of this section, we recommend consulting the works by [31, 61, 86, 163] for further insights into the varieties and equational classes of algebras.

Pocrims and Related Structures **2**

In this chapter, we study partially ordered commutative residuated integral monoids known as pocrims. Pocrims are utilized in the study of ordered algebraic structures and have applications in various fields, including algebraic logic and reasoning, representation theory, and category theory, particularly in studying enriched categories, and computer science. For instance, hoops, BL-algebras, and MV-algebras are specific subclasses of pocrims, and they play a significant role in fuzzy logic and other areas of non-classical logic. On the other hand, pocrims have a close connection with BCK-algebras, which will be studied in the last section of this chapter. It is important to note that while pocrims offer a powerful tool for modeling uncertainty and imprecision, their applications are still an active area of research.

The chapter is organized into five sections, with each section focusing on a distinct aspect, thereby offering a detailed and nuanced understanding of the algebraic structure. We begin by establishing the formal definition of pocrims and exploring a range of illustrative examples, setting the stage for a deeper investigation. The second section is dedicated to unraveling the core properties of pocrims, including boundedness, involutivity, and the double negation property (DNP), providing essential tools for their analysis. A significant contribution is the introduction of the ordinal sum of pocrims, a powerful technique for understanding their structural composition. Furthermore, we examine the relationships between filters and congruence relations, crucial for algebraic investigations. Finally, we broaden our scope to encompass BCK-algebras, demonstrating that the quasivariety of pocrims is termwise equivalent to a specific subclass of BCK-algebras with the product property, thus highlighting the connections between these algebraic structures. For those acquainted with BCK-algebras, this equivalence offers a more insightful understanding of pocrims and their applications and properties.

© The Author(s), under exclusive license to Springer Nature Switzerland AG 2026   33
A. Dvurečenskij et al., *Hoop Algebras*, Frontiers in Mathematics,
https://doi.org/10.1007/978-3-032-11736-6_2

2.1   Pocrims

This section aims to define pocrims and present several examples of this structure.

Recall that a **partially ordered monoid** is a monoid $(H; \odot, 1)$ equipped with an partially ordered relation $\leq$ which is compatible with $\odot$, that is

$$x \leq y \text{ implies } w \odot x \odot z \leq w \odot y \odot z \qquad \text{for all } x, y, w, z \in H. \tag{2.1}$$

We usually denote it by $\mathbf{H} = (H; \leq, \odot, 1)$. If 1 is the (least) greatest element of $(H, \leq)$, then the partial monoid $\mathbf{H}$ is called **(dually) integral**. A commutative partially ordered monoid $\mathbf{H}$ is **residuated** if for all $x, y \in H$, there exists a greatest element $z \in H$, denoted by $x \to y$ and called the residuum of x with respect to y, such that $z \odot x \leq y$, i.e., $\to$ is an additional binary operation on $\mathbf{H}$ satisfying the adjointness condition:

$$x \leq z \to y \text{ if and only if } z \odot x \leq y \qquad \text{for all } x, y, z \in H. \tag{2.2}$$

The binary operation $\to$ of porcims is called **residuum** or **implication**.

Definition 2.1.1 A **pocrim** is the abbreviation for partially ordered commutative residuated integral monoids.

Since 1 is assumed to be the greatest element of any pocrim $\mathbf{H}$, the adjointness condition yields

$$x \leq y \text{ if and only if } x \to y = 1, \quad \forall x, y \in H, \tag{2.3}$$

which allows us to regard every pocrim as an algebra $\mathbf{H} = (H; \odot, \to, 1)$ of type $(2, 2, 0)$. For each $x \in H$ and $n \in \mathbb{N}$, we let $x^0 := 1$, $x^1 := x$, and $x^{n+1} = x^n \odot x$.

By a **bounded pocrim** we mean an algebra $\mathbf{H} = (H; \odot, \to, 0, 1)$ of type $(2, 2, 0, 0)$ where $(H; \odot, \to, 1)$ is a pocrim with a least element 0. Every bounded pocrim satisfies the identity $0 \odot x = 0$, since $0 \odot x \leq 0 \odot 1 = 0$. Clearly, an algebra $(\{x\}; \odot, \to, x)$ where $\odot$ and $\to$ defined by $x \to x = x \odot x = x$ is a pocrim. It is called a **trivial pocrim**. A pocrim $\mathbf{H}$ is **non-trivial** if $1 < |H|$. In the sequel, we will show some important examples of pocrims:

Example 2.1.2 Every Boolean algebra $(B; \vee, \wedge, ', 0, 1)$ can be seen as a bounded pocrim $(B; \odot, \to, 0, 1)$, where $\odot = \wedge$ and $x \to y = x' \vee y$ for all $x, y \in B$. For instance, the two-element Boolean algebra $(\{0, 1\}; \vee, \wedge, ', 0, 1)$ with $0 < 1$, $0' = 1$ and $1' = 0$ induces the bounded pocrim $\mathbf{H} = (\{0, 1\}; \odot, \to, 0, 1)$, where $\odot = \wedge$, $0 \to 1 = 1$ and $1 \to 0 = 0$.

It follows that for every set X, $(\mathcal{P}(X); \cap, \to, \emptyset, X)$ is a bounded pocrim, where $Y \to Z = (X \setminus Y) \cup Z$ for all subset Y and Z of X.

The concept of a residuated lattice was introduced by Ward and Dilworth [259] as a generalization of residuation in lattices of ideals of commutative rings with an identity element (see Example 2.1.4).

Definition 2.1.3 A **commutative residuated meet-semilattice** is a pocrim $(H; \odot, \to, 1)$ such that $(H, \leq)$ is a meet-semilattice, where $\leq$ is defined by (2.3).

A **commutative residuated join-semilattice** is a pocrim $(H; \odot, \to, 1)$ such that $(H, \leq)$ is a join-semilattice, where $\leq$ is defined by (2.3).

A **commutative residuated lattice** is a pocrim $(H; \odot, \to, 1)$ such that $(H, \leq)$ is a lattice, where $\leq$ is defined by (2.3).

To be more precise, a residuated join-semilattice is an algebra $\mathbf{H} = (H; \vee, \odot, \to, 1)$ of type $(2, 2, 2, 0)$, a residuated meet-semilattice is an algebra $\mathbf{H} = (H; \wedge, \odot, \to, 1)$ of the same type, and a residuated lattice is an algebra $\mathbf{H} = (H; \vee, \wedge, \odot, \to, 1)$ of type $(2, 2, 2, 2, 0)$. A bounded residuated lattice is an algebra $\mathbf{H} = (H; \vee, \wedge, \odot, \to, 0, 1)$ of type $(2, 2, 2, 0, 0)$, where $(H; \vee, \wedge, \odot, \to, 1)$ is a residuated lattice with the least element 0.

In this book, we always speak about commutative residuated lattices and we simply call them residuated lattices (unless otherwise stated). However, non-commutative residuated lattices have become a subject of intensive research in recent years, mainly in the context of non-commutative logic. For a survey of non-commutative residuated lattices, we refer to [192], for a survey of commutative ones see, e.g., [68]. In Sect. 2.5, we will propose an equivalent definition for residuated lattices.

Example 2.1.4 ([31, 69]) Let $\mathbf{R} = (R; +, \cdot, -, 0, 1)$ be a commutative ring with an identity element 1 and $\mathrm{Id}(\mathbf{R})$ be the set of all ideals of $\mathbf{R}$. Then $(\mathrm{Id}(\mathbf{R}), \subseteq)$ is a complete lattice with $\{0\}$ as the least element and R as the top element. Suppose $\cdot$ stands for the usual ideal multiplication, i.e.,

$$I \cdot J = \{a_1 b_1 + \cdots + a_n b_n : a_1, \ldots, a_n \in I, \ b_1, \ldots, b_n \in J, \ n \in \mathbb{N}\}, \quad \forall I, J \in \mathrm{Id}(\mathbf{R}).$$

Define a binary operation $\to : \mathrm{Id}(\mathbf{R}) \times \mathrm{Id}(\mathbf{R}) \to \mathrm{Id}(\mathbf{R})$ by

$$I \to J = \{x \in R : x \cdot I \subseteq J\}.$$

Then, $(\mathrm{Id}(\mathbf{R}); \vee, \cap, \cdot, \to, \{0\}, R)$ is a bounded residuated lattice, where

$$I \vee J = \bigcap \{A \in \mathrm{Id}(\mathbf{R}) : I \cup J \subseteq A\}.$$

Therefore, $(\mathrm{Id}(\mathbf{R}); \cdot, \to, \{0\}, R)$ is a bounded pocrim.

Example 2.1.5 Let $H_1 = \{x \in \mathbb{R}: x \leq 0\}$ and $H_2 = \{x \in \mathbb{Z}: x \leq 0\}$. Define $x \odot y = x + y$ and $x \to y = \min\{y - x, 0\}$ for all $x, y \in H_1$. We can easily check that $(H_1; \odot, \to, 0)$ and $(H_2; \odot, \to, 0)$ are pocrims. Remark that $x \to y = 0$ if and only if $0 \leq y - x$ if and only if $x \leq y$ with respect to the natural ordering of real numbers.

Example 2.1.6 ([142, Example 2.1.7]) Let $\odot$ be the usual multiplication on $\mathbb{N}$ and $x \to y = y/\gcd(x, y)$ for all $x, y \in \mathbb{N}$, where $\gcd(x, y)$ is the greatest common divisor of x and y and $\mathrm{lcm}(x, y)$ is the least common multiple of x and y. Then $(\mathbb{N}; \odot, \to, 1)$ is a pocrim.

Example 2.1.7 A relatively pseudo-complemented semilattice is an algebra $\mathbf{S} = (S; \wedge, *, 1)$ of type $(2, 2, 0)$ where $(S; \wedge)$ is a meet-semilattice with the top element 1 and for all $x, y, z \in S$ we have the equivalence

$$z \leq x * y \text{ if and only if } z \wedge x \leq y.$$

Then $\mathbf{S}$ can be viewed as a residuated meet-semilattice as well as a pocrim $(S; \wedge, \odot, \to, 1)$ in which $x \odot y = x \wedge y$ and $x \to y = x * y$.

Example 2.1.8 ([70, 73]) It is known that MV-algebras can be considered as bounded residuated lattices. Indeed, given an MV-algebra $(M; \oplus, {}', 0, 1)$, we define $x \vee y = (x' \oplus y)' \oplus y$, $x \wedge y = (x' \vee y')'$, $x \odot y = (x' \oplus y')'$, and $x \to y = x' \oplus y$. Then $(M; \vee, \wedge, \odot, \to, 0, 1)$ becomes a bounded residuated lattice satisfying the identity

$$x \vee y = (x \to y) \to y, \tag{2.4}$$

in particular, $(M; \vee, \wedge, \odot, \to, 0, 1)$ satisfies (DNP):

(DNP) Double Negation Property: $x = x'' = (x \to 0) \to 0$.

For more details on MV-algebras see Sect. 1.4.1.

Example 2.1.9 ([168]) A BL-algebra is a bounded residuated lattice satisfying the identities:

(Div) Divisibility: $x \wedge y = x \odot (x \to y)$.

(Pre) Prelinearity: $(x \to y) \vee (y \to x) = 1$.

BL-algebras were invented by Hájek [168] as an algebraic semantics for his basic fuzzy logic. It is worth noting that every MV-algebra (as a residuated lattice) is a BL-algebra and a BL-algebra is an MV-algebra exactly if it satisfies the law of (DNP). A more detailed study of these structures will be carried out in Chap. 5.

Example 2.1.10 ([141, 268]) Let $H = \{0, 1, a, b, c, d\}$ be an ordered set such that $0 < a < b < c < d < 1$. Consider the binary operations $\odot$ and $\to$ defined in Table 2.1:

Then $(H; \odot, \to, 0, 1)$ is a pocrim (in particular, a bounded residuated lattice) satisfying (Pre).

Table 2.1 Operations $\odot$ and $\rightarrow$ of Example 2.1.10

Table 2.1.1

$\odot$	0	a	b	c	d	1
0	0	0	0	0	0	0
a	0	0	0	0	0	a
b	0	0	a	a	a	b
c	0	0	a	b	b	c
d	0	0	a	b	c	d
1	0	a	b	c	d	1

Table 2.1.2

$\rightarrow$	0	a	b	c	d	1
0	1	1	1	1	1	1
a	d	1	1	1	1	1
b	0	d	1	1	1	1
c	0	b	d	1	1	1
d	0	b	d	d	1	1
1	0	a	b	c	d	1

Example 2.1.11 ([137]) An MTL-algebra is a bounded residuated lattice satisfying the prelinearity identity. These algebras are an algebraic counterpart of the monoidal t-norm based logic (see [137]). Suppose that $(H, \leq)$ is a bounded linearly ordered set and $n : H \rightarrow H$ is a weak negation (see Definition 1.1.2) on L. Define $\odot : H \times H \rightarrow H$ and $\rightarrow: H \times H \rightarrow H$ by

$$x \odot y = \begin{cases} 0 & \text{if } x \leq n(y) \\ x \wedge y & \text{otherwise,} \end{cases} \qquad x \rightarrow y = \begin{cases} 1 & \text{if } x \leq y \\ n(x) \vee y & \text{otherwise.} \end{cases}$$

Then $\mathbf{H} = (H; \vee, \wedge, \odot, \rightarrow, 0, 1)$ is an MTL-algebra. Moreover, if n is an involution, then $\mathbf{H}$ fulfills the law of double negation $n(n(x)) = x$, but it need not be an MV-algebra. As an example, let H be the chain $0 < a < b < 1$ with the involution $n(0) = 1$, $n(a) = b$, $n(b) = a$, and $n(1) = 0$. Then the operations $\odot$ and $\rightarrow$ are given by Table 2.2.

Table 2.2 Operations $\odot$ and $\rightarrow$ of Example 2.1.11

Table 2.2.1

$\odot$	0	a	b	1
0	0	0	0	0
a	0	0	0	a
b	0	0	b	b
1	0	a	b	1

Table 2.2.2

$\rightarrow$	0	a	b	1
0	1	1	1	1
a	b	1	1	1
b	a	a	1	1
1	0	a	b	1

The algebra $\mathbf{H}$ does not satisfy the identity (2.4) and hence it is not induced from an MV-algebra (see Example 2.1.8). Indeed, $(b \to a) \to a = 1 \neq b = a \vee b$. Similarly, divisibility does not hold:

$$a \wedge b = a \neq 0 = b \odot a = b \odot (b \to a).$$

In Appendix B, we present all of the pocrims of order ≤ 5.

2.2   Main Properties of Pocrims

In this section, we will study and investigate the structure of pocrims to collect their main properties. These properties contribute to a deeper understanding of this structure and allow us to provide an equivalent definition for pocrims (see Theorem 2.2.7).

Lemma 2.2.1 *In every pocrim* $\mathbf{H} = (H; \odot, \to, 1)$ *the following statements hold for all* $x, y, z \in H$:

(1) $x \to 1 = 1 = x \to x.$
(2) $1 \to x = x.$
(3) $x \leq y$ *implies that* $x \odot z \leq y \odot z.$
(4) $(x \to y) \odot x \leq y$ *and* $x \odot y \leq x, y.$
(5) $x \to y \leq (x \odot z) \to (y \odot z).$
(6) $x \leq (x \to y) \to y.$
(7) $x \leq y$ *implies* $z \to x \leq z \to y$, *and* $y \to z \leq x \to z.$
(8) $x \to (y \to z) = (x \odot y) \to z = y \to (x \to z).$
(9) $x \leq y \to z$ *if and only if* $x \odot y \leq z$ *if and only if* $y \leq x \to z.$
(10) $x \leq y \to x$ *and* $y \to x = \max\{z \in H : y \odot z \leq x\}.$
(11) $x \to y \leq (z \to x) \to (z \to y).$
(12) $x \to y \leq (y \to z) \to (x \to z).$
(13) $x \to y = ((x \to y) \to y) \to y.$
(14) $(x \to y) \odot (y \to z) \leq x \to z.$
(15) $x \leq y \to (x \odot y).$

Proof The equations (1) and (2) are immediate consequences of (2.2).

(3) Let $x \leq y$. From $y \odot z \leq y \odot z$ we get $y \leq z \to (y \odot z)$ entails that $x \leq y \leq z \to (y \odot z)$, and so $x \odot z \leq y \odot z.$

(4) $x \to y \leq x \to y$ and (2.2) imply that $x \odot (x \to y) \leq y$. In addition, $x \odot y \leq 1 \odot y = y$ and $x \odot y \leq x \odot 1 = x$, since $(H; \odot, 1)$ is a partially ordered monoid.

(5) Applying (4), we have $(x \to y) \odot x \leq y$, so $(x \to y) \odot x \odot z \leq y \odot z$ which is equivalent to $x \to y \leq (x \odot z) \to (y \odot z)$, by (2.2).

(6) By (4), $x \odot (x \to y) \leq y$, so (2.2) entails that $x \leq (x \to y) \to y.$

(7) By (4) we have $(z \to x) \odot z \leq x \leq y$ which yields $z \to x \leq z \to y$. In addition, from $(y \to z) \odot x \leq (y \to z) \odot y \leq z$ we get $y \to z \leq x \to z$.

(8) For each $u \in H$, we have $u \leq x \to (y \to z)$ if and only if $u \odot x \leq y \to z$ if and only if $u \odot x \odot y \leq z$ if and only if $u \leq (x \odot y) \to z$. Thus, $(x \odot y) \to z = x \to (y \to z)$. In a similar way, we can prove that $(x \odot y) \to z = y \to (x \to z)$.

(9) The proof of this part is obvious, by (2.2), since $\odot$ is commutative.

(10) Due to (4), $x \odot y \leq x$, so by (2.2), $x \leq y \to x$. In addition, $y \to x \leq y \to x$ implies that $y \odot (y \to x) \leq x$ which means $y \to x \in \{z \in H : z \odot y \leq x\}$. Given $z \in H$ such that $y \odot z \leq x$ we have $z \leq y \to x$, by (2.2). Therefore, $y \to x = \max\{z \in H : z \odot y \leq x\}$.

(11) Using (4) twice, we have $(z \to x) \odot z \leq x$ and

$$(x \to y) \odot (z \to x) \odot z \leq (x \to y) \odot x \leq y$$

yielding $x \to y \leq ((z \to x) \odot z) \to y = (z \to x) \to (z \to y)$, by (8).

(12) It follows at once from (11) by (9).

(13) As a consequence of (6) and (7), we obtain (13).

(14) Applying (11), $y \to z \leq (x \to y) \to (x \to z)$, so by (2.2), $(x \to y) \odot (y \to z) \leq x \to z$.

(15) $x \odot y \leq x \odot y$, so by (2.2), $x \leq y \to (x \odot y)$. $\square$

Remark 2.2.2 Every finite pocrim $\mathbf{H} = (H; \odot, \to, 1)$ is bounded. Indeed, if $H = \{x_1, \ldots, x_n\}$ for some $n \in \mathbb{N}$, then by Lemma 2.2.1(4), $a := x_1 \odot \cdots \odot x_n \leq y$ for all $y \in H$, and so $\mathbf{H}$ is bounded.

Lemma 2.2.3 ([69]) *Let* $\mathbf{H} = (H; \odot, \to, 1)$ *be a pocrim. If the supremum (resp. infimum) on the left-hand side exists, then so does that on the right-hand side and the following equalities are valid:*

(i) $x \odot \bigvee \{y_i : i \in I\} = \bigvee \{x \odot y_i : i \in I\}$.

(ii) $x \to \bigwedge \{y_i : i \in I\} = \bigwedge \{x \to y_i : i \in I\}$.

(iii) $(\bigvee \{x_i : i \in I\}) \to y = \bigwedge \{x_i \to y : i \in I\}$.

Proof (i) Put $y = \bigvee \{y_i : i \in I\}$. Then $x \odot y_i \leq x \odot y$ for all $i \in I$. Having $x \odot y_i \leq z$, for all $i \in I$, then by Lemma 2.2.1(9), $y_i \leq x \to z$ for every $i \in I$, so $y \leq x \to z$ and hence $x \odot y \leq z$ proving that $x \odot y$ is the supremum of $\{x \odot y_i : i \in I\}$.

(ii) Put $y = \bigwedge \{y_i : i \in I\}$. Then due to Lemma 2.2.1(7), $x \to y \leq x \to y_i$ for all $i \in I$. If z is another lower bound of the set $\{x \to y_i : i \in I\}$, then $z \odot x \leq y_i$, for every $i \in I$ whence $z \odot x \leq y$ and consequently $z \leq x \to y$. Thus, $x \to y$ is the infimum of $\{x \to y_i : i \in I\}$.

(iii) Let $x = \bigvee \{x_i : i \in I\}$. Then by Lemma 2.2.1(7), $x \to y \leq x_i \to y$ for all $i \in I$. Assume that $z \leq x_i \to y$, for all $i \in I$. Then $x_i \leq z \to y$ for every $i \in I$, which implies $x \leq z \to y$ and so by Lemma 2.2.1(9), $z \leq x \to y$ showing that $x \to y$ is the infimum of $\{x_i \to y : i \in I\}$. $\square$

Proposition 2.2.4 ([169, Lemma 2]) *Let x and y be two elements of a pocrim* $\mathbf{H}$. *If* $x \vee y = 1$, *then* $x^k \vee y^n = 1$ *for all* $k, n \in \mathbb{N}$.

Proof Let $k, n \in \mathbb{N}$ and $x \vee y = 1$. First, we show that $x^k \vee y = 1$. We proceed by induction on k. The statement holds for $k = 1$, evidently. By induction hypothesis, assume that the statement holds for k, and let $x^{k+1}, y \leq z$, hence by (2.2), $x^k \leq x \to z$. Also, $y \leq z \leq x \to z$, by Lemma 2.2.1(10). Hence, $1 = x^k \vee y \leq x \to z$ entails that $x \leq z$. Therefore, $1 = x \vee y \leq z$, proving $z = 1$. The rest can be proved analogously by induction on n. $\qquad\square$

Remark 2.2.5 On each bounded pocrim $\mathbf{H} = (H; \odot, \to, 0, 1)$, we can always define a unary operation $' : H \to H$ by $x' := x \to 0$ for each $x \in H$. It satisfies the following properties:

   (i) $x \leq y$ implies that $y' \leq x'$ for all $x, y \in H$ (follows from Lemma 2.2.1(7)).
  (ii) $x \leq x''$ for all $x \in H$ (follows from Lemma 2.2.1(6)).
 (iii) $x''' = x'$ for all $x \in H$ (follows from Lemma 2.2.1(13)).
 (iv) $x \odot x' = x \odot (x \to 0) \leq 0$, by Lemma 2.2.1(4), so $x \odot x' = 0$ for all $x \in H$.
  (v)   Due to Lemma 2.2.1(12), $x \to y \leq y' \to x'$, and by Lemma 2.2.1(8), we get

$$x \to y' = x \to (y \to 0) = (x \odot y)' = y \to (x \to 0) = y \to x', \quad \forall x, y \in H.$$

       Similarly by (iii), $x \to y' = x \to y''' = y'' \to x'$ for all $x, y \in H$.
 (vi)   By Lemma 2.2.1(10),

$$1' = \max\{z \in H : z \odot 1 \leq 0\} = \max\{z \in H : z \leq 0\} = \max\{0\} = 0.$$

       Similarly, $0' = \max\{z \in H : z \odot 0 \leq 0\} = 1$, since $1 \odot 0 = 0$.
 (vii)   If $\mathrm{Reg}(\mathbf{H}) := \{x \in H : x'' = x\}$, then $\mathrm{Reg}(\mathbf{H}) = \{x' : x \in H\}$, see (iii). In addition, by (vi), $0'' = 1' = 0$ and $1'' = 0' = 1$, which imply $0, 1 \in \mathrm{Reg}(\mathbf{H})$. Furthermore, $\mathrm{Reg}(\mathbf{H})$ is closed under $\to$. More generally, for each $x \in H$ and $y \in \mathrm{Reg}(\mathbf{H})$ we have $x \to y \in \mathrm{Reg}(\mathbf{H})$:

$$\begin{aligned}
(x \to y)'' &= (x \to y'')'' = (x \to (y' \to 0))'' \\
&= ((x \odot y')')'', \text{ by Lemma 2.2.1 (8)} \\
&= (x \odot y')', \text{ by (iii)} \\
&= x \to y'' = x \to y.
\end{aligned}$$

       According to [239], a bounded pocrim $\mathbf{H} = (H; \odot, \to, 0, 1)$ is called **involutive** if it satisfies (DNP), i.e., $x'' = x$ for all $x \in H$, equivalently, $\mathrm{Reg}(\mathbf{H}) = H$.
(viii)   If $\mathbf{H}$ is an involutive pocrim, then $x \to y = y' \to x'$ for all $x, y \in H$. In fact,

$$x \to y = x \to y'' = x \to (y' \to 0) = (x \odot y') \to 0 = y' \to x', \text{ by Lemma 2.2.1 (8)}.$$

On each bounded pocrim $(H; \odot, \rightarrow, 0, 1)$ we can always define three binary operations:

$$x \oplus y := x' \rightarrow y'' \quad \& \quad x \star y := x' \rightarrow y \quad \& \quad x \ominus y := x \odot y' \quad \forall x, y \in H. \qquad (2.5)$$

Then $\oplus$, $\star$, and $\ominus$ satisfy the following conditions:

Proposition 2.2.6 *Let* $\mathbf{H}$ *be a bounded pocrim. Then the following assertions hold for all* $x, y, z \in H$:

  (i) $x \oplus y = (x' \odot y')'$.
 (ii) $\oplus$ *is a commutative and associative binary operation.*
(iii) $x \leq y$ *implies that* $x \oplus z \leq y \oplus z$.
(iv) $(\mathrm{Reg}(\mathbf{H}); \oplus, 0)$ *is a commutative partially ordered monoid.*
 (v)   *If* $\mathbf{H}$ *satisfies* (DNP)*, then* $\oplus = \star$.
(vi) $x \leq y$ *implies that* $x \ominus z \leq y \ominus z$ *and* $z \ominus y \leq z \ominus x$. *In addition,* $x \ominus y = 0$ *if and only if* $x \leq y''$.
(vii)   *If* $x \vee y$ *exists, then* $(x \vee y) \ominus z = (x \ominus z) \vee (y \ominus z)$.

Proof (i) Let $x, y \in H$. By Lemma 2.2.1(8),

$$x \oplus y = x' \rightarrow y'' = x' \rightarrow (y' \rightarrow 0) = (x' \odot y') \rightarrow 0 = (x' \odot y')'.$$

(ii) Choose $x, y, z \in H$. By (i), $x \oplus y = (x' \odot y')' = (y' \odot x')' = y \oplus x$. In addition,

$$\begin{aligned}
x \oplus (y \oplus z) &= (x' \odot (y \oplus z)')' = (x' \odot (y' \odot z')'')' \\
&= x' \rightarrow (y' \odot z')''', \text{ by Lemma 2.2.1(8)} \\
&= x' \rightarrow (y' \odot z')', \text{ by Remark 2.2.5 (iii)} \\
&= (x' \odot (y' \odot z'))' = (x' \odot y' \odot z')'.
\end{aligned}$$

In a similar way, it can be proved that $(x \oplus y) \oplus z = (x' \odot y' \odot z')'$. Therefore, $\oplus$ is associative.

(iii) Let $x, y, z \in H$ such that $x \leq y$. Then $y' \leq x'$ by Remark 2.2.5(i) and so $y' \odot z' \leq x' \odot z'$, see Lemma 2.2.1(3). Applying Remark 2.2.5(i), we get $(x' \odot z')' \leq (y' \odot z')'$ and so $x \oplus z \leq y \oplus z$ by (i).

(iv) By Remark 2.2.5(vii), $0 \in \mathrm{Reg}(\mathbf{H})$. For each $x, y \in \mathrm{Reg}(\mathbf{H})$, we have $x \oplus y = x' \rightarrow y'' = x' \rightarrow y \in \mathrm{Reg}(\mathbf{H})$, see Remark 2.2.5(vii) (since $y \in \mathrm{Reg}(\mathbf{H})$). In addition, $0 \oplus x = 0' \rightarrow x'' = 1 \rightarrow x'' = x$. Now, (i)–(iii) imply that $(\mathrm{Reg}(\mathbf{H}); \oplus, 0)$ is a commutative partially ordered monoid.

(v) Let $x'' = x$ for all $x \in H$. Due to Lemma 2.2.1(8),

$$y \star z = y' \rightarrow z = y' \rightarrow z'' = (y' \odot z')' = y \oplus z, \qquad \forall y, z \in H.$$

(vi) The first inequality follows from Lemma 2.2.1(3) and the second one follows from Remark 2.2.5(i) and Lemma 2.2.1(3). In addition,

$$x \ominus y = 0 \Leftrightarrow x \odot y' = 0 \Leftrightarrow (x \odot y')' = 1 \Leftrightarrow x \to y'' = 1 \Leftrightarrow x \le y''.$$

(vii) The proof is a straight consequence of Proposition 2.2.3(i). $\qquad\square$

According to Proposition 2.2.6(v), if $\mathbf{H}$ is a bounded pocrim satisfying (DNP), then $\star = \oplus$ and so by Proposition 2.2.6(ii), $\star$ is an associative binary operation. Consider the pocrim $\mathbf{H}$ in Example 2.1.10. We have

$$(a \star a) \star a = (a' \to a)' \to a = (d \to a)' \to a = b' \to a = 0 \to a = 1.$$
$$a \star (a \star a) = a' \to (a' \to a) = d \to (d \to a) = d \to b = d.$$

Therefore, Proposition 2.2.6(v) may not hold in general.

Theorem 2.2.7 *An algebra* $\mathbf{H} = (H; \odot, \to, 1)$ *is a pocrim if and only if* $(H; \odot, 1)$ *is a commutative monoid and* $\mathbf{H}$ *satisfies the following quasi-identities:*

(i) $x \to x = x \to 1 = 1.$
(ii) $(x \to y) \to ((z \to x) \to (z \to y)) = 1.$
(iii) *If* $x \to y = 1$ *and* $y \to x = 1$, *then* $x = y$.
(iv) $(x \odot y) \to z = x \to (y \to z).$

Proof By Lemma 2.2.1 and (2.3), every pocrim satisfies (i)–(iv).

Conversely, we start by showing that $1 \to x = x$. Since $(H; \odot, 1)$ is a commutative monoid by (iv), we have $x \to (1 \to x) = (x \odot 1) \to x = x \to x = 1$ and

$$(1 \to x) \to x = ((1 \to x) \odot 1) \to x = (1 \to x) \to (1 \to x) = 1,$$

and consequently, $x = 1 \to x$ by (iii). Define a relation $\le$ of H by $x \le y$ if and only if $x \to y = 1$.

We show that $\le$ is a partial order on H. The relation is reflexive and symmetric, obviously. For transitivity, let $a \le b$ and $b \le c$, i.e., $a \to b = 1$ and $b \to c = 1$. Then using (ii), $1 = (b \to c) \to ((a \to b) \to (a \to c)) = 1 \to (1 \to (a \to c)) = a \to c$, so $a \le c$. Thus, $(H, \le)$ is a poset with 1 as the greatest element.

Further, assume $a \le b$, i.e., $a \to b = 1$. Then $b \to (c \to (b \odot c)) = (b \odot c) \to (b \odot c) = 1$ by (iv) and (i), whence it follows by (ii)

$$1 = (b \to (c \to (b \odot c))) \to \Big((a \to b) \to \big(a \to (c \to (b \odot c))\big)\Big)$$

$$= 1 \to \Big(1 \to \big((a \odot c) \to (b \odot c)\big)\Big)$$

$$= (a \odot c) \to (b \odot c),$$

so that $a \odot c \leq b \odot c$.

It remains to verify the adjointness condition. If $c \leq a \to b$, then $1 = c \to (a \to b) = (c \odot a) \to b$, so $c \odot a \leq b$.

Conversely, $c \odot a \leq b$ implies $1 = (c \odot a) \to b = c \to (a \to b)$, thus $c \leq a \to b$. $\square$

Example 2.2.8 ([13, 142]) Let $\mathbf{H}_1 = (H_1; \odot_1, \to_1, 1)$ and $\mathbf{H}_2 = (H_2; \odot_2, \to_2, 1)$ be pocrims with $H_1 \cap H_2 = \{1\}$. Then, **the ordinal sum** of $\mathbf{H}_1$ and $\mathbf{H}_2$, is denoted by $\mathbf{H}_1 \oplus \mathbf{H}_2$ and is defined as follows. The underlying set of the algebra $\mathbf{H}_1 \oplus \mathbf{H}_2$ is $H_1 \cup H_2$ and

$$x \odot y = \begin{cases} x \odot_1 y & \text{if } x, y \in H_1 \\ x \odot_2 y & \text{if } x, y \in H_2 \\ y & \text{if } x \in H_2 \setminus \{1\}, \ y \in H_1 \setminus \{1\} \\ x & \text{if } x \in H_1 \setminus \{1\}, \ y \in H_2 \setminus \{1\}, \end{cases} \tag{2.6}$$

$$x \to y = \begin{cases} x \to_1 y & \text{if } x, y \in H_1 \\ x \to_2 y & \text{if } x, y \in H_2 \\ y & \text{if } x \in H_2 \setminus \{1\}, \ y \in H_1 \setminus \{1\} \\ 1 & \text{if } x \in H_1 \setminus \{1\}, \ y \in H_2 \setminus \{1\}. \end{cases} \tag{2.7}$$

If $H_1 \cap H_2 \neq \{1\}$, then we can replace $\mathbf{H}_1$ and $\mathbf{H}_2$ with isomorphic copies whose intersection is $\{1\}$. It is enough to take $\mathbf{H}_1 \times \mathbf{1}_2$ and $\mathbf{1}_1 \times \mathbf{H}_2$, respectively, where $\mathbf{1}_i$ is the trivial one-element pocrim induced from $\{1_i\}$ for $i = 1, 2$.

(i) Clearly, $\odot$ is a commutative binary operation on $H_1 \cup H_2$.

(ii) The relation $x \leq y$ if and only if $x \to y = 1$ is a partially ordered relation on $H_1 \cup H_2$. Choose $x, y \in H_1 \cup H_2$. If $x, y \in H_1$, then $x \to y = x \to_1 y$, so $x \leq y$ if and only if $x \leq_1 y$, where $\leq_1$ is the partially ordered relation on $\mathbf{H}_1$ defined in (2.2). Similarly, if $x, y \in H_2$, then $x \leq y$ if and only if $x \leq_2 y$, where $\leq_2$ is the partially ordered relation on $\mathbf{H}_2$. Now, let $x \in H_1$ and $y \in H_2 \setminus \{1\}$. Then by definition of $\to$ we have $x \leq y$. Summing up the arguments of this part, we get

$$x \leq y \Leftrightarrow \Big((x, y \in H_1, \ x \leq_1 y) \text{ or } (x, y \in H_2, \ x \leq_2 y) \text{ or } (x \in H_1 \setminus \{1\}, y \in H_2)\Big). \tag{2.8}$$

Clearly, $\leq$ is reflexive, anti-symmetric, and transitive, consequently, $\leq$ is a partially ordered relation on $H_1 \cup H_2$. In addition, $x \leq 1$ or equivalently, $x \to 1 = 1$ for all $x \in H_1 \cup H_2$.

(iii) Let $x, y, z \in H_1 \cup H_2$ such that $x \leq y$.

If $x, y \in H_1$, then we have four cases:

(1) $z \in H_1$ implies that $x \odot z = x \odot_1 z \le y \odot_1 z = y \odot z$, since $\mathbf{H}_1$ is a pocrim;

(2) $x = 1$ and $z \in H_2 \setminus \{1\}$ imply that $y = 1$ and so $x \odot z = z \le z = y \odot z$;

(3) $x \ne 1$, $z \in H_2 \setminus \{1\}$, and $y = 1$ imply that $x \odot z = x \le z = z \odot y$; and

(4) finally, $x, y \ne 1$ and $z \in H_2 \setminus \{1\}$ imply that $x \odot z = x \le y = y \odot z$.

If $x, y \in H_2$, then we have two cases: (1) $z \in H_2$, implies that $x \odot z = x \odot_2 z \le y \odot_2 z = y \odot z$, since $\mathbf{H}_2$ is a pocrim; (2) $z \in H_1 \setminus \{1\}$, then $x \odot z = z \le z = y \odot z$.

If $x \in H_1 \setminus \{1\}$ and $y \in H_2 \setminus \{1\}$ (note that by (ii), the case $y \in H_1 \setminus \{1\}$ and $x \in H_2 \setminus \{1\}$ is impossible), then we have three cases: (1) $z = 1$ implies that $x \odot z = x \le y = y \odot z$; (2) $z \in H_1 \setminus \{1\}$, then $x \odot z = x \odot_1 z \le z = y \odot z$; (3) $z \in H_2 \setminus \{1\}$ implies that $x \odot z = x \ne 1$ and $y \odot z = y \odot_2 z \in H_2$, so by (ii), $x \odot z \le y \odot z$.

It follows that $x \odot z \le y \odot z$ for all $x, y, z \in H_1 \cup H_2$.

(iv) Choose $x, y, z \in H_1 \cup H_2$. We claim that $x \odot (y \odot z) = (x \odot y) \odot z$. Evidently, the identity holds when $x, y, z \in H_1$ or $x, y, z \in H_2$. In addition, since $w \odot 1 = w$ for all $w \in H_1 \cup H_2$, we have either $x = 1$, or $y = 1$ or $z = 1$ the identity $x \odot (y \odot z) = (x \odot y) \odot z$ holds. Thus, we assume that $x, y, z \in (H_1 \cup H_2) \setminus \{1\}$.

(1) If $x, y \in H_1 \setminus \{1\}$ and $z \in H_2 \setminus \{1\}$, then $x \odot y = x \odot_1 y \in H_1 \setminus \{1\}$ and $y \odot z = y$, so

$$x \odot (y \odot z) = x \odot y = x \odot_1 y = (x \odot_1 y) \odot z = (x \odot y) \odot z.$$

(2) If $x \in H_1 \setminus \{1\}$ and $y, z \in H_2 \setminus \{1\}$, then $y \odot z = y \odot_2 z \in H_2 \setminus \{1\}$ and $x \odot y = x$, so

$$(x \odot y) \odot z = x \odot z = x = x \odot (y \odot z).$$

Since $\odot$ is commutative, the case $y \in H_1 \setminus \{1\}$ and $x, z \in H_2 \setminus \{1\}$ holds, similarly.

(3) $x, y \in H_2 \setminus \{1\}$ and $z \in H_1 \setminus \{1\}$, then $x \odot y = x \odot_2 y \in H_2 \setminus \{1\}$ and $y \odot z = z$, so

$$x \odot (y \odot z) = x \odot z = z = (x \odot_2 y) \odot z = (x \odot y) \odot z.$$

In a similar way, $x, z \in H_2 \setminus \{1\}$ and $y \in H_1 \setminus \{1\}$ imply that $x \odot y = y \odot z = y$, so $x \odot (y \odot z) = x \odot y = y = (x \odot y) \odot z$ for all $x, y, z \in H_1 \cup H_2$.

Therefore, $(H_1 \cup H_2; \odot, 1)$ is a partially ordered commutative integral monoid.

(v) It suffices to show (2.2) holds. Let $x, y, z \in H_1 \cup H_2$.

(1) If $x = 1$, then by (ii) $x \le z \to y$ implies that $z \le y$ and so by (iv), $x \odot z = z \le y$, so (2.2) holds. Similarly, either $y = 1$ or $z = 1$ we get (2.2) holds. In addition, (2.2) holds when $x, y, z \in H_1$ or $x, y, z \in H_2$, since $\mathbf{H}_1$ and $\mathbf{H}_2$ are pocrims.

(2) If $x, y \in H_1 \setminus \{1\}$ and $z \in H_2 \setminus \{1\}$, then $z \to y = y$. Thus, $x \le z \to y$ if and only if $x \le y$. Also, $x \odot z = x$, so $x \odot z \le y$ if and only if $x \le y$. Thus, (2.2) holds.

(3) If $x \in H_1 \setminus \{1\}$ and $y, z \in H_2 \setminus \{1\}$, then $z \to y = z \to_2 y \in H_2$. Furthermore, $x \odot z = x \le y$. It follows that $x \le z \to y$ and $x \odot z \le y$. Hence, (2.2) holds, evidently.

(4) If $x, z \in H_1 \setminus \{1\}$ and $y \in H_2 \setminus \{1\}$, then $z \rightarrow y = 1$ and $x \odot z = x \odot_1 z \in H_1 \setminus \{1\}$ imply that $x \odot z \leq y$. Hence, both $x \leq z \rightarrow y$ and $x \odot z \leq y$ hold and so (2.2) fulfilled.

(5) If $z, y \in H_1 \setminus \{1\}$ and $x \in H_2 \setminus \{1\}$, then $z \rightarrow y = z \rightarrow_1 y \in H_1$ and $x \odot y = y$. By (ii), we have $x \leq z \rightarrow y$ if and only if $z \rightarrow y = 1$, equivalently, $z \leq y$, also, $z = x \odot z \leq y$ if and only if $z \leq y$. Thus, (2.2) holds.

(6) If $z \in H_1 \setminus \{1\}$ and $x, y \in H_2 \setminus \{1\}$, then $z \rightarrow y = 1 \geq x$ and $x \odot z = z \leq y$. Thus, both $x \leq z \rightarrow y$ and $x \odot z \leq y$ hold, which means (2.2) holds.

(7) If $y \in H_1 \setminus \{1\}$ and $x, z \in H_2 \setminus \{1\}$, then $z \rightarrow y = y$ and $x \odot z = x \odot_2 z \in H_2 \setminus \{1\}$. It follows that both cases $x \leq z \rightarrow y$ and $x \odot z \leq y$ are impossible. Thus, (2.2) holds.

Therefore, (2.2) holds for every $x, y, z \in H_1 \cup H_2$. Summing up the above arguments, we conclude that $\mathbf{H}_1 \oplus \mathbf{H}_2 = (H_1 \cup H_2; \odot, \rightarrow, 1)$ is a pocrim.

Let $\mathbf{H} = (H; \odot, \rightarrow, 1)$ be a pocrim and $\mathbf{1} = (\{1\}; \odot, \rightarrow, 1)$ be a trivial pocrim. Then $\mathbf{H} \oplus \mathbf{1} = \mathbf{1} \oplus \mathbf{H}$, clearly.

Example 2.2.9 Let $\mathbf{H}_2 = (H; \odot_2, \rightarrow_2, 1)$ be an arbitrary pocrim and $\mathbf{H}_1 = (\{0, 1\}; \odot_1, \rightarrow_1, 0, 1)$ be the pocrim in Example 2.1.2. By the comment in Example 2.2.8, we can assume that $H_1 \cap H_2 = \{1\}$. The underlying set of the pocrim $\mathbf{H}_1 \oplus \mathbf{H}_2$ is $\{0\} \cup H_2$ and by definition, it can be easily induced that:

$$x \odot y = \begin{cases} x \odot_1 y & \text{if } x, y \in H_1 \\ 0 & \text{if } x = 0 \text{ or } y = 0, \end{cases} \qquad x \rightarrow y = \begin{cases} x \rightarrow_1 y & \text{if } x, y \in H_1 \\ 1 & \text{if } x = 0 \\ 0 & \text{if } y = 0, \ x \in H_1. \end{cases}$$

In addition, 0 is the least element of the pocrim $\mathbf{H}_1 \oplus \mathbf{H}_2$, that is $\mathbf{H}_1 \oplus \mathbf{H}_2$ is bounded.

Let $\mathbf{H}_1$ and $\mathbf{H}_2$ be two pocrims. A straightforward calculation demonstrates that $\mathbf{H}_1 \oplus \mathbf{H}_2$ is bounded if and only if $\mathbf{H}_1$ is bounded. Furthermore, the least elements of $\mathbf{H}_1$ and $\mathbf{H}_1 \oplus \mathbf{H}_2$ are identical, provided they exist.

We conclude this section with a theorem that facilitates a more profound understanding of residuated join-semilattices.

Theorem 2.2.10 *An algebra* $\mathbf{H} = (H; \vee, \odot, \rightarrow, 1)$ *of type* $(2, 2, 2, 0)$ *is a residuated join-semilattice if and only if* $(H; \vee)$ *is a join-semilattice with the greatest element* 1, $(H; \odot, 1)$ *is a commutative monoid and* $\mathbf{H}$ *satisfies the identities:*

(i) $(x \odot y) \rightarrow z = x \rightarrow (y \rightarrow z)$.

(ii) $(x \odot (x \rightarrow y)) \vee y = y$.

(iii) $x \rightarrow (x \vee y) = 1$.

Proof It is obvious that the identities (i), (ii), (iii) hold in any residuated join-semilattice.

Conversely, let **H** be an algebra satisfying the above conditions. We have to show the adjointness condition (2.2) holds, i.e., $x \odot y \leq z$ if and only if $x \leq y \rightarrow z$, where $\leq$ is the order induced by $\vee$.

If $x \leq y$, then $x \rightarrow y = x \rightarrow (x \vee y) = 1$ by (iii), and if $x \rightarrow y = 1$, then $x = x \odot (x \rightarrow y) \leq y$ by (ii). Thus, we have $x \leq y$ if and only if $x \rightarrow y = 1$. Now, $x \odot y \leq z$ implies $x \rightarrow (y \rightarrow z) = (x \odot y) \rightarrow z = 1$, so $x \leq y \rightarrow z$.

Conversely, if $x \leq y \rightarrow z$, then $(x \odot y) \rightarrow z = x \rightarrow (y \rightarrow z) = 1$, whence $x \odot y \leq z$. Thus, **H** is a residuated join-semilattice. $\square$

Remark 2.2.11 Similarly, we can characterize residuated meet-semilattices as algebras satisfying condition (i) of Theorem 2.2.10 and

(ii′) $((x \odot (x \rightarrow y)) \wedge y = x \odot (x \rightarrow y)$.
(iii′) $(x \wedge y) \rightarrow y = 1$.

In Sect. 3.1, we will see that every hoop is a meet-semilattice. Thus, hoops are examples of residuated meet-semilattices.

2.3 Filters and Deductive Systems

This section explores the connection between filters and deductive systems in the context of pocrims. We then investigate filters generated by arbitrary subsets of pocrims.

Definition 2.3.1 Given a pocrim $\mathbf{H} = (H; \odot, \rightarrow, 1)$, we call $D \subseteq H$:

(i) A **filter** if

(F1) $a \odot b \in D$ whenever $a, b \in D$.
(F2) If $a \leq b$ and $a \in D$, then $b \in D$.

(ii) A **deductive system** if $1 \in D$ and $a \rightarrow b \in D$ and $a \in D$ imply $b \in D$.

Denote by $\mathcal{D}(\mathbf{H})$ and $\mathcal{F}(\mathbf{H})$ the set of deductive systems and the set of filters of **H**, respectively.

Lemma 2.3.2 *Let* $\mathbf{H} = (H; \odot, \to, 1)$ *be a pocrim,* $D \subseteq H$ *and* $1 \in D$. *Then the following statements are equivalent:*

(i) *D is a filter.*
(ii) *D is a deductive system.*
(iii) *For all $a \in H$ and $b_1, b_2 \in D$, $((b_1 \odot b_2) \to a) \to a \in D$.*

Proof (i) $\Leftrightarrow$ (ii) Let D be a filter of $\mathbf{H}$. If $a \to b, a \in D$, then $(a \to b) \odot a \in D$. Since $(a \to b) \odot a \leq b$, it follows that $b \in D$, so D is a deductive system.

Conversely, let D be a deductive system. If $a, b \in D$, then $a \to (b \to (a \odot b)) = (a \odot b) \to (a \odot b) = 1 \in D$ whence $a \odot b \in D$. Finally, $a \leq b$ and $a \in D$ yield $a \to b = 1 \in D$, and so $b \in D$.

(ii) $\Rightarrow$ (iii) If D is a deductive system, then by (i), $b_1 \odot b_2 \in D$, so Lemma 2.2.1(6), $((b_1 \odot b_2) \to a) \to a \in D$.

(iii) $\Rightarrow$ (i) To show that (iii) yields (i), assume first that $a, b \in D$. Then $a \odot b = 1 \to (a \odot b) = ((a \odot b) \to (a \odot b)) \to (a \odot b) \in D$. Moreover, if $a \in D$ and $a \leq b$, then $a \to b = 1$, and so $b = 1 \to b = ((1 \odot a) \to b) \to b \in D$. $\qquad\square$

According to Lemma 2.3.2, the concepts of deductive systems and filters are equivalent. So, $\mathcal{F}(\mathbf{H}) = \mathcal{D}(\mathbf{H})$ and we can call a deductive system a filter and vice versa. Clearly, for each pocrim $\mathbf{H}$ we have $\{1\}, H \in \mathcal{D}(\mathbf{H})$. A filter (deductive system) F of $\mathbf{H}$ is a **proper filter** if $F \neq H$. It can be easily seen that, if $\mathbf{H}$ is a bounded pocrim, then a filter is proper if and only if it does not contain 0.

Example 2.3.3 Let $\mathbf{H}$ be a pocrim and $a \in H$. Then $X_a := \{x \in H : x \odot a = a\}$ is a filter of $\mathbf{H}$. Clearly, $1 \in X_a$. If $x, y \in X_a$, then $(x \odot y) \odot a = x \odot a = a$, that is $x \odot y \in X_a$. In addition, if $x \leq y$ and $x \in X_a$, then by Lemma 2.2.1(3), we obtain $a = x \odot a \leq y \odot a \leq 1 \odot a = a$, which implies $y \in X_a$. Therefore, X_a is a filter.

Corollary 2.3.4 *Every filter of a pocrim* $\mathbf{H} = (H; \odot, \to, 1)$ *is a subuniverse of* $\mathbf{H}$.

Proof Let F be a filter of $\mathbf{H}$. Since $F \neq \emptyset$, by (F2), $1 \in F$. By (F1), F is closed under $\odot$. In addition, for every $x, y \in F$, by Lemma 2.2.1(10), $x \to y \geq y \in F$, which implies $x \to y \in F$. Therefore, F is a subuniverse of $\mathbf{H}$. $\qquad\square$

Proposition 2.3.5 *Let* F *be a filter of a pocrim* $\mathbf{H} = (H; \odot, \to, 1)$. *If* G *is a filter of* $(F; \odot, \to, 1)$, *then* G *is also a filter of* $\mathbf{H}$.

Proof According to Corollary 2.3.4, $(F; \odot, \to, 1)$ is a pocrim. Let G be a filter of F and $y \geq x \in G$ for some $x, y \in H$. Since $F \in \mathcal{F}(\mathbf{H})$, we get $y \in F$, so by the assumption $y \in G$. In addition, G is closed under $\odot$, which means $G \in \mathcal{F}(\mathbf{H})$. $\qquad\square$

Proposition 2.3.6 *Suppose* **H** *is a pocrim and* $\emptyset \neq F \subseteq H$ *such that* $1 \in F$. *Then, for any* $x, y, z \in H$, *the next equivalence statements hold:*

  (i) $F \in \mathcal{F}(\mathbf{H})$.
  (ii) *If* $x \to y \in F$ *and* $y \to z \in F$, *then* $x \to z \in F$.
  (iii) *If* $x \to y \in F$ *and* $x \odot z \in F$, *then* $y \odot z \in F$.
  (iv) *If* $x, y \in F$ *such that* $x \leq y \to z$, *then* $z \in F$.

Proof (i) $\Rightarrow$ (ii) Suppose $x, y, z \in H$ such that $x \to y \in F$ and $y \to z \in F$. Since $F \in \mathcal{F}(\mathbf{H})$, by (F1), $(x \to y) \odot (y \to z) \in F$. Now, by Lemma 2.2.1(14), we have $(x \to y) \odot (y \to z) \leq x \to z$. Now, (F2) implies $x \to z \in F$.

(ii) $\Rightarrow$ (i) Consider $x, y \in H$ such that $x, x \to y \in F$. From (ii), $x \to y \in F$ and $1 \to x = x \in F$ it follows that $y = 1 \to y \in F$. Hence, by Lemma 2.3.2, F is a filter.

(i) $\Rightarrow$ (iii) Let $x, y, z \in H$ such that $x \to y \in F$ and $x \odot z \in F$. Then

$$z \odot y \geq z \odot (x \odot (x \to y)) = (z \odot x) \odot (x \to y) \in F,$$

entails that $z \odot y \in F$, by (F2).

(iii) $\Rightarrow$ (i) Choose $x, y \in F$. From $1 \to y = y \in F$, $1 \odot x = x \in F$ and (iii) it follows that $y \odot x \in F$. Also, for each $x \in F$ and $x \leq y$ imply that $x \to y = 1 \in F$ and $x \odot 1 \in F$. Thus, by (iii), $y \odot 1 = y \in F$.

(i) $\Rightarrow$ (iv) Suppose $x, y \in F$ and $z \in H$ such that $x \leq y \to z$. Then by (F1), $x \odot y \in F$, and by (2.2), $x \odot y \leq z$ which entails $z \in F$, by (F2).

(iv) $\Rightarrow$ (i) Let $x, y \in F$. Since $x \odot y \leq x \odot y$, by (2.2), $x \leq y \to (x \odot y)$. So, by (iv), $x \odot y \in F$. On the other hand, suppose $x \leq y$. Then from $x \leq 1 \to y$, $x, 1 \in F$, and (iv), we get that $y \in F$. $\qquad\square$

Remark 2.3.7 It is plain that the intersection of an arbitrary family $\{F_i : i \in I\}$ of filters (as well as deductive system) of a pocrim **H**, is a filter. In addition, $\mathcal{F}(\mathbf{H})$ forms a complete lattice under set inclusion, where $\{1\}$ is the least element, the top element H, and meets coincide with set-intersections. That is,

$$(\mathcal{F}(\mathbf{H}); \vee, \wedge, \{1\}, H), \qquad \bigwedge_{i \in I} F_i = \bigcap_{i \in I} F_i, \quad \bigvee_{i \in I} F_i = \bigcap \{F \in \mathcal{F}(\mathbf{H}) : \bigcup_{i \in I} F_i \subseteq F\}.$$

Hence, for every pocrim $\mathbf{H} = (H; \odot, \to, 1)$ and every $X \subseteq H$, there exists the least filter containing X, denoted by $\langle X \rangle$ or simply $\langle x \rangle$ if $X = \{x\}$. Indeed,

$$\langle X \rangle = \bigcap \{D \in \mathcal{F}(\mathbf{H}) : X \subseteq D\}.$$

It is called the filter (or deductive system) of **H generated by** X. We note that for each pocrim **H** we have $\langle \emptyset \rangle = \{1\}$.

Proposition 2.3.8 *Let* $\mathbf{H} = (H; \odot, \to, 1)$ *be a pocrim and* $X \subseteq H$. *Then*

$$\langle X \rangle = \{a \in H : a \geq x_1 \odot \cdots \odot x_n \ \textit{for some} \ x_1, \ldots, x_n \in X, \ n \in \mathbb{N}\}$$
$$= \{a \in H : x_1 \to (\cdots \to (x_n \to a) \cdots) = 1 \ \textit{for some} \ x_1, \ldots, x_n \in X, \ n \in \mathbb{N}\}.$$

Proof Suppose $\emptyset \neq X \subseteq H$. Then let

$$B = \{a \in H : a \geq x_1 \odot \cdots \odot x_n \ \text{for some} \ x_1, \ldots, x_n \in X, \ n \in \mathbb{N}\}.$$

It is enough to prove that B is the smallest filter of $\mathbf{H}$ containing X. Consider $a, b \in H$ such that $a \in B$ and $a \leq b$. According to the definition of B, there exist $n \in \mathbb{N}$ and $x_1, \ldots, x_n \in X$ such that $x_1 \odot x_2 \odot \cdots \odot x_n \leq a$. Since by the assumption, $a \leq b$ we get $x_1 \odot x_2 \odot \cdots \odot x_n \leq b$. Thus, $b \in B$. Now, suppose $a, b \in B$. According to definition of B, there exist $n, m \in \mathbb{N}$ and $x_1, \ldots, x_n, y_1, \ldots, y_m \in X$ such that $x_1 \odot x_2 \odot \cdots \odot x_n \leq a$ and $y_1 \odot y_2 \odot \cdots \odot y_m \leq b$. Then by Lemma 2.2.1(3), we get

$$(x_1 \odot x_2 \odot \cdots \odot x_n) \odot (y_1 \odot y_2 \odot \cdots \odot y_m) \leq a \odot y_1 \odot y_2 \odot \cdots \odot y_m \leq a \odot b.$$

Thus, $a \odot b \in B$, and so B is a filter of $\mathbf{H}$. On the other side, it is clear that B contains X. Now, suppose C is a filter of $\mathbf{H}$ containing X. If $a \in B$, then there are $n \in \mathbb{N}$ and $x_1, \ldots, x_n \in X$ such that $x_1 \odot x_2 \odot \cdots \odot x_n \leq a$. Since for any $1 \leq i \leq n$, we have $x_i \in X \subseteq C$ and C is a filter of $\mathbf{H}$, we get $x_1 \odot x_2 \odot \cdots \odot x_n \in C$, and so $a \in C$, which entails $B \subseteq C$. Therefore, $B = \langle X \rangle$. $\qquad\square$

Corollary 2.3.9 *Let* $\mathbf{H} = (H; \odot, \to, 1)$ *be a pocrim. For each* $x \in H$,

$$\langle x \rangle = \{a \in H : x^n \leq a \ \textit{for some} \ n \in \mathbb{N}\} = \{a \in H : x \overset{n}{\to} a = 1 \ \textit{for some} \ n \in \mathbb{N} \cup \{0\}\},$$

where $x \overset{0}{\to} y = y$ *and* $x \overset{n+1}{\longrightarrow} y = x \to (x \overset{n}{\to} y)$ *for all* $n \in \mathbb{N}$.

Proof It follows from Proposition 2.3.8. $\qquad\square$

Corollary 2.3.10 *Let* $\mathbf{H}$ *be a pocrim,* $F \in \mathcal{F}(\mathbf{H})$ *and* $x \in H \setminus F$. *Then*

(i) $\langle F \cup \{x\} \rangle = \{a \in H : y \odot x^n \leq a \ \textit{for some} \ n \in \mathbb{N} \ \textit{and} \ y \in F\}$.
(ii) $\langle F \cup \{x\} \rangle = \{a \in H : x^n \to a \in F \ \textit{for some} \ n \in \mathbb{N}\}$.

Proof (i) Suppose $B = \{a \in H : y \odot x^n \leq a \ \text{for some} \ n \in \mathbb{N} \ \text{and} \ y \in F\}$. Similarly to the proof of Proposition 2.3.8, $B \in \mathcal{F}(\mathbf{H})$. Now, assume $y \in F$ and for any $n \in \mathbb{N}$, by Lemma 2.2.1(4), we have $x^n \leq x$ and so $y \odot x^n \leq y, x$. Thus, $y, x \in B$ and so $F \cup \{x\} \subseteq B$. Now, we show that B is the smallest filter of H containing $F \cup \{x\}$. Suppose $C \in \mathcal{F}(\mathbf{H})$ such that $F \cup \{x\} \subseteq C$. Then for any $z \in B$, there exist $n \in \mathbb{N}$ and $y \in F$ such that $y \odot x^n \leq z$.

Thus, by (2.2), $y \leq x^n \to z$. Also, since $C \in \mathcal{F}(\mathbf{H})$ and $F \subseteq C$, we have $y \in C$, and so $x^n \to z \in C$. In addition, $x \in C$ and since $C \in \mathcal{F}(\mathbf{H})$, for any $n \in \mathbb{N}$ we get $x^n \in C$. Then $z \in C$ and so $B \subseteq C$. Hence, $\langle F \cup \{x\} \rangle = B$.

(ii) It follows from (i), Lemmas 2.2.1(9) and 2.3.2(ii). $\qquad\square$

2.4 Congruence Relations and Homomorphisms

This section delves into the fundamental concepts of congruence relations and homomorphisms within the framework of pocrims. We will explore the relationship between filters and congruence relations, demonstrating how congruence relations arise from filters. While not all congruence relations stem from filters, those that induce pocrim quotient structures do. We will also see that the class of all pocrims is not closed under homomorphic images, implying that it does not form a variety.

Theorem 2.4.1 *Let F be a filter of a pocrim $\mathbf{H} = (H; \odot, \to, 1)$.*

(i) *The relation Θ_F defined by*

$$(x, y) \in \Theta_F \text{ if and only if } x \to y \in F \text{ and } y \to x \in F \qquad (2.9)$$

is a congruence relation on $\mathbf{H}$.

(ii) *For each congruence relation θ the class $F := [1]_\theta$ is a filter.*

(iii) *The map $f : \mathcal{F}(\mathbf{H}) \to \mathrm{Con}(\mathbf{H})$ defined by $f(F) = \Theta_F$ is an embedding that preserves the inclusion and $[1]_{\Theta_F} = F$.*

Proof (i) Clearly, Θ_F is reflexive and symmetric. Assume that $(a, b), (b, c) \in \Theta_F$. Then $a \to b \leq (b \to c) \to (a \to c)$ (by Lemma 2.2.1(12)) yields $(b \to c) \to (a \to c) \in F$ whence $a \to c \in F$. Similarly, from $b \to a \leq (c \to b) \to (c \to a)$ it follows $c \to a \in F$. Hence, $(a, c) \in \Theta_F$ proving that Θ_F is an equivalence relation.

Compatibility with $\odot$: Take $(a, b) \in \Theta_F$ and $c \in H$. Then by Lemma 2.2.1(5) $a \to b \leq (a \odot c) \to (b \odot c)$ and $b \to a \leq (b \odot c) \to (a \odot c)$ imply $(a \odot c) \to (b \odot c) \in F$ and $(b \odot c) \to (a \odot c) \in F$, i.e., $(a \odot c, b \odot c) \in \Theta_F$.

Compatibility with $\to$: Choose $(a, b), (c, d) \in \Theta_F$. Applying Lemma 2.2.1(12), from $a \to b \leq (b \to c) \to (a \to c)$ and $b \to a \leq (a \to c) \to (b \to c)$, we conclude $(a \to c, b \to c) \in \Theta_F$, and analogously, from $c \to d \leq (b \to c) \to (b \to d)$ and $d \to c \leq (b \to d) \to (b \to c)$, we obtain $(b \to c, b \to d) \in \Theta_F$. Hence, due to transitivity, $(a \to c, b \to d) \in \Theta_F$.

(ii) Assume that $\theta \in \mathrm{Con}(\mathbf{H})$ and $F := [1]_\theta$. Clearly, $1 \in F$. If $a \to b, a \in F$, then from $(a, 1) \in \theta$ we get $((a \to b), 1 \to b) \in \theta$, also $(1, a \to b) \in \theta$, so $(1, b) = (1, 1 \to b) \in \theta$, which means $b \in F$. Therefore, F is a filter of $\mathbf{H}$.

(iii) Clearly, f is well-defined. From $x \in [1]_{\Theta_F}$ iff $x \to 1$, $1 \to x \in F$ iff $x \in F$ it follows that $[1]_{\Theta_F} = F$. In addition, $F \subseteq G$ iff $f(F) \subseteq f(G)$ for every $F, G \in \mathcal{F}(\mathbf{H})$. Therefore, f is a one-to-one map preserving $\subseteq$. $\square$

If θ is a congruence relation of a pocrim $\mathbf{H}$, we denote by x/θ the equivalence class of θ containing x, that is $x/\theta = \{y \in H : (x, y) \in \theta\}$. Set $H/\theta := \{x/\theta : x \in H\}$. By Remark 1.2.14, the quotient structure $\mathbf{H}/\theta := (H/\theta; \odot, \to, 1/\theta)$ is an algebra of type $(2, 2, 0)$, where

$$\frac{a}{\theta} \odot \frac{b}{\theta} = \frac{a \odot b}{\theta} \quad \& \quad \frac{a}{\theta} \to \frac{b}{\theta} = \frac{a \to b}{\theta}, \quad \forall a, b \in H. \tag{2.10}$$

In particular, when $\theta = \Theta_F$ for some filter F of $\mathbf{H}$, we simply denote x/Θ_F, H/Θ_F, and $\mathbf{H}/\Theta_F$ by x/F, H/F, and $\mathbf{H}/F$, respectively.

A congruence relation θ on a pocrim $\mathbf{H}$ is a **filter-congruence**[1] if the quotient structure $\mathbf{H}/\theta = (H/\theta; \odot, \to, 1/\theta)$ is a pocrim. The set of all filter-congruences of $\mathbf{H}$ is denoted by $\mathrm{Con}_F(\mathbf{H})$.

Theorem 2.4.2 *For every filter F of a pocrim $\mathbf{H}$, the quotient algebra $\mathbf{H}/F$ is a pocrim. In addition, $x/F \leq y/F$ if and only if $x \to y \in F$ for all $x, y \in H$.*

Proof We apply Theorem 2.2.7 in order to show that $(H/F; \odot, \to, 1/F)$ is a pocrim. It suffices to check the condition (iii) from Theorem 2.2.7, since identities (i), (ii), and (iv) hold, evidently. Let $a, b \in H$ such that $a/F \to b/F = 1/F$ and $b/F \to a/F = 1/F$. Then $(a \to b)/F = 1/F$ entails that $(a \to b) \to 1$, $1 \to (a \to b) \in F$, equivalently, $a \to b \in F$. In a similar way, $b/F \to a/F = 1/F$ implies that $b \to a \in F$. Hence, $a \to b, b \to a \in F$, that is $a/F = b/F$.

Now, assume that $x, y \in H$. Since $\mathbf{H}/F$ is a pocrim, we have

$$\frac{x}{F} \leq \frac{y}{F} \Leftrightarrow \frac{x}{F} \to \frac{y}{F} = \frac{1}{F} \Leftrightarrow \frac{x \to y}{F} = \frac{1}{F} \Leftrightarrow x \to y \in F.$$

$\square$

Theorem 2.4.3 *Consider the assumptions of Theorem 2.4.1. Then $\mathrm{Im}(f) = \mathrm{Con}_F(\mathbf{H})$.*

Proof By Theorem 2.4.2, $f(F) = \Theta_F \in \mathrm{Con}_F(\mathbf{H})$ for all $F \in \mathcal{F}(\mathbf{H})$. Choose a filter congruence θ of $\mathbf{H}$. Due to Theorem 2.4.1(ii), $G := 1/\theta$ is a filter of $\mathbf{H}$. We show that $f(1/\theta) = \theta$. Given $x, y \in H$. If $(x, y) \in \theta$, then $(x \to y, 1) = (x \to y, x \to x) \in \theta$ and $(y \to x, 1) = (y \to x, y \to y) \in \theta$, so $x \to y, y \to x \in 1/\theta = G$ entails that

---

[1] Filter-congruences are also known as relative congruences (see [33]).

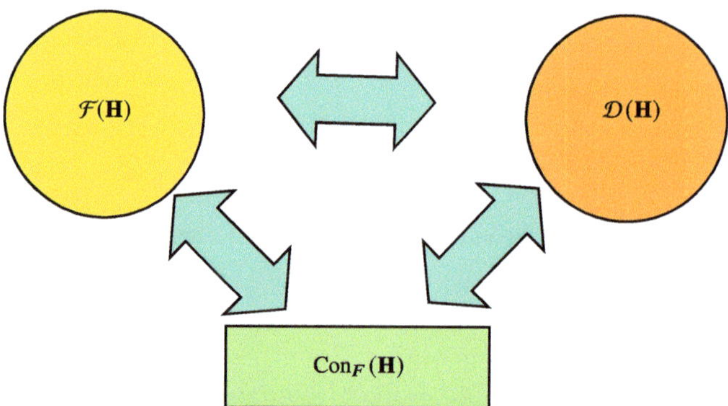

Fig. 2.1 Relation between Filters, Deductive systems, and Filter congruences

$(x, y) \in \Theta_G = f(G)$. Furthermore, if $(x, y) \in f(G) = \Theta_G$, then by definition, $x \to y, y \to x \in G = 1/\theta$, which implies

$$\frac{x}{\theta} \to \frac{y}{\theta} = \frac{x \to y}{\theta} = \frac{1}{\theta} \quad \& \quad \frac{y}{\theta} \to \frac{x}{\theta} = \frac{y \to x}{\theta} = \frac{1}{\theta}.$$

Since $\mathbf{H}/\theta$ is a pocrim, we get that $x/\theta = y/\theta$ or equivalently, $(x, y) \in \theta$. Therefore, $\theta = \Theta_G = f(G)$.   □

Theorem 2.4.3 implies that there exists a one-to-one correspondence between the set of all filters of $\mathbf{H}$ (deductive systems of $\mathbf{H}$) and $\mathrm{Con}_F(\mathbf{H})$, see Fig. 2.1.

Remark 2.4.4 Let $\mathbf{H}_1 = (H_1; \odot, \to, 1)$ and $\mathbf{H}_2 = (H_2; \odot, \to, 1)$ be two pocrims. According to Definition 1.2.10, a mapping $f : \mathbf{H}_1 \to \mathbf{H}_2$ is a homomorphism of pocrims if f preserves $\odot$, $\to$, and 1, that is

(i) $f(x \odot y) = f(x) \odot f(y)$ for all $x, y \in H_1$.
(ii) $f(x \to y) = f(x) \to f(y)$ for all $x, y \in H_1$.
(iii) $f(1) = 1$.

Furthermore, if $\mathbf{H}_1$ and $\mathbf{H}_2$ are bounded pocrims, then a homomorphism of bounded pocrims is a homomorphism f that preserves the least elements, that is $f(0) = 0$.

It is worthy of note that (1) and (2) imply (3), since $f(1) = f(1 \to 1) = f(1) \to f(1) = 1$. So, condition (iii) in Remark 2.4.4 is superfluous. Analogously to Definition 1.2.10, one can define epimorphisms and isomorphisms of pocrims.

In light of the preceding remark, we define a category $\mathcal{P}ocrim$ ($\mathcal{BP}ocrim$) whose objects are (bounded) pocrims and whose morphisms are homomorphisms of (bounded) pocrims.

Easy calculations show that for each family $\{\mathbf{H}_i : i \in I\}$ of (bounded) pocrims, the direct product of this family is a pocrim. In addition, for each pocrim $\mathbf{H} = (H; \odot, \rightarrow, 1)$ and each trivial pocrim $(\{x\}; \odot, \rightarrow, x)$ the map $f : \{x\} \rightarrow H$ defined by $f(x) = 1$ is an embedding of pocrims. Thus, all trivial pocrims are isomorphic, which means up to isomorphism there exists only one trivial pocrim denoted by $\mathbf{1} = (\{1\}; \odot, \rightarrow, 1)$. Furthermore, there exists exactly one (onto) homomorphism from $\mathbf{H}$ to $\mathbf{1}$. Thus, $\mathbf{1}$ is both initial and terminal object of $\mathcal{P}ocrim$.

While $\mathbf{1}$ can be considered as a bounded pocrim, there cannot exist a non-trivial homomorphism from $\mathbf{1}$ to any non-trivial bounded pocrim $(H; \odot, \rightarrow, 0, 1)$. This is because any such homomorphism would map 1 to 0 and 1, simultaneously, contradicting the assumption that $0 \neq 1$.

Example 2.4.5 Every pocrim with two elements is isomorphic to the pocrim $\mathbf{H} = (\{0, 1\}; \odot, \rightarrow, 0, 1)$ presented in Example 2.1.2. Consequently, up to isomorphism, there exists a unique two-element pocrim. Moreover, $(\{0, 1\}; \odot, \rightarrow, 0, 1)$ is the initial object of $\mathcal{BP}ocrim$.

Remark 2.4.6 For each homomorphism $f : \mathbf{H}_1 \rightarrow \mathbf{H}_2$ the set $I := f^{-1}(1) = \{x \in H_1 : f(x) = 1\}$ is a filter of H_1. We remark that according to Theorem 1.2.13, $\mathrm{Ker}(f)$ is indeed the congruence relation $\{(x, y) \in H_1 \times H_1 : f(x) = f(y)\}$. For each $x, y \in H_1$ we have

$$
\begin{aligned}
(x, y) \in \mathrm{Ker}(f) &\Leftrightarrow f(x) = f(y) \\
&\Leftrightarrow f(x) \rightarrow f(y) = 1 = f(y) \rightarrow f(x) \\
&\Leftrightarrow f(x \rightarrow y) = 1 = f(y \rightarrow x) \\
&\Leftrightarrow x \rightarrow y, y \rightarrow x \in I = f^{-1}(1) \\
&\Leftrightarrow (x, y) \in \Theta_I.
\end{aligned}
$$

For this reason, we also adopt the notation of $\mathrm{Ker}(f)$ to denote the filter $f^{-1}(1)$. We ensure it is treated as a congruence relation when working with congruences and as a filter when working with filters. In addition, the homomorphism f is one-to-one if and only if $\mathrm{Ker}(f) = \{1\}$.

Considering Definition 1.2.4, each subalgebra of a pocrim $\mathbf{H} = (H; \odot, \rightarrow, 1)$ is said to be a **subpocrim** of $\mathbf{H}$. Applying Remark 1.2.7(ii), every subpocrim $\mathbf{S}$ of $(H; \odot, \rightarrow, 1)$ is of the form $(S; \odot, \rightarrow, 1)$, where S is closed under $\odot$, $\rightarrow$ and 1 (equivalently, $S \neq \emptyset$ and S is a subset of H closed under $\odot$ and $\rightarrow$).

Proposition 2.4.7 *Let* $\mathbf{H}_1$ *be a subpocrim of a pocrim* $\mathbf{H}_2$.

(i) $F \cap H_1 \in \mathcal{F}(\mathbf{H}_1)$ *for each* $F \in \mathcal{F}(\mathbf{H}_2)$.

(ii) *If* $F \in \mathcal{F}(\mathbf{H}_1)$, *then* $\langle F \rangle_{H_2} = \{x \in H_2 : x \geq a,\ \exists a \in F\}$, *where* $\langle F \rangle_{H_2}$ *is the filter of* $\mathbf{H}_2$ *generated by* F.

(ii) *For each* $F \in \mathcal{F}(\mathbf{H}_1)$, *there exists* $G \in \mathcal{F}(\mathbf{H}_2)$ *such that* $F = G \cap H_1$.

Proof The proof of (i) is straightforward and the proof of (ii) is an easy consequence of Corollary 2.3.9.

(iii) Let $F = \langle F \rangle_{H_1}$. Set $G = \langle F \rangle_{H_2}$. Clearly, $F \subseteq G \cap H_1$. By (ii), for each $x \in G \cap H_1$, there exists $a \in F$ such that $x \geq a$, which implies $x \in F$, since F is a filter of $\mathbf{H}_1$. Therefore, $F = G \cap H_1$. $\qquad\square$

Theorem 2.4.8 *Let* $f : \mathbf{H}_1 \to \mathbf{H}_2$ *be a homomorphism of pocrims. Then the following statements hold:*

(i) $f(\mathbf{A})$ *is a subalgebra of* $\mathbf{H}_2$ *for each subalgebra* $\mathbf{A}$ *of the pocrim* $\mathbf{H}_1$.

(ii) $f^{-1}(\mathbf{B})$ *is a subalgebra of* $\mathbf{H}_1$ *for each subalgebra* $\mathbf{B}$ *of the pocrim* $\mathbf{H}_2$, *where* $f^{-1}(\mathbf{B}) := (f^{-1}(B); \odot, \to, 1)$.

(iii) $f^{-1}(F) \in \mathcal{F}(\mathbf{H}_1)$ *for each* $F \in \mathcal{F}(\mathbf{H}_2)$.

(iv) *If* f *is onto and* $\mathrm{Ker}(f) \subseteq F$, *then* $f(F) \in \mathcal{F}(\mathbf{H}_2)$ *for each* $F \in \mathcal{F}(\mathbf{H}_1)$.

(v) $\bar{f} : \mathbf{H}/\mathrm{Ker}(f) \to f(\mathbf{H})$, *defined by* $\bar{f}(x/\mathrm{Ker}(f)) = f(x)$, *is an isomorphism.*

Proof The proofs of (i)–(iii) are straightforward and the proof of (v) follows from Remark 2.4.6 and Theorem 1.2.16. They are left to the reader as an exercise.

(iv) Let F be a filter of $\mathbf{H}_1$ containing $\mathrm{Ker}(F)$ and f be onto. Given $y_1, y_2 \in f(F)$ there exist $x_1, x_2 \in F$ such that $f(x_i) = y_i$ for $i = 1, 2$. Clearly, $y_1 \odot y_2 = f(x_1) \odot f(x_2) = f(x_1 \odot x_2) \in f(F)$. In addition, if $z \geq f(x)$ for some $x \in F$ and $z \in H_2$, then there exists $y \in H_1$ such that $f(y) = z$. It follows that $f(x \to y) = f(x) \to f(y) = 1$, consequently, $x \to y \in \mathrm{Ker}(f) \subseteq F$. From $x \in F$ and Lemma 2.3.2, we get $y \in F$, and so $z = f(y) \in f(F)$. Therefore, $f(F)$ is a filter of $\mathbf{H}_2$. $\qquad\square$

Remark 2.4.9 ([69, 174]) (i) It should be remarked that Higgs [174] proved that pocrims do not form a variety because the class of all pocrims is not closed under homomorphic images. However, as shown in Theorem 2.2.7, they form a quasivariety (closed under the operators I (isomorphic), S (subalgebra), P (direct product), $\mathrm{P_r}$ (reduced products) and contains a trivial algebra), see also [33, 239]. In addition, Grishin [166] proved that the class of involutive pocrims also fails to form a variety. On the contrary, the class of all residuated semilattices is a variety.

(ii) By (i), the class of all pocrims does not form a variety, as it is not closed under homomorphic images. This means that the quotient of a pocrim $\mathbf{H}$ by a congruence relation

θ may not be a pocrim. Equivalently, a congruence θ is not necessarily determined by $K := 1/\theta$, i.e., in general, Θ_K can differ from θ since $\mathbf{H}/\Theta_K$ is a pocrim, while $\mathbf{H}/\theta$ need not necessarily be. Note that we always have $\theta \subseteq \Theta_K$.

2.5 BCK-Algebras

BCK-algebras introduced by Iséki and Imai [176, 177, 183, 184] from two primary perspectives: set theory and propositional calculi.

Set-Theoretic Perspective: Union, intersection, and difference are fundamental set operations. Boolean algebras generalize these operations and their properties. Distributive lattices generalize union and intersection, while upper and lower semilattices focus on either operation alone. In contrast, the set-theoretic difference, along with its properties, inspired K. Iséki 's definition of BCK-algebras.

Propositional Calculus Perspective: BCK-algebras can be viewed as generalizations of implication-based logic systems. They provide an algebraic framework for studying propositional calculi, particularly those focusing on implication.

Pocrims are closely related to BCK-algebras. Relationship between pocrims and BCK-algebras has been studied in [33, 34, 145, 231, 232]. In this section, we will see some basic results about this relation.

Definition 2.5.1 A **BCK-algebra** is an algebra $\mathbf{H} = (H; \rightarrow, 1)$ of type $(2, 0)$ satisfying the following axioms for all $x, y, z \in H$.

> (BCK1) $(x \rightarrow y) \rightarrow ((y \rightarrow z) \rightarrow (x \rightarrow z)) = 1$.
> (BCK2) $x \rightarrow ((x \rightarrow y) \rightarrow y) = 1$.
> (BCK3) $x \rightarrow x = 1$.
> (BCK4) $x \rightarrow 1 = 1$.
> (BCK5) If $x \rightarrow y = 1$ and $y \rightarrow x = 1$, then $x = y$.

We note that there exists another definition of BCK-algebras which is dual to the above one (see [214, 266]):

A BCK-algebra is an algebra with a binary operation $*$ and a constant 0 that satisfies the quasi-identities:

> (BCK1$'$) $((z * x) * (z * y)) * (y * x) = 0$.
> (BCK2$'$) $(y * (y * x)) * x = 0$.
> (BCK3$'$) $x * x = 0$.
> (BCK4$'$) $0 * x = 0$.
> (BCK5$'$) If $x * y = 0$ and $y * x = 0$, then $x = y$.

Thus, $x * y$ is replaced by $y \to x$ and 0 by 1. The operation $*$ is an analog of the set-theoretical difference, while $\to$ can be understood as the truth function of implication in a certain propositional calculus.

Given a BCK-algebra $\mathbf{H} = (H; \to, 1)$, the relation $\leq$ defined via $x \leq y$ if and only if $x \to y = 1$ is a partial order on H with 1 at the top (the proof can be achieved just as in the proof of Theorem 2.2.7). If the poset $(H, \leq)$ is a $\wedge$-semilattice, then $\mathbf{H}$ is called a **BCK-$\wedge$** bf -semilattice. Similarly, a $\vee$-semilattice can be defined. In addition, a BCK-lattice is a BCK-algebra $\mathbf{H}$ where it is both $\vee$-semilattice and $\wedge$-semilattice.

On the other hand, if $(H, \leq)$ is a poset with a greatest element 1, then upon defining

$$x \to y = \begin{cases} 1 & \text{if } x \leq y \\ y & \text{otherwise,} \end{cases}$$

the structure $(H; \to, 1)$ becomes a BCK-algebra; even more, this is a Hilbert algebra, i.e., a BCK-algebra that fulfills the identity

$$x \to (y \to z) = (x \to y) \to (x \to z).$$

A well-known example of BCK-algebras is based on the powerset of a set X as follows:

Example 2.5.2 Let X be a set and $\mathcal{P}(X)$ be the set of all subsets of X. Define the binary operation $\to: \mathcal{P}(X) \times \mathcal{P}(X) \to \mathcal{P}(X)$ by $A \to B = B \setminus A$ for all $A, B \in \mathcal{P}(X)$, where $\setminus$ is the set theoretic difference. Then $(\mathcal{P}(X); \to, \emptyset)$ is a BCK-algebra. In addition,

$$A \leq B \Leftrightarrow B \setminus A = \emptyset \Leftrightarrow B \subseteq A.$$

Therefore, X is the least element of the BCK-algebra $(\mathcal{P}(X); \to, \emptyset)$.

Example 2.5.3 ([260]) Let $A = \{a_n : n \in \mathbb{N}_0\}$, $B = \{b_n : n \in \mathbb{N}_0\}$ and $\mathbb{N}$ be three pairwise disjoint sets and $H = A \cup B \cup \mathbb{N}_0$. Define the binary operation $\to: H \times H \to H$ as follows:

$$n \to m = \max\{0, m - n\} \quad a_n \to m = b_n \to m = 0$$
$$n \to a_m = a_{m+n} \quad n \to b_m = b_{m+n}$$
$$a_n \to a_m = b_n \to b_m = m \to n \quad b_n \to a_m = a_n \to b_m = m \to (n + 1).$$

Then $(H, \to, 0)$ is a BCK-algebra, with its Hasse diagram is depicted in Fig. 2.2. This structure is known as **Wronski's BCK-algebra**.

The basic properties of BCK-algebras are listed in Lemma 2.2.1. For the remainder of this text, we will primarily employ Definition 2.5.1 for BCK-algebras. We will highlight a few key properties that will be essential in our subsequent discussions.

Fig. 2.2 Hasse diagram of
Wroński's BCK-algebra

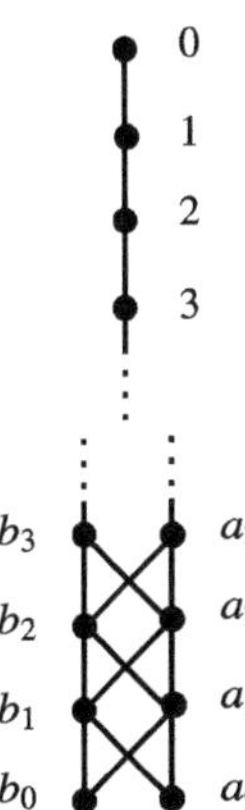

Lemma 2.5.4 *Let* $\mathbf{H} = (H; \rightarrow, 1)$ *be a BCK-algebra. Then for each* $x, y, z \in H$ *we have:*

  (i) $1 \rightarrow x = x$.

  (ii) $x \leq y$ *implies* $y \rightarrow z \leq x \rightarrow z$ *and* $z \rightarrow x \leq z \rightarrow y$.

 (iii) $y \leq x \rightarrow y$.

 (iv) $x \rightarrow (y \rightarrow z) = y \rightarrow (x \rightarrow z)$.

  (v) $(x \rightarrow y) \rightarrow ((z \rightarrow x) \rightarrow (z \rightarrow y)) = 1$.

 (vi) $x \leq y \rightarrow z$ *if and only if* $y \leq x \rightarrow z$.

Proof (i) Using (BCK3) and (BCK2), we get $x \rightarrow (1 \rightarrow x) = x \rightarrow ((x \rightarrow x) \rightarrow x) = 1$. But $(1 \rightarrow x) \rightarrow x = ((x \rightarrow x) \rightarrow x) \rightarrow x \geq x \rightarrow x = 1$, so $(1 \rightarrow x) \rightarrow x = 1$, whence we conclude $1 \rightarrow x = x$ by (BCK5).

(ii) Let $x \leq y$, i.e., $x \rightarrow y = 1$. Then by (BCK1) and (i), we have $(y \rightarrow z) \rightarrow (x \rightarrow z) = 1$, so $y \rightarrow z \leq x \rightarrow z$. The second inequality follows from (BCK1) as well. Indeed, $(z \rightarrow x) \rightarrow (z \rightarrow y) = (z \rightarrow x) \rightarrow ((x \rightarrow y) \rightarrow (z \rightarrow y)) = 1$, so $z \rightarrow x \leq z \rightarrow y$.

(iii) This is a consequence of (i) and (ii), since $x \leq 1$ yields $y = 1 \rightarrow y \leq x \rightarrow y$.

(iv) By (BCK1), we have $x \rightarrow (y \rightarrow z) \leq ((y \rightarrow z) \rightarrow z) \rightarrow (x \rightarrow z)$. Since $y \leq (y \rightarrow z) \rightarrow z$, it follows $x \rightarrow (y \rightarrow z) \leq ((y \rightarrow z) \rightarrow z) \rightarrow (x \rightarrow z) \leq y \rightarrow (x \rightarrow z)$ by (ii). By interchanging x and y, we obtain the converse inequality.

(v) Using the exchange identity (iv), we have

$$(x \rightarrow y) \rightarrow ((z \rightarrow x) \rightarrow (z \rightarrow y)) = (z \rightarrow x) \rightarrow ((x \rightarrow y) \rightarrow (z \rightarrow y)) = 1, \text{ by (BCK1).}$$

(vi) Applying (iv), we have $x \leq y \rightarrow z$ iff $x \rightarrow (y \rightarrow z) = 1$ iff $y \rightarrow (x \rightarrow z) = 1$ iff $y \leq x \rightarrow z$. $\qquad\square$

Theorem 2.5.5 *For every pocrim* $(H; \odot, \rightarrow, 1)$, *the reduct* $(H; \rightarrow, 1)$ *is a BCK-algebra.*

Proof It follows from Lemma 2.2.1 and Theorem 2.2.7. $\square$

More generally, every subalgebra of $(H; \rightarrow, 1)$ is a BCK-algebra, where $(H; \odot, \rightarrow, 1)$ is a pocrim. A partial converse of the last statement was independently obtained by Fleischer [145], Ono and Komori [231], and Palasiński [232]:

Theorem 2.5.6 *Every BCK-algebra is isomorphic to a subalgebra of a $\{\rightarrow, 1\}$-reduct of a residuated lattice.*[2]

We will sketch the proof instead of providing a full proof for Theorem 2.5.6.

Let $(X; \rightarrow, 1)$ be an arbitrary BCK-algebra and let W be the set of all words over X. Every element of W is a sequence $\mathbf{x} = x_1 x_2 \ldots x_n$ for some $n \in \mathbb{N} \cup \{0\}$ and $x_1, \ldots x_n \in X$, where n is termed the length of the word $\mathbf{x}$ and the empty word is the word with length 0. In addition, for every elements $\mathbf{x} = x_1 \ldots x_n$ and $\mathbf{y} = y_1 \ldots y_m$ of W, $\mathbf{x}^r$ and $\mathbf{xy}$ stand for the reverse word of $\mathbf{x}$ and for the concatenation of $\mathbf{x}$ and $\mathbf{y}$, respectively:

$$\mathbf{x}^r = x_n \ldots x_1 \quad \& \quad \mathbf{xy} = x_1 \ldots x_n y_1 \ldots y_m. \tag{2.11}$$

For every $\mathbf{x} \in W$ and $a \in X$, $\mathbf{x} \rightarrow a$ denotes $x_1 \rightarrow (\cdots \rightarrow (x_n \rightarrow a) \cdots)$. Due to Lemma 2.5.4(iv), $\mathbf{x} \rightarrow a = 1$ if and only if $\mathbf{x}^r \rightarrow a = 1$ for every $\mathbf{x} \in W$ and $a \in X$. Let Q be the set of all finite non-empty subsets of W. Direct calculations demonstrate that the binary relation $\sim$ on W defined by $\{\mathbf{x}_1, \ldots \mathbf{x}_n\} \sim \{\mathbf{y}_1, \ldots, \mathbf{y}_m\}$ if and only if for all $\mathbf{w} \in W$ and $a \in X$:

$$\left(\mathbf{w} \rightarrow (\mathbf{x}_i \rightarrow a) = 1, \ \forall i = 1, \ldots, n\right) \Leftrightarrow \left(\mathbf{w} \rightarrow (\mathbf{y}_j \rightarrow a) = 1, \ \forall j = 1, \ldots, m\right) \tag{2.12}$$

is an equivalence on Q. We use $[\mathbf{x}_1, \ldots, \mathbf{x}_n]$ to denote the equivalence class of $\sim$ containing $\{\mathbf{x}_1, \ldots, \mathbf{x}_n\}$. As usual, $Q/\sim$ is the set of all equivalence classes of Q with respect to $\sim$. Define $\sqcap : (Q/\sim)^2 \rightarrow Q/\sim$ and $* : (Q/\sim)^2 \rightarrow Q/\sim$ by

$$[\mathbf{x}_1, \ldots, \mathbf{x}_n] \sqcap [\mathbf{y}_1, \ldots, \mathbf{y}_m] := [\mathbf{x}_1, \ldots, \mathbf{x}_n, \mathbf{y}_1, \ldots, \mathbf{y}_m]$$

$$[\mathbf{x}_1, \ldots, \mathbf{x}_n] * [\mathbf{y}_1, \ldots, \mathbf{y}_m] := [\mathbf{x}_i \mathbf{y}_j : i = 1, \ldots, n, \ j = 1 \ldots, m].$$

Then both $\sqcap$ and $*$ are well-defined binary operations on $Q/\sim$ (for more detail see [201, p. 8]). Furthermore, $*$ is commutative, since $\mathbf{w} \rightarrow (\mathbf{xy} \rightarrow a) = \mathbf{w} \rightarrow (\mathbf{yx} \rightarrow a)$ for all $\mathbf{x}, \mathbf{y}, \mathbf{w} \in W$ and $a \in X$.

Lemma 2.5.7 *For every BCK-algebra $(X; \rightarrow, 1)$, the structure $(Q/\sim; \sqcap, *, [1])$ is a dually integral meet-semilattice-ordered monoid.*

---

[2] In other words, every BCK-algebra is isomorphic to a $\{\rightarrow, 1\}$-subreduct of a residuated lattice.

If $\sqsubseteq$ is the partially ordered relation associated with the meet-semilattice $(Q/\sim; \sqcap)$, then we have $[\mathbf{x}_1, \ldots, \mathbf{x}_n] \sqsubseteq [\mathbf{y}_1, \ldots, \mathbf{y}_m]$ if and only if for every $a \in X$ and $\mathbf{w} \in W$ we have the following implication:

$$\left(\mathbf{w} \to (\mathbf{x}_i \to a) = 1, \ \forall i = 1, \ldots, n\right) \Rightarrow \left(\mathbf{w} \to (\mathbf{y}_j \to a) = 1, \ \forall j = 1, \ldots, m\right). \quad (2.13)$$

Henceforth, we denote the dually integral meet-semilattice-ordered monoid $(Q/\sim; \sqcap, *, [1])$ by $\mathbf{M}(X)$.

Recall that a filter of a meet-semilattice is defined analogously to a filter on a lattice; that is, F is a filter of a meet-semilattice $(L; \wedge)$ if it is an upper set closed under $\wedge$.

Remark 2.5.8 Let $(M; \wedge, \cdot, e)$ be a dually integral meet-semilattice-ordered monoid and $\mathbf{F}(M)$ be the set of all filters of $(M; \wedge)$ containing the empty subsets. We define the following additional operations of $\mathbf{F}(M)$:

$$X \vee Y := \{a \in M : a \geq x \wedge y, \ \text{for some } x, y \in X \cup Y\}$$
$$X \cdot Y := \{a \in M : a \geq x \cdot y, \ \text{for some } x, y \in X \cup Y\}$$
$$X \to Y := \{a \in M : \{a\} \cdot X \subseteq Y\},$$

where $\{a\} \cdot X = \{a \cdot x : x \in X\}$. Then $(\mathbf{F}(M); \vee, \cap, \cdot, \to, \emptyset, M)$ is a complete integral residuated lattice.

Proposition 2.5.9 ([201, Proposition 1.2.4]) *Let $(X; \to, 1)$ be a BCK-algebra and $\mathbf{M}_X$ be the dually integral meet-semilattice-ordered monoid $(Q/\sim; \sqcap, *, [1])$ associated to $(X; \to, 1)$. Then the map $f : X \to \mathbf{F}(\mathbf{M}_X)$ defined by*

$$f(a) := \{[\mathbf{x}_1, \ldots, \mathbf{x}_n] \in Q/\sim : \mathbf{x}_i \to a = 1, \ \text{for all } i = 1, \ldots, n\}$$

is a one-to-one homomorphism of type $\{\to, 1\}$ into $(\mathbf{F}(\mathbf{M}_X); \to, Q/\sim)$. In addition, if $a \vee b$ exists in X, then $f(a \vee b) = f(a) \vee f(b)$.

The proof of Theorem 2.5.6 follows from Lemma 2.5.7, Remark 2.5.8, and Proposition 2.5.9.

Definition 2.5.10 ([178, 179]) A BCK-algebra $(X; \to, 1)$ satisfies the product condition if
(P) $x \odot y := \min\{z \in X : x \leq y \to z\}$ exists for all $x, y \in X$.

In such a case, this BCK-algebra is called **BCK-algebra with (P) condition** or **BCK(P)-algebra**.

Note that any bounded linearly ordered BCK-algebra satisfies condition (P).

Proposition 2.5.11 ([79, 179, 181]) *Let $(X; \to, 1)$ be a BCK(P)-algebra with product $\odot$. Then it satisfies (2.2). In addition, the algebra $(X; \odot, \to, 1)$ is a pocrim.*

Proof Let $(X; \to, 1)$ be a BCK(P)-algebra and $x, y, z \in X$. If $x \odot y \le z$, then by Lemma 2.5.4(ii), $y \to (x \odot y) \le y \to z$ entails that $x \le y \to z$. Similarly, if $x \le y \to z$, then by definition of $\odot$, we get that $x \odot y \le z$. Thus, (2.2) holds.

Clearly, $(X; \odot, \to, 1)$ is an algebra of type $(2, 2, 0)$, where $x \odot y := \min\{z \in X : x \le y \to z\}$ for all $x, y \in X$. From Lemma 2.5.4(iv) we can easily deduce that $x \odot y = y \odot x$. In addition, $x \odot 1 = \min\{z \in X : x \le 1 \to z\} = \min\{z \in X : x \le z\} = \min(\uparrow x) = x$. Now, we show that $\odot$ is associative. Given $w \in X$, by (2.2), we have $(x \odot y) \odot z \le w$ iff $x \odot y \le z \to w$ iff $x \le y \to (z \to w)$ iff $y \le x \to (z \to w) = z \to (x \to w)$, by Lemma 2.5.4(iv) and (vi), iff $y \odot z \le x \to w$ iff $x \odot (y \odot z) \le w$. Hence, $(x \odot y) \odot z = x \odot (y \odot z)$, consequently, $(X; \odot, 1)$ is a commutative monoid.

To show that $(X; \odot, \to, 1)$ is a pocrim, we apply Theorem 2.2.7:

Theorem 2.2.7(i) follows from (BCK3) and (BCK4).

Theorem 2.2.7(ii) follows from Lemma 2.5.4(v).

Theorem 2.2.7(iii) follows from (BCK5).

Suppose that $u \in X$. Then $u \le x \to (y \to z)$ iff $x \odot u \le y \to z$ iff $y \odot (x \odot u) \le z$ iff $(x \odot y) \odot u \le z$ iff $u \le (x \odot y) \to z$. Hence, $(x \odot y) \to z = x \to (y \to z)$, so Theorem 2.2.7(iv) holds.

Therefore, $(X; \odot, \to, 1)$ is a pocrim. $\qquad\qquad\square$

Theorem 2.5.12 ([181, Theorem 2.1.6]) *BCK(P)-algebras are termwise equivalent to pocrims.*

Proof Given a pocrim $(H; \odot, \to, 1)$, the reduct $(H; \to, 1)$ is a BCK-algebra by Theorem 2.5.5. Choose $x, y, z \in H$. If $x \le y \to z$, then $x \odot y \le z$. In addition, $y \to (x \odot y) \ge x$ (by Lemma 2.2.1(15)) entails that $x \odot y = \min\{z \in H : x \le y \to z\}$. Hence, $(H; \to, 1)$ is a BCK(P)-algebra. Conversely, if $(X; \to, 1)$ is a BCK(P)-algebra, then by Proposition 2.5.11, $(X; \odot, \to, 1)$ is a pocrim, where $x \odot y = \min\{z \in X : x \le y \to z\}$ for all $x, y \in X$. Therefore, the map φ from the class of pocrims to the class of BCK(P)-algebras sending each pocrim $(H; \odot, \to, 1)$ to the reduct $(H; \to, 1)$ is a bijection map. $\quad\square$

Definition 2.5.13 A BCK-algebra **H** is called **commutative** if the following identity holds:

$$(x \to y) \to y = (y \to x) \to x, \quad \forall x, y \in H \qquad \text{(quasi-commutativity).}$$

A straightforward verification yields that if $\mathbf{H} = (H; \to, 1)$ is a commutative BCK-algebra, then $(H, \le)$ is a join-semilattice, where

$$x \vee y = (x \to y) \to y.$$

On the other hand, in a BCK-algebra $\mathbf{H} = (H; \rightarrow, 1)$, if $(H, \leq)$ is a join-semilattice, then $x \vee y$ is not necessarily equal to $(x \rightarrow y) \rightarrow y$, so $\mathbf{H}$ need not be commutative.

For instance, let $(H; \rightarrow, 1)$ be the reduct of the MTL-algebra from Example 2.1.11. Then $(\mathbf{H}, \rightarrow, 1)$ is a BCK-algebra, and $(\mathbf{H}, \leq)$ is a join-semilattice (specifically, a chain). However, $(b \rightarrow a) \rightarrow a = 1$, whereas $(a \rightarrow b) \rightarrow b = b$.

Remark 2.5.14 Similar to pocrims, the class of BCK-algebras forms a quasivariety. However, as shown by Wroński [260], this class does not form a variety. In particular, it fails to be closed under homomorphic images (see [214, Chap. 6] or [266, Example 1.5.1]). Consider the Wroński's BCK-algebra $\mathbf{H} = (H; \rightarrow, 1)$ in Example 2.5.3. Then $\{\mathbb{N}_0, A, B\}$ is a partition for H and easy calculations show that the relation θ defined by

$$(x, y) \in \theta \Leftrightarrow \text{ both of } x \text{ and } y \text{ are in one of the sets } \mathbb{N}_0, A, \text{ or } B,$$

is a congruence relation on $\mathbf{H}$. Additionally, $\mathbf{H}/\theta$ has exactly three classes $\{A, B, \mathbb{N}_0\}$, and $\mathbf{H}/\theta = (\{A, B, \mathbb{N}_0\}; \rightarrow, \mathbb{N}_0)$ is an algebra of type $(2, 0)$, where the binary operation $\rightarrow$ is defined by Table 2.3:

Clearly, $\mathbf{H}/\theta$ is not a BCK-algebra, since $A \rightarrow B = B \rightarrow A = \mathbb{N}_0$, but $A \neq B$. Therefore, the homomorphic image of the natural projection map $\pi_\theta : \mathbf{H} \rightarrow \mathbf{H}/\theta$ is not a BCK-algebra.

In contrast, Yutani [267] proved that the class of commutative BCK-algebras forms a variety.

Theorem 2.5.15 ([267]) *An algebra* $(H; \rightarrow, 1)$ *of type* $(2, 0)$ *is a commutative BCK-algebra if and only if it satisfies the equations:*

(i) $(x \rightarrow y) \rightarrow y = (y \rightarrow x) \rightarrow x.$
(ii) $x \rightarrow (y \rightarrow z) = y \rightarrow (x \rightarrow z).$
(iii) $x \rightarrow x = 1.$
(iv) $1 \rightarrow x = x.$

Table 2.3 BCK-algebra in Remark 2.5.14

$\rightarrow$	A	B	$\mathbb{N}_0$
A	$\mathbb{N}_0$	$\mathbb{N}_0$	$\mathbb{N}_0$
B	$\mathbb{N}_0$	$\mathbb{N}_0$	$\mathbb{N}_0$
$\mathbb{N}_0$	A	B	$\mathbb{N}_0$

Proof Assume that $(H; \to, 1)$ is an algebra of type $(2, 0)$ satisfying (i)–(iv). Choose $x, y, z \in H$. From (i), (iv), and (iii) it follows that $(x \to 1) \to 1 = (1 \to x) \to x = x \to x = 1$, and so

$$x \to 1 = x \to ((x \to 1) \to 1) = (x \to 1) \to (x \to 1) = 1, \text{ by (ii) and (iii).} \quad (2.14)$$

$$\begin{aligned}
(x \to y) \to ((y \to z) \to (x \to z)) &= (x \to y) \to (x \to ((y \to z) \to z)), \text{ by (ii)} \\
&= (x \to y) \to (x \to ((z \to y) \to y)), \text{ by (i)} \\
&= (x \to y) \to ((z \to y) \to (x \to y)), \text{ by (ii)} \\
&= (z \to y) \to ((x \to y) \to (x \to y)), \text{ by (ii)} \\
&= (z \to y) \to 1 = 1, \text{ by (2.14).}
\end{aligned}$$

Thus, (BCK1) holds. In addition, (BCK2) follows from (ii) and (iii), directly. The conditions (BCK3) and (BCK4) are (iii) and (iv), exactly. Now, let $x \to y = 1 = y \to x$. Then by (i) and (iv) we get

$$x = 1 \to x = (y \to x) \to x = (x \to y) \to y = 1 \to y = y,$$

which means (BCK5) holds. Therefore, $(H; \to, 1)$ is a BCK-algebra. Conversely, if $(H; \to, 1)$ is a commutative BCK-algebra, then by Lemma 2.5.4, it satisfies (i)–(iv), evidently. $\qquad\square$

A **bounded BCK-algebra** is an algebra $\mathbf{H} = (H; \to, 0, 1)$, where $(H; \to, 1)$ is a BCK-algebra with a least element 0, i.e., $0 \to x = 1$ for all $x \in H$. Mundici [222] proved that MV-algebras and bounded commutative BCK-algebras are termwise equivalent:

Theorem 2.5.16 *Let* $(H; \oplus, {}', 0, 1)$ *be an MV-algebra. Define* $x \to y = x' \oplus y$ *for all* $x, y \in H$. *Then* $(H; \to, 0, 1)$ *is a bounded commutative BCK-algebra.*

Conversely, if $(H; \to, 0, 1)$ *is a bounded commutative BCK-algebra, define*

$$x \oplus y = (x \to 0) \to y \quad \& \quad \neg x = x \to 0, \quad \forall x, y \in H.$$

Then $(H; \oplus, \neg, 0, 1)$ *is an MV-algebra.*

Proof The proof is left to the readers. $\qquad\square$

By applying Theorems 2.5.12 and 2.5.16, we can establish the relationships between BCK-algebras, pocrims, and MV-algebras as depicted in Fig. 2.3.

For further information on BCK-algebras, we suggest the monographs [180, 182, 201, 214, 266].

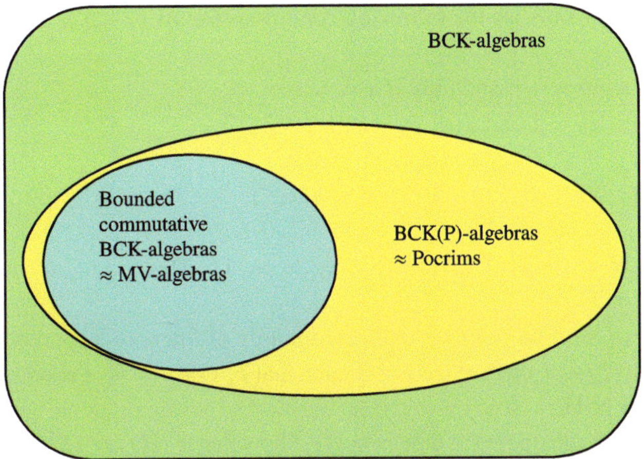

Fig. 2.3 Relation between BCK-algebras, pocrims, and MV-algebras

Bibliographical Remarks and Suggestions for Further Study

This chapter provided a comprehensive study of pocrims and their relationships with other algebraic structures, particularly residuated lattices and BCK-algebras. This chapter meticulously gathered various properties and examples of pocrims from diverse sources (those were cited in the text), facilitating a deeper understanding for the reader. A thorough understanding of the properties, concepts, and results presented in this chapter is essential for comprehending the subsequent chapters. While the scope of pocrims is significantly broader, the collected material adequately supports our main goal: investigating hoops.

It is noteworthy that the classes of BCK-$\wedge$-semilattices, pocrim-$\wedge$-semilattices, and pocrim-$\vee$-semilattices are arithmetical varieties, and so strictly locally affine complete varieties [193, Sect. 3.4]. In [217], the quasivariety of involutive pocrims was studied. It was proven that if a variety V is finitely generated by involutive pocrims of finite type, then it has a finitely based equational theory. Furthermore, relatively compatible operations in pocrims and BCK-algebras were investigated in [207]. For a deeper understanding of the relationship between BCK-algebras and pocrims, we recommend consulting [179, 181].

2.6 Exercises

2.6.1 Let $(H; \odot, \rightarrow, 1)$ be a pocrim. Define the mapping $d : H^2 \rightarrow H$ by

$$d(x, y) = (x \rightarrow y) \odot (y \rightarrow x) \qquad \forall x, y \in H. \tag{2.15}$$

Show that d satisfies the following conditions for all $x, y, z \in H$:

(i) $d(x, y) = 1$ if and only if $x = y$.
(ii) $d(x, x) = 1$ and $d(x, 1) = x$.
(iii) $d(y, z) \odot d(x, y) \odot d(y, z) \leq d(x, z)$.
(iv) $d(x, y) \odot d(y, z) \odot d(x, y) \leq d(x, z)$.

2.6.2 Let F be a filter of a pocrim $(H; \odot, \rightarrow, 1)$. Then $(x, y) \in \theta_F$ if and only if $d(x, y) \in F$.

2.6.3 Let $\{F_i : i \in I\}$ be an upward directed family of filters of a pocrim $\mathbf{H}$. It means that for each $i, j \in I$ there exists $k \in I$ such that $F_i \cup F_j \subseteq F_k$. Prove that $\bigcup\{F_i : i \in I\}$ is a filter of $\mathbf{H}$.

2.6.4 Let $\mathbf{H}_1$ be a subalgebra of a pocrim $\mathbf{H}_2$. Show that $\mathcal{F}(\mathbf{H}_1) = \{F \cap H_1 : F \in \mathcal{F}(\mathbf{H}_2)\}$ (Hint: Show that $\uparrow F \in \mathcal{F}(\mathbf{H}_2)$ and $H_1 \cap \uparrow F = F$ for each $F \in \mathcal{F}(\mathbf{H}_1)$).

2.6.5 Let $\{\mathbf{H}_i : i \in I\}$ be a family of pocrim. Prove that the following statements hold on the direct product $\prod_{i \in I} \mathbf{H}_i$ (see Sect. 1.2):

(i) $\mathcal{F}(\prod_{i \in I} \mathbf{H}_i) = \{\prod_{i \in I} F_i : F_i \in \mathcal{F}(\mathbf{H}_i), \ i \in I\}$.
(ii) $\mathbf{A}$ is a subalgebra of $\prod_{i \in I} \mathbf{H}_i$ if and only if $\pi_i(\mathbf{A})$ is a subalgebra of $\mathbf{H}_i$ for all $i \in I$.

2.6.6 Let $\mathbf{H}$ be a pocrim. Prove that if $\mathbf{H}$ is a lattice, then every filter of $\mathbf{H}$ is a lattice filter of $(H; \vee, \wedge)$ (see Definition 1.3.17).

2.6.7 Complete the proof of Theorem 2.4.8.

2.6.8 Find a bounded pocrim $\mathbf{H}$ such that $\mathrm{Reg}(\mathbf{H})$ is not closed under $\odot$.

2.6.9 Complete the proof of Theorem 2.5.16.

2.6.10 Applying the ordinal sum of pocrims, prove that every pocrim can be seen as a subalgebra of a $\{\odot, \rightarrow, 1\}$-reduct of a bounded pocrim $(H; \odot, \rightarrow, 0, 1)$.

2.6.11 Let $\mathbf{H}$ be a bounded pocrim and $\ominus$ and $\star$ be the binary operations were defined in (2.5). Show that if $\mathbf{H}$ satisfies (DNP), then $(x \ominus y) \ominus z = x \ominus (y \star z)$ for all $x, y, z \in H$.

2.6.12 Let $(X; \rightarrow, 0, 1)$ be a bounded BCK-algebra such that $(x \rightarrow 0) \rightarrow 0 = x$ for all $x \in X$. Prove that the following assertions are equivalent, where $\leq$ is the partially ordered relation on the BCK-algebra $(X; \rightarrow, 1)$.

(i) $(X, \leq)$ is a $\wedge$-semilattice.
(ii) $(X, \leq)$ is a $\vee$-semilattice.
(iii) $(X, \leq)$ is a lattice.

2.6.13 Let $(X; \to, 1)$ be a BCK-algebra such that $(X, \leq)$ is a lattice. Show that:

(i) $(x \vee y) \to z = (x \to z) \wedge (y \to z)$ for every $x, y, z \in X$.

$$(x \vee y) \to z = (x \to z) \wedge (y \to z), \quad \forall x, y, z \in X.$$

(ii) If $(x \wedge y) \to z = (x \to z) \vee (y \to z)$ for every $x, y, z \in X$, then $(X, \leq)$ is a distributive lattice.

Structure of Hoops 3

Hoop-algebras were initially introduced by Bosbach [48, 49] under the name "complementary semigroups" and later explored by Büchi and Owens [59], an unpublished paper. These algebras are particularly significant in the fields of universal algebra and algebraic logic. Hoop algebras have various applications, especially in fuzzy logic and non-classical logics. Esteva et al. [137] provided axiomatizations for logics corresponding to hoops, which are known as hoop logics. More information is available in B. Bosbach recent papers [52, 53]. Over the years, many researchers have studied hoops, enriching the theory with deep structure theorems.

This chapter introduces hoops, a significant class of algebraic structures derived from pocrims, and explores their rich properties and diverse subclasses. We begin by defining hoops through pocrims, indeed, a hoop is a naturally ordered pocrims, and then demonstrate their variety identified via four identities. Establishing that every hoop is a meet-semilattice, we proceed to collect and analyze its main properties. The fundamental properties of hoops, along with those inherited from pocrims, and particularly the relationships between the operations $\wedge$, $\odot$, $\rightarrow$, and $\vee$ (whenever $\vee$ exists), are essential for the subsequent sections. While every finite hoop is a lattice, this property does not hold in general, even for bounded hoops. We introduce and investigate $\vee$-hoops and $\sqcup$-hoops. It is shown that every $\sqcup$-hoop is a $\vee$-hoop, and we delineate the conditions under which these concepts coincide. Utilizing the join-center or cancellative-center of a hoop, we show how bounded hoops can be derived. We then delve into special subclasses, including cancellative, Wajsberg, basic, idempotent, and product hoops, detailing their properties and interrelationships, and revealing the categorical equivalence between cancellative hoops and Abelian ℓ-groups. All five of the aforementioned subclasses of hoops form varieties. Finally, we provide a comprehensive collection of diverse examples, illustrating the breadth and depth of hoop theory.

© The Author(s), under exclusive license to Springer Nature Switzerland AG 2026

A. Dvurečenskij et al., *Hoop Algebras*, Frontiers in Mathematics,
https://doi.org/10.1007/978-3-032-11736-6_3

3.1　Hoops

In this section, we familiarize ourselves with the definitions, fundamental concepts, basic properties, and several examples of Hoops.

We say that a pocrim $\mathbf{H} = (H; \odot, \rightarrow, 1)$ is **naturally ordered** if for all $a, b \in H$, $a \leq b$ if and only if $a = b \odot c$ for some $c \in H$. A **hoop-algebra** or simply **hoop** is a naturally ordered pocrim.

Definition 3.1.1 ([59]) An algebra $\mathbf{H} = (H; \odot, \rightarrow, 1)$ of type $(2, 2, 0)$ is called a **hoop** if it is a naturally ordered pocrim. A **bounded hoop** is an algebra $\mathbf{H} = (H; \odot, \rightarrow, 0, 1)$ of type $(2, 2, 0, 0)$ such that $(H; \odot, \rightarrow, 1)$ is a hoop with the least element 0.

A **linear (or totally ordered) hoop** is a hoop in which $\leq$ is a total order.

We note that the trivial pocrim $\mathbf{1} = (\{1\}; \odot, \rightarrow, 1)$ is evidently a hoop. A hoop $\mathbf{H}$ is called **trivial** if $|H| = 1$, otherwise, is called **non-trivial**.

In what follows, we will adopt the convention that $\odot$ has priority over the operation $\rightarrow$. Sometimes, for the sake of clarity, we shall put parentheses even if this is not necessary.

Of course, every relatively pseudo-complemented semilattice $(S; \wedge, *, 1)$ (see Example 2.1.7) can be viewed as a hoop in which $x \odot y = x \wedge y$ and $x \rightarrow y = x * y$ since $x \wedge y = (x * y) \wedge x = (y * x) \wedge y$, hence $x \leq y$ if and only if $x = (y * x) \wedge y$.

We shall denote the set of natural numbers by $\mathbb{N}$. We define $x^0 = 1$, $x \overset{0}{\rightarrow} y = y$ and

$$x^n = x^{n-1} \odot x, \quad \text{and} \quad x \overset{n}{\rightarrow} y = x \rightarrow (x \overset{n-1}{\longrightarrow} y), \quad \forall n \in \mathbb{N}. \tag{3.1}$$

There are two equivalent definitions of hoops, as shown in Theorems 3.1.2 and 3.1.3.

Theorem 3.1.2 ([31]) *An algebra* $\mathbf{H} = (H; \odot, \rightarrow, 1)$ *of type* $(2, 2, 0)$ *is a hoop if and only if it satisfies the following assertions:*

(H0)　$(H; \odot, 1)$ *is a commutative monoid.*
(H1)　$x \rightarrow x = x \rightarrow 1 = 1$.
(H2)　$(x \odot y) \rightarrow z = x \rightarrow (y \rightarrow z)$.
(H3)　$x \odot (x \rightarrow y) = y \odot (y \rightarrow x)$.

Proof Suppose that $\mathbf{H}$ is a hoop. Then (H0) holds, evidently. Lemma 2.2.1(1) and (8) imply (H1) and (H2). Choose $x, y \in H$. By Lemma 2.2.1(4), $x \odot (x \rightarrow y) \leq y$ entails that $y \odot z = x \odot (x \rightarrow y) \leq x$ for some $z \in H$. Thus, $z \leq y \rightarrow x$ and so $x \odot (x \rightarrow y) = y \odot z \leq y \odot (y \rightarrow x)$. In a similar way, we can show that $y \odot (y \rightarrow x) \leq x \odot (x \rightarrow y)$. Therefore, (H3) holds.

Conversely, assume that $\mathbf{H} = (H; \odot, \rightarrow, 1)$ is an algebra of type $(2, 2, 0)$ satisfying (H0)–(H3). We will apply Theorem 2.2.7 in order to show that $\mathbf{H}$ is a hoop. Choose $x, y, z \in H$. If $x \rightarrow y = y \rightarrow x = 1$, then by (H3), $x = y$. It suffices to show that $(x \rightarrow y) \rightarrow ((z \rightarrow x) \rightarrow (z \rightarrow y)) = 1$. We have

$$
\begin{aligned}
(x \rightarrow y) \rightarrow \big((z \rightarrow x) \rightarrow (z \rightarrow y)\big) &= (x \rightarrow y) \rightarrow \big((z \odot (z \rightarrow x)) \rightarrow y\big), \text{ by (H2)} \\
&= (x \rightarrow y) \rightarrow \big((x \odot (x \rightarrow z)) \rightarrow y\big), \text{ by (H3)} \\
&= (x \rightarrow y) \rightarrow \big((x \rightarrow z) \rightarrow (x \rightarrow y)\big), \text{ by (H2)} \\
&= \big((x \rightarrow y) \odot (x \rightarrow z)\big) \rightarrow (x \rightarrow y), \text{ by (H2)} \\
&= (x \rightarrow z) \rightarrow \big((x \rightarrow y) \rightarrow (x \rightarrow y)\big), \text{ by (H2)} \\
&= (x \rightarrow z) \rightarrow 1 = 1, \text{ by (H1)}.
\end{aligned}
$$

Hence, by Theorem 2.2.7, $\mathbf{H}$ is a pocrim. Furthermore, $x \leq y$ implies that $x \rightarrow y = 1$, consequently, by (H3), $x = x \odot (x \rightarrow y) = y \odot (y \rightarrow x)$. Therefore, $\mathbf{H}$ is a naturally ordered pocrim, or equivalently, $\mathbf{H}$ is a hoop. $\qquad\square$

The concepts of homomorphism, embedding, and isomorphism for hoops are defined analogously to those for pocrims (see Remark 2.4.4). As a straightforward consequence of Theorem 3.1.2, the class of hoops forms a variety. We shall denote by H this variety. In addition, the category whose objects are hoops and whose morphisms are homomorphisms of hoops is called the category of hoops and denoted by $\mathcal{H}oop$. There exists another reduced form of Theorem 3.1.2 as follows:

Theorem 3.1.3 *An algebra* $\mathbf{H} = (H; \odot, \rightarrow, 1)$ *of type* $(2, 2, 0)$ *is a hoop if and only if it satisfies the following assertions:*

(i) $\odot$ *is commutative such that* $1 \odot x = x$.
(ii) $x \rightarrow x = 1 = x \rightarrow 1$.
(iii) $(x \odot y) \rightarrow z = x \rightarrow (y \rightarrow z)$.
(iv) $x \odot (x \rightarrow y) = y \odot (y \rightarrow x)$.

Proof Suppose that $\mathbf{H}$ is a hoop. Then by Theorem 3.1.2, (i)–(iv) hold.

Conversely, assume that $\mathbf{H} = (H; \odot, \rightarrow, 1)$ is an algebra of type $(2, 2, 0)$ satisfying (i)–(iv). By Theorem 3.1.2, it suffices to show that $\odot$ is associative. First, define a binary relation $\leq$ in H by $x \leq y$ if and only if $x \rightarrow y = 1$.

(1) By (ii), $\leq$ is reflexive. Let $x, y \in H$ be such that $x \leq y \leq x$. Then by (iv) and (i),

$$
x = x \odot 1 = x \odot (x \rightarrow y) = y \odot (y \rightarrow x) = y \odot 1 = y,
$$

entails that $\leq$ is anti-symmetric.

(2) Let $x \le y$ and $y \le z$ for some $x, y, z \in H$. Then $x \to y = y \to z = 1$ and so

$$
\begin{aligned}
x \to z &= (x \odot 1) \to z = (x \odot (x \to y)) \to z \\
&= (y \odot (y \to x)) \to z, \text{ by (iv)} \\
&= (y \to x) \to (y \to z), \text{ by (iii)} \\
&= (y \to x) \to 1 = 1, \text{ by (ii)}
\end{aligned}
$$

that is, $x \le z$, so $\le$ is transitive, consequently, it is a partially ordered relation on H.

Now, let $x, y, z \in H$ and $(x \odot y) \odot z \le u$ for some $u \in H$. Then, by (iii), we have

$$
\begin{aligned}
(x \odot y) \odot z \le u &\Leftrightarrow \big((x \odot y) \odot z\big) \to u = 1 \\
&\Leftrightarrow (x \odot y) \to (z \to u) = 1, \text{ by (iii)} \\
&\Leftrightarrow x \to \big(y \to (z \to u)\big) = 1, \text{ by (iii)} \\
&\Leftrightarrow x \to \big((y \odot z) \to u\big) = 1, \text{ by (iii)} \\
&\Leftrightarrow \big(x \odot (y \odot z)\big) \to u = 1, \text{ by (iii)} \\
&\Leftrightarrow x \odot (y \odot z) \le u.
\end{aligned}
$$

Thus, $x \odot (y \odot z) = (x \odot y) \odot z$, i.e., $(H; \odot, 1)$ is a commutative monoid. Therefore, by Theorem 3.1.2, $(H; \odot, \to, 1)$ is a hoop. $\square$

Proposition 3.1.4 ([48, 49, 69]) *Let* $\mathbf{H} = (H; \odot, \to, 1)$ *be a hoop. For any* $x, y, z, w \in H$, *the following statements hold:*

(i) $(H, \le)$ *is a meet-semilattice, with* $x \wedge y = x \odot (x \to y) = y \odot (y \to x)$ *with the top element* 1. *In addition,* $x \odot y \le x \wedge y$ *for all* $x, y \in H$.

(ii) *For any* $n \in \mathbb{N}$, $x \overset{n}{\to} y = x^n \to y$.

(iii) $y \to ((x \to y) \to x) = y \to x$.

(iv) $\big(((y \to x) \to x) \to y\big) \to (y \to x) = y \to x$.

(v) $(x \to y) \odot (z \to w) \le (x \wedge z) \to (y \wedge w)$.

(vi) $(x \to y) \to (y \to x) = y \to x$.

(vii) $(x \to y) \to (x \to z) = (y \to x) \to (y \to z)$.

Proof (i) By Lemma 2.2.1(4), $x \odot (x \to y) \le x, y$. Suppose that $z \le x, y$ for some $z \in H$. So, by Lemma 2.2.1(7), $x \to z \le x \to y$ entails that

$$
z = z \odot 1 = z \odot (z \to x) = x \odot (x \to z) \le x \odot (x \to y),
$$

implying $x \odot (x \to y)$ is the greatest lower bound for x and y, that is $x \wedge y = x \odot (x \to y)$. In a similar way, it can be proved that $x \wedge y = y \odot (y \to x)$. In addition, $x \odot y \le x, y$ implies that $x \odot y \le x \wedge y$, by Lemma 2.2.1(4).

(ii) A straightforward inductive argument, utilizing Lemma 2.2.1(8), establishes that

$$x \xrightarrow{n} y = x^n \to y, \quad \forall n \in \mathbb{N}.$$

(iii) Let us denote $(x \to y) \to x$ by z. We have to prove that $y \to z = y \to x$. By Lemma 2.2.1(10), $x \leq z$. Hence, applying Lemma 2.2.1(7), we get that $y \to x \leq y \to z$. Also, $x \leq z$ implies that $z \to y \leq x \to y$ and, by Lemma 2.2.1(6), $x \to y \leq ((x \to y) \to x) \to x = z \to x$. Thus, we have $z \to y \leq z \to x$, consequently, $(y \to z) \odot y = y \wedge z = (z \to y) \odot z \leq (z \to x) \odot z = z \wedge x \leq x$. Hence, by (i), $y \to z \leq y \to x$.

(iv) Let $z = (y \to x) \to x$. By (i), Lemma 2.2.1(6) and (13), we have that $y \leq z$ and $z \to x = y \to x$, so

$$(z \to y) \to (y \to x) = (z \to y) \to (z \to x) = ((z \to y) \odot z) \to x$$
$$= (z \wedge y) \to x = y \to x.$$

(v) By Lemma 2.2.1(7), we have that $x \to y \leq (x \wedge z) \to y$ and $z \to w \leq (x \wedge z) \to w$. It follows that

$$(x \to y) \odot (z \to w) \leq ((x \wedge z) \to y) \odot ((x \wedge z) \to w) \leq ((x \wedge z) \to y) \wedge ((x \wedge z) \to w).$$

Applying Lemma 2.2.3(ii), we get that $((x \wedge z) \to y) \wedge ((x \wedge z) \to w) = (x \wedge z) \to (y \wedge w)$.

(vi) First, we note that $y \to x \leq (x \to y) \to (y \to x)$, by Lemma 2.2.1(10). Set $w := (x \to y) \to x$. By Lemma 2.2.1(10), $x \leq w$, and

$$w \to y \leq x \to y \leq ((x \to y) \to x) \to x = w \to x, \quad \text{by Lemma 2.2.1(6) and (7)},$$

entails from (i) that $y \odot (y \to w) = w \odot (w \to y) \leq w \odot (w \to x) \leq x$. Hence, by (2.2), $y \to w \leq y \to x$, consequently, $(x \to y) \to (y \to x) = y \to w \leq y \to x$. Therefore, $(x \to y) \to (y \to x) = y \to x$.

(vii) By (H3) and Lemma 2.2.1(8), we obtain

$$(x \to y) \to (x \to z) = \big(x \odot (x \to y)\big) \to z$$
$$= \big(y \odot (y \to x)\big) \to z$$
$$= (y \to x) \to (y \to z).$$

$\square$

Theorem 3.1.5 ([31, p. 237]) *A pocrim* $\mathbf{H} = (H; \odot, \to, 1)$ *is a hoop if and only if* $\mathbf{H}$ *satisfies* (H3).

Proof Let $\mathbf{H} = (H; \odot, \rightarrow, 1)$ be a pocrim satisfying $x \odot (x \rightarrow y) = y \odot (y \rightarrow x)$. Choose two elements $x \leq y$ of H. From $x = x \odot 1 = x \odot (x \rightarrow y) = y \odot (y \rightarrow x)$ it follows that $\mathbf{H}$ is naturally ordered and so by definition, $\mathbf{H}$ is a hoop.

The proof of the converse follows from Proposition 3.1.4(i). $\qquad\qquad\qquad\square$

Theorem 3.1.5 and Proposition 3.1.4(i) imply that if a pocrim $\mathbf{H}$ satisfies (H3), then it is meet-semilattice, indeed, $x \wedge y = x \odot (x \rightarrow y) = y \odot (y \rightarrow x)$.

At the end of this section, we will verify some properties of linear hoops.

Proposition 3.1.6 ([111]) *Let* $\mathbf{H}$ *be a linear hoop.*

   (i) *If* $x \leq z \leq y$ *and* $z \odot z = z$, *then* $x \odot y = x$ *for each* $x, y, z \in H$.
  (ii) *If* $z \odot z = z$ *and* $x < z \leq y$, *then* $y \rightarrow x = x$ *for each* $x, y, z \in H$.
 (iii) *There exist no elements* $x, y, z \in H$ *such that* $x < z < y$, $x \odot y = x$, $x \odot z < x$, *and* $y \odot z < z$.
 (iv) *If* $x \odot x < x$ *and* $x \odot y = x$, *then* $y \rightarrow x = x$ *for each* $x, y \in H$.

Proof (i) Suppose that $x, y, z \in H$ such that $x \leq z \leq y$ and $z \odot z = z$. Then

$$x = x \wedge z = z \odot (z \rightarrow x), \text{ by Proppsition } 3.1.4(i)$$
$$= z \odot z \odot (z \rightarrow x)$$
$$= z \odot x \leq x \odot y, \text{ by Lemma } 2.2.1(3)$$
$$\leq x.$$

Therefore, $x \odot y = x$.

(ii) Let $x, y, z \in H$ such that $z \odot z = z$ and $x < z \leq y$. If $z \leq y \rightarrow x$, then

$$y \odot z \leq y \odot (y \rightarrow x) = y \wedge x = x$$

entails that $z = z \odot z \leq y \odot z \leq x$ which is absurd. Hence, $y \rightarrow x < z$ and so $x \leq y \rightarrow x < z \leq y$. If $x < y \rightarrow x$, then by (i), from $y \rightarrow x < z \leq y$ it follows that $x < y \rightarrow x = (y \rightarrow x) \odot y = y \wedge x = x$, which is a contradiction. Therefore, $y \rightarrow x = x$.

(iii) Let $x, y, z \in H$ satisfying (iii). Set $u := x \rightarrow (z \odot x)$. Then $u < y$. Otherwise, $y \leq u$ implies that $z \odot x < x = y \odot x \leq u \odot x = (x \rightarrow (z \odot x)) \odot x = x \wedge (z \odot x) = z \odot x$, which is a contradiction. That is, $u < y$. By Lemma 2.2.1(8),

$$y \rightarrow u = y \rightarrow (x \rightarrow (z \odot x)) = (y \odot x) \rightarrow (z \odot x) = x \rightarrow (z \odot x) = u,$$

consequently, $u = u \wedge y = y \odot (y \rightarrow u) = y \odot u$.

From $z \odot x \le z \odot x$ and Lemma 2.2.1(9) we get $z \le x \to (x \odot z) = u$. Hence,

$$z = z \wedge u = u \odot (u \to z) = (u \odot y) \odot (u \to z)$$
$$= (u \odot (u \to z)) \odot y = (u \wedge z) \odot y = z \odot y < z,$$

which is a contradiction. Therefore, there exist no elements $x, y, z \in H$ satisfying (iii).

(iv) Let $x, y \in H$ such that $x \odot x < x$ and $y \odot x = x$. Then $x \le y$. Also, $y \ne x$, since $y \odot x = x$ and $x \odot x < x$. That is, $x < y$. Consider the filter $X_x = \{z \in H : z \odot x = x\}$, clearly, $y \in X_x$ (see Example 2.3.3). We show that $y \to x < y$, otherwise, $y \le y \to x$ implies that $y \to x \in X_x$, and so by (F1), $x = y \wedge x = (y \to x) \odot y \in X_x$, which contradicts with the assumption. Hence, we have $x \le y \to x < y$. Similar proof shows that $y \to x \notin X_x$, consequently, $(y \to x) \odot x < x$.

We claim that $y \to x = x$. On the contrary, $x < y \to x$ implies that

$$x < y \to x < y \quad \& \quad y \odot x = x,$$
$$y \odot (y \to x) = y \wedge x = x < y \to x \quad \& \quad (y \to x) \odot x < x,$$

This directly contradicts part (iii). Therefore, the claim holds true. $\square$

Each subalgebra $(S; \odot, \to, 1)$ of a hoop $\mathbf{H} = (H; \odot, \to, 1)$ is called a **subhoop** of $\mathbf{H}$. Due to Definition 1.2.4 and Remark 1.2.7, a non-empty subset S of a hoop $\mathbf{H} = (H; \odot, \to, 1)$ is a subuniverse of $\mathbf{H}$ if it is closed under $\to$ and $\odot$. We note that $S \ne \emptyset$ implies that $1 = x \to x \in S$ for each $x \in S$, which means $(S; \odot, \to, 1)$ is a hoop. Consequently, by Theorem 3.1.2, if a pocrim $\mathbf{H}$ is a hoop, then its subhoops are precisely its subpocrims. In addition, if $\mathbf{H} = (H; \odot, \to, 0, 1)$ is bounded, then a bounded subhoop of $\mathbf{H}$ is a subhoop $\mathbf{S}$ such that $0 \in S$.

Filters and deductive systems, as defined for pocrims, carry over directly to the context of hoops. Specifically, a subset F of a hoop $(H; \odot, \to, 1)$ is a filter if and only if it is a filter when viewed as a pocrim. Consequently, the results established for deductive systems and filters in Chap. 2 remain valid in the setting of hoops. A more in-depth exploration of these concepts within the framework of hoops is the focus of Chap. 4.

3.2   ⊔-Hoops and ∨-Hoops

As established in Proposition 3.1.4(i), every hoop forms a $\wedge$-semilattice. This section introduces an additional binary operation $\sqcup$ on hoops. We will first explore the hoop $\mathbf{H}$, where $x \sqcup y$ represents the least upper bound of x and y for all $x, y \in H$. Subsequently, we will delve into hoops where the partially ordered relation (2.2) exhibits a lattice structure in the general case.

According to [11], on each hoop $\mathbf{H}$ one can define a binary operation $\sqcup$ as follows:

$$x \sqcup y = \big((x \to y) \to y\big) \wedge \big((y \to x) \to x\big) \quad \text{for all } x, y \in H. \tag{3.2}$$

We note that by Proposition 3.1.4(i), $\sqcup$ is well-defined.

Proposition 3.2.1 ([11]) *Let $\mathbf{H}$ be a hoop. For any $x, y \in H$, the following assertions hold:*

(i) $x \sqcup y = y \sqcup x$.
(ii) $x, y \leq x \sqcup y$.
(iii) $x \leq y$ *if and only if* $x \sqcup y = y$.
(iv) $x \sqcup x = x$ *and* $x \sqcup 1 = 1$.
(v) *If $\mathbf{H}$ is bounded, then* $0 \sqcup x = x$.
(vi) $(x \to y) \sqcup (y \to x) = 1$.

Proof (i) The proof is obvious.

(ii) By Lemma 2.2.1(6), $x \leq (x \to y) \to y$ and by Lemma 2.2.1(10) $x \leq (y \to x) \to x$. Hence, $x \leq x \sqcup y$. In a similar way, $y \leq x \sqcup y$.

(iii) If $x \leq y$, then by (2.3), $(x \to y) \to y = 1 \to y = y$. Hence,

$$x \sqcup y = y \wedge ((y \to x) \to x) = y, \text{ by Lemma } 2.2.1(6).$$

Conversely, suppose that $x \sqcup y = y$. It follows that $x \wedge y = x \wedge (x \sqcup y) = x$, by (ii). That is, $x \leq y$.

(iv) $x \sqcup x = ((x \to x) \to x) \wedge ((x \to x) \to x) = (1 \to x) \wedge (1 \to x) = x \wedge x = x$. In addition, $x \sqcup 1 = ((x \to 1) \to 1) \wedge ((1 \to x) \to x) = 1 \wedge (x \to x) = 1$.

(v) $0 \sqcup x = ((0 \to x) \to x) \wedge ((x \to 0) \to 0) = (1 \to x) \wedge x'' = x \wedge x'' = x$, by Remark 2.2.5(ii).

(vi) By Proposition 3.1.4(vi),

$$((x \to y) \to (y \to x)) \to (y \to x) = (y \to x) \to (y \to x) = 1.$$

Similarly, $((y \to x) \to (x \to y)) \to (x \to y) = 1$. Therefore, $(x \to y) \sqcup (y \to x) = 1$. $\qquad\square$

Proposition 3.2.2 ([11, Proposition 2.4]) *Let $\mathbf{H}$ be a hoop. The following are equivalent:*

(i) $\sqcup$ *is associative.*
(ii) $x \leq y$ *implies* $x \sqcup z \leq y \sqcup z$ *for all* $x, y, z \in H$.
(iii) $x \sqcup (y \wedge z) \leq (x \sqcup y) \wedge (x \sqcup z)$ *for all* $x, y, z \in H$.
(iv) $\sqcup$ *is the join operation on H (that is $x \sqcup y = x \vee y$ for all $x, y \in H$).*

Proof (i) ⇔ (iv) Assume that ⊔ is associative. By Proposition 3.2.1(ii), $x, y \leq x \sqcup y$. Let u be an upper bound for x and y. By (i) and Lemma 2.2.1, we get

$$(x \sqcup y) \sqcup u = x \sqcup (y \sqcup u) = x \sqcup \left(\big((u \to y) \to y\big) \wedge \big((y \to u) \to u\big) \right)$$

$$= x \sqcup \left(\big((u \to y) \to y\big) \wedge (1 \to u) \right) = x \sqcup \big(((u \to y) \to y) \wedge u\big)$$

$$= x \sqcup u, \text{ by Lemma } 2.2.1(6)$$

$$= ((x \to u) \to u) \wedge ((u \to x) \to x) = (1 \to u) \wedge ((u \to x) \to x)$$

$$= u \wedge ((u \to x) \to x) = u, \text{ by Lemma } 2.2.1(6).$$

Now, Proposition 3.2.1(iii) implies that $x \sqcup y \leq u$, which means $x \sqcup y$ is the least upper bound for x and y. Conversely, suppose that $x \sqcup y = x \vee y$ for all $x, y \in H$. Then $x \sqcup (y \sqcup z) = x \vee (y \vee z) = (x \vee y) \vee z = (x \sqcup y) \sqcup z$ for all $x, y, z \in H$.

(ii) ⇔ (iv) Suppose that (ii) holds. Let $x, y \leq z$ be elements of H. By (ii), $x \sqcup y \leq z \sqcup y$ and so

$$((x \to y) \to y) \wedge ((y \to x) \to x) \leq ((z \to y) \to y) \wedge ((y \to z) \to z)$$

$$= ((z \to y) \to y) \wedge z = z.$$

Hence, by Proposition 3.2.1(ii), $x \sqcup y$ is the least upper bound for x and y. The proof of the converse is clear.

(iii) ⇔ (iv) Let (iii) hold and $x, y \leq z$ be elements of H. From

$$x \sqcup z = ((x \to z) \to z) \wedge ((z \to x) \to x) = z \wedge ((z \to x) \to x) = z$$

and (iii) it follows that

$$x \sqcup y = x \sqcup (y \wedge z) \leq (x \sqcup y) \wedge (x \sqcup z) = (x \sqcup y) \wedge z,$$

consequently, $x \sqcup y \leq z$. Hence, Proposition 3.2.1(ii) implies that $x \sqcup y = x \vee y$. Conversely, let (iv) hold and $x, y, z \in H$ be given. Then $x \sqcup (y \wedge z) = x \vee (y \wedge z)$, $x \sqcup y = x \vee y$, and $x \sqcup z = x \vee z$. Clearly, $x, y \wedge z \leq x \vee y$ and $x, y \wedge z \leq x \vee z$, which means $(x \vee y) \wedge (x \vee z)$ is an upper bound for x and $y \wedge z$, that is,

$$x \sqcup (y \wedge z) = x \vee (y \wedge z) \leq (x \vee y) \wedge (x \vee z) = (x \sqcup y) \wedge (x \sqcup z).$$

$$\square$$

Applying Proposition 3.2.2, we define ⊔-hoops as follows:

Definition 3.2.3 By a ⊔-**hoop** we mean a hoop algebra **H** satisfying one of the equivalent conditions of Proposition 3.2.2.

In each hoop $\mathbf{H}$ the subset $\{x \in H : x \vee y \text{ exists for all } y \in H\}$ is called the **join-center**[1] of $\mathbf{H}$ and denoted by $C_\vee(\mathbf{H})$. Clearly, $1 \in C_\vee(\mathbf{H})$ and $x \vee y \in C_\vee(\mathbf{H})$ for every $x, y \in C_\vee(\mathbf{H})$. In addition, if $\mathbf{H}$ is bounded, then $0 \in C_\vee(\mathbf{H})$.

Definition 3.2.4 A hoop $\mathbf{H}$ is said to be a **join hoop** or $\vee$ **-hoop** if $C_\vee(\mathbf{H}) = H$.

Evidently, every $\sqcup$-hoop is a $\vee$-hoop. In the next section, we will see that basic hoops and Wajsberg hoops are examples of $\vee$-hoops.

The following proposition, originally stated for bounded hoops in [79], holds true for any hoop because the proof relies solely on properties common to all hoops, regardless of boundedness.

Proposition 3.2.5 ([79, Theorem 2.2]) *Let* $\mathbf{H} = (H; \odot, \rightarrow, 1)$ *be a hoop and* $a \in C_\vee(\mathbf{H})$. *Then* $\mathbf{H}_a := ([a, 1]; \odot^a, \rightarrow, a, 1)$ *is a bounded hoop where the implication of* $\mathbf{H}_a$ *is the restriction of* $\rightarrow$ *to the set* $[a, 1] \times [a, 1]$ *(which we continue to denote simply by* $\rightarrow$*), and*

$$x \odot^a y := (x \odot y) \vee a, \qquad \forall x, y \in [a, 1].$$

In addition, if $\leq_a$ *is the partially ordered relation of* $\mathbf{H}_a$*, then it satisfies the following conditions for all* $x, y \in [x, y]$:

(i) $x \leq_a y$ *if and only if* $x \leq y$.
(ii) $x \wedge_a y = x \wedge y$, *where* $x \wedge_a y$ *is the meet operation in the hoop* $\mathbf{H}_a$.
(iii) *If* $x \vee_a y$ *exists if and only if* $x \vee y$, *where* $x \wedge_a y$ *is the join of* x *and* y *in* $\mathbf{H}_a$. *Furthermore,* $x \vee_a = x \vee y$.

Proof Note that for each $x, y \in [a, 1]$ from $a \leq y \leq x \rightarrow y$ and $a \leq (x \odot y) \vee a$ it follows that $[a, 1]$ is closed under $\rightarrow$ and $\odot^a$. By Theorem 3.1.2 we need to verify that (H0)–(H3) hold. Choose $x, y, z \in [a, 1]$.

(1) Clearly, $\odot^a$ is a commutative binary operation and $x \odot^a 1 = (x \odot 1) \vee a = x \vee a = x$. In addition,

$$\begin{aligned}
x \odot^a (y \odot^a z) &= (x \odot (y \odot^a z)) \vee a \\
&= \bigl(x \odot ((y \odot z) \vee a)\bigr) \vee a \\
&= \bigl((x \odot y \odot z) \vee (x \odot a)\bigr) \vee a, \text{ by Lemma 2.2.3(i)} \\
&= (x \odot y \odot z) \vee ((x \odot a) \vee a) \\
&= (x \odot y \odot z) \vee a, \text{ by Lemma 2.2.1(4).}
\end{aligned}$$

---

[1] Join-center of a pocrim is defined similarly.

Similarly, $(x \odot^a y) \odot^a z = (x \odot y \odot z) \vee a$, so, $([a, 1]; \odot^a, a)$ is a commutative monoid such that (H0) holds. In addition, (H1) holds clearly.

(2) By Lemma 2.2.3(iii), we have

$$(x \odot^a y) \rightarrow z = ((x \odot y) \vee a) \rightarrow z = ((x \odot y) \rightarrow z) \wedge (a \rightarrow z)$$
$$= ((x \odot y) \rightarrow z) \wedge 1 = (x \odot y) \rightarrow z = x \rightarrow (y \rightarrow z).$$

(3) $x \odot^a (x \rightarrow y) = \left(x \odot (x \rightarrow y)\right) \vee a = \left(y \odot (y \rightarrow x)\right) \vee a = y \odot^a (y \rightarrow x)$.

From (1)–(3) and Theorem 3.1.2, it follows that $([a, 1]; \odot^a, \rightarrow, 1)$ is a hoop with the least element a. Therefore, $\mathbf{H}_a = ([a, 1]; \odot^a, \rightarrow, a, 1)$ is a bounded hoop.

Now, let $x, y \in [a, 1]$.

(i) The proof is straightforward, by definition of $\rightarrow$ on $\mathbf{H}_a$.

(ii) By Proposition 3.1.4(i), $x \wedge_a y = x \odot_a (x \rightarrow y) = (x \odot (x \rightarrow y)) \vee a = (x \wedge y) \vee a = x \wedge y$, since $a \leq x, y$.

(iii) The proof is evident by (i). We note that if $x \vee y$ exists, then $x \vee y \in [a, 1]$. $\square$

Example 3.2.6 Let $H = \{0, a, b, c, 1\}$ be a poset where $0 < a, b < c < 1$. Define two binary operations $\odot$ and $\rightarrow$ on H in Table 3.1:

Easy calculation show that $(H; \odot, \rightarrow, 1)$ is an algebra of type $(2, 2, 0)$ satisfying (H0)–(H2) (see [67, Exm 3.3]). Choose $x, y \in H$.

(1) If $x = 0$, then $x \odot (x \rightarrow y) = y \odot (y \rightarrow x) = 0$. Also, $y = 1$ implies that

$$x \odot (x \rightarrow y) = x = y \odot (y \rightarrow x).$$

(2) If $x = y$, then $x \odot (x \rightarrow x) = x = x \wedge x$.

(3) $a \odot (a \rightarrow b) = a \odot b = 0 = a \wedge b = b \odot a = b \odot (b \rightarrow a)$.

(4) $a \odot (a \rightarrow c) = a = a \wedge c = c \odot a = c \odot (c \rightarrow a)$.

(5) $b \odot (b \rightarrow c) = b \odot 1 = b = b \wedge c = c \odot b = c \odot (c \rightarrow b)$.

Table 3.1 Operations $\odot$ and $\rightarrow$ of Example 3.2.6

Table 3.1.1

$\odot$	0	a	b	c	1
0	0	0	0	0	0
a	0	a	0	a	a
b	0	0	b	b	b
c	0	a	b	c	c
1	0	a	b	c	1

Table 3.1.2

$\rightarrow$	0	a	b	c	1
0	1	1	1	1	1
a	b	1	b	1	1
b	a	a	1	1	1
c	0	a	b	1	1
1	0	a	b	c	1

(1)–(5) imply that (H3) holds. Hence, by Theorem 3.1.2, $(H; \odot, \rightarrow, 0, 1)$ is a bounded hoop. Clearly, $a \vee b = c$ and

$$a \sqcup b = ((a \rightarrow b) \rightarrow b) \wedge ((b \rightarrow a) \rightarrow a) = (b \rightarrow b) \wedge (a \rightarrow a) = 1 \wedge 1 = 1 > c = a \vee b.$$

Therefore, $(H; \odot, \rightarrow, 0, 1)$ is a bounded $\vee$-hoop while it is not a $\sqcup$-hoop, by Proposition 3.2.2(iv).

Proposition 3.2.7 *Every finite hoop is a $\vee$-hoop.*

Proof Let $\mathbf{H}$ be a finite hoop, $x, y \in H$, and $\mathcal{U}\{x, y\}$ be the set of all upper bounds of x and y. Since H is finite, by Proposition 3.1.4(i), $a := \bigwedge \mathcal{U}\{x, y\}$ exists. Clearly, $x, y \leq a$ and $a \leq z$ for all $z \in \mathcal{U}\{x, y\}$, that is, $x \vee y = a$. Therefore, $\mathbf{H}$ is a $\vee$-hoop. $\square$

Proposition 3.2.8 ([96, Proposition 3.4]) *Let $\mathbf{H}$ be a $\vee$-hoop. Then $(x \vee y) \odot (x \vee z) \leq x \vee (y \odot z)$ for all $x, y, z \in H$.*

Proof Choose $x, y, z \in H$. By Lemma 2.2.3(i), we get that

$$\begin{aligned}
(x \vee y) \odot (x \vee z) &= \big((x \vee y) \odot x\big) \vee \big((x \vee y) \odot z\big) \\
&= \big((x \odot x) \vee (y \odot x)\big) \vee \big((x \odot z) \vee (y \odot z)\big) \\
&= \big((x \odot x) \vee (y \odot x) \vee (x \odot z)\big) \vee (y \odot z) \\
&\leq x \vee (y \odot z), \text{ by Lemma } 2.2.1(4).
\end{aligned}$$

$\square$

Proposition 3.2.9 ([153, Proposition 2.8]) *Let $\mathbf{H}$ be a hoop and $x, y, z \in H$ such that $x \vee y$ exists.*

(i) $(x \vee y) \rightarrow z = (x \rightarrow z) \wedge (y \rightarrow z)$.
(ii) *For any* $n \in \mathbb{N}$, $(x \vee y)^n \rightarrow z = \bigwedge\{(a_n \odot \cdots \odot a_1) \rightarrow z \colon a_i \in \{x, y\}\}$.

Proof (i) Since $x, y \leq x \vee y$, by Lemma 2.2.1(7), we get that $(x \vee y) \rightarrow z \leq x \rightarrow z$ and $(x \vee y) \rightarrow z \leq y \rightarrow z$, so $(x \vee y) \rightarrow z \leq (x \rightarrow z) \wedge (y \rightarrow z)$. If $a \leq (x \rightarrow z) \wedge (y \rightarrow z)$ for some $a \in H$, then $a \leq x \rightarrow z$ and $a \leq y \rightarrow z$, so $a \odot x \leq z$ and $a \odot y \leq z$, hence $x, y \leq a \rightarrow z$. But this implies $x \vee y \leq a \rightarrow z$, that is $a \odot (x \vee y) \leq z$, so $a \leq (x \vee y) \rightarrow z$. Hence, $a \leq (x \rightarrow z) \wedge (y \rightarrow z)$ implies $a \leq (x \vee y) \rightarrow z$. It follows that $(x \rightarrow z) \wedge (y \rightarrow z) \leq (x \vee y) \rightarrow z$.

(ii) For $n = 1$, the proof follows from (i). Assume that the equality holds for $n \in \mathbb{N}$. Then

$$(x \vee y)^{n+1} \rightarrow z = \big((x \vee y) \odot (x \vee y)^n\big) \rightarrow z$$

$$= (x \vee y) \rightarrow \big((x \vee y)^n \rightarrow z\big), \text{ by (H2)}$$

$$= (x \vee y) \rightarrow \bigwedge\{(a_n \odot \cdots \odot a_1) \rightarrow z : a_i \in \{x, y\}\}$$

$$= \Big(x \rightarrow \bigwedge\{(a_n \odot \cdots \odot a_1) \rightarrow z : a_i \in \{x, y\}\}\Big)$$

$$\wedge \Big(y \rightarrow \bigwedge\{(a_n \odot \cdots \odot a_1) \rightarrow z : a_i \in \{x, y\}\}\Big), \text{ by (i)}$$

$$= \bigwedge\{x \rightarrow ((a_n \odot \cdots \odot a_1) \rightarrow z) : a_i \in \{x, y\}\}$$

$$\wedge \bigwedge\{y \rightarrow ((a_n \odot \cdots \odot a_1) \rightarrow z) : a_i \in \{x, y\}\}, \text{ by Lemma 2.2.3(ii)}$$

$$= \bigwedge\{(x \odot a_n \odot \cdots \odot a_1) \rightarrow z : a_i \in \{x, y\}\}$$

$$\wedge \bigwedge\{(y \odot a_n \odot \cdots \odot a_1) \rightarrow z : a_i \in \{x, y\}\}, \text{ by (H2)}$$

$$= \bigwedge\{(a_{n+1} \odot a_n \odot \cdots \odot a_1) \rightarrow z : a_i \in \{x, y\}\}.$$

Therefore, $(x \vee y)^n \rightarrow z = \bigwedge\{(a_n \odot \cdots \odot a_1) \rightarrow z : a_i \in \{x, y\}\}$ for all $n \in \mathbb{N}$. $\qquad\square$

Lemma 3.2.10 *Let* $\mathbf{H}$ *be a hoop. Then* $x \wedge \bigvee_{i \in I} y_i = \bigvee_{i \in I}(x \wedge y_i)$, *whenever the join of the left side of the equation exists. In addition, every* $\vee$-*hoop is a distributive lattice.*

Proof Let $\{y_i : i \in I\}$ be a subset of H such that $\bigvee_{i \in I} y_i$ exists. Choose $x \in H$. For each $i \in I$, the inequality $x \wedge y_i \leq x \wedge \bigvee_{i \in I} y_i$ is obvious. Thus, $x \wedge \bigvee_{i \in I} y_i$ is an upper bound for the set $\{x \wedge y_i : i \in I\}$. Let $z \in H$ be an arbitrary upper bound for $\{x \wedge y_i : i \in I\}$. We have

$$\Big(\bigvee_{i \in I} y_i\Big) \wedge x = \Big(\bigvee_{i \in I} y_i\Big) \odot \Big(\Big(\bigvee_{i \in I} y_i\Big) \rightarrow x\Big) = \bigvee_{i \in I}\Big(y_i \odot \Big(\Big(\bigvee_{i \in I} y_i\Big) \rightarrow x\Big)\Big), \text{ by Lemma 2.2.3(i)}.$$

Applying Lemma 2.2.1(7), we get $\big(\bigvee_{i \in I} y_i\big) \rightarrow x \leq y_i \rightarrow x$ for any $i \in I$, so

$$y_i \odot \Big(\Big(\bigvee_{i \in I} y_i\Big) \rightarrow x\Big) \leq y_i \odot (y_i \rightarrow x) = y_i \wedge x \leq z, \quad \forall i \in I.$$

It follows that

$$x \wedge \bigvee_{i \in I} y_i = \Big(\bigvee_{i \in I} y_i\Big) \odot \Big(\Big(\bigvee_{i \in I} y_i\Big) \rightarrow x\Big) = \bigvee_{i \in I}\Big(y_i \odot \Big(\Big(\bigvee_{i \in I} y_i\Big) \rightarrow x\Big)\Big) \leq z.$$

Therefore, $x \wedge \bigvee_{i \in I} y_i$ is the least upper bound of the set $\{x \wedge y_i : i \in I\}$, i.e., $x \wedge \bigvee_{i \in I} y_i = \bigvee_{i \in I}(x \wedge y_i)$. Now, let $\mathbf{H}$ be a $\vee$-hoop. Then $y \vee z$ and $(x \wedge y) \vee (x \wedge z)$ exist for each

$x, y, z \in H$. Therefore, by the first part of the proof, we have $x \wedge (y \vee z) = (x \wedge y) \vee (x \wedge z)$, which means $\mathbf{H}$ is a distributive lattice. $\qquad\square$

The following proposition was proved in [96, 97] for pseudo BL-algebra. The proof works for hoops as well.

Proposition 3.2.11 ([153, Proposition 2.15]) *Let $\mathbf{H}$ be a hoop and $x, y \in H$ such that $x \vee y = 1$. Then*

(i) $x \odot y = x \wedge y$.
(ii) $x^n \vee y^n = 1$ *for all $n \in \mathbb{N}$.*

Proof (i) By Lemma 2.2.1(4), $x \odot y \leq x \wedge y$. Also, by Proposition 3.2.1(ii), $x \sqcup y$ is a upper bound for x and y, this leads us

$$((x \to y) \to y) \wedge ((y \to x) \to x) = 1,$$

hence $(x \to y) \to y = (y \to x) \to x = 1$. It follows that $y \to x \leq x$, so

$$x \wedge y = (y \to x) \odot y \leq x \odot y.$$

(ii) By Lemma 3.2.10,

$$x = x \odot 1 = x \odot (x \vee y) = x^2 \vee (x \odot y),$$

so

$$1 = x \vee y = x^2 \vee (x \odot y) \vee y = x^2 \vee y.$$

Similarly, we get that

$$y = y \odot 1 = y \odot (x^2 \vee y) = (y \odot x^2) \vee y^2.$$

Hence,

$$x^2 \vee y^2 = x^2 \vee (y \odot x^2) \vee y^2 = x^2 \vee y = 1.$$

Similar arguments show that $x^{2n} \vee y^{2n} = 1$ for all $n \in \mathbb{N}$. Now, $x^{2n} \leq x^n$ and $y^{2n} \leq y^n$, imply that $x^n \vee y^n = 1$. $\qquad\square$

Remark 3.2.12 Suppose that $\mathbf{H}$ is a hoop such that $\sqcup$ is associative. By Proposition 3.2.2, we get that $\vee = \sqcup$ on H. Applying Lemma 3.2.10, it follows that $(H; \wedge, \vee)$ is a distributive lattice.

Proposition 3.2.13 *Let $\mathbf{H}$ be a $\vee$-hoop satisfying prelinearity. Then for each $x, y \in H$, if $x \vee y$ exists, then $x \vee y = x \sqcup y$.*

Proof Choose $x, y \in H$ such that $x \vee y$ exists. Clearly, $x \vee y \le x \sqcup y$. Let $a \in H$ such that $a \le (x \to y) \to y$ and $a \le (y \to x) \to x$. From $a \le (x \to y) \to y$, Lemma 2.2.1(13), and (2.2) it follows that $x \to y \le a \to y$. Similarly, $y \to x \le a \to x$, so

$$1 = (x \to y) \vee (y \to x) \le (a \to y) \vee (a \to x) \le a \to (y \vee x), \text{ by Lemma 2.2.1(7),}$$

entails that $a \le x \vee y$. Since a was an arbitrary lower bound of $(x \to y) \to y$ and $(y \to x) \to x$, it follows that $x \sqcup y = ((x \to y) \to y) \wedge ((y \to x) \to x) \le x \vee y$. $\qquad\square$

By Proposition 3.2.1, every ⊔-hoop is a ∨-hoop. The converse is true when the hoop satisfies prelinearity.

Corollary 3.2.14 *A ∨-hoop* **H** *satisfying prelinearity is a ⊔-hoop.*

Proof It follows from Proposition 3.2.13. $\qquad\square$

Applying Propositions 3.2.1(vi), 3.2.2, and Corollary 3.2.14 we get:

$$\boxed{⊔ - hoops} \Leftrightarrow \boxed{\vee - hoops + (Pre)} \tag{3.3}$$

We end this section with an interesting property of ∨-hoops satisfying prelinearily.

Proposition 3.2.15 *Let* **H** *be a ∨-hoop satisfying prelinearity. Then*

$$x \odot (y \wedge z) = (x \odot y) \wedge (x \odot z), \qquad \forall x, y, z \in H. \tag{3.4}$$

Proof Choose $x, y, z \in H$. Due to Lemma 3.2.10, $(H; \wedge, \vee)$ is distributive, so by Lemma 3.2.10, we get that

$$
\begin{aligned}
(x \odot y) \wedge (x \odot z) &= \big((x \odot y) \wedge (x \odot z)\big) \odot 1 \\
&= \big((x \odot y) \wedge (x \odot z)\big) \odot \big((y \to z) \vee (z \to y)\big), \text{ by (Pre)} \\
&= \big(((x \odot y) \wedge (x \odot z)) \odot (y \to z)\big) \vee \big(((x \odot y) \wedge (x \odot z)) \odot (z \to y)\big), \\
&\qquad\qquad\qquad\qquad\qquad\qquad\qquad\qquad\qquad\qquad \text{by Lemma 2.2.3(i)} \\
&\le \big(x \odot y \odot (y \to z)\big) \vee \big(x \odot z \odot (z \to y)\big) \\
&= \big(x \odot (y \wedge z)\big) \vee \big(x \odot (y \wedge z)\big), \text{ by Proposition 3.1.4(i)} \\
&= x \odot (y \wedge z).
\end{aligned}
$$

The converse inequality is obvious (see Lemma 2.2.1(3)). $\qquad\square$

3.3   Some Classes of Hoops

This section delves into the realm of hoops, focusing on several important classes: cancellative hoops, Wajsberg hoops, basic hoops, idempotent hoops, and product hoops. We will explore their relationships and properties. Notably, we will demonstrate that every basic hoop is a ⊔-hoop, and every Wajsberg hoop is a basic hoop. Furthermore, cancellative hoops are a special type of Wajsberg hoops, and bounded Wajsberg hoops are in one-to-one correspondence with MV-algebras.

3.3.1   Cancellative Hoops

Let $\mathbf{H}$ be a hoop. Define the **cancellative center** of $\mathbf{H}$, $C_c(\mathbf{H})$, by

$$C_c(\mathbf{H}) = \{a \in H : a \odot x = a \odot y \text{ implies that } x = y, \ \forall x, y \in H\}.$$

Clearly, $1 \in C_c(\mathbf{H})$, for each hoop $\mathbf{H}$. In addition, if $\mathbf{H}$ is bounded, then $0 \in C_c(\mathbf{H})$ if and only if $H = \{0\}$.

Example 3.3.1 (i) Consider the binary operations $\odot$ and $\rightarrow$ on the interval $(-\infty, 0]$ of the real numbers that are defined by $x \odot y = x + y$ and $x \rightarrow y = \min\{0, y - x\}$. Then $((-\infty, 0]; \odot, \rightarrow, 0)$ is a hoop. We can easily verify that $C_c((-\infty, 0]; \odot, \rightarrow, 0) = (-\infty, 0]$.

(ii) Let $\mathbf{H}$ be the hoop was defined in Example 3.2.6. Then $C_c(\mathbf{H}) = \{1\}$.

Proposition 3.3.2 *Let $\mathbf{H}$ be a hoop and $a \in H$. Then the following assertions are equivalent:*

(i) $a \in C_c(\mathbf{H})$.
(ii) $a \rightarrow (a \odot x) = x$ *for all* $x \in H$.
(iii) $x \rightarrow y = (x \odot a) \rightarrow (y \odot a)$ *for all* $x, y \in H$.
(iv) $x \odot a \leq y \odot a$ *if and only if* $x \leq y$.

Proof (i) $\Rightarrow$ (ii) Let $a \in C_c(\mathbf{H})$ and $x \in H$. By Proposition 3.1.4(i), $a \odot (a \rightarrow (a \odot x)) = a \wedge (a \odot x) = a \odot x$, so by the assumption, $a \rightarrow (a \odot x) = x$.

(ii) $\Rightarrow$ (iii) By (ii) and Theorem 3.1.2(H2), we have

$$x \rightarrow y = x \rightarrow (a \rightarrow (a \odot y)) = (a \odot x) \rightarrow (a \odot y).$$

(iii) $\Rightarrow$ (iv) Applying (iii), $x \odot a \leq y \odot a$ if and only if $(x \odot a) \rightarrow (y \odot a) = 1$ if and only if $x \rightarrow y = 1$ if and only if $x \leq y$.

(iv) $\Rightarrow$ (i) Let $x \odot a = y \odot a$ for some $x, y \in H$. From $x \odot a \leq y \odot a$ and (iv), we get $x \leq y$. Similarly, $y \odot a \leq x \odot a$ implies that $y \leq x$, which entails $x = y$. Therefore, $a \in C_c(\mathbf{H})$. $\square$

Corollary 3.3.3 *If* $\mathbf{H}$ *is a hoop and* $a \in C_c(\mathbf{H})$*, then* $a^n \to a^{n+1} = a$ *for all* $n \in \mathbb{N}$.

Proof Let $n \in \mathbb{N}$. Then by Proposition 3.3.2(ii), we have $a^n \to a^{n+1} = a^n \to (a^n \odot a) = a$. $\square$

Definition 3.3.4 ([15, 30]) A hoop $\mathbf{H} = (H; \odot, \to, 1)$ is called **cancellative** if the monoid $(H; \odot, 1)$ is cancellative i.e., $x \odot z = y \odot z$ implies that $x = y$ for all $x, y, z \in H$, equivalently, $C_c(\mathbf{H}) = H$.

No bounded hoop $\mathbf{H}$ with more than one element is cancellative, since $0 \odot x = 0$ for all $x \in H$. It follows from Remark 2.2.2 that non-trivial finite hoops are not cancellative.

The next proposition will provide us with a representation of cancellative hoops.

Proposition 3.3.5 ([30] [153, Proposition 4.1]) *A hoop* $\mathbf{H}$ *is cancellative if and only if* $y \to (x \odot y) = x$.

Proof Suppose that $\mathbf{H}$ is cancellative. Since $x \odot y \le y$, we have

$$x \odot y = y \wedge (x \odot y) = (y \to (x \odot y)) \odot y,$$

hence, $x = y \to (x \odot y)$. The proof of the converse follows from Proposition 3.3.2(ii). $\square$

Corollary 3.3.6 *The class of cancellative hoops forms a variety.*

Proof It follows from Proposition 3.3.5 and Theorem 1.6.4 directly. $\square$

Proposition 3.3.7 *Let* $\mathbf{H}$ *be a cancellative hoop. For all* $x, y, z \in H$,

(i) $z \to x = (z \odot y) \to (x \odot y)$.
(ii) $x \odot y \le z \odot y$ *if and only if* $x \le z$.
(iii) $(x \to y) \to y = (y \to x) \to x$.

Proof The proofs of (i) and (ii) follow directly from Proposition 3.3.2.
(iii) Let $x, y \in H$.

$$\begin{aligned}
(x \to y) \to y &= \big((x \to y) \odot x\big) \to (y \odot x), \text{ by (i)} \\
&= \big((y \to x) \odot y\big) \to (x \odot y), \text{ by (H2)} \\
&= (y \to x) \to x, \text{ by (i)}.
\end{aligned}$$

$\square$

Example 3.3.8 [141, Sect. 2.4.2] Let $\mathbf{G} = (G; \vee, \wedge, +, -, 0)$ be an arbitrary Abelian ℓ-group and $N(\mathbf{G})$ be the negative cone of $\mathbf{G}$, that is $N(\mathbf{G}) = \{a \in G : a \leq 0\}$. Define the binary operations $\odot$ and $\rightarrow$ on $N(\mathbf{G})$ as follows for all $x, y \in N(\mathbf{G})$:

$$x \odot y = x + y \quad \& \quad x \rightarrow y = (y - x) \wedge 0. \tag{3.5}$$

Then $\mathbf{N(G)} = (N(\mathbf{G}); \odot, \rightarrow, 0)$ is a hoop. We shall verify the identities from Theorem 3.1.2. Let $a, b, c \in N(\mathbf{G})$.

(H0) $a \odot 0 = a + 0 = a$.

(H1) $a \rightarrow a = 0 \wedge 0 = 0$.

(H2) $(a \odot b) \rightarrow c = \big(c - (a + b)\big) \wedge 0 = (c - b - a) \wedge 0$ and

$$a \rightarrow (b \rightarrow c) = \big(((c - b) \wedge 0) - a\big) \wedge 0 = (c - b - a) \wedge (-a) \wedge 0$$
$$= (c - b - a) \wedge 0.$$

(H3) $(a \rightarrow b) \odot a = (b - a) \wedge 0 + a = b \wedge a$. Also, $a \odot (a \rightarrow b) = a + (-a + b) \wedge 0 = b \wedge a$.

In addition, the hoop $N(\mathbf{G})$ is a cancellative, evidently. It is worth noting that some mathematical literature denotes the concept we represent as $N(\mathbf{G})$ by $\mathbf{G}^{-}$. We will adopt both notations throughout this book for clarity and consistency with existing literature.

The cancellative hoop $\mathbf{N(\mathbb{Z})}$ is denoted by $\mathbf{L}_{\infty}$. In Proposition 3.3.19, we will see that every cancellative hoop is isomorphic to $N(\mathbf{G})$ for some ℓ-group $\mathbf{G}$.

3.3.2   Wajsberg Hoops

A hoop $\mathbf{H} = (H; \odot, \rightarrow, 1)$ is called **Wajsberg**[2] if satisfies the condition (3.6).

$$(x \rightarrow y) \rightarrow y = (y \rightarrow x) \rightarrow x, \quad \forall x, y \in H. \tag{3.6}$$

Proposition 3.3.9 ([153, Proposition 4.3]) *Every Wajsberg hoop* $\mathbf{H}$ *satisfies the following statements:*

(i) $x \sqcup y = (x \rightarrow y) \rightarrow y = (y \rightarrow x) \rightarrow x$ *for all* $x, y \in H$.

(ii) $\sqcup$ *is associative.*

(iii) $x \vee y = (x \rightarrow y) \rightarrow y$.

---

[2] Wajsberg hoops are called Łukasiewicz hoops in [141].

Proof (i) The proof of (i) follows from the definition of $\sqcup$ and Wajsberg hoops.

(ii) If $x \le y$ and $z \in H$, then applying twice Lemma 2.2.1(7), we get that $y \to z \le x \to z$ and $(x \to z) \to z \le (y \to z) \to z$, that is $x \sqcup z \le y \sqcup z$. By Proposition 3.2.2, $\sqcup$ is associative.

(iii) It follows from (i), (ii), and Proposition 3.2.2. $\qquad\square$

Corollary 3.3.10 *Every Wajsberg hoop is a $\sqcup$-hoop satisfying the prelinearity condition. In addition, $x \vee y = ((x \to a) \wedge (y \to a)) \to a$ for each $x, y, a \in H$ satisfying $a \le x, y$.*

Proof It follows from Proposition 3.3.9 and Proposition 3.2.2. Hence, $x \vee y = x \sqcup y$ for all $x, y \in H$. In addition, by Proposition 3.3.9 and Proposition 3.2.2, $x \vee y = (x \to y) \to y$ for all $x, y \in H$. From Proposition 3.1.4(vi) it follows that

$$(x \to y) \vee (y \to x) = \big((x \to y) \to (y \to x)\big) \to (y \to x) = (y \to x) \to (y \to x) = 1.$$

Now, assume that $x, y, a \in H$ such that $a \le x, y$. From $(x \to a) \wedge (y \to a) \le x \to a$ and Lemma 2.2.1(7) it follows that $x = x \vee a = (x \to a) \to a \le ((x \to a) \wedge (y \to a)) \to a$. Similarly, $y \le ((x \to a) \wedge (y \to a)) \to a$ entails that $x \vee y \le ((x \to a) \wedge (y \to a)) \to a$. Also, $(x \vee y) \to a \le x \to a, y \to a$, so $(x \vee y) \to a \le (x \to a) \wedge (y \to a)$. By applying Lemma 2.2.1(7), we obtain

$$((x \to a) \wedge (y \to a)) \to a \le ((x \vee y) \to a) \to a = (x \vee y) \vee a = x \vee y.$$

Therefore, $((x \to a) \wedge (y \to a)) \to a = x \vee y$. $\qquad\square$

Example 3.3.11 Let $\odot$ and $\to$ be binary operations on the unit real interval $[0, 1]$ defined as follows:

$$x \odot y = (x + y - 1) \vee 0 \quad \& \quad x \to y = (1 - x + y) \wedge 1 \quad \forall x, y \in [0, 1].$$

Then $([0, 1]; \odot, \to, 0, 1)$ is a bounded Wajsberg hoop.

Furthermore, for each $n \in \mathbb{N}$, $L_n := \{0, 1/n, 2/n, \ldots, (n-1)/n, 1\}$ (see Example 1.4.2) is a subhoop of the bounded hoop $([0, 1]; \odot, \to, 0, 1)$. We denote the bounded hoop algebra $(L_n; \odot, \to, 0, 1)$ by $\mathbf{L}_n$. Furthermore, $\mathbf{L}_0$ denotes the trivial hoop $\mathbf{1}$.

Example 3.3.12 ([227]) Let $H = \{0, a, b, c, 1\}$ be a poset such that $0 < c < a, b < 1$. Define two binary operations $\odot$ and $\to$ on H as Table 3.2:

Then $(H; \odot, \to, 0, 1)$ is a bounded hoop. Clearly, it is not a Wajsberg hoop, since $(a \to 0) \to 0 = 1 \ne a = (0 \to a) \to a$.

Theorem 3.3.13 *Bounded Wajsberg hoops are termwise equivalent to MV-algebras.*

Table 3.2 Operations $\odot$ and $\to$ of Example 3.3.12

Table 3.2.1

$\odot$	0	c	a	b	1
0	0	0	0	0	0
c	0	c	c	c	c
a	0	c	a	c	a
b	0	c	c	b	b
1	0	c	a	b	1

Table 3.2.2

$\to$	0	c	a	b	1
0	1	1	1	1	1
c	0	1	1	1	1
a	0	b	1	b	1
b	0	a	a	1	1
1	0	a	b	c	1

Proof Let BWH and MV be the class of all bounded Wajsberg hoops and the class of all MV-algebras, respectively. First, assume that $(M; \oplus, ', 0, 1)$ be an MV-algebra. Consider the binary operations $\odot$ and $\to$ on M are defined by

$$x \odot y := (x' \oplus y')' \quad \& \quad x \to y := x' \oplus y, \quad \forall x, y \in M.$$

Applying properties of MV-algebra, we can show that $(M; \odot, \to, 0, 1)$ is a bounded Wajsberg hoop with $x \vee y = (x \to y) \to y$ and $x \wedge y = (x' \vee y')'$ for all $x, y \in M$.

$$\alpha : \text{MV} \to \text{BWH}, \quad \alpha((M; \oplus, ', 0, 1)) := (M; \odot, \to, 0, 1).$$

Conversely, for each bounded Wajsberg hoop $(H; \odot, \to, 0, 1)$, the algebra $(H; \oplus, ', 0, 1)$ is an MV-algebra, where $x' = x \to 0$ and $x \oplus y = (x' \odot y')'$ for all $x, y \in H$.

$$\beta : \text{BWH} \to \text{MV} \quad \& \quad \beta((H; \odot, \to, 0, 1)) := (H; \oplus, ', 0, 1).$$

Then $\beta \circ \alpha = \text{Id}_{\text{MV}}$ and $\alpha \circ \beta = \text{Id}_{\text{BWH}}$. $\qquad\square$

Given a bounded Wajsberg hoop $\mathbf{H} = (H; \odot, \to, 0, 1)$ the MV-algebra $\beta(\mathbf{H}) = (H; \oplus, ', 0, 1)$ is called the **MV-algebra corresponding** to the bounded Wajsberg hoop $\mathbf{H}$, similarly, for each MV-algebra $\mathbf{M} = (M; \oplus, ', 0, 1)$, the bounded Wajsberg hoop $\alpha(\mathbf{M}) = (M; \odot, \to, 0, 1)$ is called the **bounded Wajsberg hoop corresponding** to the MV-algebra $\mathbf{M}$.

Corollary 3.3.14 *For every bounded Wajsberg hoop* $\mathbf{H} = (H; \odot, \to, 0, 1)$ *there exists a unital ℓ-group* $(\mathbf{G}, u)$ *such that* $\mathbf{H} \cong \mathbf{G}[\mathbf{u}]$. *In addition,* $\alpha(\Gamma(\mathbf{G}, u)) = \mathbf{G}[\mathbf{u}]$ *and* $\beta(\mathbf{G}[\mathbf{u}]) = \Gamma(\mathbf{G}, u)$, *for every unital ℓ-group* $(\mathbf{G}, u)$.

Proof The proof is an immediate consequence of Theorems 3.3.13 and 1.4.15. $\qquad\square$

Theorem 3.3.15 *A bounded hoop $(H; \odot, \rightarrow, 0, 1)$ is a Wajsberg hoop if and only if $x'' = x$ for all $x \in H$.*

Proof Let $\mathbf{H}$ be a bounded hoop satisfying $x'' = x$ for all $x \in H$. Choose $x, y \in H$. By Lemma 2.2.1(8),

$$x \rightarrow y = x \rightarrow y'' = x \rightarrow (y' \rightarrow 0) = y' \rightarrow (x \rightarrow 0) = y' \rightarrow x'. \tag{3.7}$$

In addition,

$$(x \rightarrow y) \rightarrow y = (y' \rightarrow x') \rightarrow y'' = (y' \rightarrow x') \rightarrow (y' \rightarrow 0)$$
$$= (y' \odot (y' \rightarrow x')) \rightarrow 0 = (y' \wedge x')'.$$

In a similar way, we can prove that $(y \rightarrow x) \rightarrow x = (y' \wedge x')'$. Hence, $(x \rightarrow y) \rightarrow y = (y \rightarrow x) \rightarrow x$, which means $\mathbf{H}$ is a Wajsberg hoop. Conversely, assume that $\mathbf{H}$ is a bounded Wajsberg hoop. Due to Corollary 3.3.10, prelinearity holds. Also, for each $x \in H$ we have $x'' = (x \rightarrow 0) \rightarrow 0 = (0 \rightarrow x) \rightarrow x = 1 \rightarrow x = x$. Therefore, $\mathbf{H}$ is an MV-algebra. $\square$

Proposition 3.3.16 ([13, p. 108]) *Every cancellative hoop $\mathbf{H}$ is a Wajsberg hoop.*

Proof It follows from Proposition 3.3.7(iii). $\square$

From Corollary 3.3.10 and Proposition 3.3.16 it follows that every cancellative hoop is a $\vee$-hoop satisfying $(x \rightarrow y) \vee (y \rightarrow x) = 1$.

Before we state an important theorem about the relation between cancellative hoops and ℓ-group, we recall some fundamental results in the theory of ℓ-groups.

Proposition 3.3.17 ([147, Proposition X.1] (Nakada)) *A necessary and sufficient condition that a partially ordered monoid $\mathbf{S} = (S; +, 0)$ be the positive cone of a partially ordered group $\mathbf{G} = (G; +, -, 0)$ is that:*

(i) *S is cancellative (i.e., it satisfies cancellation laws on both sides).*
(ii) *$x \leq y$ if and only if $y = x + a = b + x$ for some $a, b \in S$.*

Evidently, in Proposition 3.3.17, S is a commutative if and only if $\mathbf{G}$ is Abelian. As a consequence of Proposition 3.3.17, if $\mathbf{S}$ is a commutative ℓ-monoid, then $\mathbf{G}$ is an Abelian ℓ-group. In addition, for every $x \in G$, there exist $x_1, x_2 \in S$ such that $x = x_1 - x_2$, where $x_1 = x \vee 0$ and $x_2 = -x \vee 0$. Furthermore,

$$x \wedge y = ((x_1 + y_2) \wedge (y_1 + x_2)) - x_2 - y_2, \quad \forall x, y \in G.$$
$$x \vee y = ((x_1 + y_2) \vee (y_1 + x_2)) - x_2 - y_2, \quad \forall x, y \in G.$$

For more details see [29, 147, Chap. XIII] and [141, p. 35].

To facilitate a better understanding of the subsequent lemma, we find it beneficial to recall some properties of ℓ-groups in more detail. Consider an Abelian ℓ-group $\mathbf{G} = (G; \vee, \wedge, +, -, 0)$. For every $g \in G$, $g^+ := g \vee 0$ and $g^- := -g \vee 0$ are elements of $\mathbf{G}^+$ and by Proposition 1.4.10, $g + g^- = g + (-g \vee 0) = (g - g) \vee g = g^+$, which implies $g = g^+ - g^-$. Similarly, we can show that $g = -g^- + g^+$.

$$g^+ \wedge g^- = (g + g^-) \wedge g^- = (g \wedge 0) + g^- = -(-g \vee 0) + g^- = -g^- + g^- = 0, \text{ by Proposition 1.4.10.}$$

From Proposition 1.4.10(v) we conclude that $g^+ + g^- = g^+ \vee g^-$.

Lemma 3.3.18 ([29]) *Let $\mathbf{G}_1 = (G_1; \vee, \wedge, +, -, 0)$ and $\mathbf{G}_2 = (G_2; \vee, \wedge, +, -, 0)$ be two ℓ-groups with the positive cone $\mathbf{G}_1^+$ and $\mathbf{G}_2^+$, respectively. Then every ℓ-monoid morphism $f : (\mathbf{G}_1^+; \vee, \wedge, +, 0) \to (\mathbf{G}_2^+; \vee, \wedge, +, 0)$ can be uniquely extended to an ℓ-homomorphism F from $\mathbf{G}_1$ to $\mathbf{G}_2$. In addition, F is an ℓ-isomorphism if and only if f is an isomorphism of ℓ-monoids.*

Proof Let $f : (\mathbf{G}_1^+; \vee, \wedge, +, 0) \to (\mathbf{G}_2^+; \vee, \wedge, +, 0)$ be a homomorphism of ℓ-monoids. Define $F : G_1 \to G_2$ by $f(g) = f(g^+) - f(g^-)$ for all $g \in G_1$. The rest of the proof follows from the note right after Proposition 3.3.17. $\qquad\square$

Proposition 3.3.19 ([30, 32, 153]) *For any cancellative hoop $\mathbf{H}$ there exists an Abelian ℓ-group $\mathbf{G}$ such that $\mathbf{H} = \mathbf{N}(\mathbf{G})$.*

Proof Let $\mathbf{H} = (H; \odot, \to, 1)$ be a cancellative hoop. By Proposition 3.3.16, $\mathbf{H}$ is a Wajsberg hoop and so it is a $\vee$-hoop, see Proposition 3.3.9. Consider the monoid $\mathbf{S} = (H; \odot, 1)$ endowed with the dual order relation $\leq_d$, that is $x \leq_d y$ if and only if $y \leq x$ if and only if $y = x \odot z$ for some $z \in H$. Evidently, $\mathbf{S}$ is a cancellative partially ordered monoid satisfying the conditions of Proposition 3.3.17. In addition, by Proposition 2.2.3(i), $\odot$ distributes on $\vee_d$. Furthermore, due to Corollary 3.3.10 and Proposition 3.2.15, $\odot$ distributes over $\wedge_d$, i.e., $(S; \vee_d, \wedge_d, \odot, 1)$ is an ℓ-monoid (note that $\vee_d$ and $\wedge_d$ in $\mathbf{S}$ are $\wedge$ and $\vee$ in $\mathbf{H}$, respectively). Applying Nakada's Theorem (Proposition 3.3.17 and the note after the proposition), there exists an ℓ-ordered group $\mathbf{G} = (G; \vee_d, \wedge_d, +, -, 0)$ such that $\mathbf{G}^+ = \mathbf{S}$ and $x + y = x \odot y$ for all $x, y \in H$.

Consider the hoop $\mathbf{N}(\mathbf{G}) = (\mathbf{N}(\mathbf{G}); +, \to, 0)$, where $x \to y = (y - x) \wedge_d 0$. Define $f : \mathbf{H} \to \mathbf{N}(\mathbf{G})$ by $f(x) = -x$ for all $x \in H$. Clearly, f is a bijection. For each $x, y \in H$ we have

$$(f(x) \to f(y)) + f(x) = \big((f(y) - f(x)) \wedge_d 0\big) + f(x)$$
$$= f(y) \wedge_d f(x), \text{ by Proposition } 1.4.10\text{(iii)}$$
$$= -y \wedge_d -x$$
$$= -(y \vee_d x), \text{ by Proposition } 1.4.10\text{(i)}$$
$$= -(x \wedge y)$$
$$= f(x \wedge y)$$
$$= f((x \to y) \odot x)$$
$$= f((x \to y) + x)$$
$$= f(x \to y) + f(x).$$

Since $\mathbf{N}(\mathbf{G})$ is cancellative, it follows that $f(x) \to f(y) = f(x \to y)$. Hence, f is a hoop isomorphism from $\mathbf{H}$ to $\mathbf{N}(\mathbf{G})$. $\qquad\square$

Proposition 3.3.19 provides us with another proof for Proposition 3.3.16 as follows:

Proof Let $\mathbf{H}$ be a cancellative hoop. By Proposition 3.3.19, without loss of generality, we assume that $\mathbf{H} = (\mathbf{N}(\mathbf{G}); \odot, \to, 0)$ for some Abelian ℓ-group $(G; \vee, \wedge; +, -, 0)$. Choose $x, y \in \mathbf{N}(\mathbf{G})$. Applying properties of Abelian ℓ-groups, we get

$$(x \to y) \to y = \big((y - x) \wedge 0\big) \to y = \Big(y - \big((y - x) \wedge 0\big)\Big) \wedge 0$$
$$= \Big(y + ((x - y) \vee 0)\Big) \wedge 0, \text{ by Proposition } 1.4.10\text{(i)}$$
$$= \Big((y - y + x) \vee y\Big) \wedge 0, \text{ by Proposition } 1.4.10\text{(ii)}$$
$$= (x \vee y) \wedge 0 = x \vee y, \text{ since } x, y \leq 0.$$

In a similar way, $(y \to x) \to x = x \vee y$ entails that $\mathbf{H}$ is a Wajsberg hoop. $\qquad\square$

Let $\mathcal{CH}$ be the category whose objects are cancellative hoops and whose morphisms are hoop homomorphisms between cancellative hoops. Furthermore, suppose that $\mathcal{A\ell G}$ is the category whose objects are Abelian ℓ-groups and morphisms between objects are ℓ-homomorphisms (see Sect. 1.4.2). The next theorem shows that the categories $\mathcal{A\ell G}$ and $\mathcal{CH}$ are categorically equivalent (for the definition of categorically equivalent see [208, Chap. 4]).

Theorem 3.3.20 ([31, Theorem 1.17]) *For each ℓ-group $\mathbf{G}$ and each ℓ-homomorphism* $f : \mathbf{G}_1 \to \mathbf{G}_2$*, we define*

$$\mathcal{N}(\mathbf{G}) = \mathbf{N}(\mathbf{G}), \quad \mathcal{N}(f : \mathbf{G}_1 \to \mathbf{G}_2) = f|_{\mathbf{G}_1^-} : \mathbf{N}(\mathbf{G}_1) \to \mathbf{N}(\mathbf{G}_2). \tag{3.8}$$

Then $N : \mathcal{A}\ell\mathcal{G} \to C\mathcal{H}$ *is a covariant functor which establishes a categorically equivalence between* $\mathcal{A}\ell\mathcal{G}$ *and* $C\mathcal{H}$.

Proof The proof is straightforward by Propositions 3.3.19 and 3.3.17. $\square$

Corollary 3.3.21 ([31, Corollary 1.18]) *The variety* CH *of cancellative hoops is generated by* $N(\mathbb{Z})$.

Proof By Proposition 1.5.22, the variety A of Abelian ℓ-groups is generated by the group of integers $\mathbb{Z}$. By applying Theorem 1.5.21 and the functor N in Theorem 3.3.20, we can easily deduce that $CH = HSP(N(\mathbb{Z})) = V(N(\mathbb{Z}))$. $\square$

Example 3.3.22 Let $H = (H; \odot, \to, 0, 1)$ be a bounded Wajsberg hoop. Every proper filter F of H is a Wajsberg hoop. Indeed, by Corollary 2.3.4, F is closed under $\odot$ and $\to$. Since H satisfies (H0)–(H3), we get that these conditions hold on the hoop $(F; \odot, \to, 1)$, evidently. Hence, by Theorem 3.1.2, $(F; \odot, \to, 1)$ is a hoop. In addition, $(F; \odot, \to, 1)$ is a Wajsberg hoop, since (3.6) holds on H.

In [75], R. Cignoli and A. Torrens gave a description of the free cancellative hoops in terms of linear functions. In Sect. 6.1, we will represent unbounded linear Wajsberg hoops.

3.3.3 Basic Hoops

In this section, we present the main properties of basic hoops as an important subclass within the class of hoops. We will show that every basic hoop is a $\sqcup$-hoop that satisfies property (Pre). Additionally, each Wajsberg hoop is also a basic hoop.

Definition 3.3.23 A hoop $H = (H; \odot, \to, 1)$ is called **basic** if it satisfies (3.9)

$$(x \to y) \to z \leq ((y \to x) \to z) \to z, \qquad \forall x, y, z \in H. \tag{3.9}$$

Proposition 3.3.24 *Every linear hoop* H *is a basic hoop.*

Proof Let H be a linear hoop and $x, y \in H$. If $x \leq y$, then

$$(x \to y) \to z = 1 \to z = z \leq ((y \to x) \to z) \to z, \text{ by Lemma 2.2.1(10).}$$

The inequality $y \leq x$ implies that $((y \to x) \to z) \to z = z \to z = 1 \geq (x \to y) \to z$. Therefore, H is a basic hoop. $\square$

Evidently, by Theorem 1.6.4, the class of basic hoops, BH, forms a variety, indeed, it is the class of algebra $(H; \odot, \to, 1)$ of type $(2, 2, 0)$ satisfying the conditions (H0)–(H3) and the following additional identity:

$$\big((x \to y) \to z\big) \to \big(((y \to x) \to z) \to z\big) = 1 \quad \forall x, y, z \in H. \tag{3.10}$$

Lemma 3.3.25 ([153, Lemma 4.5]) *Let* **H** *be a basic hoop. For any* $x, y, z \in H$, *the following statements hold:*

(i) $(x \to y) \vee (y \to x) = 1 = (x \to y) \sqcup (y \to x)$.
(ii) $x \to y = (x \sqcup y) \to y$.
(iii) $(x \sqcup y) \to z = (x \to z) \wedge (y \to z)$.
(iv) $x \vee y = x \sqcup y$.

Proof (i) Let $x \to y, y \to x \le a$. By definition of basic hoops, we get that $(x \to y) \to a \le ((y \to x) \to a) \to a$. Also, $(x \to y) \to a = (y \to x) \to a = 1$, so, $1 \le 1 \to a = a$ entails that, $a = 1$. Therefore, $(x \to y) \vee (y \to x) = 1$. The proof of the second part follows from Proposition 3.2.1(ii).

(ii) Using $x \le x \sqcup y$ and Lemma 2.2.1(7), we get that $(x \sqcup y) \to y \le x \to y$. Let us prove the converse inequality. Since

$$x \sqcup y = ((x \to y) \to y) \wedge ((y \to x) \to x) \le (x \to y) \to y.$$

by Lemmas 2.2.1(13) and (7), it follows that

$$x \to y = ((x \to y) \to y) \to y \le (x \sqcup y) \to y.$$

(iii) The inequality $(x \sqcup y) \to z \le (x \to z) \wedge (y \to z)$ is obvious, by Lemma 2.2.1(7) and $x, y \le x \sqcup y$. Set $a = \big((x \to z) \wedge (y \to z)\big) \to \big((x \sqcup y) \to z\big)$. It suffices to prove that $a = 1$. From

$$
\begin{aligned}
\big((x \to z) \wedge (y \to z)\big) \odot \big((x \sqcup y) \to y\big) \odot (x \sqcup y) &= \big((x \to z) \wedge (y \to z)\big) \odot \big((x \sqcup y) \wedge y\big) \\
&= \big((x \to z) \wedge (y \to z)\big) \odot y \\
&\le (y \to z) \odot y \\
&= y \wedge z \le z,
\end{aligned}
$$

it follows that $\big((x \to z) \wedge (y \to z)\big) \odot \big((x \sqcup y) \to y\big) \le (x \sqcup y) \to z$, so by (2.2) and Lemma 2.2.1(8), $(x \sqcup y) \to y \le a$. Applying (ii), we get $x \to y \le a$, that is $(x \to y) \to a = 1$. In a similar way, we can show that $(y \to x) \to a = 1$. By the definition of basic hoops, we get

$$1 = (x \to y) \to a \le ((y \to x) \to a) \to a = 1 \to a = a.$$

(iv) We apply Proposition 3.2.2(ii). Let $x, y, z \in H$ such that $x, y \leq z$. By (iii), $(x \sqcup y) \to z = (x \to z) \wedge (y \to z) = 1 \wedge 1 = 1$ which entails that $x \sqcup y \leq z$. Hence, by Proposition 3.2.1(ii), $x \sqcup y$ is the least upper bound of x and y, i.e., $x \vee y = x \sqcup y$ for all $x, y \in H$. $\qquad\square$

Proposition 3.3.26 *Let* **H** *be a hoop. Then* **H** *is a basic hoop if and only if* $\sqcup$ *is associative and* $(x \to y) \sqcup (y \to x) = 1$ *for all* $x, y \in H$.

Proof Let **H** be a basic hoop. Then the proof follows from Lemma 3.3.25(i) and (iv). Conversely, applying Proposition 3.1.4(i) and Proposition 3.2.9(i), it follows that

$$((x \to y) \to z) \odot ((y \to x) \to z) \leq ((x \to y) \to z) \wedge ((y \to x) \to z)$$
$$= ((x \to y) \vee (y \to x)) \to z$$
$$= 1 \to z = z.$$

Hence, $(x \to y) \to z \leq ((y \to x) \to z) \to z$, and so **H** is a basic hoop. $\qquad\square$

Theorem 3.3.27 *Every basic hoop* **H** *is a distributive lattice, where* $\vee = \sqcup$.

Proof Let **H** be a basic hoop. Due to Proposition 3.1.4(i), $(H, \leq)$ is a meet-semilattice. By Proposition 3.3.26, it is a lattice with $\vee = \sqcup$. To get that $(H; \wedge, \vee)$ is a distributive lattice, apply Lemma 3.2.10. $\qquad\square$

Theorem 3.3.28 *A hoop* **H** *is a basic hoop if and only if it is a* $\vee$*-hoop satisfying prelinearity if and only if it is a* $\sqcup$*-hoop.*

Proof Assume that **H** is a $\vee$-hoop satisfying the prelinearity condition. Then by Corollary 3.2.14, it is a $\sqcup$-hoop and $\vee = \sqcup$. So, by Proposition 3.3.26, it is a basic hoop. If **H** is a basic hoop, then by Proposition 3.3.26 it is a $\vee$-hoop satisfying prelinearity.

The rest of the proof follows from Theorem 3.3.27, Proposition 3.2.1(vi) and (3.3). $\qquad\square$

From Theorem Proposition 3.2.1(vi), Theorem 3.3.28 and Proposition 3.3.26 it follows that Fig. 3.1 presents equivalent classes of hoops:

Remark 3.3.29 Applying Proposition 3.2.5 and Theorem 3.3.28 it follows that if $\mathbf{H} = (H; \odot, \to, 1)$ is a basic hoop, then $([a, 1]; \odot^a, \to, a, 1)$ is a bounded basic hoop. In addition, if **H** is a Wajsberg hoop, then so is $([a, 1]; \odot^a, \to, a, 1)$.

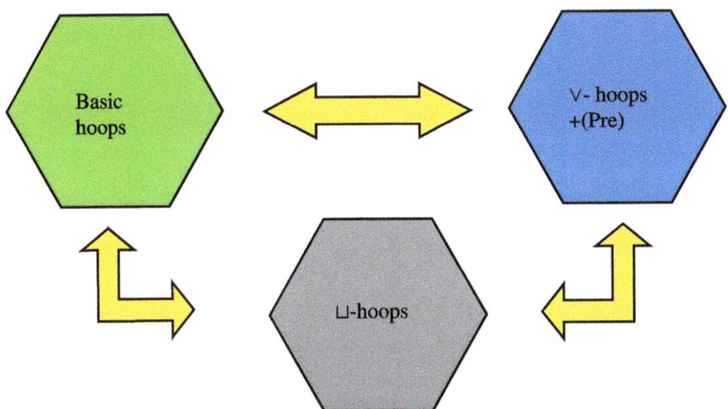

Fig. 3.1 Relation between basic hoops, ∨-hoops, and ⊔-hoops

Proposition 3.3.30 *Let* **H** *be a basic hoop. Then for all* $x, y, z \in H$,

(i) $x \odot (y \wedge z) = (x \odot y) \wedge (x \odot z)$.
(ii) $x \odot (y \vee z) = (x \odot y) \vee (x \odot z)$.

Proof (i) By Theorem 3.3.28, **H** is a ∨-hoop satisfying preliniearity, so Proposition 3.2.15 implies that $x \odot (y \wedge z) = (x \odot y) \wedge (x \odot z)$ for all $x, y, z \in H$.

(ii) The proof follows from Lemma 2.2.3(i), since **H** is ∨-hoop. □

Proposition 3.3.31 *Let* **H** *be a basic hoop such that for all* $x, y \in H$, $y \to x = x$ *implies* $x = 1$ *or* $y = 1$. *Then* **H** *is a linear Wajsberg hoop.*

Proof Let $x, y \in H$. Applying Proposition 3.1.4(vi), we get that

$$(x \to y) \to (y \to x) = y \to x.$$

Hence, by the hypothesis, $x \to y = 1$ or $y \to x = 1$, that is $x \leq y$ or $y \leq x$. Thus, **H** is a linear hoop. Let us prove now that **H** is Wajsberg. Let $x, y \in H$. If $x = y$, then the proof is obvious. Assume $x \neq y$. Since **H** is linear, we can suppose that $x < y$. It follows that $(x \to y) \to y = 1 \to y = y$, so it suffices to show that $(y \to x) \to x = y$. By Proposition 3.1.4(iv),

$$\big(((y \to x) \to x) \to y\big) \to (y \to x) = y \to x,$$

so by the hypothesis and the fact that $y \to x \neq 1$, it follows that $((y \to x) \to x) \to y = 1$, that is $(y \to x) \to x \leq y$. But, from Lemma 2.2.1(6), we have also that $y \leq (y \to x) \to x$. Hence, we have obtained that $(y \to x) \to x = y$. □

Proposition 3.3.32 ([153, Proposition 4.9]) *Any Wajsberg hoop is a basic hoop.*

Proof Let $x, y \in H$.

$$
\begin{aligned}
y \to x &= \big(((y \to x) \to x) \to y\big) \to (y \to x), \text{ by Proposition 3.1.4(iv)} \\
&= \big(((x \to y) \to y) \to y\big) \to (y \to x), \text{ since } \mathbf{H} \text{ is a Wajsberg hoop} \\
&= (x \to y) \to (y \to x), \text{ by Lemma 2.2.1(13).}
\end{aligned}
$$

So, by Proposition 3.3.9(i),

$$
(x \to y) \vee (y \to x) = ((x \to y) \to (y \to x)) \to (y \to x) = (y \to x) \to (y \to x) = 1.
$$

In addition, due to Proposition 3.3.9(ii), $\vee$ is associative. Applying now Proposition 3.3.26, we obtain that $\mathbf{H}$ is a basic hoop. $\square$

Remark 3.3.33 By Propositions 3.3.16 and 3.3.32, $\mathbf{N(G)}$ is a basic hoop for each ℓ-group $\mathbf{G} = (G; \vee, \wedge, +, -, 0)$. A direct proof for this statement is as follows:

By Example 3.3.8, $\mathbf{N(G)}$ is a hoop. Let $x, y, z \in H$. Firstly, we prove that $\mathbf{N(G)}$ is basic. We have

$$
\begin{aligned}
(x \to y) \to z &= \Big(z - \big((y - x) \wedge 0\big)\Big) \wedge 0 \\
&= \Big(z + \big((-y + x) \vee 0\big)\Big) \wedge 0, \text{ by Proposition 1.4.10(i)} \\
&= \big((z - y + x) \vee z\big) \wedge 0, \text{ by Proposition 1.4.10(ii).}
\end{aligned}
$$

Similarly, $(y \to x) \to z = \big((z - x + y) \vee z\big) \wedge 0$. It follows from Proposition 1.4.10 that

$$
\begin{aligned}
((y \to x) \to z) \to z &= \Big(z - \big(((z - x + y) \vee z) \wedge 0\big)\Big) \wedge 0 \\
&= \Big(\big(z - (z - x + y) \vee z\big) \vee z\Big) \wedge 0 \\
&= \Big(\big((z - y + x - z) \wedge 0) \vee z\big) \vee z\Big) \wedge 0 \\
&= \big((z - y + x - z) \vee z\big) \wedge (0 \vee z) \wedge 0 \\
&= \big((z - y + x - z) \vee z\big) \wedge 0.
\end{aligned}
$$

Since $z \leq 0$, it follows that $0 \leq -z$, so $z - y + x \leq z - y + x - z$, consequently,

$$
(x \to y) \to z = ((z - y + x) \vee z) \wedge 0 \leq ((z - y + x - z) \vee z) \wedge 0 = ((y \to x) \to z) \to z.
$$

Therefore, $\mathbf{N(G)}$ is a basic hoop.

In Remark 2.2.5(vii), we showed that if $\mathbf{H}$ is a bounded hoop, then $\mathrm{Reg}(\mathbf{H}) = \{x \in H : x'' = x\}$ is closed under $\rightarrow$. In the next theorem, we prove that $\mathrm{Reg}(\mathbf{H})$ is a subhoop of $\mathbf{H}$ when $\mathbf{H}$ is a bounded basic hoop.

Proposition 3.3.34 *Let $\mathbf{H}$ be a basic hoop. Then*

(i) $(x \wedge y) \rightarrow z = (x \rightarrow z) \vee (y \rightarrow z)$ *for all* $x, y, z \in H$.
(ii) $(x \vee y) \rightarrow z = (x \rightarrow z) \wedge (y \rightarrow z)$ *for all* $x, y, z \in H$.

Proof First, we note that due to Theorem 3.3.28, $\mathbf{H}$ is a $\vee$-hoop satisfying (Pre). Let $x, y, z \in H$. By Lemma 2.2.3(ii), $x \rightarrow (x \wedge y) = (x \rightarrow x) \wedge (x \rightarrow y) = x \rightarrow y$, so Lemma 2.2.1(12) implies that $x \rightarrow y \leq ((x \wedge y) \rightarrow z) \rightarrow (x \rightarrow z)$, consequently, $(x \rightarrow y) \odot ((x \wedge y) \rightarrow z) \leq x \rightarrow z$. In a similar way, $(y \rightarrow x) \odot ((x \wedge y) \rightarrow z) \leq y \rightarrow z$. It follows that

$$
\begin{aligned}
(x \wedge y) \rightarrow z &= ((x \wedge y) \rightarrow z) \odot 1 \\
&= ((x \wedge y) \rightarrow z) \odot ((x \rightarrow y) \vee (y \rightarrow x)), \text{ by Theorem 3.3.28} \\
&= (((x \wedge y) \rightarrow z) \odot (x \rightarrow y)) \vee (((x \wedge y) \rightarrow z) \odot (y \rightarrow x)) \\
&\leq (x \rightarrow z) \vee (y \rightarrow z), \text{ by Lemma 2.2.1(3) and (7).}
\end{aligned}
$$

In addition, from $x \wedge y \leq x, y$ and Lemma 2.2.1(7) we get that $(x \rightarrow z) \vee (y \rightarrow z) \leq (x \wedge y) \rightarrow z$. Therefore, $(x \wedge y) \rightarrow z = (x \rightarrow z) \vee (y \rightarrow z)$. The proof of the other part follows from Lemma 2.2.3(iii). $\square$

Corollary 3.3.35 ([79, Proposition 4.1]) *Let $\mathbf{H}$ be a bounded basic hoop. Then*

$$(x \wedge y)' = x' \vee y' \quad \& \quad (x \vee y)' = x' \wedge y', \quad x, y \in H. \tag{3.11}$$

Proof The proof is straightforward by Proposition 3.3.34. $\square$

Theorem 3.3.36 ([79, Proposition 4.30]) *Let $\mathbf{H}$ be a bounded basic hoop. Then $\mathrm{Reg}(\mathbf{H})$ is a subhoop of $\mathbf{H}$ as well as a Wajsberg hoop.*

Proof Due to Remark 2.2.5(vii), it suffices to show that $\mathrm{Reg}(\mathbf{H})$ is closed under $\odot$. For this purpose, we claim that $(x \odot y)'' = x'' \odot y''$ for all $x, y \in H$. Choose $x, y \in H$.

$$(x \odot y)'' = (x \odot y)'' \wedge y'', \text{ since} x \odot y \leq y$$
$$= y'' \odot \left(y'' \to (x \odot y)''\right)$$
$$= y'' \odot \left(y'' \to ((x \odot y) \to 0)'\right)$$
$$= y'' \odot \left(y'' \to (x \to y')'\right), \text{ by Lemma 2.2.1(8)}$$
$$= y'' \odot \left(y'' \to (y'' \to x')'\right), \text{ by Remark 2.2.5(v)}$$
$$= y'' \odot \left(y'' \odot (y'' \to x')\right)', \text{ by Lemma 2.2.1(8)}$$
$$= y'' \odot (y'' \wedge x')', \text{ by Proposition 3.1.4(i)}$$
$$= y'' \odot (y''' \vee x'') = y'' \odot (y' \vee x''), \text{ by Corollary 3.3.35}$$
$$= (y'' \odot y') \vee (y'' \odot x'') = 0 \vee (y'' \odot x'') = y'' \odot x''.$$

It follows that $\mathrm{Reg}(\mathbf{H})$ is a subhoop of $\mathbf{H}$. In addition, $x'' = x$ for all $x \in \mathrm{Reg}(\mathbf{H})$, so by Theorem 3.3.15, $(\mathrm{Reg}(\mathbf{H}); \odot, \to, 0, 1)$ is a bounded Wajsberg hoop, equivalently, an MV-algebra, by Theorem 3.3.13. $\qquad\square$

Proposition 3.3.37 *If $\mathbf{H} = (H; \odot, \to, 0, 1)$ is a bounded Wajsberg hoop, then $x \to (y \vee z) = (x \to y) \vee (x \to z)$ for all $x, y, z \in H$.*

Proof By Proposition 3.3.32 and Theorem 3.3.15, $\mathbf{H}$ is a bounded basic hoop satisfying (DNP). Choose $x, y, z \in H$.

$$x \to (y \vee z) = x \to (y \vee z)'' = x \to (y' \wedge z')', \text{ by Corollary 3.3.35}$$
$$= \left(x \odot (y' \wedge z')\right)', \text{ by Lemma 2.2.1(8)}$$
$$= \left((x \odot y') \wedge (x \odot z')\right)', \text{ by Proposition 3.3.30(i)}$$
$$= (x \odot y')' \vee (x \odot z')', \text{ by Corollary 3.3.35}$$
$$= (x \to y) \vee (x \to z), \text{ by Lemma 2.2.1(8)}.$$

$\qquad\square$

Corollary 3.3.38 *Every Wajsberg hoop $\mathbf{H}$ satisfies the following identity:*

$$x \to (y \vee z) = (x \to y) \vee (x \to z), \quad \forall x, y, z \in H. \tag{3.12}$$

Proof Let $\mathbf{H}$ be a Wajsberg hoop and $x, y, z \in H$. Consider the bounded Wajsberg hoop $\mathbf{H}_a := ([a, 1]; \odot^a, \to, a, 1)$, where $a := x \odot y \odot z$. Let $\vee_a$ and $\wedge_a$ be the join and meet operations of $\mathbf{H}_a$. Clearly, $x, y, z \in [a, 1]$. By Proposition 3.2.5, the join and meet of each pair of elements of $\mathbf{H}_a$ coincide with those on $\mathbf{H}$. Therefore, by Proposition 3.3.36, we obtain

$$x \to (y \vee z) = x \to (y \vee_a z) = (x \to y) \vee_a (x \to z) = (x \to y) \vee (x \to z).$$

$\qquad\square$

3.3.4 Idempotent Elements and Gödel Hoops

Definition 3.3.39 In each hoop $\mathbf{H} = (H; \odot, \rightarrow, 1)$, the set of idempotent elements of H with respect to $\odot$ is denoted by $\mathrm{Idm}(\mathbf{H})$, that is $\mathrm{Idm}(\mathbf{H}) = \{x \in H : x \odot x = x\}$. If $\mathrm{Idm}(\mathbf{H}) = H$, then $\mathbf{H}$ is said to be a **Gödel** hoop or an **idempotent hoop**.

Clearly, for every cancellative Gödel hoop $\mathbf{H}$, we have $H = \{1\}$, since $x \odot x = x = x \odot 1$ implies that $x = 1$ for all $x \in H$.

Proposition 3.3.40 ([120]) *Let b be an idempotent element of a hoop $\mathbf{H} = (H; \odot, \rightarrow, 1)$. Then $b \odot x = b \wedge x$ for all $x \in H$.*

Proof Choose $x \in H$. By Lemma 2.2.1(4), $x \odot b \leq x \wedge b$. Also, $b \wedge x = b \odot (b \rightarrow x) = (b \odot b) \odot (b \rightarrow x) = b \odot (b \wedge x) \leq b \odot x$, so $b \wedge x = b \odot x$. $\qquad\square$

Proposition 3.3.41 *Let $\mathbf{H} = (H; \odot, \rightarrow, 0, 1)$ be a bounded Wajsberg hoop and $a \in H$. Then the following assertions are equivalent:*

  (i) $a \in \mathrm{Idm}(\mathbf{H})$.
 (ii) $a \vee a' = 1$.
(iii) $a \wedge a' = 0$.
 (iv) $a' \odot a' = a'$.

Proof First, we note that by Proposition 3.3.9(iii), $\mathbf{H}$ is a $\vee$-hoop. In addition, it satisfies (DNP), see Theorem 3.3.15.

(i) $\Rightarrow$ (ii) Let $a \in H$ be an idempotent element. Applying Lemma 2.2.1(8), we have $a \vee a' = (a \rightarrow a') \rightarrow a' = (a \odot a)' \rightarrow a' = a' \rightarrow a' = 1$.

(ii) $\Rightarrow$ (iii) Due to Lemma 2.2.3(iii), $0 = 1' = (a \vee a')' = (a' \wedge a'') = a' \wedge a$.

(iii) $\Rightarrow$ (i) By Corollary 3.3.35, we get $a \vee a' = (a' \wedge a)' = 0' = 1$, consequently, by Proposition 3.3.30(ii), $a = a \odot 1 = a \odot (a \vee a') = (a \odot a) \vee (a \odot a') = (a \odot a) \vee 0 = a \odot a$.

(ii) $\Rightarrow$ (iv) Applying Proposition 3.3.30(ii),

$$a' = a' \odot 1 = a' \odot (a \vee a') = (a' \odot a) \vee (a' \odot a') = a' \odot a'.$$

(iv) $\Rightarrow$ (ii) $a' \odot a'$ implies that $a' \in \mathrm{Idm}(\mathbf{H})$, so by "(i) $\Rightarrow$ (ii)" we get $1 = a' \vee a'' = a' \vee a$. $\qquad\square$

It follows from Proposition 3.3.40 that $\odot = \wedge$ on each Gödel hoop $\mathbf{H}$.

Before we state the next proposition, we recall that for the definition of Heyting algebras, see Definition 1.3.36.

Theorem 3.3.42 *Gödel bounded $\vee$-hoops and Heyting algebras are termwise equivalent.*

Proof For each Heyting algebra $(L; \vee, \wedge, \to, 0, 1)$, the algebraic structure $(L; \wedge, \to, 0, 1)$ is a Gödel bounded $\vee$-hoop. Conversely, for each Gödel bounded $\vee$-hoops $(H; \odot, \to, 0, 1)$, the algebraic structure $(H; \odot, \to, 0, 1)$ is a Heyting algebra, by Proposition 3.3.40, Proposition 3.2.9(i) and Lemma 2.2.3(ii). $\qquad\square$

Using idempotent elements of a hoop, we can obtain new hoops which are bounded as follows:

Example 3.3.43 Let **H** be a hoop and $a, b \in \mathrm{Idm}(\mathbf{H})$.

(i) Then $([a, 1]; \odot, \to, a, 1)$ is a bounded hoop. Indeed, for each $x, y \in [a, 1]$ we have $a = a \odot a \le x \odot y$ and $a \le y \le x \to y$, so $[a, 1]$ is closed under $\odot$ and $\to$. Hence, $([a, 1]; \odot, \to, 1)$ is a hoop, while a is the least element of $[a, 1]$, which entails that $([a, 1]; \odot, \to, a, 1)$ is a bounded hoop. Considering the notation in Proposition 3.2.5, $x \odot^a y = x \odot y$ for all $x, y \in [a, 1]$, so $\mathbf{H}_a = ([a, 1]; \odot, \to, a, 1)$.

(ii) $([a, b]; \odot, \to_b, a, b)$ is a bounded hoop, where $x \to_b y := (x \to y) \wedge b$. In order to prove it, we apply Theorem 3.1.2. Clearly, $[a, b]$ is closed under $\odot$ and $\to_b$. Choose $x, y, z \in [a, b]$. By Proposition 3.3.40, $x \odot b = x \wedge b = x$, so (H1) holds, evidently. Also, $x \to_b x = (x \to x) \wedge b = 1 \wedge b = b$ and $x \to_b b = (x \to b) \wedge b = 1 \wedge b = b$, since $x \le b$, that implies, (H2). Furthermore, from

$$
\begin{aligned}
(x \to_b (y \to_b z)) &= \bigl(x \to ((y \to z) \wedge b)\bigr) \wedge b \\
&= \Bigl(\bigl((x \to (y \to z))\bigr) \wedge (x \to b)\Bigr) \wedge b, \text{ by Lemma 2.2.3(ii)} \\
&= \bigl((x \to (y \to z))\bigr) \wedge b \\
&= \bigl((x \odot y) \to z\bigr) \wedge b \\
&= (x \odot y) \to_b z,
\end{aligned}
$$

it follows that (H2) holds. Finally, $x \odot (x \to_b y) = x \odot ((x \to y) \wedge b) = x \odot (x \to y) \odot b = (x \odot b) \odot (x \to y) = (x \wedge b) \odot (x \to y) = x \odot (x \to y)$. In a similar way, we can show that $y \odot (y \to_b x) = y \odot (y \to x)$, which entails that (H3) holds. Therefore, $([a, b]; \odot, \to_b, a, b)$ is a bounded hoop.

Example 3.3.44 ([152]) Let $G_n := \{0, x_1, \ldots, x_{n-1}, 1\}$ be equipped with a partial ordering $\le$ such that $0 < x_1 < \cdots < x_{n-1} < 1$. Consider the binary operations $\odot$ and $\to$ are defined as follows:

$$
x \odot y = \min\{x, y\} \quad \& \quad x \to y = \begin{cases} 1 & \text{if } x \le y \\ y & \text{otherwise.} \end{cases}
$$

Then $(G_n; \odot, \rightarrow, 0, 1)$ is a bounded hoop. Evidently, $x \odot x = x$ for all $x \in G_n$, which means it is a Gödel bounded hoop. Note that for each $0 < x \in G_n$, we have $x'' = (x \rightarrow 0) \rightarrow 0 = 0 \rightarrow 0 = 1$.

At the end of this section, we will study the set of Boolean elements of bounded hoops. Let $\mathbf{H} = (H; \odot, \rightarrow, 0, 1)$ be a bounded $\vee$-hoop. An element $x \in H$ is complemented if there exists $y \in H$ such that $x \vee y = 1$ and $x \wedge y = 0$. Applying Lemma 3.2.10, we can easily deduce that the element y is unique (see [35, Theorem 6.2]). It is called the complement of x. Indeed, if $z \in H$ such that $x \vee z = 1$ and $x \wedge z = 0$, then by Lemma 3.2.10 we have $z = z \wedge 1 = z \wedge (x \vee y) = (z \wedge x) \vee (z \wedge y) = 0 \vee (z \wedge y) = z \wedge y$, which implies $z \leq y$. Similar proof shows that $y \leq z$, consequently, $z = y$. Note that x is also the complement of y. Clearly, 0 and 1 are complemented.

The set of all complemented elements of a bounded $\vee$-hoop $\mathbf{H} = (H; \odot, \rightarrow, 0, 1)$ is denoted by $B(\mathbf{H})$, i.e.,

$$B(\mathbf{H}) := \{x \in H : x \vee y = 1, \ x \wedge y = 0, \ \exists y \in H\}.$$

Since $\mathbf{H}$ is a bounded distributive lattice, due to Example 1.3.33(iii), $(B(\mathbf{H}); \vee, \wedge, ', 0, 1)$ is a Boolean algebra. The set $B(\mathbf{H})$ is also called the **Boolean part** of $\mathbf{H}$.

Proposition 3.3.45 *Let $\mathbf{H}$ be a bounded basic hoop. Then* $\mathrm{Reg}(\mathbf{H}) \cap \mathrm{Idm}(\mathbf{H}) = B(\mathbf{H})$. *In addition, $B(\mathbf{H})$ is a subuniverse of a bounded subhoop of $\mathbf{H}$ and a Boolean algebra.*

Proof Let $x \in B(\mathbf{H})$. Then there exists $y \in H$ such that $x \vee y = 1$ and $x \wedge y = 0$. From $x \odot y \leq x \wedge y = 0$ it follows that $x \leq y'$. Also, by Lemma 2.2.3(i), $y' = y' \odot 1 = y' \odot (x \vee y) = (y' \odot x) \vee (y' \odot y) = y' \odot x \leq x$ yields that $y' = x$. On the other hand, by Corollary 3.3.35, $x' \wedge x = x' \wedge y' = (x \vee y)' = 0$ and $x' \vee x = x' \vee y' = (x \wedge y)' = 1$ that entail x' is the unique complement of x and $x' \in B(\mathbf{H})$. Similarly, x'' is the complemented of x' which implies $x = x''$ (since both x and x'' are complements of x'), which means $x \in \mathrm{Reg}(\mathbf{H})$. In addition, $x = x \odot 1 = x \odot (x \vee x') = (x \odot x) \vee (x \odot x') = (x \odot x) \vee 0 = x \odot x$, i.e., $x \in \mathrm{Idm}(\mathbf{H})$. Therefore, $B(\mathbf{H}) \subseteq \mathrm{Reg}(\mathbf{H}) \cap \mathrm{Idm}(\mathbf{H})$.

Conversely, let $x \in \mathrm{Reg}(\mathbf{H}) \cap \mathrm{Idm}(\mathbf{H})$. By Proposition 3.3.40, $x \wedge x' = x \odot x' = 0$ and so $x \vee x' = x'' \vee x' = (x' \wedge x)' = 0' = 1$, see Corollary 3.3.35. Hence, $x \in B(\mathbf{H})$.

Therefore, by the comment right before the proposition, $\mathrm{Reg}(\mathbf{H}) \cap \mathrm{Idm}(\mathbf{H})$ forms a Boolean algebra, where $x' := x \rightarrow 0$ is the complement of x for all $x \in B(\mathbf{H})$. Now, we show that $B(\mathbf{H})$ is a closed under $\odot$ and $\rightarrow$. Choose $x, y \in B(\mathbf{H})$.

By Proposition 3.3.40, $x \odot y = x \wedge y \in B(\mathbf{H})$. From $y \in B(\mathbf{H})$ we have $y' \in B(\mathbf{H})$, so $x \odot y' \in B(\mathbf{H})$, consequently, $x \rightarrow y = x \rightarrow y'' = (x \odot y')' \in B(\mathbf{H})$. Therefore, $B(\mathbf{H})$ is closed under the operations $\odot, \rightarrow, 0$, and 1, equivalently, $(B(\mathbf{H}); \odot, \rightarrow, 0, 1)$ is a bounded subhoop of $\mathbf{H}$. Evidently, $(B(\mathbf{H}); \vee, \wedge, ', 0, 1)$ is a Boolean algebra. $\qquad\square$

Lemma 3.3.46 *Let a be a Boolean element of a bounded basic hoop $\mathbf{H}$. Then for all $x, y \in H$:*

$$(a \vee x) \odot (a \vee y) = (x \odot y) \vee a \quad \& \quad (a \wedge x) \odot (a \wedge y) = (x \odot y) \wedge a.$$

Proof Let $x, y \in H$. By Lemma 2.2.3(i), we have

$$
\begin{aligned}
(a \vee x) \odot (a \vee y) &= (a \odot a) \vee (a \odot y) \vee (x \odot a) \vee (x \odot y) \\
&= \big(a \vee (a \odot y) \vee (x \odot a)\big) \vee (x \odot y), \text{ by Proposition 3.3.45} \\
&= a \vee (x \odot y).
\end{aligned}
$$

In addition, due to Proposition 3.3.41, $a \wedge (x \odot y) = a \odot (x \odot y) = a \odot a \odot x \odot y = (a \odot x) \odot (a \odot y) = (a \wedge x) \odot (a \wedge y)$. $\qquad\square$

Theorem 3.3.47 *Let $\mathbf{H} = (H; \odot, \to, 0, 1)$ be a bounded basic hoop and $a \in B(\mathbf{H})$. Then $f_a : \mathbf{H} \to \mathbf{H}_a$ defined by $f_a(x) = x \vee a$ is an onto homomorphism of bounded hoops. In addition, $f : \mathbf{H} \to \mathbf{H}_a \times \mathbf{H}_{a'}$ defined by $f(x) = (x \vee a, x \vee a')$ is an isomorphism of bounded hoops.*

Proof Clearly, f_a is well-defined, $f_a(0) = a$ and $f_a(1) = 1$. Choose $x, y \in H$. By Lemma 3.3.46,

$$
\begin{aligned}
f_a(x) \odot f_a(y) &= (a \vee x) \odot (a \vee y) \\
&= \big((a \vee x) \odot a\big) \vee \big((a \vee x) \odot y\big) \\
&= (a \odot a) \vee (x \odot a) \vee (a \odot y) \vee (x \odot y) \\
&= \big(a \vee (x \odot a) \vee (a \odot y)\big) \vee (x \odot y) \\
&= a \vee (x \odot y) = f_a(x \odot y).
\end{aligned}
$$

Due to Lemma 3.2.10, $y = (y \vee a) \wedge (y \vee a')$, so by Lemma 2.2.3(ii),

$$x \to y = (x \to (y \vee a)) \wedge (x \to (y \vee a')),$$

consequently,

$$
\begin{aligned}
(x \to y) \vee a &= \big((x \to (y \vee a)) \vee a\big) \wedge \big((x \to (y \vee a')) \vee a\big) \\
&= (x \to (y \vee a)) \wedge \big((x \to (y \vee a')) \vee a\big), \text{ since } x \to (y \vee a) \geq y \vee a \geq a \\
&= (x \to (y \vee a)) \wedge 1, \text{ since} (x \to (y \vee a')) \vee a \geq (y \vee a') \vee a \geq a' \vee a = 1 \\
&= x \to (y \vee a) = (x \vee a) \to (y \vee a), \text{ by Lemma 2.2.3(iii)}.
\end{aligned}
$$

Hence, $f_a(x \to y) = f_a(x) \to f_a(y)$, i.e., f_a is a homomorphism of bounded hoops, which is onto, clearly.

Furthermore, $f(x) = (f_a(x), f_{a'}(x))$ for every $x \in H$, which means f is a homomorphism of bounded hoops. Since a is a Boolean element of H, we have $f(u) = f(v)$ implies that $u = (u \vee a) \wedge (u \vee a') = (v \vee a) \wedge (v \vee a') = v$, which entails f is one-to-one. Easy calculations show that $f(x \vee y) = (x, y)$ for every $(x, y) \in [a, 1] \times [a', 1]$, i.e., f is onto. Therefore, f is an isomorphism of bounded hoops and $\mathbf{H} \cong \mathbf{H}_a \times \mathbf{H}_{a'}$. $\square$

Corollary 3.3.48 *A bounded basic hoop* $\mathbf{H} = (H; \odot, \rightarrow, 0, 1)$ *is directly indecomposable if and only if* $B(\mathbf{H}) \setminus \{0, 1\} \neq \emptyset$.

3.3.5   Product Hoops

In this section, we will follow the definition of product hoops that was introduced in [11, p. 90].

Definition 3.3.49   A basic hoop $\mathbf{H} = (H; \odot, \rightarrow, 1)$ is called a **product** hoop if it satisfies the following conditions for all $x, y, z \in H$:

$$(x \rightarrow z) \vee \big((x \rightarrow (x \odot y)) \rightarrow y\big) = 1 \qquad \forall x, y, z \in H. \qquad (PH)$$

Since BH is a variety, and (PH) is an identity of type $\{\odot, \rightarrow, 1\}$, by Theorem 1.6.4, we get that the class of all product algebra forms a variety (note that $x \vee y = ((x \rightarrow y) \rightarrow y) \wedge ((y \rightarrow x) \rightarrow x)$ and $x \wedge y = x \odot (x \rightarrow y)$ for all $x, y \in H$).

Every cancellative hoop is a Wajsberg hoop and so it is a basic hoop (see Proposition 3.3.32 and Proposition 3.3.16), consequently, we can propose the following proposition:

Proposition 3.3.50 *If* $\mathbf{H}$ *is a cancellative hoop, then* $\mathbf{H}$ *is a product hoop.*

Proof Let $\mathbf{H}$ be a cancellative hoop and $x, y, z \in H$. Then by Proposition 3.3.5, $x \rightarrow (x \odot y) = y$, it follows that

$$(x \rightarrow z) \vee \big((x \rightarrow (x \odot y)) \rightarrow y\big) = (x \rightarrow z) \vee (y \rightarrow y) = 1.$$

$\square$

Example 3.3.51   ([210, p. 520]) Let $[0, 1] \subseteq \mathbb{R}$ equipped with the binary operations $\odot$ and $\rightarrow$ be defined as follows:

$$x \odot y = x \cdot y \text{(the product of real numbers)}, \quad x \rightarrow y = \begin{cases} 1 & \text{if } x \leq y \\ y/x & \text{otherwise.} \end{cases}$$

Then $[0, 1]_P := ([0, 1]; \odot, \rightarrow, 0, 1)$ is a bounded hoop, where the implication $\rightarrow$ is called Goguen's implication. Clearly, the partial order relation on the hoop $[0, 1]_P$ coincides with the natural partial order relation of the real numbers.

Let $x, y, z \in [0, 1]$.

(i) If $x \le z$ or $y = 1$, then by definition of $\rightarrow$, we have $(x \rightarrow z) \vee ((x \rightarrow (x \odot y)) \rightarrow y) = 1$.

(ii) If $z < x$ and $y \ne 1$, then $x \odot y = x \cdot y < 1$, and so $x \rightarrow (x \odot y) = x \rightarrow (x \cdot y) = (x \cdot y)/x = y$, which implies $(x \rightarrow (x \cdot y)) \rightarrow y = y \rightarrow y = 1$. Thus, $(x \rightarrow z) \vee ((x \rightarrow (x \odot y)) \rightarrow y) = 1$.

Therefore, $[0, 1]_P$ is a product hoop.

Remark 3.3.52 An easy calculation shows that if $\mathbf{H} = (H; \odot, \rightarrow, 0, 1)$ is a bounded basic hoop, then it is a product hoop if and only if $x' \vee ((x \rightarrow (x \odot y)) \rightarrow y) = 1$ for all $x, y \in H$. Evidently, (PH) implies $x' \vee ((x \rightarrow (x \odot y)) \rightarrow y) = 1$ for all $x, y \in H$ (it suffices to choose $z = 0$). The converse follows from the following inequality:

$$x' \vee ((x \rightarrow (x \odot y)) \rightarrow y) \le (x \rightarrow z) \vee ((x \rightarrow (x \odot y)) \rightarrow y) \quad \forall x, y, z \in H.$$

Example 3.3.53 ([153, Proposition 5.5]) For each Abelian ℓ-group $(G; \vee, \wedge, +, -, 0)$ the hoop $\mathbf{N(G)}$ is a product hoop. Indeed, by Example 3.3.8, $\mathbf{N(G)}$ is cancellative and so by Proposition 3.3.50, $\mathbf{N(G)}$ is a product hoop.

Linearly ordered product hoops have some interesting properties as follows:

Proposition 3.3.54 ([167]) *Let* $\mathbf{H} = (H; \odot, \rightarrow, 1)$ *be a linear product hoop. Then the following statements hold:*

(i) *If* $\mathbf{H}$ *is unbounded, then* $x \rightarrow (x \odot y) = y$ *for all* $x, y \in H$. *If* $\mathbf{H}$ *is bounded, then* $x \rightarrow (x \odot y) = y$ *and* $x' = 0$ *for all* $x \in H \setminus \{0\}$ *and* $y \in H$.
(ii) *If* $\mathbf{H}$ *is unbounded, then* $\mathbf{H}$ *is cancellative. In addition, if* $\mathbf{H}$ *is bounded and* $x \ne 0$, *then* $x \odot y = x \odot z$ *implies that* $y = z$ *for all* $x, y \in H$.
(iii) *If* $\mathbf{H}$ *is bounded and* $x \ne 0$, *then* $x \odot y < x \odot z$ *implies that* $y < z$.

Proof (i) First, assume that $\mathbf{H}$ is not bounded and $x, y \in H$. Choose $z \in H$ such that $z < x$. Then $x \rightarrow z \ne 1$, so by (PH), $1 = (x \rightarrow z) \vee ((x \rightarrow (x \odot y)) \rightarrow y)$ implies that $(x \rightarrow (x \odot y)) \rightarrow y = 1$, equivalently, $x \rightarrow (x \odot y) \le y$. On the other hand,

$$y \rightarrow (x \rightarrow (x \odot y)) = (x \odot y) \rightarrow (x \odot y) = 1, \quad \text{by Lemma 2.2.1(5)}$$

consequently, $y \le x \rightarrow (x \odot y)$. It follows that $y = x \rightarrow (x \odot y)$.

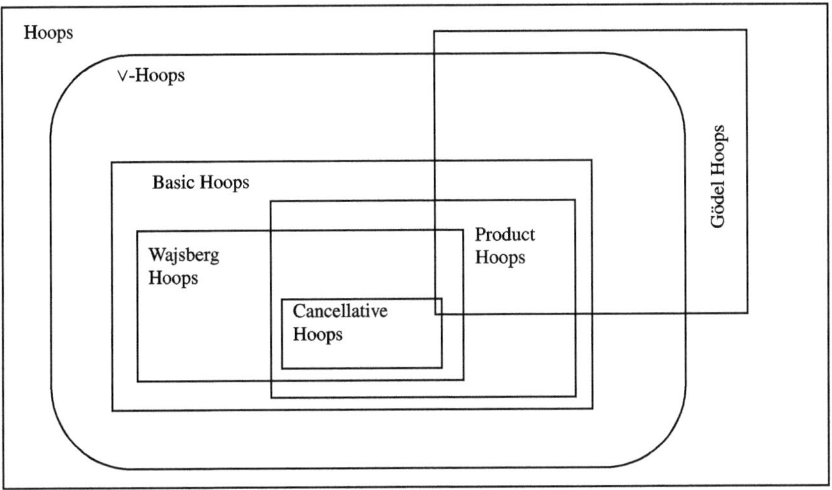

Fig. 3.2 Relation between various classes of hoops

Now, assume that $\mathbf{H}$ is bounded, $x \in H \setminus \{0\}$ and $y \in H$. Since $0 < x$, then $x' \neq 1$, similarly to the first part, since $\mathbf{H}$ is linearly ordered, from $x' \vee \big((x \to (x \odot y)) \to y\big) = 1$, we conclude that $x \to (x \odot y) = y$. In addition, for $y = 0$ we get $0 = x \to (x \odot 0) = x'$.

(ii) Let x be an element of H such that $x \neq 0$. Choose $y, z \in H$ such that $x \odot y = x \odot z$. By (i), $y = x \to (x \odot y) = x \to (x \odot z) = z$. Note that if $\mathbf{H}$ is not unbounded, then the proof works for every $x \in H$. Therefore, $\mathbf{H}$ is cancellative.

(iii) Let $x \in H \setminus \{0\}$ and $y, z \in H$ such that $x \odot y < x \odot z$. If $z \leq y$, then by Lemma 2.2.1(3), $x \odot z \leq x \odot y$ which is a contradiction. Therefore, $y < z$. Similarly to part (ii), if $\mathbf{H}$ is not unbounded, then the proof works for every $x \in H$. $\square$

In Fig. 3.2, we represent a relation between different types of hoops that were introduced in Sect. 3.3. Note that the intersection of the classes of cancellative and Gödel hoops consists solely of trivial hoops.

3.3.6 Some Examples

This section presents several examples of hoops, inspired by [11, 31, 79, 151, 152]. These examples will deepen our understanding of the structure of hoops.

Example 3.3.55 Consider the bounded pocrim $(B; \odot, \to, 0, 1)$ induced from a Boolean algebra $(B; \vee, \wedge, ', 0, 1)$ in Example 2.1.2. Then $(B; \odot, \to, 0, 1)$ is a bounded Wajsberg hoop. Indeed, $x \odot (x \to y) = x \wedge (x' \vee y) = x \wedge y$, so by Theorem 3.1.5, $(B; \odot, \to, 0, 1)$

is a hoop. Also, $(x \rightarrow 0) \rightarrow 0 = (x' \vee 0)' \vee 0 = x'' = x$, so $(B; \odot, \rightarrow, 0, 1)$ satisfies (DNP), consequently, it is a bounded Wajsberg hoop, see Theorem 3.3.15.

Example 3.3.56 Due to Proposition 1.3.37, for each Heyting algebra $(A; \vee, \wedge, \rightarrow, 0, 1)$, the structure $(A; \wedge, \rightarrow, 0, 1)$ is a bounded $\vee$-hoop.

Example 3.3.57 Let $\mathbf{G} = (G; \vee, \wedge, +, -, 0)$ be an arbitrary Abelian ℓ-group. For an arbitrary element $u \in G$, $u \geq 0$ define on the set $G[u] = [0, u]$ the following operations:

$$a \odot b = (a - u + b) \vee 0 \quad \& \quad a \rightarrow b = (b - a + u) \wedge u. \tag{3.13}$$

$\mathbf{G[u]} = (G[u]; \odot, \rightarrow, u)$ is a bounded Wajsberg hoop. For the sake of completeness, we shall give here proof of this fact. Firstly, let us note that $G[u]$ is closed under the operations $\odot, \rightarrow$. We shall verify the identities from Theorem 3.1.2. Let $a, b, c \in [0, u]$.

(H0) $a \odot u = (a - u + u) \vee 0 = a \vee 0 = a$, since $a \geq 0$.

(H1) $a \rightarrow a = u \wedge u = u$.

(H2)

$$\begin{aligned}
(a \odot b) \rightarrow c &= \big((a - u + b) \vee 0\big) \rightarrow c \\
&= \big(c - (a - u + b) \vee 0 + u\big) \wedge u \\
&= (c - b + u - a + u) \wedge (c + u) \wedge u \\
&= (c - b + u - a + u) \wedge u,
\end{aligned}$$

and

$$\begin{aligned}
a \rightarrow (b \rightarrow c) &= \big((b \rightarrow c) - a + u\big) \wedge u \\
&= \big((c - b + u) \wedge u - a + u\big) \wedge u \\
&= (c - b + u - a + u) \wedge (u - a + u) \wedge u \\
&= (c - b + u - a + u) \wedge u.
\end{aligned}$$

(H3)

$$\begin{aligned}
(a \rightarrow b) \odot a &= \big((a \rightarrow b) - u + a\big) \vee 0 \\
&= \big((b - a + u) \wedge u - u + a\big) \vee 0 \\
&= \big((b - a + u - u + a) \wedge (u - u + a)\big) \vee 0 \\
&= (b \wedge a) \vee 0 \\
&= b \wedge a,
\end{aligned}$$

similarly, $(b \rightarrow a) \odot b = a \wedge b$.

Hence, $\mathbf{G}[\mathbf{u}]$ is a hoop. It is obvious that $\mathbf{G}[\mathbf{u}]$ is bounded. It remains to prove the Wajsberg hoop condition. Let $a, b \in G[u]$. We have that

$$
\begin{aligned}
(a \to b) \to b &= \big(u - (a \to b) + b\big) \wedge u \\
&= \big(u - (b - a + u) \wedge u + b\big) \wedge u \\
&= \big((u - u + a - b + b) \vee (u - u + b)\big) \wedge u \\
&= (a \vee b) \wedge u \\
&= a \vee b,
\end{aligned}
$$

$$
\begin{aligned}
(b \to a) \to a &= \big(u - (b \to a) + a\big) \wedge u \\
&= \big(u - (a - b + u) \wedge u + a\big) \wedge u \\
&= \big((u - u + b - a + a) \vee (u - u + a)\big) \wedge u \\
&= (b \vee a) \wedge u = a \vee b.
\end{aligned}
$$

Remark 3.3.58 (i) The bounded hoop $([0, 1]; \odot, \to, 0, 1)$ in Example 3.3.11 is indeed $\mathbb{R}[1]$. In addition, similarly to Example 1.4.14(ii), we can show that $\mathbb{Z}[n] \cong \mathbf{L}_n$ for all $n \in \mathbb{N}$. Indeed, $f : \mathbb{Z}[n] \to \mathbf{L}_n$ defined by $f(m) = m/n$ for each $m \in \{0, 1, \dots, n\}$ is a hoop isomorphism.

(ii) It is worth noting that $\mathbf{G}[\mathbf{u}]$ is precisely the hoop corresponding to the MV-algebra $\Gamma(\mathbf{G}, u)$, i.e., $\mathbf{G}[\mathbf{u}] = \alpha(\Gamma(\mathbf{G}, u))$, where α is the map in Theorem 3.3.13.

Example 3.3.59 Let $\mathbf{R} = (R; +, \cdot, -, 0, 1)$ be a commutative integral domain with 1. Then $\mathbf{R}$ is said to be a **Dedekind domain** if every non-zero ideal of $\mathbf{R}$ is uniquely representable as the product of finitely many prime ideals (see [174, Sect. 8.6]). Due to Example 2.1.4, $(\mathrm{Id}(\mathbf{R}); \cdot, \to, \{0\}, R)$ is a pocrim. Choose $I, J \in \mathrm{Id}(\mathbf{R})$.

If $I \subseteq J$, there are prime ideals $P_1, \dots, P_n$ of $\mathbf{R}$ such that $I = P_1 \cdot P_2 \cdots P_n$ and $J = P_{j_1} \cdot P_{j_2} \cdots P_{j_m}$ where $P_{j_1}, P_{j_2} \dots P_{j_m} \in \{P_1, \dots P_n\}$. Thus, $I = J \cdot K$, where $K = P_{i_1} \cdot P_{i_2} \cdots P_{i_t}$ and $\{P_{i_1} \dots, P_{i_t}\} = \{P_1, \dots P_n\} \setminus \{P_{j_1}, \dots, P_{j_m}\}$. Furthermore, if $I = J \cdot K$ for some $K \in \mathrm{Id}(\mathbf{R})$, then $I \subseteq J$. Therefore, $I \subseteq J$ if and only if $I = J \cdot K$ for some $K \in \mathrm{Id}(\mathbf{R})$, which means $(\mathrm{Id}(\mathbf{R}); \cdot, \to, \{0\}, R)$ is a naturally ordered pocrim, equivalently, a hoop.

Example 3.3.60 For each Abelian ℓ-group $\mathbf{G} = (G; \vee, \wedge, +, -, 0)$, the hoop $\mathbf{N}(\mathbf{G})$ is a product hoop. Indeed, by Example 3.3.8, $\mathbf{N}(\mathbf{G})$ is a cancellative hoop and so by Proposition 3.3.16, it is a Wajsberg hoop. Now, Proposition 3.3.50 implies that $\mathbf{N}(\mathbf{G})$ is a product hoop.

Example 3.3.61 ([141, Example 2.1.7]) Consider the pocrim $(\mathbb{N}; \odot, \to, 1)$ in Example 2.1.6. We claim that $x \wedge y = \mathrm{lcm}(x, y)$ for each $x, y \in \mathbb{N}$. Let $x, y \in \mathbb{N}$ and

$a = \mathrm{lcm}(x, y)$. Then $a \to x = x/\gcd(a, x) = x/x = 1$. Similarly, $a \to y = 1$. In addition, if $z \in \mathbb{N}$ such that $z \to x = 1$ and $z \to y = 1$, then $x/\gcd(x, z) = 1 = y/\gcd(y, z)$, which entails $\gcd(x, z) = x$ and $\gcd(y, z) = y$. It follows that $\gcd(a, z) = a$, consequently, $z \to a = 1$, which implies $x \wedge y = \mathrm{lcm}(x, y)$. In addition, for each $x, y \in \mathbb{N}$ we have

$$x \odot (x \to y) = x(y/\gcd(x, y)) = xy/\gcd(x, y) = y \odot (y \to x).$$

Due to $\mathrm{lcm}(x, y) = xy/\gcd(x, y)$, we get $x \odot (x \to y) = y \odot (y \to x) = x \wedge y$ for all $x, y \in \mathbb{N}$. Therefore, by Theorem 3.1.5, $(\mathbb{N}; \odot, \to, 1)$ is a hoop.

Example 3.3.62 ([13, 141]) Let $\mathbf{H}_1 = (H_1; \odot_1, \to_1, 1)$ and $\mathbf{H}_2 = (H_2; \odot_2, \to_2, 1)$ be hoops with $H_1 \cap H_2 = \{1\}$. According to Example 2.2.8, $\mathbf{H}_1 \oplus \mathbf{H}_2$ is a pocrim. We can easily verify that $\mathbf{H}_1 \oplus \mathbf{H}_2$ is a hoop (Applying Theorem 3.1.5). Similarly to pocrims, **the ordinal sum** of the hoops $\mathbf{H}_1$ and $\mathbf{H}_2$, is denoted by $\mathbf{H}_1 \oplus \mathbf{H}_2$.

Proposition 3.3.63 *Let* $\mathbf{H}_1 = (H_1; \odot_1, \to_1, 1)$ *and* $\mathbf{H}_2 = (H_2; \odot_2, \to_2, 1)$ *be hoops with* $H_1 \cap H_2 = \{1\}$. *Then:*

(i) *If* $\mathbf{H}_1$ *and* $\mathbf{H}_2$ *are linear hoops, then* $\mathbf{H}_1 \oplus \mathbf{H}_2$ *is also linear.*
(ii) $\mathbf{H}_1$ *and* $\mathbf{H}_2$ *are subalgebras of* $\mathbf{H}_1 \oplus \mathbf{H}_2$.
(iii) $\mathbf{H}_2$ *is a filter of* $\mathbf{H}_1 \oplus \mathbf{H}_2$.
(iv) *If* $\mathbf{H}_1$ *is a linear hoop and* $\mathbf{H}_2$ *is a basic hoop, then* $\mathbf{H}_1 \oplus \mathbf{H}_2$ *is a basic hoop.*

Proof Consider the hoop $\mathbf{H}_1 \oplus \mathbf{H}_2 = (H_1 \cup H_2; \odot, \to, 1)$.

(i) Let $x, y \in H_1 \cup H_2$. If $x, y \in H_1$ or $x, y \in H_2$, then use the fact that $\mathbf{H}_1$ and $\mathbf{H}_2$ are linear to get that $x \leq y$ or $y \leq x$. Suppose that $x \in H_1 \setminus \{1\}$ and $y \in H_2 \setminus \{1\}$. Then $x \to y = 1$, so $x \leq y$. Similarly, if $x \in H_2 \setminus \{1\}$ and $y \in H_1 \setminus \{1\}$, then $y \to x = 1$, so $y \leq x$. Thus, we have proved that $\mathbf{H}_1 \oplus \mathbf{H}_2$ is a linear hoop.

(ii) It is obvious.

(iii) According to definition of $\odot$ on the hoop $\mathbf{H}_1 \oplus \mathbf{H}_2$, we have $x \odot y = x \odot_2 y \in H_2$ for all $x, y \in H_2$. So, (F1) holds. Now, assume that y is an element of $H_1 \cup H_2$ such that $y \geq x \in H_2$. If $y \notin H_2$, then by definition of $\to$ on $\mathbf{H}_1 \oplus \mathbf{H}_2$ (see Example 2.2.8) we have $x \to y = y \neq 1$ and $y \to x = 1$ which implies $y < x$, a contradiction. So, $y \in H_2$, that entails (F2). Therefore, H_2 is a filter of $\mathbf{H}_1 \oplus \mathbf{H}_2$.

(iv) Let $\mathbf{H}_1$ be a linear hoop and $\mathbf{H}_2$ be a basic hoop. We claim that $\mathbf{H}_1 \oplus \mathbf{H}_2$ is a $\vee$-hoop. To establish this, it suffices to demonstrate that $x \vee y$ exists for all $x, y \in (H_1 \cup H_2) \setminus 1$. Consider $x, y \in (H_1 \cup H_2) \setminus \{1\}$. Using (2.7) and (2.8), we analyze the following cases, where $\vee_1$ and $\vee_2$ denote the join operations on $\mathbf{H}_1$ and $\mathbf{H}_2$, respectively:

(1) If $x, y \in H_1$, then $x \vee_1 y \in H_1 \setminus \{1\}$ and by (2.8), we can easily verify that $x \vee y = x \vee_1 y$. Note that $\mathbf{H}_1$ is linear and $x, y \in H_1 \setminus \{1\}$.

(2) If $x, y \in H_2$, then $x \vee_2 y \in H_2$ and by (2.8), we have $x \vee y = x \vee_2 y$.

(3) If $x \in H_1$ and $y \in H_2$, then $x \leq y$, which entails $x \vee y = y$.

It follows that $\mathbf{H}_1 \oplus \mathbf{H}_2$ is a $\vee$-hoop. To prove that $\mathbf{H}_1 \oplus \mathbf{H}_2$ is a basic hoop, we apply Theorem 3.3.28. It suffices to demonstrate that $\mathbf{H}_1 \oplus \mathbf{H}_2$ satisfies (Pre). Since both $\mathbf{H}_1$ and $\mathbf{H}_2$ are basic hoops, by the definition of $\rightarrow$ on $\mathbf{H}_1 \oplus \mathbf{H}_2$ (see (2.7)), it is sufficient to verify that $(x \rightarrow y) \vee (y \rightarrow x) = 1$ for all $x \in H_1 \setminus \{1\}$ and $y \in H_2 \setminus \{1\}$. This condition holds, as $x \rightarrow y = 1$ and $y \rightarrow x = x$. Therefore, $\mathbf{H}_1 \oplus \mathbf{H}_2$ is a basic hoop. $\qquad\square$

Let us consider the two-element Boolean algebra $\mathbf{L}_1 = \{0, 1\}$. Then $\mathbf{L}_1$ is a hoop that is not cancellative, since $0 \odot 1 = 0 \odot 0 = 0$ and $1 \neq 0$. From now on, we use $\mathbf{2}$ to denote $\mathbf{L}_1$, for simplicity. Indeed, $\mathbf{2} := (\{0, 1\}; \odot, \rightarrow, 1)$ is the unique hoop with two elements, where

$$x \odot y = \begin{cases} 1 & \text{if } x = y = 1 \\ 0 & \text{otherwise,} \end{cases} \qquad x \rightarrow y = \begin{cases} 0 & \text{if } x = 1,\ y = 0 \\ 1 & \text{otherwise.} \end{cases}$$

Theorem 3.3.64 *Let $\mathbf{H}_1 = (H_1; \odot_1, \rightarrow_1, 1)$ be a non-linear hoop and $\mathbf{H}_2 = (H_2; \odot_2, \rightarrow_2, 1)$ be a non-trivial hoop. Then $\mathbf{H}_1 \oplus \mathbf{H}_2$ is not a basic hoop. In particular, $\mathbf{H}_1 \oplus \mathbf{2}$ is not a basic hoop.*

Proof Since $\mathbf{H}_1$ is not linear, there exist $x, y \in H_1$ such that x, y are incomparable, so $x \rightarrow_1 y \neq 1$ and $y \rightarrow_1 x \neq 1$. Choose $z \in H_2 \setminus \{1\}$. Now, by (2.7), in the hoop $\mathbf{H}_1 \oplus \mathbf{H}_2$ we have:

$$(x \rightarrow y) \rightarrow z = (x \rightarrow_1 y) \rightarrow z = 1, \text{since} x \rightarrow_1 y \neq 1.$$
$$((y \rightarrow x) \rightarrow z) \rightarrow z = ((y \rightarrow_1 x) \rightarrow z) \rightarrow z = 1 \rightarrow z = z, \text{since} x \rightarrow_1 y \neq 1.$$

Therefore, by (3.9), $\mathbf{H}_1 \oplus \mathbf{H}_2$ is not a basic hoop.

The proof of the other part is a straight consequence of the first part. $\qquad\square$

Applying Theorem 3.3.64, we can prove the converse of Proposition 3.3.63(iv) as follows:

Corollary 3.3.65 *Let $\mathbf{H}_1$ and $\mathbf{H}_2$ be non-trivial hoops. If $\mathbf{H}_1 \oplus \mathbf{H}_2$ is a basic hoop, then $\mathbf{H}_1$ is linear and $\mathbf{H}_2$ is a basic hoop.*

Proof Assume that $\mathbf{H}_1 \oplus \mathbf{H}_2$ is a basic hoop. By Theorem 3.3.64, $\mathbf{H}_1$ must be linear. Let $\vee_2$ ($\leq_2$) and $\vee$ ($\leq$) denote the join operations (partial order relations) on $\mathbf{H}_2$ and $\mathbf{H}$, respectively. If $x_1, x_2 \in H_2$, then by (2.8), $x \vee_2 y$ is an upper bound for x and y in $\mathbf{H}_1 \oplus \mathbf{H}_2$, implying that $x \vee y \leq x \vee_2 y$. On the other hand, if $u \in H_1 \cup H_2$ is an upper bound for x and y

in $\mathbf{H}_1 \oplus \mathbf{H}_2$, then $x, y \le u$, so $u \in H_2$, and by (2.8) we get, $x, y \le_2 u$, which implies, $x \vee_2 y \le_2 u$, in conclusion, $x \vee_2 y \le u$. Therefore, $x \vee_2 y = x \vee y$. In addition,

$$(x \to_2 y) \vee_2 (y \to_2 x) = (x \to y) \vee (y \to x) = 1, \text{ since} \mathbf{H}_1 \oplus \mathbf{H}_2 \text{is a basic hoop.}$$

Therefore, by Theorem 3.3.28, $\mathbf{H}_2$ is a basic hoop. $\qquad\square$

By Proposition 3.3.32, it follows that the hoop $\mathbf{H}$ in Theorem 3.3.64 is not a Wajsberg hoop. Hence, as in the case of hoops (see [11, Page 12]), the ordinal sum construction allows us to obtain various examples of hoops that are not basic.

Example 3.3.66 Let $\mathbf{H} = (H; \odot, \to, 1)$ be an arbitrary hoop. Evidently, $\mathbf{2}$ is a bounded hoop. By definition of the ordinal sum, $\mathbf{2} \oplus \mathbf{H}$ is a bounded hoop. The result demonstrates that ordinal sums provide a method for constructing bounded hoops.

Ordinal sum is a useful tool for studying finite hoops and pocrims. For the definitions of $\mathbf{L}_n$ and $\mathbf{G}_n$ see Examples 1.4.2 and 3.3.44, respectively.

Example 3.3.67 ([22]) It can be shown that there are 7 pocrims with 4 elements:
$\mathbf{2} \times \mathbf{2}$, $\mathbf{L}_3$, $\mathbf{G}_4$, $\mathbf{2} \oplus \mathbf{L}_2$, $\mathbf{L}_2 \oplus \mathbf{2}$, $\mathbf{P}_4$, and $\mathbf{Q}_4$. Additionally, $\mathbf{P}_4$ and $\mathbf{Q}_4$ are the smallest pocrims that are not hoops:
$\mathbf{P}_4$ comprises the chain $0 < q < p < 1$. The operation for $\mathbf{P}_4$ are as Table 3.3:
(where, for future reference, we additionally provide a tabulation of the double negation mapping, δ). In $\mathbf{P}_4$, $\delta(q) = p$, so $\mathbf{P}_4$ is not involutive. Moreover, $\mathbf{P}_4$ is not a hoop since it is not naturally ordered: there is no z with $p \odot z = q$. However, the image of double negation is a subpocrim with universe $\{0, p, 1\}$ isomorphic to the involutive hoop $\mathbf{L}_2$.
$\mathbf{Q}_4$ comprises the chain $0 < v < u < 1$ and has the operation tables as Table 3.4:
Like $\mathbf{P}_4$, $\mathbf{Q}_4$ is not naturally ordered and hence not a hoop, because there is no z with $u \odot z = v$. $\mathbf{Q}_4$ is involutive.

Table 3.3 Operations $\odot$, $\to$, and δ of $\mathbf{P}_4$

Table 3.3.1

$\odot$	0	p	q	1
0	0	0	0	0
p	0	0	0	p
q	0	0	0	q
1	0	p	q	1

Table 3.3.2

$\to$	0	p	q	1
0	1	1	1	1
p	p	1	p	1
q	p	1	1	1
1	1	1	1	1

Table 3.3.3

δ	
0	0
p	p
q	p
1	1

Table 3.4 Operations $\odot$, $\rightarrow$ and δ of $\mathbf{Q}_4$

Table 3.4.1

$\odot$	0	u	v	1
0	0	0	0	0
u	0	u	0	u
v	0	0	0	v
1	0	u	v	1

Table 3.4.2

$\rightarrow$	0	u	v	1
0	1	1	1	1
u	v	1	v	1
v	u	1	1	1
1	0	u	v	1

Table 3.4.3

δ	
0	0
u	u
v	v
1	1

In Sect. 6.1, we will further discuss the ordinal sums of hoops. We will demonstrate that the ordinal sum is a highly useful tool for studying the class of hoops.

Bibliographical Remarks and Suggestions for Further Study

The notion of a hoop was initially introduced by Büchi and Owens [59], an unpublished paper. Subsequently, it has garnered attention from numerous researchers across diverse fields of algebra. Filters, congruences, and deductive systems of hoops have been investigated in a series of papers by Borzooei et al. [37, 42, 44, 46, 196]. Subvarieties and subquasivarieties of hoops have been extensively studied by Aglianó et al. [8, 9, 13, 14, 18, 218, 256]. The logical aspects of hoops have been explored in [137, 155]. Notable representations of various classes of hoops have been provided in [6, 157, 210]. Furthermore, topological hoops have been studied in [3, 5, 38]. Consequently, the theory of hoops is now quite rich. Many algebraic structures associated with logic, such as MV-algebras and BL-algebras, can be considered subclasses of hoops. Therefore, a thorough understanding of these algebraic structures is not only necessary but essential for a deeper understanding of hoop algebras. For the theory of MV-algebras, we highly recommend consulting the monographs [73, 99, 103, 123, 225], and for BL-algebras, we recommend [63, 167, 204, 229].

Recently, a non-commutative generalization of hoops, termed pseudo-hoops, was introduced and studied in [153]. These structures generalize both pseudo MV-algebras and pseudo BL-algebras. For those interested in this area, we recommend studying [76, 79, 80, 111, 116]. In Appendix A, we will briefly mention some results associated with pseudo-hoops and related structures.

Another generalization of MV-algebras, termed quasi-MV algebras, motivated by investigations into the structure of quantum logical gates, was introduced in [156, 203]. The logic associated with these structures was investigated in [58, 234]. Further advanced studies related to quasi-MV algebras can be found in [57, 135, 189, 198, 233].

Furthermore, EMV-algebras, introduced in [128, 129] as a common generalization of MV-algebras and generalized Boolean algebras, can be viewed as hoops with specific properties. Studying EMV-algebras can enhance our understanding of hoop theory.

3.4   Exercises

3.4.1 Let $\mathbf{H} = (H; \odot, \rightarrow, 1)$ be a pocrim and

$$X := \{(x \odot (x \rightarrow y)) \rightarrow (y \odot (y \rightarrow x)) : x, y \in H\}.$$

Prove that $\mathbf{H}/F$ is a hoop, where $F = \langle X \rangle$.

3.4.2 Let $\mathbf{H}$ be a linear hoop and $a \in H$ be an idempotent element. Prove that $a \rightarrow x = x$ for all $x \in H$ such that $x < a$. Deduce that if $\mathbf{H}$ is a linear Wajsberg hoop, then either $\mathbf{H}$ is bounded and $x \in \{0, 1\}$ or $\mathbf{H}$ is unbounded and $x = 1$.

3.4.3 Let $\mathbf{H}_1$ and $\mathbf{H}_2$ be two hoops such that $H_1 \cap H_2 = \{1\}$ and $a \in C_c(\mathbf{H}_2)$. Prove that $C_c(\mathbf{H}_1 \oplus \mathbf{H}_2) = C_c(\mathbf{H}_2)$

3.4.4 A bounded hoop $\mathbf{H}$ is **special** if $(x \rightarrow y)' = (y \rightarrow x)'$ for all $x, y \in H$. Prove that:

(i) $\mathbf{H}$ is special if and only if $x' = 0$ for all $x \in H \setminus \{0\}$.
(ii) If $\mathbf{H}$ is special, then $x \odot y \neq 0$ for all $x, y \in H \setminus \{0\}$.
(iii) If $\mathbf{H}$ is special and $x, y \in H$ are covers of 0, then $x = y$.

The hoop in Example 3.3.12 is special (special hoops were defined in [227]).

3.4.5 Let $\mathbf{H}$ be a non-trivial bounded hoop and $H^* := H \setminus \{0\}$. Prove that:

(i)   If $\mathbf{H}$ is a special hoop, then H^* is a subuniverse of $\mathbf{H}$.
(ii) $\mathbf{H}$ is special if and only if $H = \mathbf{2} \oplus \mathbf{H}^*$.
(iii) Let C be the category whose objects are non-trivial special hoops and whose morphisms are homomorphisms of bounded hoops. Show that categories C and $\mathcal{H}oop$ are isomorphic.
      Hint: Let $\mathcal{F} : \mathcal{H}oop \rightarrow C$ be defined by $\mathcal{F}(\mathbf{H}) = \mathbf{2} \oplus \mathbf{H}$ and $\mathcal{G} : C \rightarrow \mathcal{H}oop$ be defined by $\mathcal{G}(\mathbf{H}) = \mathbf{H}^*$. Verify that $\mathcal{F}$ and $\mathcal{G}$ are functors, $\mathcal{F} \circ \mathcal{G}$ is the identity functor on C, and $\mathcal{G} \circ \mathcal{F}$ is the identity functor on $\mathcal{H}oop$, which means categories C and $\mathcal{H}oop$ are isomorphic.

3.4.6 Let $(L, \leq)$ be a poset. Then element $x \in L$ is called a **node**, if for each $y \in L, x \leq y$ or $y \leq x$. Let $\mathbf{H}$ be a bounded hoop, and denote by Node($\mathbf{H}$) the set of all nodes of $\mathbf{H}$. Prove the following statements:

(i)   If $B(\mathbf{H}) = \{x \in H : \exists y \in H \, \text{st.} \, x \vee y = 1 \, \text{and} \, x \wedge y = 0\}$, then Node($\mathbf{H}$) $\cap$ $B(\mathbf{H}) = \{0, 1\}$.
(ii) If $\mathbf{H}$ is a $\vee$-hoop, then (Node($\mathbf{H}$), $\vee, \wedge, 0, 1$) is a bounded distributive lattice.
(iii) If $x^2 = x$ for all $x \in H$, then Node($\mathbf{H}$) $\cap$ Reg($\mathbf{H}$) $= \{0, 1\}$.

3.4.7 Prove that generalized Boolean algebras (see Definition 1.3.32) and idempotent Wajsberg hoops are termwise equivalent.

3.4.8 Let $\mathbf{H}$ be a $\vee$-hoop. Show that the following statements hold:

  (i) $x \odot y = (x \vee y) \odot (x \wedge y)$ for all $x, y \in H$.
  (ii) $(x \to y) \odot (z \to w) \leq (x \vee z) \to (y \vee w)$ for all $x, y, z, w \in H$.
  (iii) $(x \to y) \to z \leq ((y \to x) \to z) \to z$ for all $x, y, z \in H$.
  (iv) If $w \leq (x \odot y) \vee z$, then $(x \vee z) \odot (y \vee z \vee (x \to w)) \leq (x \odot y) \vee z$ for all $x, y, z, w \in H$.
  (v) $(x \vee z) \odot (y \vee z \vee x') \leq (x \odot y) \vee z$ for all $x, y, z \in H$.

3.4.9 Let $a < b$ be elements of $\mathbb{R}$. Define the binary operations $\odot$ and $\to$ on the real interval $[a, b]$ as follows:

$$x \odot y := \max(x + y - b, a) \quad \& \quad x \to y := \min(b - x + y, b).$$

Prove that $([a, b]; \odot, \to, a, b)$ is a Wajsberg hoop. In addition, the map $f : [0, 1] \to [a, b]$ defined by $f(x) = a + (b - a)x$ is an hoop isomorphism from $([0, 1]; \odot, \to, 0, 1)$ to $([a, b]; \odot, \to, a, b)$ (see Example 3.3.11). The map $g : [a, b] \to [0, 1]$ defined by $g(x) = (x - a)/(b - a)$ is the inverse of f.

3.4.10 Prove that $\mathbf{L}_n \cong \mathbb{Z}[\mathbf{n}]$ for all $n \in \mathbb{N}$ (Hint: see Remark 3.3.58). In addition, show that for every $m, n \in \mathbb{N}$, $\mathbf{L}_m$ is a subhoop of $\mathbf{L}_n$ if and only if $m \mid n$.

3.4.11 Prove that Boolean algebras are termwise equivalence to bounded $\vee$-hoops with (DNP) satisfying $x^2 = x$ for any $x \in H$.

3.4.12 Let $\mathbf{H} = (H; \odot, \to, 0, 1)$ be a bounded $\vee$-hoop and $a, b \in \mathrm{B}(\mathbf{H})$. Show that:

  (i) If $a \leq x$, then $a' \to x = x$ for all $x \in H$.
  (ii) $a \to x = a \to (a \to x)$ for all $x \in H$.
  (iii) $a \to (x \to y) = (a \to x) \to (a \to y)$ for all $x, y \in H$.
  (iv) $a' \to x = a \vee x$ for all $x \in H$.

3.4.13 Let X be a subset of a hoop $\mathbf{H}$. The **right stabilizer** and the **left stabilizer** of X are defined as follows, respectively (see [38]):

$$\mathrm{St}_r(X) := \{a \in H : a \to x = x, \ \forall x \in X\} \quad \& \quad \mathrm{St}_l(X) := \{a \in H : x \to a = a, \ \forall x \in X\}.$$

Also, the **product stabilizer** of X is defined by

$$\mathrm{St}_\odot(X) := \{a \in A : a \odot x = x, \ \forall x \in X\}.$$

Prove that:

(i) If $X \in \mathcal{F}(\mathbf{H})$, then $\mathrm{St}_\odot(X) \subseteq X$.

(ii) If $\mathbf{H}$ is a basic hoop and $a \in A$ such that $a \vee x = 1$ for all $x \in X$, then $a \in \mathrm{St}_\mathrm{l}(X) \cap \mathrm{St}_\mathrm{r}(X)$.

(iii) If $\mathbf{H}$ is a bounded hoop with (DNP), then $\mathrm{St}_\mathrm{r}(X) = \mathrm{St}_\mathrm{l}(X)$.

(iv) $\mathrm{St}_\odot(X)$ and $\mathrm{St}_\mathrm{r}(X)$ are filters of $\mathbf{H}$. Show that $\mathrm{St}_\mathrm{l}(X)$ may fail to be a filter.

Filters and Congruences of Hoops

4

This chapter focuses on the crucial role of congruences in the analysis of hoop structures, exploring their close relationship with quotient algebras and homomorphisms. The investigation of hoop congruences is organized into four sections. Initially, we establish the connection between filters and congruences in hoops, highlighting their one-to-one correspondence, a property distinct from pocrims. The second section delves into the properties of prime, maximal, and ultrafilters, laying the groundwork for later investigations into Wajsberg, basic, simple, semisimple, and perfect hoops. Prime filters are also tied to the representation of these algebras. We define two distinct types of prime filters, which are equivalent in basic hoops. We examine the hierarchical relationship between these filters and address the existence of maximal filters, particularly in bounded hoops. The third section explores implicative, positive implicative, fantastic, and involutive filters, revealing their connections and the nature of their associated quotient algebras. The study of these filters is important, as each corresponds to a specific type of congruence. Quotient hoops associated with implicative filters are idempotent hoops, while quotient hoops associated with fantastic filters are Wajsberg hoops. In addition, the notions of an involutive filter and a fantastic filter are equivalent within the context of bounded hoops. The study of these filters is important since each of them corresponds to a special type of congruence. Finally, we investigate ideals in bounded hoops. It is well-established that filters and ideals are dual notions in lattices; however, this duality does not extend to hoops. Nevertheless, in bounded hoops with the Double Negation Property (DNP), a subset F is a filter if and only if F' is an ideal. We then present several examples and compile the principal properties of ideals in bounded hoops.

© The Author(s), under exclusive license to Springer Nature Switzerland AG 2026

A. Dvurečenskij et al., *Hoop Algebras*, Frontiers in Mathematics,
https://doi.org/10.1007/978-3-032-11736-6_4

4.1   Filters

Building on the results from Sect. 2.3, which explored filters and congruence relations in pocrims, we observed that the quotient structure induced by a general congruence relation on a pocrim (denoted **H**) is not necessarily a pocrim itself. Consequently, we introduced the concept of filter-congruence relations, establishing a one-to-one correspondence with filters of **H**. Now, we turn our attention to filters and congruence relations within the context of hoops. In this section, we will demonstrate that for hoops, the notions of congruence and filter-congruence coincide.

Definition 4.1.1 A non-empty subset F of a hoop $\mathbf{H} = (H; \odot, \rightarrow, 1)$ is a filter if it satisfies the conditions (F1) and (F2) of Definition 2.3.1.

By definition, F is a filter of the hoop **H** if and only if it is a filter of the pocrim **H**. Consequently, for each subset $X \subseteq H$ the least filter of the hoop $(H; \odot, \rightarrow, 1)$ generated by X coincides with the least filter of the pocrim $(H; \odot, \rightarrow, 1)$ generated by X. Therefore, by Proposition 2.3.8, we have

$$\langle X \rangle = \bigcap \{F \in \mathcal{F}(\mathbf{H}) : X \subseteq F\} \tag{4.1}$$

$$= \{y \in H : x_1 \odot \cdots \odot x_n \leq y, \text{ for some } n \in \mathbb{N}, \ x_1, \ldots, x_n \in X\}. \tag{4.2}$$

Moreover, all the results from Sect. 2.3 are applicable to hoops in place of pocrims.

By Remark 2.3.7, $(\mathcal{F}(\mathbf{H}), \subseteq)$ is a complete lattice. The following example shows that the union of filters in a hoop does not necessarily form a filter (Fig. 4.1).

Example 4.1.2 Let $H = \{0, a, b, 1\}$ be a set with the following Hasse diagram. Define the operations $\odot$ and $\rightarrow$ on H in Table 4.1.

Then $(H; \odot, \rightarrow, 0, 1)$ is a bounded hoop. Obviously, $F_1 = \{a, 1\}$ and $F_2 = \{b, 1\} \in \mathcal{F}(\mathbf{H})$, but $F_1 \cup F_2 = \{a, b, 1\} \notin \mathcal{F}(\mathbf{H})$, because $a \odot b = 0 \notin F_1 \cup F_2$.

Example 4.1.3 ([229, Example 3.2]) Define the binary operations $\odot$ and $\rightarrow$ on the set $H = \{0, a, b, c, d, 1\}$ according to Table 4.2.

Fig. 4.1 Hasse diagram of the hoop in Example 4.1.2

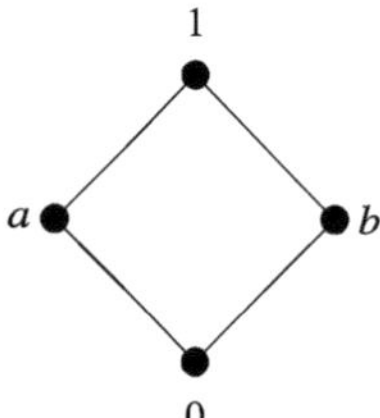

Table 4.1 Operations $\odot$ and $\rightarrow$ of Example 4.1.2

Table 4.1.1

$\odot$	0	a	b	1
0	0	0	0	0
a	0	a	0	a
b	0	0	b	b
1	0	a	b	1

Table 4.1.2

$\rightarrow$	0	a	b	1
0	1	1	1	1
a	b	1	b	1
b	a	a	1	1
1	0	a	b	1

Table 4.2 Operations $\odot$ and $\rightarrow$ of Example 4.1.3

Table 4.2.1

$\odot$	0	a	b	c	d	1
0	0	0	0	0	0	0
a	0	b	b	d	0	a
b	0	b	b	0	0	b
c	0	b	0	c	d	c
d	0	0	0	d	0	d
1	0	a	b	c	d	1

Table 4.2.2

$\rightarrow$	0	a	b	c	d	1
0	1	1	1	1	1	1
a	d	1	a	c	c	1
b	c	1	1	c	c	1
c	b	a	b	1	a	1
d	a	1	a	1	1	1
1	0	a	b	c	d	1

Then $\mathbf{H} = (H; \odot, \rightarrow, 0, 1)$ is a bounded hoop, where $0 < b < a < 1, 0 < d < a < 1$ and $0 < d < c < 1$. An easy calculation shows that $F = \{1, a, b\}$ is a filter of $\mathbf{H}$.

Example 4.1.4 Considering the proof of Theorem 3.3.13, we can easily check that a subset I of M is an ideal of the MV-algebra $\mathbf{M} = (M; \oplus, ', 0, 1)$ (see Sect. 1.4.1) if and only if $I' := \{x': x \in I\}$ is a filter of it corresponding bounded hoop $\mathbf{H} := (M; \odot, \rightarrow, 0, 1)$ and vise versa. In addition,

$$\mathcal{F}(\mathbf{H}) = \{I': I \in \mathcal{I}(\mathbf{M})\} \quad \& \quad \mathcal{I}(\mathbf{M}) = \{F': F \in \mathcal{F}(\mathbf{H})\}. \tag{4.3}$$

In addition, $A \subseteq \mathbf{H}$ is closed under $\odot$ and $\rightarrow$ if and only if A is closed under $\oplus$ and $'$, which means subalgebras of the hoop $\mathbf{H}$ are exactly subalgebras of the MV-algebra $\mathbf{M}$. In a similar way, $f : (H_1; \odot, \rightarrow, 0, 1) \rightarrow (H_2; \odot, \rightarrow, 0, 1)$ is a homomorphism between bounded hoops if and only if $f : (H_1; \oplus, ', 0, 1) \rightarrow (H_2; \oplus, ', 0, 1)$ is a homomorphism between MV-algebras.

Example 4.1.5 Consider the hoop $\mathbf{H} = \mathbf{N}(\mathbf{G})$ where $\mathbf{G} = (G; \vee, \wedge, +, -, 0)$ is an Abelian ℓ-group. By Example 3.3.8, it is a cancellative hoop (note that due to Proposition 3.3.19, every cancellative hoop is of the form $\mathbf{H} = \mathbf{N}(\mathbf{G})$ for some Abelian ℓ-group $\mathbf{G}$). Then

$$\mathcal{F}(\mathbf{H}) = \{\mathbf{K}^- : \mathbf{K} \text{ is a convex } \ell \text{-subgroup of } \mathbf{G}\}.$$

Indeed, choose a convex ℓ-subgroup $\mathbf{K}$ of $\mathbf{G}$. Given $x, y \in H$. If $x, y \in \mathbf{K}^-$, then $x \odot y = x + y \in K$ and $x + y \le 0$, entails that (F1) holds. In addition, if $y \ge x \in \mathbf{K}^-$, then $x \le y \le 0$, and so $y \in \mathbf{K}^-$ (since $\mathbf{K}$ is convex), consequently, (F2) holds. It follows that $\mathbf{K}^-$ is a filter of $\mathbf{H}$.

In addition, for each filter F of $\mathbf{H}$, $(F; +, 0)$ is a cancellative lattice ordered submonoid of the monoid of $\mathbf{G}$. It suffices to set $K := \{x - y : x, y \in F\}$. We can easily verify that $\mathbf{K}$ is a convex ℓ-subgroup of G and $\mathbf{K}^- = F$. Therefore, $F(\mathbf{H}) = \{\mathbf{K}^- : \mathbf{K} \text{ is a convex } \ell \text{-subgroup of } \mathbf{G}\}$.

Proposition 4.1.6 *Let F be a filter of $\mathbf{H}$. Then F is closed with respect to $\wedge$.*

Proof It follows from (F2) and Proposition 3.1.4(i). $\qquad\qquad\qquad\qquad\qquad\square$

Example 4.1.7 Consider an Abelian ℓ-group $(G; \vee, \wedge, +, -, 0)$ and $u \in \mathbf{G}^+$. By Example 3.3.57, $\mathbf{G}[\mathbf{u}]$ is a hoop. If K is a convex ℓ-subgroup of G, then $F = \{x \in G[u] : u - x \in K\}$ is a filter of $\mathbf{G}[\mathbf{u}]$. Indeed, firstly, let us remark that, since K is normal, if $u - x \in K$, then $u = (u - x) + x \in K + x = x + K$, hence $-x + u \in K$. That is, $F = \{x \in G[u] : -x + u \in K\}$. We have $u \in F$, since $u - u = 0 \in K$. Let $x \in F$, $y \in G[u]$ such that $x \le y$. Since K is convex, $0 \le u - y \le u - x$, and $0, u - x \in K$ it follows that $u - y \in K$, hence $y \in F$. Let $x, y \in F$, that is $u - x, u - y \in K$. We get that

$$0 \le u - (x \odot y) = u - ((x - u + y) \vee 0) = (u - y + u - x) \wedge u \le (u - y) + (u - x) \in K,$$

so $u - (x \odot y) \in K$, hence $x \odot y \in F$. Therefore, F is a filter of $\mathbf{G}[\mathbf{u}]$.

Proposition 4.1.8 ([154, Proposition 3.4]) *Let $\mathbf{H}$ be a hoop and $x, y \in H$. If $x \vee y$ exists, then $\langle x \vee y \rangle = \langle x \rangle \cap \langle y \rangle$.*

Proof It is obvious that $x \in \langle x \rangle$ and $y \in \langle y \rangle$. Since $x, y \le x \vee y$, it follows that $x \vee y \in \langle x \rangle$ and $x \vee y \in \langle y \rangle$, so $x \vee y \in \langle x \rangle \cap \langle y \rangle$. Hence, $\langle x \vee y \rangle \subseteq \langle x \rangle \cap \langle y \rangle$. Conversely, let $z \in \langle x \rangle \cap \langle y \rangle$. Due to Corollary 2.3.9, there are $n, m \in \mathbb{N}$ such that $x^n \le z$ and $y^m \le z$. By Proposition 3.2.9(ii), we get that

$$(x \vee y)^{n+m} \to z = \bigwedge \{(a_{n+m} \odot \cdots \odot a_1) \to z : a_i \in \{x, y\}\}. \tag{4.4}$$

Consider $a_1, \ldots, a_{n+m} \in \{x, y\}$. Denote by r the number of occurrences of x in the sequence $a_1, \ldots, a_{n+m}$ and by s the number of occurrences of y in the sequence $a_1, \ldots, a_{n+m}$. Of course, $r + s = n + m$. By Lemma 2.2.1(4), we have $a_1 \odot \cdots \odot a_{n+m} \le x^r$ and $a_1 \odot \cdots \odot a_{n+m} \le y^s$. Applying Lemma 2.2.1(7), it follows that $x^r \to z \le (a_1 \odot \cdots \odot a_{n+m}) \to z$ and $y^s \to z \le (a_1 \odot \cdots \odot a_{n+m}) \to z$. If $r \le n$, then $s \ge m$, so $y^s \le y^m$. Applying again Lemma 2.2.1(7), we get that $y^s \to z \ge y^m \to z = 1$, since $y^m \le z$. Hence, $y^s \to z = 1$, so $(a_1 \odot \cdots \odot a_{n+m}) \to z = 1$. Similarly, if $r > n$, then $x^r \le x^n$, so $x^r \to z \ge x^n \to z = 1$, since $x^n \le z$. It follows that $x^r \to z = 1$, hence $(a_1 \odot \cdots \odot a_{n+m}) \to z = 1$. Therefore, $(a_1 \odot \cdots \odot a_{n+m}) \to z = 1$ for any $a_1, \ldots, a_{n+m} \in \{x, y\}$, hence by (4.4), $(x \vee y)^{n+m} \to z = 1$. It follows that $(x \vee y)^{n+m} \le z$, consequently, $z \in \langle x \vee y \rangle$. Therefore, $\langle x \vee y \rangle = \langle x \rangle \cap \langle y \rangle$. $\qquad \square$

Proposition 4.1.9 ([154, Proposition 3.4]) *Let* $\mathbf{H}$ *be a* $\vee$-*hoop,* $F \in \mathcal{F}(\mathbf{H})$ *and* $x, y \in H$. *Then*

$$\langle F \cup \{x \vee y\} \rangle = \langle F \cup \{x\} \rangle \cap \langle F \cup \{y\} \rangle. \tag{4.5}$$

Proof Let $a \in \langle F \cup \{x \vee y\} \rangle$. By Corollary 2.3.10(ii), there exists $k \in \mathbb{N}$ such that $(x \vee y)^k \to a \in F$. Since $x, y \le x \vee y$ and $(H; \odot, 1)$ is a commutative partially ordered monoid, $x^k \le (x \vee y)^k$ and $y^k \le (x \vee y)^k$, so by Lemma 2.2.1(7), $(x \vee y)^k \to a \le x^k \to a$ and $(x \vee y)^k \to a \le y^k \to a$, entails that $(x \vee y)^k \to a \in F$ (since $F \in \mathcal{F}(\mathbf{H})$), consequently, $x^k \to a \in F$ and $y^k \to a \in F$. Hence, $a \in \langle F \cup \{x\} \rangle \cap \langle F \cup \{y\} \rangle$, which implies $\langle F \cup \{x \vee y\} \rangle \subseteq \langle F \cup \{x\} \rangle \cap \langle F \cup \{y\} \rangle$.

Conversely, choose $z \in \langle F \cup \{x\} \rangle \cap \langle F \cup \{y\} \rangle$. Then by Corollary 2.3.10(ii), there exists $m, n \in \mathbb{N}$ such that $x^n \to z \in F$ and $y^m \to z \in F$. By Proposition 3.2.9(ii), we get that $(x \vee y)^n \to z = \bigwedge \{(a_n \odot \cdots \odot a_1) \to z : a_i \in \{x, y\}\}$. Consider $a_1, \ldots, a_{n+m} \in \{x, y\}$. Denote by r the number of occurrences of x in the sequence $a_1, \ldots, a_{n+m}$ and by s the number of occurrences of y in the sequence $a_1, \ldots, a_{n+m}$. Of course, $r + s = n + m$. We have that $a_1 \odot \cdots \odot a_{n+m} \le x^r$ and $a_1 \odot \cdots \odot a_{n+m} \le y^s$. Applying Lemma 2.2.1(7), it follows that $x^r \to z \le (a_1 \odot \cdots \odot a_{n+m}) \to z$ and $y^s \to z \le (a_1 \odot \cdots \odot a_{n+m}) \to z$. If $r \le n$, then $s \ge m$, so $y^s \le y^m$. Applying again Lemma 2.2.1(7), we get that $y^m \to z \le y^s \to z$, since $y^m \to z \in F$ and $F \in \mathcal{F}(\mathbf{H})$. Hence, $y^s \to z \in F$, so $(a_1 \odot \cdots \odot a_{n+m}) \to z \in F$. Similarly, if $r > n$, then $x^r \le x^n$, so $x^n \to z \le x^r \to z$, since $x^n \to z \in F$ and $F \in \mathcal{F}(\mathbf{H})$. It follows that $x^r \to z \in F$, hence $(a_1 \odot \cdots \odot a_{n+m}) \to z \in F$, consequently, $(a_1 \odot \cdots \odot a_{n+m}) \to z \in F$ for any $a_1, \ldots, a_{n+m} \in \{x, y\}$, which implies $(x \vee y)^{n+m} \to z \in F$. Thus, $z \in \langle F \cup \{x \vee y\} \rangle$. Therefore, $\langle F \cup \{x\} \rangle \cap \langle F \cup \{y\} \rangle \subseteq \langle F \cup \{x \vee y\} \rangle$. $\qquad \square$

Definition 4.1.10 Let $\mathbf{H}$ be a hoop. For each $x \in H$, we define the **polar** $x^{\perp} := \{y \in H : x \vee y = 1\}$.

Remark 4.1.11 Note that the concept of polarity can be extended to pocrims in a similar manner. Moreover, for any pocrim $(H; \odot, \to, 1)$ and elements $x, y \in H$, we have $x, y \le (x \to y) \to y$. Consequently, if $y \in x^{\perp}$, then $(x \to y) \to y = 1 = (y \to x) \to x$.

Example 4.1.12 Given an element x of a $\vee$-hoop $\mathbf{H}$, we have $1 \in x^{\perp}$. If $y, z \in x^{\perp}$, then by Proposition 3.2.8, $1 = (x \vee y) \odot (x \vee z) \leq x \vee (y \odot z)$, entails that $y \odot z \in x^{\perp}$, so (F1) holds. In addition, for each $z \geq y \in x^{\perp}$ we have $1 = y \vee x \leq z \vee x$, so $z \in x^{\perp}$, i.e., (F2) holds. Therefore, $x^{\perp}$ is a filter of $\mathbf{H}$.

Proposition 4.1.13 ([154, Proposition 3.14]) *Let $\mathbf{H}$ be a hoop and $\theta \in \mathrm{Con}(\mathbf{H})$. Then $\Theta_F = \theta$ where $F = 1/\theta$.*

Proof First, we note that by Proposition 2.4.1(ii), $F = 1/\theta$ is a filter of $\mathbf{H}$. Choose $\theta \in \mathrm{Con}(\mathbf{H})$. Let $(x, y) \in \theta$. Then $(x \to y, 1) = (x \to y, x \to x) \in \theta$ and $(y \to x, 1) = (y \to x, y \to y) \in \theta$, which means $x \to y, y \to x \in 1/F$. Hence, $(x, y) \in \Theta_F$. Conversely, assume that $(x, y) \in \Theta_F$. Then $(x \to y, 1) \in \theta$ and $(y \to x, 1) \in \theta$. It follows that

$$y = y \odot 1 = y \odot (y \to x) = x \wedge y = \big(x \odot (x \to y)\big)\theta(x \odot 1) = x,$$

entails that $(x, y) \in \theta$. Therefore, $\Theta_F = \theta$. $\qquad\square$

Proposition 4.1.13 shows that if $\mathbf{H}$ is a hoop, then $\mathrm{Con}(\mathbf{H}) = \mathrm{Con}_F(\mathbf{H})$.

Theorem 4.1.14 ([154, Proposition 3.15]) *The map $f : \mathcal{F}(\mathbf{H}) \to \mathrm{Con}(\mathbf{H})$ defined by $f(F) = \Theta_F$ is an isomorphism between the lattice of filters of $\mathbf{H}$ and the lattice of congruences of $\mathbf{H}$. In addition, the map $g : \mathrm{Con}(\mathbf{H}) \to \mathcal{F}(\mathbf{H})$ defined by $g(\theta) = 1/\theta$ is its inverse.*

Proof The proof follows from Theorem 2.4.1(iii) and Proposition 4.1.13. In addition, $F_1 \subseteq F_2$ if and only if $\Theta_{F_1} \subseteq \Theta_{F_2}$ for all $F_1, F_2 \in \mathcal{F}(\mathbf{H})$. $\qquad\square$

Let $\mathbf{H}$ be a hoop and $F \in \mathcal{F}(\mathbf{H})$. Recall from Theorem 2.4.2 that $\mathbf{H}/F := (H/F; \odot, \to, 1/F)$ is a pocrim. The next theorem will show that $\mathbf{H}/F$ is a hoop.

Theorem 4.1.15 ([43, Lemma 3.1]) *Let F be a filter of a hoop $\mathbf{H}$. Then $\mathbf{H}/F$ is a hoop. In addition, $x \to y \in F$ if and only if $x/F \leq y/F$ for all $x, y \in H$.*

Proof Considering Theorem 2.4.2, it suffices to show that $\mathbf{H}/F$ is a naturally ordered pocrim. Choose $x, y \in H$ such that $x/F \leq y/F$. By Theorem 2.4.2, $x \to y = z \in F$, entails that $z \odot x \leq y$. By the assumption, there exists $w \in H$ such that $z \odot x = y \odot w$. It follows that

$$\frac{x}{F} = \frac{1}{F} \odot \frac{x}{F} = \frac{z}{F} \odot \frac{x}{F} = \frac{z \odot x}{F} = \frac{y \odot w}{F} = \frac{y}{F} \odot \frac{w}{F}.$$

$\qquad\square$

Theorem 4.1.16 *Let F be a filter of a $\vee$-hoop $\mathbf{H}$. Then $\mathbf{H}/F$ is a $\vee$-hoop, and $x/F \vee y/F = (x \vee y)/F$ for all $x, y \in H$.*

Proof Choose $x, y \in H$. From $x \to (x \vee y) = 1 \in F$, $y \to (x \vee y) \in F$ and Theorem 2.4.2 it follows that $(x \vee y)/F$ is an upper bound for x/F and y/F. Let $x/F, y/F \leq z/F$ for some $z \in H$. Then $x \to z = w_1 \in F$ and $y \to z = w_2 \in F$. Hence, $(x \to z) \wedge (y \to z) = w_1 \wedge w_2 \in F$, by Proposition 4.1.6. On the other hand, by Proposition 3.2.9(i), $(x \to z) \wedge (y \to z) = (x \vee y) \to z$, consequently, $(x \vee y)/F \leq z/F$. Thus, $(x \vee y)/F$ is the least upper bound of x/F and y/F. Therefore, $\mathbf{H}/F$ is a $\vee$-hoop. $\qquad\square$

Let $\mathbf{H}$ be a $\vee$-hoop. Theorem 4.1.16 also shows that the operation $\vee$ on $\mathbf{H}/F$ defined by $x/F \vee y/F = (x \vee y)/F$ is well-defined. Of course, there is another way to show that $\vee$ is well-defined on $\mathbf{H}/F$. It suffices to show that Θ_F is compatible with $\vee$. Assume that $(x_1, x_2), (y_1, y_2) \in \Theta_F$ for some $x_1, x_2, y_1, y_2 \in H$. Applying Proposition 3.2.9(i), we get that

$$\begin{aligned}
(x_1 \vee y_1) \to (x_2 \vee y_2) &= \big(x_1 \to (x_2 \vee y_2)\big) \wedge \big(y_1 \to (x_2 \vee y_2)\big) \\
&\geq \big((x_1 \to x_2) \vee (x_1 \to y_2)\big) \wedge \big((y_1 \to x_2) \vee (y_1 \to y_2)\big), \text{ by Lemma 2.2.1 (7)} \\
&\geq (x_1 \to x_2) \wedge (y_1 \to y_2) \in F, \text{ by Proposition 4.1.6.}
\end{aligned}$$

Hence, $(x_1 \vee y_1) \to (x_2 \vee y_2) \in F$. In a similar way, we can show that $(x_2 \vee y_2) \to (x_1 \vee y_1) \in F$, which implies $(x_1 \vee y_1, x_2 \vee y_2) \in \Theta_F$. Therefore, Θ_F is compatible with $\vee$.

Theorem 4.1.17 *Let F be a filter of a hoop $\mathbf{H}$.*

(i) $\mathcal{F}(\mathbf{H}/F) = \{G/F : G \in \mathcal{F}(\mathbf{H}), \ F \subseteq G\}$ *where* $X/F = \{x/F : x \in X\}$ *for all subsets X of H.*

(ii) $G_1/F = G_2/F$ *if and only if $G_1 = G_2$ for each filters G_1, G_2 of $\mathbf{H}$ containing F.*

(iii) $(\mathbf{H}/F)/(G/F) \cong \mathbf{H}/G$ *for each $F \subseteq G \in \mathcal{F}(\mathbf{H})$.*

Proof (i) The proof is left to the reader as an exercise.

(ii) Let $G_1/F = G_2/F$. Choose $x \in G_1$, then there exists $y \in G_2$ such that $x/F = y/F$, equivalently, $x \to y, y \to x \in F \subseteq G_2$. Now, by Lemma 2.3.2, it follows that $x \in G_2$, entails that $G_1 \subseteq G_2$. A similar proof shows the reverse inclusion. Therefore, $G_1 = G_2$.

(iii) By (i), G/F is a filter of $\mathbf{H}/F$. Define $f : \mathbf{H}/F \to \mathbf{H}/G$ by $f(x/F) = x/G$. Since $F \subseteq G$, f is well-defined. Also, f preserves $\odot$ and $\to$, clearly, so by the first isomorphism theorem (see Theorem 1.2.16), $(\mathbf{H}/F)/\operatorname{Ker}(f) \cong f(\mathbf{H}) = \mathbf{H}/G$. An easy calculation shows that $\operatorname{Ker}(f) = G/F$, which means $(\mathbf{H}/F)/(G/F) \cong \mathbf{H}/G$. $\qquad\square$

Theorem 4.1.17 remarkably extends to hold true for general pocrims as well.

4.2   Prime and Maximal Filters

Prime and maximal filters are crucial types of filters in hoops. They play a significant role in representing and characterizing various subclasses of hoops.

4.2.1   Prime Filters

Definition 4.2.1 Let **H** be a hoop. A proper filter P of **H** is called a **prime filter** if $x \to y \in P$ or $y \to x \in P$ for each $x, y \in H$. The set of all prime filters of H is denoted by Spec(**H**).

Example 4.2.2 (i) According to Example 4.1.2, $P_1 = \{a, 1\}$ and $P_2 = \{b, 1\}$ are prime filters of **H** but $P_3 = \{1\} \notin \text{Spec}(\mathbf{H})$ since $a \to b = b \notin P_3$.

(ii) If **H** is a linear hoop, then every filter of **H** is prime, since $x \to y = 1$ or $y \to x = 1$ for all $x, y \in H$.

(iii) Let **H** be an arbitrary hoop. Then $F = H$ is a prime filter of $2 \oplus \mathbf{H}$ (see Example 3.3.66). An easy calculation shows that F is a proper filter of $2 \oplus \mathbf{H}$. It suffices to show that $x \to y \in F$ or $y \to x \in F$ for all $x, y \in H$. Choose $x, y \in H \cup \{0\}$. If $x, y \in H$, then $x \to y, y \to x \in F$. If $x = 0 \in 2$, then $x \to y = 1 \in F$ for all $y \in H$. Clearly, $x = y = 0$ entails that $x \to y \in F$. Therefore, F is a prime filter.

Example 4.2.3 Consider $H = \{0, a, b, c, d, 1\}$ is a lattice with the Hasse diagram in Fig. 4.2. Define the operations $\odot$ and $\to$ on H as Table 4.3.

Then $(H; \odot, \to, 0, 1)$ is a bounded hoop and $F = \{d, 1\}$ is a filter of **H**. In addition, $c \to d = a \to b = a \to d = d \in F$ and $a \to a = a \to c = a \to 1 = b \to d = b \to 1 = b \to b = b \to c = c \to c = c \to 1 = d \to d = 0 \to x = 1 \in F$ for all $x \in H$. Therefore, $x \to y \in F$ or $y \to x \in F$ for all $x, y \in H$, which implies F is prime.

Remark 2.3.7 established that the poset $(\mathcal{F}(\mathbf{H}), \subseteq)$ forms a complete lattice for any hoop **H**. The following proposition demonstrates that every prime filter of **H** is a meet-irreducible element of this lattice (see Definition 1.3.13).

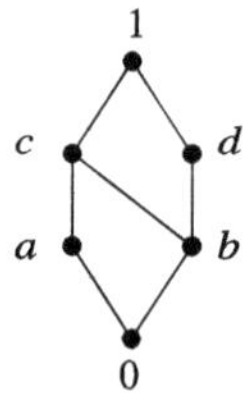

Fig. 4.2 Hasse diagram of the hoop in Example 4.2.3

Table 4.3 Operations $\odot$ and $\to$ of Example 4.2.3

Table 4.3.1

$\odot$	0	a	b	c	d	1
0	0	0	0	0	0	0
a	0	0	a	a	a	a
b	0	a	b	b	b	b
c	0	a	b	c	b	c
d	0	a	b	b	d	d
1	0	a	b	c	d	1

Table 4.3.2

$\to$	0	a	b	c	d	1
0	1	1	1	1	1	1
a	a	1	d	1	d	1
b	0	a	1	1	1	1
c	0	a	d	1	d	1
d	0	a	c	c	1	1
1	0	a	b	c	d	1

Proposition 4.2.4 ([169]) *Let* $\mathbf{H}$ *be a hoop and* $P \in \mathrm{Spec}(\mathbf{H})$. *Then*

(i) $\Lambda = \{F \in \mathcal{F}(\mathbf{H}) : P \subseteq F \neq H\}$ *is a chain in the poset* $(\mathrm{Spec}(\mathbf{H}), \subseteq)$.

(ii) P *is a meet-irreducible element in the lattice* $(\mathcal{F}(\mathbf{H}), \subseteq)$, *that is* $I \cap J = P$ *implies that* $I = P$ *or* $J = P$ *for each* $I, J \in \mathcal{F}(\mathbf{H})$.

Proof (i) Let G and F be two filters of $\mathbf{H}$ containing P. Clearly, by definition, $G, F \in \mathrm{Spec}(\mathbf{H})$. Let $x \in F \setminus G$. For each $y \in G$ we have $y \to x \in H \setminus P$, otherwise, $y \to x \in G$, Lemma 2.3.2(ii), implies that $x \in G$, which is absurd. It follows that $x \to y \in P$ for all $y \in G$, consequently, $x \to y \in F$, and so $y \in F$ for all $y \in G$, that is $G \subseteq F$. Therefore, Λ is a chain.

(ii) Let $I, J \in \mathcal{F}(\mathbf{H})$ such that $I \cap J = P$. Assume that $x \in I \setminus P$. For each $y \in J$ we have $y \to x \notin P$, otherwise, $y \to x \in P \subseteq J$ implies that $x \in J$ (by Lemma 2.3.2(ii)), consequently, $x \in I \cap J = P$, a contradiction. It follows from the assumption that $x \to y \in P \subseteq I$ for all $y \in J$, which entails $y \in I$ for all $y \in J$, i.e., $J \subseteq I$. Therefore, $J = I \cap J = P$. $\qquad\square$

Definition 4.2.5 Let P be a proper filter of a $\vee$-hoop $\mathbf{H}$. Then P is called a $\vee$**-prime filter** if $x \vee y \in P$ implies $x \in P$ or $y \in P$ for any $x, y \in H$. The set of all $\vee$-prime filters of H is denoted by $\mathrm{Spec}_\vee(\mathbf{H})$.

Example 4.2.6 Assume $(H = \{0, a, b, c, 1\}, \leq)$ is a poset where $0 < c < a, b < 1$. Define the operations $\odot$ and $\to$ on H by Table 4.4.

Then $(H; \odot, \to, 0, 1)$ is a bounded hoop. Obviously, $F = \{b, 1\}$ is a $\vee$-prime filter of H but $\{1\}$ is not, since $a \vee b = 1 \in \{1\}$ but $a \notin \{1\}$ and $b \notin \{1\}$.

Table 4.4 Operations $\odot$ and $\rightarrow$ of Example 4.2.6

Table 4.4.1

$\odot$	0	c	a	b	1
0	0	0	0	0	0
c	0	c	c	c	c
a	0	c	a	c	a
b	0	c	c	b	b
1	0	c	a	b	1

Table 4.4.2

$\rightarrow$	0	c	a	b	1
0	1	1	1	1	1
c	0	1	1	1	1
a	0	b	1	b	1
b	0	a	a	1	1
1	0	c	a	b	1

Proposition 4.2.7 ([37, Proposition 3.9]) *Let* $\mathbf{H}$ *be a hoop and* $P \in \mathrm{Spec}(\mathbf{H})$.

(i) *If* $x \vee y$ *exists and* $x \vee y \in P$ *for some* $x, y \in H$, *then* $x \in P$ *or* $y \in P$.

(ii) *If* $\mathbf{H}$ *is a* $\vee$-*hoop, then* $\mathrm{Spec}(\mathbf{H}) \subseteq \mathrm{Spec}_\vee(\mathbf{H})$.

Proof (i) Let $x, y \in H$ such that $x \vee y$ exists and $x \vee y \in P$. Due to Lemma 2.2.1 parts (6) and (10), $x \vee y \le (x \rightarrow y) \rightarrow y, (y \rightarrow x) \rightarrow x$, then by (F2), $(x \rightarrow y) \rightarrow y \in P$ and $(y \rightarrow x) \rightarrow x \in P$. Moreover, by the assumption, $x \rightarrow y$ or $y \rightarrow x \in P$ for all $x, y \in H$. Since $(x \rightarrow y) \rightarrow y \in P$, if $x \rightarrow y \in P$, then by Proposition 2.3.6(ii), $y \in P$. In a similar way, since $(y \rightarrow x) \rightarrow x \in P$, if $y \rightarrow x \in P$, then $x \in P$.

(ii) The proof follows from (i) immediately. $\square$

Proposition 4.2.8 ([37, Proposition 3.8]) *Let* $\mathbf{H}$ *be a* $\vee$-*hoop and* P *be a proper filter of* $\mathbf{H}$. *Then* $P \in \mathrm{Spec}_\vee(\mathbf{H})$ *if and only if for any two filters* I *and* J *of* $\mathbf{H}$, *if* $I \cap J \subseteq P$, *then* $I \subseteq P$ *or* $J \subseteq P$.

Proof Let $P \in \mathrm{Spec}_\vee(\mathbf{H})$. If $J \nsubseteq P$ and $I \nsubseteq P$, then there are $x \in I$ and $y \in J$ such that $x, y \notin P$. Since $x, y \le x \vee y$ and $I, J \in \mathcal{F}(\mathbf{H})$, then by (F2), $x \vee y \in I \cap J \subseteq P$, and so $x \vee y \in P$. Also, $P \in \mathrm{Spec}_\vee(\mathbf{H})$, so $x \in P$ or $y \in P$, which is a contradiction.

Conversely, let P be a proper filter of $\mathbf{H}$ such that $x \vee y \in P$ for some $x, y \in H$. By Proposition 4.1.8, $\langle x \rangle \cap \langle y \rangle = \langle x \vee y \rangle \subseteq P$. Then by assumption, $\langle x \rangle \subseteq P$ or $\langle y \rangle \subseteq P$. Since $x \in \langle x \rangle$ and $y \in \langle y \rangle$, then $x \in P$ or $y \in P$. Therefore, $P \in \mathrm{Spec}_\vee(\mathbf{H})$. $\square$

Corollary 4.2.9 ([37, Corollary 3.10]) *Let* $\mathbf{H}$ *be a* $\vee$-*hoop satisfying* (Pre). *Then* $\mathrm{Spec}_\vee(\mathbf{H}) = \mathrm{Spec}(\mathbf{H})$.

Proof By Proposition 4.2.7(ii), $\mathrm{Spec}(\mathbf{H}) \subseteq \mathrm{Spec}_{\vee}(\mathbf{H})$. Suppose that $F \in \mathrm{Spec}_{\vee}(\mathbf{H})$. Choose $x, y \in F$. From $(x \to y) \vee (y \to x) = 1 \in F$, we get that $x \to y \in F$ or $y \to x \in F$, which entails that $F \in \mathrm{Spec}(\mathbf{H})$. Therefore, $\mathrm{Spec}(\mathbf{H}) = \mathrm{Spec}_{\vee}(\mathbf{H})$. $\square$

Corollary 4.2.10 *Let* $\mathbf{H}$ *be a hoop. Then* $\mathbf{H}$ *is linearly ordered if and only if* $\{1\}$ *is a prime filter of* $\mathbf{H}$*. In addition, if* $\mathbf{H}$ *is linear, then* $\mathcal{F}(\mathbf{H}) = \mathrm{Spec}(\mathbf{H}) = \mathrm{Spec}_{\vee}(\mathbf{H})$*.*

Proof First, assume that $\mathbf{H}$ is a linear hoop. Since $\mathbf{H}$ satisfies prelinearity, Corollary 4.2.9 implies that $\mathrm{Spec}(\mathbf{H}) = \mathrm{Spec}_{\vee}(\mathbf{H})$. On the other hand, for each $F \in \mathcal{F}(\mathbf{H})$, we have $x \to y = 1 \in F$ or $y \to x = 1 \in F$, so $F \in \mathrm{Spec}(\mathbf{H})$. Therefore, $\mathcal{F}(\mathbf{H}) = \mathrm{Spec}(\mathbf{H}) = \mathrm{Spec}_{\vee}(\mathbf{H})$. The converse of this corollary also holds, since $\{1\}$ is a filter of $\mathbf{H}$.

The proof of the other part is straightforward. $\square$

Theorem 4.2.11 *Let* F *be a proper filter of a hoop* $\mathbf{H}$*. Then* F *is prime if and only if* $\mathbf{H}/F$ *is linearly ordered.*

Proof First, assume that $\mathbf{H}/F$ is linearly ordered. Choose $x, y \in H$. Then $x/F \leq y/F$ or $y/F \leq x/F$. By Theorem 4.1.15, $x \to y \in F$ or $y \to x \in F$, respectively, which means F is prime. The proof of the converse follows from Theorem 4.1.15 similarly. $\square$

Theorem 4.2.12 *Let* $\mathbf{H}$ *be a* $\vee$*-hoop and* $F \in \mathcal{F}(\mathbf{H})$*. Then for each* $x \in H \setminus F$*, there exists a* $\vee$*-prime filter* P *containing* F *such that* $x \notin P$*.*

Proof Let $\Omega = \{G \in \mathcal{F}(\mathbf{H}) : F \subseteq G \neq H, \ x \notin G\}$. Since $F \in \Omega$, we have $\Omega \neq \emptyset$. In addition, any chain of elements in Ω has an upper bound in Ω, it is the union of the elements (see Exercise 2.6). By Zorn's lemma for $(\Omega, \subseteq)$, there exists a maximal element $P \in \Omega$ such that $F \subseteq P$ and $x \notin P$. Now, we prove that P is a $\vee$-prime filter of H. Let $y \vee z \in P$ and $y, z \notin P$. Then $P \subseteq \langle P \cup \{y\}\rangle \cap \langle P \cup \{z\}\rangle$. Since P is maximal of Ω, we get $\langle P \cup \{y\}\rangle, \langle P \cup \{z\}\rangle \notin \Omega$ and so $x \in \langle P \cup \{y\}\rangle \cap \langle P \cup \{z\}\rangle$. By Proposition 4.1.9, we have $x \in \langle P \cup \{y \vee z\}\rangle = P$, which is a contradiction. Hence, $y \in P$ or $z \in P$. Therefore, there exists a $\vee$-prime filter P containing F such that $x \notin P$. $\square$

Corollary 4.2.13 *Every* $\vee$*-hoop with more than* 1 *element has at least one prime filter.*

Proof Choose $x \in H \setminus \{1\}$. It suffices to set $F = \{1\}$ in Theorem 4.2.12. $\square$

Corollary 4.2.14 *Let* $\mathbf{H}$ *be a* $\vee$*-hoop. Every proper filter of* $\mathbf{H}$ *is equal to the intersection of all* $\vee$*-prime filters containing it. In particular,* $\bigcap\{P : P \in \mathrm{Spec}_{\vee}(\mathbf{H})\} = \{1\}$ *for each non-trivial hoop* $\mathbf{H}$*.*

Proof First, assume that $H = \{1\}$. Then $\mathrm{Spec}_\vee(\mathbf{H}) = \emptyset$ and $\mathcal{F}(\mathbf{H}) = \{\{1\}\}$. Clearly, $\bigcap\{P : P \in \mathrm{Spec}_\vee(\mathbf{H})\} = \bigcap\{P : P \in \emptyset\} = H$. Assume that $\mathbf{H}$ is a $\vee$-hoop with more than 1 elements and $F \in \mathcal{F}(\mathbf{H})$ be a proper filter of $\mathbf{H}$. Let $x \in H \setminus F$. By Theorem 4.2.12, there exists $Q \in \mathrm{Spec}_\vee(\mathbf{H})$ containing F. Clearly, $F \subseteq \bigcap\{P \in \mathrm{Spec}_\vee(\mathbf{H}) : F \subseteq P\}$. Assume that $\bigcap\{P \in \mathrm{Spec}_\vee(\mathbf{H}) : F \subseteq P\} \not\subseteq F$. Then there exists $a \in \bigcap\{P \in \mathrm{Spec}_\vee(\mathbf{H}) : F \subseteq P\} \setminus F$. By Theorem 4.2.12, there is $G \in \mathrm{Spec}_\vee(\mathbf{H})$ such that $F \subseteq G$ and $a \notin G$, which is a contradiction with $a \in \bigcap\{P \in \mathrm{Spec}_\vee(\mathbf{H}) : F \subseteq P\} \subseteq G$. Hence, $\bigcap\{P \in \mathrm{Spec}_\vee(\mathbf{H}) : F \subseteq P\} \subseteq F$. Therefore, $F = \bigcap\{P \in \mathrm{Spec}_\vee(\mathbf{H}) : F \subseteq P\}$. Suppose that $\mathbf{H}$ is not trivial, i.e., $1 < |H|$. Then $\{1\}$ is a proper filter of $\mathbf{H}$ and so $\bigcap\{P \in \mathrm{Spec}_\vee(\mathbf{H}) : \{1\} \subseteq P\} = \{1\}$. Obviously, $\mathrm{Spec}_\vee(\mathbf{H}) = \{P \in \mathrm{Spec}_\vee(\mathbf{H}) : \{1\} \subseteq P\}$, so we have $\bigcap\{P : P \in \mathrm{Spec}_\vee(\mathbf{H})\} = \{1\}$. $\square$

The following corollary is a direct consequence of Corollaries 4.2.9 and 4.2.14.

Corollary 4.2.15 *Let* $\mathbf{H}$ *be a* $\vee$*-hoop satisfying* (Pre). *Then*

$$F = \bigcap\{P \in \mathrm{Spec}(\mathbf{H}) : F \subseteq P\}, \quad \forall F \in \mathcal{F}(\mathbf{H}) \setminus \{H\}.$$

4.2.2   Maximal and Ultrafilters

In this section, we will investigate maximal filters and ultrafilters in the context of hoops. We begin by introducing the concept of ultrafilters, a specific type of maximal filter. We will then demonstrate that every proper filter of a bounded hoop is contained within a maximal filter, thus establishing the existence of at least one maximal filter in every bounded hoop.

Definition 4.2.16 A filter U of hoop $\mathbf{H}$ is called an **ultrafilter** if $\mathbf{H}/U \cong \mathbf{L}_1$.

Remark 4.2.17 (i) Clearly, every ultrafilter U of a hoop $\mathbf{H}$ is proper, otherwise, $|\mathbf{H}/U| = 1$, a contradiction.

(ii) Recall that, by Example 3.3.11, $\mathbf{L}_1 = (\{0, 1\}; \odot, \rightarrow, 0, 1)$ is a bounded hoop. In addition, up to isomorphism, there is exactly one hoop with two elements. Therefore, a filter U of a hoops $\mathbf{H}$ is an ultrafilter if and only if $\mathbf{H}/U$ contains exactly two elements.

(iii) Let U be an ultrafilter of a hoop $\mathbf{H}$. Then there exists an isomorphism $f : \mathbf{H}/U \rightarrow \mathbf{L}_1$. We have $x \in H \setminus U$, if and only if $x/U \neq 1/U$ if and only if $H = (x/U) \cup (1/U) = (x/U) \cup U$, which implies U is an ultrafilter of $\mathbf{H}$ if and only if $(x/U) \cup U = H$ for all $x \in H \setminus U$. Note that $(x/U) \cap U = \emptyset$ for all $x \in H \setminus U$.

Proposition 4.2.18 *Let* U *be a proper filter of a hoop* $\mathbf{H} = (H; \odot, \rightarrow, 0, 1)$.

(i) *If* U *is an ultrafilter, then* $U \in \mathrm{Spec}(\mathbf{H})$.

(ii) U *is an ultrafilter of* $\mathbf{H}$ *if and only if* $x \rightarrow y \in U$ *for each* $x, y \in H \setminus U$.

(iii) *If* **H** *is bounded, then* U *is an ultrafilter if and only if it satisfies the following condition:*

$$x \in U \iff x' \notin U, \quad \forall x \in H. \tag{4.6}$$

Proof (i) The proof is evident by Theorem 4.2.11.

(ii) Suppose that U is an ultrafilter of **H**. Then by Remark 4.2.17(ii), $|\mathbf{H}/U| = 2$, so for each $x, y \in H \setminus U$ we have $x/U = y/U$ (since $x, y \notin 1/U$), consequently, $x \to y \in U$. Conversely, let $x \to y \in U$ for all $x, y \in H \setminus U$. If $x \in U$, then $x/U = 1/U$. In addition, if $x, y \in H \setminus U$, then by the assumption, $x \to y, y \to x \in U$, which means $x/U = y/U$. Therefore, $|\mathbf{H}/U| = 2$.

(iii) Let **H** be a bounded hoop satisfying (4.6). Choose $x \in H$. If $x \in U$, then $x/U = 1/U$, so $x \in 1/U = U$. Clearly, $x \notin U$ implies $x' \in U$. So $x'/U = 1/U$, consequently,

$$x''/U = (x'/U)' = 1/U \to 0/U = (1 \to 0)/U = 0/U,$$

which implies $0/U \leq x/U \leq x''/U = 0/U$, see Remark 2.2.5(ii). Thus, $x/U = 0/U$, equivalently, $x \in 0/U$. It follows that $H = (1/U) \cup (0/U)$. Therefore, by Remark 4.2.17(ii), U is an ultrafilter of **H**.

Conversely, assume that **H** is a bounded hoop such that $\mathbf{H}/U \cong \mathbf{L}_1$. If $x \in U$, then $x' \notin U$, clearly (since $x' \in U$ implies $0 = x \odot x' \in U$). On the other hand, if $x' \notin U$, then $x'/U \neq 1/U$, which implies $x'/U = 0/U$, consequently, $x/U = 1/U$, otherwise, $x \in 0/U$. Hence, $x'/U = (x/U) \to (0/U) = 1/U$, which is a contradiction. Hence, $x' \notin U$ implies that $x/U = 1/U$, equivalently $x \in U$. Therefore, U is an ultrafilter of **H**. $\qquad\square$

We note that in [37], the definition of ultrafilters was restricted to bounded hoops. Proposition 4.2.18(iii) demonstrates that our Definition 4.2.16 coincides with Definition 3.12 in [37] within the context of bounded hoops.

Example 4.2.19 (i) According to Example 4.1.2, $U_1 = \{a, 1\}$ and $U_2 = \{b, 1\}$ are two ultrafilters of **H**.

(ii) Assume $H = \{0, a, b, 1\}$ is a chain such that $0 < a < b < 1$. Then $(H; \odot, \to, 0, 1)$ is a bounded hoop where the operation are defined by Table 4.5:

In addition, $\{a, b, 1\}$ is an ultrafilter of **H**.

(iii) Let $H = \{0, a, b, c, 1\}$ be a set with the following Hasse diagram in Fig. 4.3. Define two operations $\to$ and $\odot$ on H as Table 4.6:

Then $(H; \odot, \to, 0, 1)$ is a bounded hoop and $F = \{a, b, c, 1\}$ is an ultrafilter of **H**.

(iv) Let $\mathbf{G} = (G; \vee, \wedge, +, -, 0)$ be an Abelian ℓ-group. Consider the Abelian ℓ-group $\mathbf{K} = \mathbb{Z} \overrightarrow{\times} \mathbf{G}$ and the element $u = (1, 0) \in \mathbf{K}^+$. Due to Example 3.3.57, $\mathbf{K}[u]$ is a bounded Wajsberg hoop. Furthermore,

$$K[u] = \{(x, y) \in \mathbb{Z} \times \mathbf{G} : (0, 0) \leq (x, y) \leq (1, 0)\} = (\{0\} \times \mathbf{G}^+) \cup (\{1\} \times \mathbf{G}^-),$$

Table 4.5 Operations $\odot$ and $\rightarrow$ of Example 4.2.19(ii)

Table 4.10.1

$\odot$	0	a	b	1
0	0	0	0	0
a	0	a	a	a
b	0	a	a	b
1	0	a	b	1

Table 4.10.2

$\rightarrow$	0	a	b	1
0	1	1	1	1
a	0	1	1	1
b	0	b	1	1
1	0	a	b	1

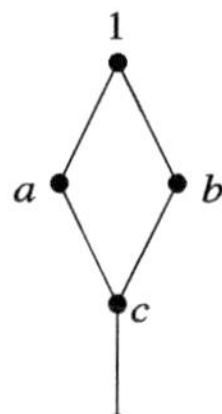

Fig. 4.3 The Hasse diagram of **H**

Table 4.6 Operations $\odot$ and $\rightarrow$ of Example 4.2.19(iii)

Table 4.5.1

$\odot$	0	c	a	b	1
0	0	0	0	0	0
c	0	c	c	c	c
a	0	c	a	c	a
b	0	c	c	b	b
1	0	c	a	b	1

Table 4.5.2

$\rightarrow$	0	c	a	b	1
0	1	1	1	1	1
c	0	1	1	1	1
a	0	b	1	b	1
b	0	a	a	1	1
1	0	c	a	b	1

and $(0, x)' = (1, 0) - (0, x) = (1, -x)$ and $(1, y)' = (1, 0) - (1, y) = (0, -y)$ for all $x \in \mathbf{G}^+$ and $y \in \mathbf{G}^-$. We can easily verify that $F := \{1\} \times \mathbf{G}^-$ is the unique maximal filter of $\mathbf{K}[u]$. In addition, F is an ultrafilter. Note that if D is a filter of $\mathbf{K}[u]$ such that $(0, x) \in D$ for some $x \in \mathbf{G}^+$, then $\{1\} \times \mathbf{G}^- \subseteq D$, by (F2). In addition, by (F1),

$$(0, 0) = (0, -x) \vee (0, 0) = \big((0, x) - (1, 0) + (1, -x)\big) \vee (0, 0) = (0, x) \odot (1, -x) \in D,$$

that is $D = K[u]$.

Lemma 4.2.20 *Let U be an ultrafilter of a bounded hoop $\mathbf{H}$. Then $x'' \in U$ if and only if $x \in U$. In addition, if $\mathbf{H}$ satisfies* (DNP), *then $H = U \cup U'$.*

Proof Choose $x \in H$. If $x \in U$, then by Remark 2.2.5(ii), $x \leq x''$, which entails $x'' \in U$. Conversely, assume that $x'' \in U$. Then by the definition of ultrafilter and Remark 2.2.5(iii), $x' = x''' \notin U$, consequently, $x \in U$. Since U is a proper filter of $\mathbf{H}$, we have $U \cap U' = \emptyset$. Let $x \in H$. If $x \notin U$, then $x' \in U$, consequently, $x = x'' \in U'$. Therefore, $H = U \cup U'$.$\square$

Theorem 4.2.21 *Let $\mathbf{H}$ be a bounded $\vee$-hoop. Then every ultrafilter is a $\vee$-prime filter.*

Proof **First proof**: Let U be an ultrafilter of $\mathbf{H}$. Since $1 \in U$, then $1' = 0 \notin U$, and so U is a proper filter. Suppose $x \vee y \in U$ for some $x, y \in H$. Since $x \vee y \leq (x \to y) \to y$, then by (F2), $(x \to y) \to y \in U$. Let $x \notin U$. By Lemma 2.2.1(7), $x \to y \geq x' \in U$, so $x \to y \in U$. From $(x \to y) \to y \in U$ and Lemma 2.3.2(ii) it follows that $y \in U$. Therefore, $U \in \mathrm{Spec}_\vee(\mathbf{H})$.

Second proof: It follows from Propositions 4.2.18 and 4.2.7(ii). $\square$

Definition 4.2.22   A **maximal filter** is a proper filter M of $\mathbf{H}$ that is not included in any other proper filter. Equivalently, M is maximal if $M \subseteq G \subseteq H$ implies that $M = G$ or $G = H$ for all $G \in \mathcal{F}(\mathbf{H})$. Denote by $\mathrm{Max}(\mathbf{H})$ the set of all maximal filters of $\mathbf{H}$.

Example 4.2.23   (i) According to Example 4.2.19(ii) and (iii), both filters are maximal filters of H.

(ii) The filter $F = \{b, c, d, 1\}$ in Example 4.2.3 is a maximal filter of $\mathbf{H}$ which is not an ultrafilter of H, since $a, a' \notin F$.

(iii) Let $\mathbf{H} = ([0, 1]; \odot, \to, 1)$ be a hoop where

$$x \odot y = (x + y - 1) \vee 0, \quad x \to y = (1 - x + y) \wedge 1.$$

Indeed, $\mathbf{H} = \mathbb{R}[1]$. Then $\{1\}$ is a maximal filter of the bounded $\mathbf{H}$ which is not an ultrafilter.

Proposition 4.2.24 *A proper filter M of a hoop $\mathbf{H}$ is a maximal filter if and only if $\langle M \cup \{x\}\rangle = H$ for all $x \in H \setminus M$.*

Proof Assume that M is a maximal filter of $\mathbf{H}$ and $x \in H \setminus M$. From $M \subseteq \langle M \cup \{x\}\rangle$ and $M \neq \langle M \cup \{x\}\rangle$ it follows that $H = \langle M \cup \{x\}\rangle$.

Conversely, suppose that there exists a filter G of H such that $M \subseteq G \subseteq H$. For each $x \in G \setminus M$ we have $H = \langle M \cup \{x\}\rangle \subseteq G$. Hence, M is a maximal filter of $\mathbf{H}$. $\square$

Proposition 4.2.25 *Let* **H** *be a hoop and* M *be a proper filter of* H. *The following conditions are equivalent:*

(i) M *is a maximal filter.*

(ii)   *For all* $x \in H$, *if* $x \notin M$, *then for any* $a \in H$, $x^n \to a \in M$ *for some* $n \in \mathbb{N}_0$.

Proof (i) $\Rightarrow$ (ii) Let $x \in H \setminus M$. By Proposition 4.2.24, we have $\langle M \cup \{x\} \rangle = H$. Applying Corollary 2.3.10, we get that for all $a \in H$ there is $n \in \mathbb{N}_0$ and $m \in M$ such that $m \odot x^n \leq a$, so $m \leq x^n \to a$. Since M is a filter of H, it follows that $x^n \to a \in M$.

(ii) $\Rightarrow$ (i) Let $x \in H \setminus M$. For any $a \in H$, there is $n \in \mathbb{N}_0$ such that $x^n \to a \in M$, so by Corollary 2.3.10, $a \in \langle M \cup \{x\} \rangle$. Hence, $\langle M \cup \{x\} \rangle = H$. Proposition 4.2.24 implies that M is maximal. $\qquad\qquad\qquad\square$

Theorem 4.2.26 ([37, Theorem 3.15]) *Every ultrafilter of a hoop* **H** *is a maximal filter.*

Proof Suppose U is an ultrafilter of a hoop **H**. By Remark 4.2.17(i), U is a proper filter of **H**. Let $a \in H \setminus U$. By Proposition 4.2.18(ii), for each $x \in H \setminus U$, $a \to x \in U$, so Proposition 4.2.25 implies that U is a maximal filter. $\qquad\qquad\qquad\square$

Corollary 4.2.27 *A proper filter* M *of a bounded hoop* **H** *is maximal if and only if for each* $x \in H \setminus M$ *there exists* $n \in \mathbb{N}$ *such that* $(x^n)' \in M$.

Proof Let M be a proper filter of a bounded hoop **H**. If M is a maximal filter of **H** and $x \in H \setminus M$, then by Proposition 4.2.24, $0 \in H = \langle H \cup \{x\} \rangle$, and so by Corollary 2.3.10(ii), there exists $n \in \mathbb{N}$ such that $(x^n)' = x^n \to 0 \in M$. The proof of the converse follows from Corollary 2.3.10(ii) and Proposition 4.2.25, similarly. $\qquad\qquad\qquad\square$

Proposition 4.2.28 *Every proper filter of a bounded hoop* **H** *is contained in a maximal filter. In addition,* $\mathrm{Max}(\mathbf{H}) \neq \emptyset$ *for all bounded hoop* **H**.

Proof Let F be a proper filter of **H**. Zorn's lemma states that a partially ordered set containing upper bounds for every chain (that is, every totally ordered subset) necessarily contains at least one maximal element. Consider $\Omega = \{P \in \mathcal{F}(\mathbf{H}) : F \subseteq P \neq H\}$. Since $F \in \Omega$, we get $\Omega \neq \emptyset$. Let $\{P_i\}_{i \in I}$ be a chain in the partially ordered set $(\Omega, \subseteq)$. Set $P = \bigcup_{i \in I} P_i$. Clearly, $P \in \mathcal{F}(\mathbf{H})$ and $F \subseteq \bigcup_{i \in I} P_i$. Obviously, $P \neq H$ (otherwise, $0 \in P$ implies that $0 \in P_i$ for some $i \in I$, which is absurd) and so $P \in \Omega$. Hence, P is an upper bound of Ω and by using Zorn's Lemma, there exists a maximal element $M \in \Omega$ which is a maximal filter of H and $F \subseteq M$.

The proof of the second part is straightforward. For each bounded hoop $(H; \odot, \to, 0, 1)$, the filter $\{1\}$ is proper, so there exists a maximal filter of $(H; \odot, \to, 0, 1)$ containing $\{1\}$. Therefore, $\mathrm{Max}(\mathbf{H}) \neq \emptyset$. $\qquad\square$

Corollary 4.2.29 *Let P be a prime filter of a bounded hoop $\mathbf{H}$. Then there exists a unique maximal filter of H containing P.*

Proof The proof follows from Propositions 4.2.28 and 4.2.4. $\qquad\square$

Theorem 4.2.30 *Let P be a proper filter of a bounded hoop H satisfying the following condition. Then P is maximal.*

$$x, y \in H \setminus P \text{ implies } x \to y \in P \text{ and } y \to x \in P. \tag{4.7}$$

Proof Assume that $P \in \mathcal{F}(\mathbf{H})$ satisfying (4.7) and $P \subseteq M \subseteq H$. If $P \neq M$, then there is $x \in M \setminus P$. Clearly, $0 \notin P$, so by the assumption $0 \to x \in P$ and $x \to 0 \in P$, which entails that $x \to 0 \in M$, consequently, $0 = x \odot (x \to 0) \in M$, since $M \in \mathcal{F}(\mathbf{H})$, which means $M = H$. Therefore, $P \in \mathrm{Max}(\mathbf{H})$. $\qquad\square$

The next example shows that the assumption of boundedness is necessary in Proposition 4.2.28.

Example 4.2.31 Let $H = \{-n : n \in \mathbb{N}\}$. Consider the following operations on H:

$$x \odot y = \min\{x, y\}, \quad (x \to y = 1, \text{ if } x \le y \ \& \ x \to y = y \text{ if } x > y).$$

Easy calculation shows that $\mathbf{H} = (H; \odot, \to, -1)$ is a hoop. We claim that it has no maximal filter. Indeed, for every $m \in \mathbb{N}$, we have $F_n := \{x \in H : -m \le x\}$ satisfies (F1) and (F2) which means it is a filter of $\mathbf{H}$. In addition, if F is a proper filter of $\mathbf{H}$, then there exists $n \in \mathbb{N}$ such that $-n \notin F$, consequently, $F \subsetneq F_n \subsetneq H$. Therefore, $\mathbf{H}$ has no maximal filter.

Proposition 4.2.32 *Let $\mathbf{H}$ be a $\vee$-hoop. Then every maximal filter is a $\vee$-prime filter.*

Proof Let M be a maximal filter of $\mathbf{H}$. Then by definition, M is proper. Assume that $x \vee y \in M$ for some $x, y \in H$. If $x \in H \setminus M$ and $y \in H \setminus M$, then by Proposition 4.2.24, $\langle M \cup \{x\}\rangle = H = \langle M \cup \{y\}\rangle$. Now, Proposition 4.1.9 implies that $M = \langle M \cup \{x \vee y\}\rangle = \langle M \cup \{x\}\rangle \cap \langle M \cup \{y\}\rangle = H$, which is a contradiction. Hence, $x \in M$ or $y \in M$, consequently, $M \in \mathrm{Spec}_{\vee}(\mathbf{H})$. $\qquad\square$

4.3   (Positive) Implicative Filters

Positive implicative and implicative filters are connected to positive implicational calculus and weak positive implicational calculus, respectively, within the framework of implicational functors in logical systems. We initiate this section with a definition of positive implicative filters.

Definition 4.3.1 ([197]) A non-empty subset F of a hoop $\mathbf{H}$ is called a **positive implicative filter** of $\mathbf{H}$ if for any $x, y, z \in H$, it satisfies in the following conditions:

(PIF1) $1 \in F$.

(PIF2) If $z \to ((x \to y) \to x) \in F$ and $z \in F$, then $x \in F$.

Example 4.3.2 (i) Let $H = \{0, a, b, c, d, 1\}$. We define two operations $\odot$ and $\to$ on H as Table 4.7:

Then $\mathbf{H} = (H; \odot, \to, 0, 1)$ is a bounded hoop and $F = \{b, c, 1\}$ is a positive implicative filter of $\mathbf{H}$.

(ii) Let $\mathbf{H}$ be the hoop of Example 4.2.19(ii). Then $F = \{a, b, 1\}$ is a positive implicative filter of $\mathbf{H}$.

Proposition 4.3.3 *Every positive implicative filter of a hoop $\mathbf{H}$ is a filter.*

Proof Suppose F is a positive implicative filter of $\mathbf{H}$ and $x, y \in H$ such that $x, x \to y \in F$. Then

$$x \to ((y \to 1) \to y) = x \to (1 \to y) = x \to y \in F \quad \& \quad x \in F.$$

So, (PIF2) implies that $y \in F$. Thus, by Lemma 2.3.2, $F \in \mathcal{F}(\mathbf{H})$. □

The following example shows that the converse of Proposition 4.3.3 may not be true.

Table 4.7 Operations $\odot$ and $\to$ of Example 4.3.2(i)

Table 4.6.1

$\odot$	0	a	b	c	d	1
0	0	0	0	0	0	0
a	0	a	d	0	d	a
b	0	d	c	c	0	b
c	0	0	c	c	0	c
d	0	d	0	0	0	d
1	0	a	b	c	d	1

Table 4.6.2

$\to$	0	a	b	c	d	1
0	1	1	1	1	1	1
a	c	1	b	c	b	1
b	d	a	1	b	a	1
c	a	a	1	1	a	1
d	b	1	1	b	1	1
1	0	a	b	c	d	1

Table 4.8 Operations $\odot$ and $\rightarrow$ of Example 4.3.4

Table 4.7.1

$\odot$	0	a	b	1
0	0	0	0	0
a	0	0	a	a
b	0	a	b	b
1	0	a	b	1

Table 4.7.2

$\rightarrow$	0	a	b	1
0	1	1	1	1
a	a	1	1	1
b	0	a	1	1
1	0	a	b	1

Example 4.3.4 Let $H = \{0, a, b, 1\}$ be a chain such that $0 < a < b < 1$. Define the binary operations $\odot$ and $\rightarrow$ on H by Table 4.8.

Then $\mathbf{H} = (H; \odot, \rightarrow, 0, 1)$ is a bounded hoop. Clearly, $F = \{1, b\}$ is a filter of $\mathbf{H}$. However, it is not a positive implicative filter of $\mathbf{H}$, since $b \rightarrow ((a \rightarrow 0) \rightarrow a) = 1 \in F$ and $b \in F$, but $a \notin F$.

The following theorem unveils several equivalent characterizations of positive implicative filters, providing a richer understanding of this concept.

Theorem 4.3.5 ([37, Theorem 4.3]) *Suppose that F is a subset of a hoop $\mathbf{H}$ containing* 1. *The following equivalent statements hold:*

(i)   *F is a positive implicative filter.*
(ii) *$F \in \mathcal{F}(\mathbf{H})$ and for any $(x \rightarrow y) \rightarrow x \in F$ implies that $x \in F$.*
(iii) *$F \in \mathcal{F}(\mathbf{H})$ and $((x \rightarrow y) \rightarrow x) \rightarrow x \in F$ for any $x, y \in H$.*

Proof (i) $\Rightarrow$ (ii) Let F be a positive implicative filter of H. Then by Proposition 4.3.3, $F \in \mathcal{F}(\mathbf{H})$. Choose $x, y \in H$ such that $(x \rightarrow y) \rightarrow x \in F$. By Lemma 2.2.1(2), we have $1 \rightarrow ((x \rightarrow y) \rightarrow x) \in F$. So, by the assumption, $x \in F$.

(ii) $\Rightarrow$ (i) Let $x, y, z \in H$ such that $z \rightarrow ((x \rightarrow y) \rightarrow x) \in F$ and $z \in F$. Then $(x \rightarrow y) \rightarrow x \in F$, since $F \in \mathcal{F}(\mathbf{H})$. Hence, by (ii), $x \in F$.

(ii) $\Rightarrow$ (iii) By Lemma 2.2.1(10), $x \leq ((x \rightarrow y) \rightarrow x) \rightarrow x$, also, by using Lemma 2.2.1(7), $(((x \rightarrow y) \rightarrow x) \rightarrow x) \rightarrow y \leq x \rightarrow y$, so

$$1 = ((x \rightarrow y) \rightarrow x) \rightarrow ((x \rightarrow y) \rightarrow x)$$
$$= (x \rightarrow y) \rightarrow (((x \rightarrow y) \rightarrow x) \rightarrow x)$$
$$\leq \Big((((x \rightarrow y) \rightarrow x) \rightarrow x) \rightarrow y \Big) \rightarrow (((x \rightarrow y) \rightarrow x) \rightarrow x).$$

Consequently, $\big((((x \to y) \to x) \to x) \to y \big) \to (((x \to y) \to x) \to x) \in F$. Now, due to (ii), $((x \to y) \to x) \to x \in F$.

(iii) $\Rightarrow$ (ii) Let $(x \to y) \to x \in F$ for some $x, y \in H$. Due to (iii), $((x \to y) \to x) \to x \in F$, so Proposition 2.3.6(ii) implies that $x \in F$. $\qquad\qquad\square$

The next theorem provides us with more equivalent conditions for positive implicative filters on bounded hoops.

Theorem 4.3.6 ([37]) *Let F be a filter of a bounded hoop $\mathbf{H}$ containing 1. The following assertions are equivalent:*

(i) *F is a positive implicative filter.*
(ii) *$(x' \to x) \to x \in F$ for all $x \in H$.*
(iii) *If $x \to (z' \to y) \in F$ and $y \to z \in F$, then $x \to z \in F$ for all $x, y, z \in H$.*
(iv) *If $y' \to (x \to y) \in F$, then $x \to y \in F$ for all $x, y \in H$.*

Proof (i) $\Rightarrow$ (ii) It is enough to let $y = 0$ in Theorem 4.3.5(iii).

(ii) $\Rightarrow$ (i) Let $x, y \in H$. Since $0 \leq y$, by Lemma 2.2.1(7), we have $x \to 0 \leq x \to y$ and so $(x \to y) \to x \leq x' \to x$, consequently, $(x' \to x) \to x \leq ((x \to y) \to x) \to x$. Since $(x' \to x) \to x \in F$ and $F \in \mathcal{F}(\mathbf{H})$, we get $((x \to y) \to x) \to x \in F$ for any $x, y \in H$. Hence, by Theorem 4.3.5(iii), F is positive implicative.

(i) $\Rightarrow$ (iv) Given $x, y \in H$ such that $y' \to (x \to y) \in F$. By Lemma 2.2.1(10), $y \leq x \to y$ and by Lemma 2.2.1(7), we get $(x \to y)' = (x \to y) \to 0 \leq y \to 0 = y'$, so

$$y' \to (x \to y) \leq (x \to y)' \to (x \to y).$$

(F2) entails that $(x \to y)' \to (x \to y) \in F$. On the other words,

$$1 \to \big(((x \to y) \to 0) \to (x \to y)\big) \in F.$$

Since F is a positive implicative, and $1 \in F$, we get $x \to y \in F$.

(iv) $\Rightarrow$ (i) Let $x, y \in H$ such that $(x \to y) \to x \in F$. Since $x' \leq x \to y$, by Lemma 2.2.1(7), $(x \to y) \to x \leq x' \to x$ and so by (F2), $x' \to x \in F$. It follows that

$$x' \to (1 \to x) = x' \to x \in F, \quad \text{by Lemma 2.2.1 (2).}$$

Now, (iv) implies that $x = 1 \to x \in F$. Therefore, by Theorem 4.3.5(ii), F is positive implicative.

(iv) $\Rightarrow$ (iii) Given $x, y, z \in H$ such that $x \to (z' \to y) \in F$ and $y \to z \in F$. By Lemma 2.2.1(8), $(x \odot z') \to y \in F$ and $y \to z \in F$. So, (F1) and Lemma 2.2.1(14) imply that $(x \odot z') \to z \geq ((x \odot z') \to y) \odot (y \to z) \in F$, consequently, $z' \to (x \to z) = (x \odot z') \to z \in F$. Thus, by (iv), $x \to z \in F$.

(iii) $\Rightarrow$ (iv) Let $y' \to (x \to y) \in F$ for some $x, y, z \in H$. Then by Lemma 2.2.1(8), $x \to (y' \to y) \in F$. Now, from $y \to y = 1 \in F$ and (iii), we get $x \to y \in F$. $\qquad\square$

Corollary 4.3.7 ([37]) *Let* **H** *be a bounded hoop and* F *be a positive implicative filter of* **H**. *Then* $x'' \to x \in F$ *for all* $x \in H$.

Proof Let F be a positive implicative filter of a bounded hoop **H**. By Lemma 2.2.1(7), $x' \to 0 \le x' \to x$, and so $x'' \to (x' \to x) = 1 \in F$ for all $x \in H$. Since F is a positive implicative filter and $x \to x \in F$, by Theorem 4.3.6(iii), $x'' \to x \in F$ for all $x \in H$. $\qquad\square$

Corollary 4.3.8 *Let* F *and* G *be two filters of a bounded hoop* **H** *such that* $F \subseteq G$. *If* F *is a positive implicative filter, so is* G.

Proof By Theorem 4.3.6(ii), the proof is straightforward. $\qquad\square$

Now, we introduce the notion of implicative filters on hoops and investigate some of their properties.

Definition 4.3.9 ([197]) Let **H** be a hoop and $\emptyset \ne F \subseteq H$. Then F is called an **implicative filter** of **H** if for any $x, y, z \in H$, $x \to (y \to z) \in F$ and $x \to y \in F$ imply $x \to z \in F$.

Example 4.3.10 (i) Let **H** be the hoop in Example 4.1.2. Then every filter of **H** is implicative.
(ii) Let **H** be the hoop of Example 4.2.19(ii). Then $F = \{a, b, 1\}$ is an implicative filter of **H**.

Theorem 4.3.11 *Every implicative filter of* **H** *is a filter of* **H**.

Proof If F is an implicative filter of **H** such that $x \in F$ and $x \to y \in F$, then by Lemma 2.2.1(1) and (2), $1 \to x \in F$ and $1 \to (x \to y) \in F$. Since F is an implicative filter of **H**, we get $y = 1 \to y \in F$. Hence, $F \in \mathcal{F}(\mathbf{H})$. $\qquad\square$

Theorem 4.3.12 ([37]) *Suppose* F *is a non-empty subset of a hoop* **H**. *The next statements are equivalent:*

(i) F *is an implicative filter of* **H**.
(ii) $F \in \mathcal{F}(\mathbf{H})$ *and if* $y \to (y \to x) \in F$, *then* $y \to x \in F$ *for all* $x, y \in H$.
(iii) $F \in \mathcal{F}(\mathbf{H})$ *and if* $z \to (y \to x) \in F$, *then* $(z \to y) \to (z \to x) \in F$ *for all* $x, y, z \in H$.

(iv) $1 \in F$ *and if* $z \to (y \to (y \to x)) \in F$ *and* $z \in F$, *then* $y \to x \in F$ *for all* $x, y, z \in H$.

(v) $F \in \mathcal{F}(\mathbf{H})$ *and* $x \to x^2 \in F$ *for all* $x \in H$.

Proof (i) $\Rightarrow$ (ii) By Theorem 4.3.11, $F \in \mathcal{F}(\mathbf{H})$. Consider $x, y \in H$ such that $y \to (y \to x) \in F$. Since $y \to y = 1 \in F$ and F is an implicative filter of $\mathbf{H}$, we have $y \to x \in F$.

(ii) $\Rightarrow$ (iii) Let $x, y, z \in H$ such that $z \to (y \to x) \in F$. Then Lemma 2.2.1(11) implies that $y \to x \leq (z \to y) \to (z \to x)$, so by Lemma 2.2.1(7) and (8), we get

$$z \to (y \to x) \leq z \to \big((z \to y) \to (z \to x)\big) = z \to \big(z \to ((z \to y) \to x)\big).$$

From $F \in \mathcal{F}(\mathbf{H})$ and $z \to (y \to x) \in F$, we obtain $z \to (z \to ((z \to y) \to x)) \in F$. Now, by (ii), $z \to ((z \to y) \to x) \in F$ and by Lemma 2.2.1(8), we get $(z \to y) \to (z \to x) \in F$.

(iii) $\Rightarrow$ (iv) Suppose $x, y, z \in H$ such that $z \to (y \to (y \to x)) \in F$ and $z \in F$. Since $F \in \mathcal{F}(\mathbf{H})$, we have $y \to (y \to x) \in F$, and so using (iii), we get

$$y \to x = 1 \to (y \to x) = (y \to y) \to (y \to x) \in F.$$

(iv) $\Rightarrow$ (v) First, we claim that $F \in \mathcal{F}(\mathbf{H})$. Let $x, y \in H$ such that $x, x \to y \in F$. Then from $1, x \in F, x \to (1 \to (1 \to y)) = x \to y \in F$ and (iv) it follows that $y = 1 \to y \in F$. Hence, $F \in \mathcal{F}(\mathbf{H})$. For any $x \in L$, by Lemma 2.2.1(8), we have

$$1 \to (x \to (x \to x^2)) = 1 \to ((x \odot x) \to x^2)$$
$$= 1 \to (x^2 \to x^2) = 1 \to 1 = 1 \in F.$$

Now, $1 \in F$ and (iv) imply that $x \to x^2 \in F$.

(v) $\Rightarrow$ (ii) Let $x, y \in H$ such that $x \to (x \to y) \in F$. By Lemma 2.2.1(8), $x^2 \to y \in F$. Also, by the assumption, $x \to x^2 \in F$, so (F1) implies that $(x \to x^2) \odot (x^2 \to y) \in F$. Due to Lemma 2.2.1(14), $(x \to x^2) \odot (x^2 \to y) \leq x \to y$, so by (F2), $x \to y \in F$.

(ii) $\Rightarrow$ (i) Let $x, y, z \in H$ such that $x \to (y \to z) \in F$ and $x \to y \in F$. Then by Lemma 2.2.1(8), $y \to (x \to z) \in F$ and by using (F1), $(x \to y) \odot (y \to (x \to z)) \in F$. In addition, $(x \to y) \odot (y \to (x \to z)) \leq x \to (x \to z)$, by Lemma 2.2.1(14), so $x \to (x \to z) \in F$. Hence, by (ii), $x \to z \in F$. Therefore, F is an implicative filter of $\mathbf{H}$. $\qquad \square$

Corollary 4.3.13 *Let* $\mathbf{H}$ *be a hoop. Then* $x^2 = x$ *for all* $x \in H$ *if and only if every filter of* $\mathbf{H}$ *is an implicative filter.*

Proof Suppose for any $x \in H$ we have $x^2 = x$ and F is a filter of $\mathbf{H}$. Then $x \to x^2 = 1 \in F$, so by Theorem 4.3.12(v), F is an implicative filter of $\mathbf{H}$. Conversely, let every filter of $\mathbf{H}$ be an implicative filter. Then $\{1\}$ is implicative and by Theorem 4.3.12(v) for any $x \in H$ we have $x \to x^2 \in \{1\}$. Thus, $x \leq x^2$. On the other side, $x^2 \leq x$, by Lemma 2.2.1(4). Therefore, $x = x^2$ for all $x \in H$. $\qquad \square$

Corollary 4.3.14 *Let F and G be two filters of a hoop $\mathbf{H}$. If F is an implicative filter and $F \subseteq G$, then so is G.*

Proof By Theorem 4.3.12(v), the proof is clear. $\qquad\square$

Corollary 4.3.15 *Let F be a filter of a hoop $\mathbf{H}$. Then F is implicative if and only if*

$$z \to (y \to x) \in F \iff (z \to y) \to (z \to x) \in F, \quad \forall x, y, z \in H.$$

Proof Let $x, y, z \in H$. Applying Lemma 2.2.1(7), from $y \le z \to y$ we obtain $(z \to y) \to x \le y \to x$, consequently, $(z \to y) \to (z \to x) = z \to ((z \to y) \to x) \le z \to (y \to x)$. Thus, if F is a filter, then by (F2), $(z \to y) \to (z \to x) \in F$ implies that $z \to (y \to x) \in F$. The remainder of the proof follows from Proposition 4.3.12(iii). $\qquad\square$

Proposition 4.3.16 *A filter F of a hoop $\mathbf{H}$ is implicative if and only if $\mathbf{H}/F$ is a Gödel hoop.*

Proof If F is implicative, then by Theorem 4.3.12(v), $x \to x^2 \in F$ for all $x \in H$. On the other hand, $x^2 \to x = 1 \in F$ for all $x \in H$, so $x/F = x^2/F = x/F \odot x/F$ for all $x \in H$. Hence, $\mathbf{H}/F$ is a Gödel hoop. Conversely, let $\mathbf{H}/F$ be a Gödel hoop. Then $x^2/F = x/F$ for all $x \in H$, equivalently, $x \to x^2 \in F$ for all $x \in H$. Hence, by Theorem 4.3.12(v), F is implicative. $\qquad\square$

Theorem 4.3.17 *Every positive implicative filter of a hoop $\mathbf{H}$ is an implicative filter.*

Proof Let F be a positive implicative filter of a hoop $\mathbf{H}$. Given $x, y \in H$, let $y \to (y \to x) \in F$. By Lemma 2.2.1(13), $((y \to x) \to x) \to x = y \to x$, so Lemma 2.2.1(12) implies that

$$y \to \big(((y \to x) \to x) \to x\big) = y \to (y \to x) \le ((y \to x) \to x) \to (y \to x).$$

Then, $y \to (y \to x) \in F$ and $F \in \mathcal{F}(\mathbf{H})$ imply that $((y \to x) \to x) \to (y \to x) \in F$. In addition, due to Theorem 4.3.5(ii), $y \to x \in F$. Hence, by Theorem 4.3.12(ii), F is an implicative filter of $\mathbf{H}$. $\qquad\square$

Theorem 4.3.18 *Let $\mathbf{H}$ be a bounded hoop. A proper filter F of $\mathbf{H}$ is a maximal and implicative filter if and only if $(x \to y) \wedge (y \to x) \in F$ for all $x, y \in H \setminus F$.*

Proof Let $\mathbf{H}$ be a bounded hoop and F be a maximal and implicative filter of $\mathbf{H}$. Choose $x, y \in H \setminus F$, and by Proposition 4.2.24, $\langle F \cup \{x\}\rangle = H$. By Proposition 4.2.25, there exists $n \in \mathbb{N}$ such that, $x^n \to 0 \in F$. Lemma 2.2.1(8) entails that $x^{n-1} \to (x \to 0) \in F$. Since

$x^{n-1} \rightarrow x = 1 \in F$ and F is an implicative filter, then $x^{n-1} \rightarrow 0 \in F$. Repeating this argument inductively, we eventually obtain $x \rightarrow 0 \in F$. Moreover, due to Lemma 2.2.1(7), $x' \leq x \rightarrow y$ for any $y \in H$. It follows from (F2) that $x \rightarrow y \in F$. In a similar way, since $y \in H \setminus F$ we have $\langle F \cup \{y\} \rangle = H$. Hence, $y \rightarrow x \in F$. Applying Proposition 4.1.6, we conclude that $(x \rightarrow y) \wedge (y \rightarrow x) \in F$.

Conversely, let F be a filter of $\mathbf{H}$ such that $(x \rightarrow y) \wedge (y \rightarrow x) \in F$ for all $x, y \in F$. Choose $x \in H \setminus F$. Since F is a proper filter, then $0 \notin F$, so by the assumption, $x \rightarrow 0, 0 \rightarrow x \in F$, entails that $x' \in F$. Furthermore, by Lemma 2.2.1(6), $x \leq (x \rightarrow 0) \rightarrow 0$, consequently, $0 = x \odot x' \in \langle F \cup \{x\} \rangle$, which means $\langle F \cup \{x\} \rangle = H$. Hence, by Proposition 4.2.24, F is a maximal filter. Now, let $x \rightarrow (x \rightarrow y) \in F$ for some $x, y \in H$. Since F is a filter, $x \in F$ implies that $x \rightarrow y \in F$. If $x \in H \setminus F$, then by the assumption, $x \rightarrow 0 = x' \in F$. From (F2) and $x' \leq x \rightarrow y$ it follows that $x \rightarrow y \in F$ for every $y \in H$. Therefore, Theorem 4.3.12(ii) implies that F is an implicative filter. $\qquad\square$

Theorem 4.3.19 *Let $\mathbf{H}$ be a bounded hoop with* (DNP). *Then every implicative filter is a positive implicative filter.*

Proof Let F be an implicative filter of $\mathbf{H}$ and $x'' = x$ for all $x \in H$. Suppose $x \rightarrow ((y \rightarrow z) \rightarrow y) \in F$ and $x \in F$ for some $x, y, z \in H$. Since F is an implicative filter, by Proposition 4.3.11, F is a filter. Thus, $(y \rightarrow z) \rightarrow y \in F$. Moreover, since $\mathbf{H}$ satisfies (DNP), then $(y \rightarrow z) \rightarrow y'' = (y \rightarrow z) \rightarrow y \in F$. Hence, by Lemma 2.2.1(8), $y' \rightarrow ((y \rightarrow z) \rightarrow 0) \in F$. Since F is an implicative filter, by Theorem 4.3.12(iii), $(y' \rightarrow (y \rightarrow z)) \rightarrow (y' \rightarrow 0) \in F$. Applying Lemma 2.2.1(8) and (DNP), we get $((y' \odot y) \rightarrow z) \rightarrow y \in F$. Moreover,

$$y = 1 \rightarrow y = (0 \rightarrow z) \rightarrow y = ((y' \odot y) \rightarrow z) \rightarrow y \in F,$$

so, $y \in F$. Therefore, F is a positive implicative filter of $\mathbf{H}$. $\qquad\square$

Theorem 4.3.20 *Let F be a filter of a hoop $\mathbf{H}$. Then F is an implicative filter if and only if $\langle F \cup \{a\} \rangle = \{x \in H : a \rightarrow x \in F\}$ for any $a \in H$.*

Proof Let F be an implicative filter, $a \in H$ and $F_a = \{x \in H : a \rightarrow x \in F\}$. Clearly, $1, a \in F_a$. First, we show that F_a is a filter of $\mathbf{H}$. Suppose that $x, x \rightarrow y \in F_a$ for some $x, y \in H$. Then $a \rightarrow x, a \rightarrow (x \rightarrow y) \in F$. Since F is an implicative filter, then $a \rightarrow y \in F$, and so $y \in F_a$. Hence, by Lemma 2.3.2, F_a is a filter. Moreover, by Lemmas 2.2.1(10), $x \leq a \rightarrow x$ for all $x \in F$, entail that $F \subseteq F_a$, by (F2), consequently, $F \cup \{a\} \subseteq F_a$. Now, we show that F_a is the smallest filter containing $\{a\}$ and F. Let $G \in \mathcal{F}(\mathbf{H})$ such that $F \cup \{a\} \subseteq G$. For each $x \in F_a$, we have $a \rightarrow x \in F \subseteq G$ and $a \in G$, so $x \in G$, that is $F_a \subseteq G$. Therefore, F_a is the least filter of $\mathbf{H}$ containing $F \cup \{a\}$.

Conversely, if $x \rightarrow (x \rightarrow y) \in F$ for any $x, y \in H$, then $x \rightarrow y \in F_x$. Also, $x \rightarrow x = 1 \in F$, so $x \in F_x$, consequently, by the assumption and Lemma 2.3.2, $y \in F_x$, i.e., $x \rightarrow y \in F$. Therefore, Theorem 4.3.12(ii) implies that F is an implicative filter. $\qquad\square$

Theorem 4.3.21 *Every ultrafilter of a bounded hoop* **H** *is an implicative filter.*

Proof Choose an ultrafilter M of a bounded **H**. Suppose that $x \to (y \to z) \in M$ and $x \to y \in M$ for some $x, y, z \in H$. Since M is a filter of **H**, by Theorem 3.1.2(H2) and (F1), $(x \to y) \odot (y \to (x \to z)) \in M$. By Lemma 2.2.1(14), $(x \to y) \odot (y \to (x \to z)) \leq x \to (x \to z)$, and so by (F2), $x \to (x \to z) \in M$.

If $x \in M$, then by Lemma 2.3.2, $x \to z \in M$.

If $x \notin M$, then $x' \in M$. Since $x' \leq x \to z$ for any $z \in H$ (see Lemma 2.2.1(7)), (F2) implies that $x \to z \in M$.

Therefore, M is an implicative filter of **H**. $\square$

Theorem 4.3.22 ([37]) *A subset F of a hoop* $\mathbf{H} = (H; \odot, \to, 1)$ *is a positive implicative filter of* **H** *if and only if F is an implicative filter of* **H** *and $(x \to y) \to y \in F$ implies that $(y \to x) \to x \in F$ for all $x, y \in H$.*

Proof Let F be a positive implicative filter of **H**. By Theorem 4.3.17, F is an implicative filter of **H**. Assume that $(x \to y) \to y \in F$ for some $x, y \in H$. Due to Lemma 2.2.1(6), $x \leq (y \to x) \to x$ and by Lemma 2.2.1(7), we get

$$((y \to x) \to x)' = ((y \to x) \to x) \to 0 \leq x \to 0 = x'. \tag{4.8}$$

Similarly, $x' \leq x \to y$ and so $(x \to y) \to y \leq x' \to y$. Applying (4.8), we obtain $x' \to y \leq ((y \to x) \to x)' \to y$. Thus,

$$(x \to y) \to y \leq x' \to y \leq ((y \to x) \to x)' \to y. \tag{4.9}$$

Furthermore, by Lemma 2.2.1(6) and (7), we get

$$((y \to x) \to x)' \to y \leq ((y \to x) \to x)' \to ((y \to x) \to x). \tag{4.10}$$

Using (4.9), (4.10), (F2) and assumption, $((y \to x) \to x)' \to ((y \to x) \to x) \in F$. Since F is a positive implicative filter of **H**, Theorem 4.3.5(ii) implies that $(y \to x) \to x \in F$.

Conversely, suppose that $(x \to y) \to x \in F$ for some $x, y \in H$. By Lemma 2.2.1(7), $(x \to y) \to x \leq (x \to 0) \to x$ which implies $x' \to x \in F$, by (F2). Also, by Lemma 2.2.1(6), $x \leq x''$, so $x' \to x \leq x' \to x''$, consequently, $(x \to 0) \to ((x \to 0) \to 0) = x' \to x'' \in F$. On the other words, $(x \to 0) \to (x \to 0) = 1 \in F$ and F is implicative, so $(x \to 0) \to 0 \in F$. Thus, by the assumption, $x = (0 \to x) \to x \in F$. Therefore, by Theorem 4.3.5(ii), F is a positive implicative filter of L. $\square$

4.3.1 Fantastic Filters

At the end of this section, we will study fantastic filters on hoops.

Definition 4.3.23 ([197]) A subset F of a hoop $\mathbf{H}$ is called a **fantastic filter** of $\mathbf{H}$ if $1 \in F$ and for any $x, y, z \in H$, $z \to (y \to x) \in F$ and $z \in F$ imply $((x \to y) \to y) \to x \in F$.

Example 4.3.24 (i) In Example 4.1.2, the set $F = \{1\}$ is a fantastic filter of H.
(ii) Consider the hoop in Example 4.3.2(i). Then $F = \{b, c, 1\}$ is a fantastic filter of H.
(iii) The set $F = \{b, 1\}$ of the hoop in Example 4.3.4 is a fantastic filter.

Theorem 4.3.25 ([197]) *Every fantastic filter of a hoop* $\mathbf{H}$ *is a filter of* $\mathbf{H}$.

Proof Suppose F is a fantastic filter of $\mathbf{H}$ and $x, x \to y \in F$ for some $x, y \in H$. Then $x \in F$ and $x \to (1 \to y) = x \to y \in F$ and so by the assumption, and Lemma 2.2.1(1) and (2), $y = ((y \to 1) \to 1) \to y \in F$. Therefore, $F \in \mathcal{F}(\mathbf{H})$. $\square$

Theorem 4.3.26 *Let* F *be a filter of a hoop* $\mathbf{H}$. *The following statements are equivalent:*

(i) F *is a fantastic filter of* $\mathbf{H}$.
(ii) $y \to x \in F$ *implies that* $((x \to y) \to y) \to x \in F$ *for each* $x, y \in H$.
(iii) $((x \to y) \to y) \to ((y \to x) \to x) \in F$ *for all* $x, y \in H$.

Proof (i) $\Rightarrow$ (ii) Suppose that $y \to x \in F$ for some $x, y \in H$. Then $1 \to (y \to x) = y \to x \in F$ and $1 \in F$, so by (i), $((x \to y) \to y) \to x \in F$.

(ii) $\Rightarrow$ (i) Assume that $z \to (y \to x) \in F$ and $z \in F$ for some $x, y, z \in H$. By the assumption, $F \in \mathcal{F}(\mathbf{H})$, entails that $y \to x \in F$ and so by (ii), $((x \to y) \to y) \to x \in F$. Thus, F is a fantastic filter of $\mathbf{H}$.

(i) $\Rightarrow$ (iii) Choose $x, y \in H$. By Lemma 2.2.1(6), $x \le (y \to x) \to x$, so,

$$(x \to y) \to y \le (((y \to x) \to x) \to y) \to y, \quad \text{by Lemma 2.2.1(7).}$$

Applying Lemma 2.2.1(7) again, we get

$$\Big((((y \to x) \to x) \to y) \to y \Big) \to ((y \to x) \to x) \le ((x \to y) \to y) \to ((y \to x) \to x). \tag{4.11}$$

On the other hand, $(y \to x) \to (y \to x) = 1 \in F$ and Lemma 2.2.1(8) imply that

$$1 \to \big(y \to ((y \to x) \to x) \big) = y \to ((y \to x) \to x) \in F.$$

Now, since F is a fantastic filter of $\mathbf{H}$ and $1 \in F$, we have

$$\Big(\big(((y \to x) \to x) \to y \big) \to y \Big) \to ((y \to x) \to x) \in F.$$

The assumption, (F2), and (4.11) imply

$$((x \to y) \to y) \to ((y \to x) \to x) \in F.$$

(iii) $\Rightarrow$ (ii) Choose $x, y \in H$ such that $y \to x \in F$. By (iii) and Lemma 2.2.1(8) we obtain $(y \to x) \to (((x \to y) \to y) \to x) \in F$. It follows from $F \in \mathcal{F}(\mathbf{H})$ and $y \to x \in F$ that $((x \to y) \to y) \to x \in F$. $\qquad\square$

Corollary 4.3.27 ([197]) *If $F, G \in \mathcal{F}(\mathbf{H})$ such that $F \subseteq G$ and F is a fantastic filter of $\mathbf{H}$, then G is too.*

Proof By Theorem 4.3.26(iii), the proof is clear. $\qquad\square$

Corollary 4.3.28 ([80, Theorem 4.20]) *A filter F of a hoop $\mathbf{H}$ is fantastic if and only if $\mathbf{H}/F$ is a Wajsberg hoop. In addition, if $\mathbf{H}$ is bounded, then F is fantastic if and only if $\mathbf{H}/F$ is a bounded Wajsberg hoop, equivalently, an MV-algebra.*

Proof The proof of the first part is straightforward by Theorem 4.3.26(iii). The second part follows similarly from Theorem 3.3.13. $\qquad\square$

Theorem 4.3.29 ([197]) *Every positive implicative filter of a hoop $\mathbf{H}$ is fantastic.*

Proof Let F be a positive implicative filter of $\mathbf{H}$ and $y \to x \in F$ for some $x, y \in H$. Due to Proposition 4.3.3, $F \in \mathcal{F}(\mathbf{H})$. By Lemma 2.2.1(10), $x \le ((x \to y) \to y) \to x$, so by part (7) of that lemma,

$$\big(((x \to y) \to y) \to x \big) \to y \le x \to y.$$

Similarly,

$$(x \to y) \to x \le \Big(\big((((x \to y) \to y) \to x) \to y \big) \to x \Big). \tag{4.12}$$

and

$$\Big(\big((((x \to y) \to y) \to x) \to y \big) \to \Big(((x \to y) \to y) \to x \Big)$$

$$= ((x \to y) \to y) \to \Big((((((x \to y) \to y) \to x) \to y) \to x \Big), \quad \text{by Lemma 2.2.1(8)}$$

$$\ge ((x \to y) \to y) \to ((x \to y) \to x), \quad \text{by (4.12)}. \tag{4.13}$$

In addition, by Lemma 2.2.1(6) we have $x \rightarrow y \leq ((x \rightarrow y) \rightarrow y) \rightarrow y$, which implies $(x \rightarrow y) \odot ((x \rightarrow y) \rightarrow y) \leq y$ and so

$$y \rightarrow x \leq \big((x \rightarrow y) \odot ((x \rightarrow y) \rightarrow y)\big) \rightarrow x, \text{ by Lemma 2.2.1(7)}$$
$$= ((x \rightarrow y) \rightarrow y) \rightarrow ((x \rightarrow y) \rightarrow x), \text{ by Lemma 2.2.1 (8)}.$$

Since $F \in \mathcal{F}(\mathbf{H})$ and $y \rightarrow x \in F$, we get $((x \rightarrow y) \rightarrow y) \rightarrow ((x \rightarrow y) \rightarrow x) \in F$, and by (4.13), we have

$$((((x \rightarrow y) \rightarrow y) \rightarrow x) \rightarrow y) \rightarrow (((x \rightarrow y) \rightarrow y) \rightarrow x) \in F.$$

Since F is a positive implicative filter, we have $((x \rightarrow y) \rightarrow y) \rightarrow x \in F$. Theorem 4.3.26(ii) yields F is a fantastic filter of $\mathbf{H}$. $\square$

The next theorem delves into characterizing positive implicative, implicative, and fantastic filters in hoops.

Theorem 4.3.30 ([197]) *F is a positive implicative filter of $\mathbf{H}$ if and only if F is an implicative and fantastic filter of $\mathbf{H}$.*

Proof If F is a positive implicative filter of $\mathbf{H}$, then by Theorems 4.3.29 and 4.3.17, the proof is clear.

Conversely, assume that F is an implicative and fantastic filter of $\mathbf{H}$. Let $x, y \in H$ such that $(x \rightarrow y) \rightarrow x \in F$. By Theorem 4.3.25, F is a filter, so Lemma 2.2.1(10) implies that $(x \rightarrow y) \rightarrow ((x \rightarrow y) \rightarrow y) \in F$. Since F is an implicative filter of $\mathbf{H}$, we have $(x \rightarrow y) \rightarrow y \in F$. In addition, from $y \leq x \rightarrow y$ and Lemma 2.2.1(7), it follows that $(x \rightarrow y) \rightarrow x \leq y \rightarrow x$, consequently, $y \rightarrow x \in F$. Since F is fantastic, by Theorem 4.3.26(ii), we obtain $((x \rightarrow y) \rightarrow y) \rightarrow x \in F$. Now, from $F \in \mathcal{F}(\mathbf{H})$ and $(x \rightarrow y) \rightarrow y \in F$, we get $x \in F$. Therefore, by Theorem 4.3.5(ii), F is a positive implicative filter of $\mathbf{H}$. $\square$

Figure 4.4 presents the relationship between implicative, fantastic, and positive implicative filters of hoops.

Definition 4.3.31 ([80, Definition 4.1]) A filter F of a bounded hoop $(H; \odot, \rightarrow, 0, 1)$ is said to be an **involutive** filter if $x'' \rightarrow x \in F$ for all $x \in H$.

We note that every ultrafilter F of a bounded hoop $\mathbf{H}$ is involutive. Indeed, if $x \in F$, then due to Lemma 2.2.1(10) and (F2), $x'' \rightarrow x \in F$. In addition, if $x \in H \setminus F$, then by Lemma 4.2.20, $x'' \in H \setminus F$, this leads us to $x'' \rightarrow x \in F$, see Proposition 4.2.18(ii). Therefore, ultrafilters are a special kind of involutive filters on bounded hoops.

Fig. 4.4 Relation between filters of hoops

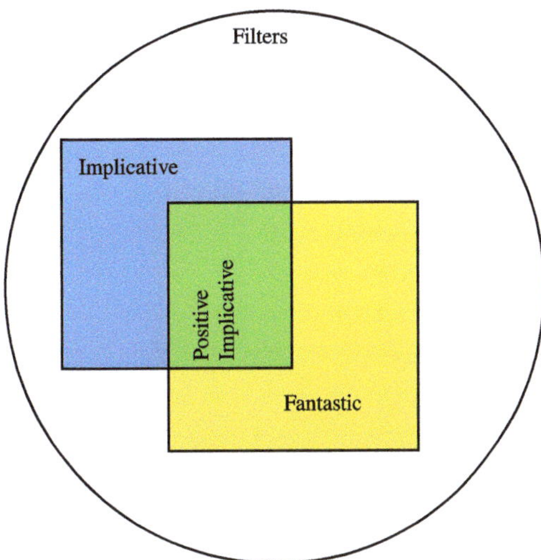

Proposition 4.3.32 ([80, Theorem 4.22]) *Let* $\mathbf{H}$ *be a bounded hoop and* $F \in \mathcal{F}(\mathbf{H})$. *Then* F *is an involutive filter if and only if* F *is a fantastic filter.*

Proof Assume that F is fantastic and $x \in H$. Then by Theorem 4.3.26(iii),

$$((x \to 0) \to 0) \to ((0 \to x) \to x) \in F,$$

entails that $x'' \to x \in F$, i.e., F is involutive. Conversely, let F be involutive. Consider the hoop $\mathbf{H}/F$, which is a bounded hoop with the least element $0/F$, clearly. Let $x \in H$. since $x'' \to x \in F$, we have $x''/F \to x/F = (x'' \to x)/F = 1/F$. Also, $x/F \to x''/F = (x \to x'')/F = 1/F$, since $x \le x''$, so $x/F = x''/F = (x/F)''$ for all $x \in H$, which means $\mathbf{H}/F$ is a bounded hoop satisfying (DNP). Now, Theorem 3.3.15 implies that $\mathbf{H}$ is a Wajsberg hoop. Therefore, by Corollary 4.3.28, F is a fantastic filter. $\qquad\square$

4.4 Ideals of Hoops

In this section, we are going to define and investigate the concept of ideals in hoops. According to Sect. 1.3, the concepts of ideals and filters are dual on lattices. But they can be different on hoops.

Definition 4.4.1 ([1]) A non-empty subset $I \subseteq H$ of a bounded hoop $\mathbf{H} = (H; \odot, \to, 0, 1)$ is called an **ideal** of $\mathbf{H}$ if it satisfies the following conditions:

Table 4.9 Operations $\odot$ and $\rightarrow$ of Example 4.4.2

Table 4.8.1

$\odot$	0	a	b	c	d	1
0	0	0	0	0	0	0
a	0	a	0	a	0	a
b	0	0	0	0	b	b
c	0	a	0	a	b	c
d	0	0	b	b	d	d
1	0	a	b	c	d	1

Table 4.8.2

$\rightarrow$	0	a	b	c	d	1
0	1	1	1	1	1	1
a	d	1	d	1	d	1
b	c	c	1	1	1	1
c	b	c	d	1	d	1
d	a	a	b	c	1	1
1	0	a	b	c	d	1

(I1) $0 \in I$.

(I2) $x' \rightarrow y \in I$ for all $x, y \in I$.

(I3) If $x, y \in H$, $x \leq y$ and $y \in I$, then $x \in I$.

It is clear that H and $\{0\}$ are the trivial ideals of **H**. The set of all ideals of **H** is denoted by $\mathcal{I}(\mathbf{H})$. An ideal I is **proper** if $I \neq H$. It can be easily seen that an ideal I is proper if and only if it does not contain 1 (Fig. 4.5).

Example 4.4.2 ([1]) Let $H = \{0, a, b, c, d, 1\}$. We define two operations $\odot$ and $\rightarrow$ on H by Table 4.9.

Routine calculations show that $\mathbf{H} = (H; \odot, \rightarrow, 0, 1)$ is a bounded hoop. It is easy to see that $I = \{0, a\} \in \mathcal{I}(\mathbf{H})$.

Given a family $\{I_i : i \in I\}$ of ideals of a bounded hoop **H**, it is easy to see that $\bigcap_{i \in I} I_i$ is an ideal of **H**. It follows that for each subset X of a bounded hoop **H** the least ideal of **H** containing X exists. It is called the **ideal is generated by** X in **H**, and it is denoted by $(X]$. Evidently, it is the intersection of all ideals of **H** containing X. Note that $\bigcup_{i \in I} I_i$ may not be an ideal of **H**, in general.

Fig. 4.5 The Hasse diagram of the hoop in Example 4.4.2

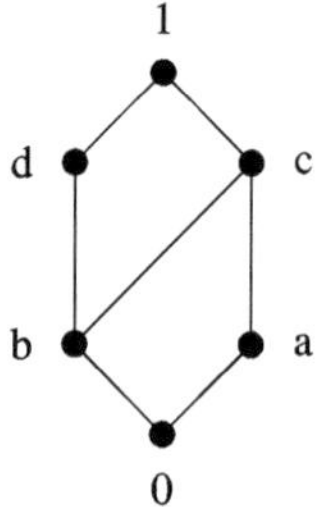

Clearly, if $\mathbf{H}$ is a bounded hoop and $X = \emptyset$, then $\{0\}$ is the least ideal of $\mathbf{H}$ containing X, that is, $(\emptyset] = \{0\}$.

Example 4.4.3 Assume that $\mathbf{H}$ is the bounded hoop in Example 4.1.2. It is easy to see that $I_1 = \{0, a\}$ and $I_2 = \{0, b\}$ are two ideals of $\mathbf{H}$, but $I_3 = I_1 \cup I_2 = \{0, a, b\}$ is not an ideal of $\mathbf{H}$, since $a' \to b = b \to b = 1 \notin I_3$.

Recall that the operation $x \star y := x' \to y$ is a binary operation on each bounded hoop $\mathbf{H}$. In addition, due to Proposition 2.2.6(v), if $\mathbf{H}$ satisfies (DNP) then the operation "$\star$" is associative.

Proposition 4.4.4 ([1]) *Let I be an ideal of a bounded hoop $\mathbf{H}$. Then $x \in I$ if and only if $x'' \in I$ for all $x \in H$.*

Proof Choose $x \in H$. If $x'' \in I$, then by Remark 2.2.5(ii) and (I3), $x \in I$. Conversely, if $x \in I$, then by (I2), $x'' = x' \to 0 \in I$. $\qquad\qquad\square$

Proposition 4.4.5 ([1]) *Let I be a non-empty subset of a bounded hoop $\mathbf{H} = (H; \odot, \to, 0, 1)$. Then, for any $x, y \in H$, the following statements are equivalent:*

  (i) $I \in \mathcal{I}(\mathbf{H})$.
 (ii) (1) $0 \in I$ and I is closed under $\star$; and (2) $x' \odot y \in I$ and $x \in I$ imply that $y \in I$.
(iii) (1) $0 \in I$ and I is closed under $\star$; and (2) $(x' \to y')' \in I$ and $x \in I$ imply that $y \in I$.

Proof (i) $\Rightarrow$ (ii) Let $I \in \mathcal{I}(\mathbf{H})$. Then by (I1), $0 \in I$ and by (I2) $x \star y \in I$ for each $x, y \in I$. Now, suppose that $x, y \in H$ such that $x' \odot y, x \in I$. Since $x' \odot y \leq x' \odot y$, by Lemma 2.2.1(9), $y \leq x' \to (x' \odot y)$. From $x' \odot y \in I$, $x \in I$, and (I2) it follows that $x' \to (x' \odot y) \in I$. Thus, by (I3), $y \in I$.

(ii) $\Rightarrow$ (iii) First we show that I satisfies (I3). Suppose $x \in H$ and $x \leq y \in I$. Then by Remark 2.2.5(i), $y' \leq x'$, and so $(y' \to x')' = 0$. Hence, by (H2) and Remark 2.2.5(ii), $y' \odot x \leq (y' \odot x)'' = 0$ entails that $y' \odot x = 0 \in I$. Since $y \in I$, by (ii), $x \in I$. Now, let $(x' \to y')' \in I$ and $x \in I$ for some $x, y \in H$. By Remark 2.2.5(ii), $x' \odot y \leq (x' \odot y)'' = ((x' \odot y)')' = (x' \to y')'$, consequently $x' \odot y \in I$. From $x \in I$ and (ii) it follows that $y \in I$.

(iii) $\Rightarrow$ (i) It is clear that conditions (I1) and (I2) hold. Let $x \in H$ and $x \leq y \in I$. By Remark 2.2.5(i), $y' \leq x'$, equivalently, $y' \to x' = 1$, and so $(y' \to x')' = 0 \in I$. Since $y \in I$, by (iii), $x \in I$. Therefore, $I \in \mathcal{I}(\mathbf{H})$. $\qquad\qquad\square$

In the next proposition, we investigate the relation between filters and ideals in bounded hoops. For this purpose, given a non-empty subset X of a bounded hoop $\mathbf{H}$, we define $X' = \{x' : x \in X\}$. Clearly, if $\mathbf{H}$ satisfies (DNP), then $x \in X'$ if and only if $x' \in X$.

Proposition 4.4.6 ([1]) *If* $\mathbf{H} = (H; \odot, \rightarrow, 0, 1)$ *is a bounded hoop satisfying* (DNP) *and* $I \subseteq H$, *then* $I \in \mathcal{I}(\mathbf{H})$ *if and only if* $I' \in \mathcal{F}(\mathbf{H})$.

Proof Let $I \in \mathcal{I}(\mathbf{H})$ and $F := I'$. We have $1 = 0' \in I' = F$. Choose $x, y \in F$. Then $x', y' \in I$ and so by (I2), $x \rightarrow y' = x'' \rightarrow y' \in I$. It follows from Lemma 2.2.1(8) that $(x \odot y)' = x \rightarrow y' \in I$, equivalently, $x \odot y = (x \odot y)'' \in F$. Now, suppose $x \leq y$ and $x \in F$. Then $x' \in I$ and by Remark 2.2.5(i), $y' \leq x'$. Since $I \in \mathcal{I}(\mathbf{H})$ and $x' \in I$, we get $y' \in I$, entails that $y \in F$.

Conversely, assume that $F \in \mathcal{F}(\mathbf{H})$ and $I := F'$. Then $0 = 1' \in F' = I$. Let $x \leq y$ and $y \in I$. Due to Remark 2.2.5(i), $y' \leq x'$ and $y' \in F$, thus, $x' \in F$ and so $x = x'' \in I$. Suppose $x, y \in I$. Then $x', y' \in F$. By (F1), $x' \odot y' \in F$, equivalently $(x' \odot y')' \in F' = I$. It follows from Lemma 2.2.1(8) that $x' \rightarrow y = x' \rightarrow y'' = (x' \odot y')' \in I$. Hence, (I2) holds, and $I \in \mathcal{I}(\mathbf{H})$. $\square$

The following example shows that the condition (DNP) is necessary in Proposition 4.4.6 (Fig. 4.6).

Example 4.4.7 ([1]) Let $H = \{0, a, b, c, d, e, f, 1\}$. Define two operations $\odot$ and $\rightarrow$ on H by Table 4.10.

Routine calculations confirm that $\mathbf{H} = (H; \odot, \rightarrow, 0, 1)$ is a bounded hoop. Moreover, $I = \{0, a, b, c\} \in \mathcal{I}(\mathbf{H})$, but $I' = \{1, d\} \notin \mathcal{F}(\mathbf{H})$, since $d \leq e, f$ and $e, f \notin F$.

Proposition 4.4.8 *Let* $\mathbf{H}$ *be a bounded* $\vee$-*hoop and* $I \in \mathcal{I}(\mathbf{H})$. *Then the following statements hold for any* $x, y \in H$:

(i) $x, y \in I$ *if and only if* $x \vee y \in I$.
(ii) *If* $x, y \in I$, *then* $x \wedge y \in I$.

Proof Let $I \in \mathcal{I}(\mathbf{H})$ and $x, y \in H$.

(i) If $x \vee y \in I$, then by (I3), $x, y \in I$, evidently. Suppose that $x, y \in I$. Then by Lemma 2.2.3(i), $x' \odot (x \vee y) = (x' \odot x) \vee (x' \odot y) = x' \odot y$, see Remark 2.2.5(iv). Since

Fig. 4.6 Hasse diagram of $\mathbf{H}$ in
Example 4.4.7

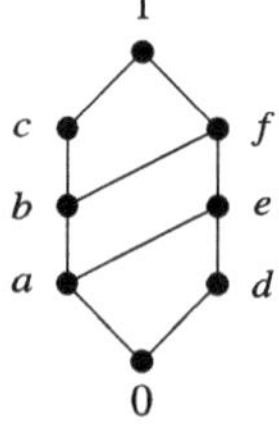

Table 4.10 Operations $\odot$ and $\to$ of Example 4.4.7

Table 4.9.1

$\odot$	0	a	b	c	d	e	f	1
0	0	0	0	0	0	0	0	0
a	0	a	a	a	0	a	a	a
b	0	a	a	b	0	a	a	b
c	0	a	b	c	0	a	b	c
d	0	0	0	0	d	d	d	d
e	0	a	a	a	d	e	e	e
f	0	a	a	b	d	e	e	f
1	0	a	b	c	d	e	f	1

Table 4.9.2

$\to$	0	a	b	c	d	e	f	1
0	1	1	1	1	1	1	1	1
a	d	1	1	1	d	1	1	1
b	d	f	1	1	d	f	1	1
c	d	e	f	1	d	e	f	1
d	c	c	c	c	1	1	1	1
e	0	c	c	c	d	1	1	1
f	0	b	c	c	d	f	1	1
1	0	a	b	c	d	e	f	1

$x' \odot y \le y$, $y \in I$ and $I \in \mathcal{I}(\mathbf{H})$, we get $x' \odot y \in I$, entails that $x' \odot (x \vee y) \in I$. Now, Proposition 4.4.5 implies that $x \vee y \in I$.

(ii) The proof is clear by (I3), since $x \wedge y \le x \in I$. $\qquad\square$

In the following example, we show that the converse of Proposition 4.4.8(ii), may not be true, in general. Indeed, according to Example 4.4.2, $I = \{0, a\} \in \mathcal{I}(\mathbf{H})$ and $a \wedge b = a \odot (a \to b) = a \odot d = 0 \in I$ but $b \notin I$.

Proposition 4.4.9 ([1]) *Let I be a subset of a bounded hoop $\mathbf{H}$ such that $0 \in I$. Then the following conditions are equivalent:*

(i) $I \in \mathcal{I}(\mathbf{H})$.
(ii) $L(x, y) = \{z \in H : z \odot x' \le y\} \subseteq I$ *for any* $x, y \in I$.
(iii) *If* $(z \odot x') \odot y' = 0$, *then* $z \in I$ *for any* $z \in H$ *and* $x, y \in I$.

Proof (i) $\Rightarrow$ (ii) Let $a \in L(x, y)$. Then $a \odot x' \le y$. Since $y \in I$ and $I \in \mathcal{I}(\mathbf{H})$, $a \odot x' \in I$. From $x \in I$ and Proposition 4.4.5 it follows that $a \in I$.

(ii) $\Rightarrow$ (iii) Let $x, y \in I$ and $(z \odot x') \odot y' = 0$. Since $0, y \in I$ and $(z \odot x') \odot y' \le 0$, by (ii), we get $z \odot x' \in L(0, y) \subseteq I$, and so $z \odot x' \in I$. Moreover, since $z \odot x' \le z \odot x'$ and $z \odot x', x \in I$, by (ii), $z \in L(z \odot x', x) \subseteq I$. Hence, $z \in I$.

(iii) $\Rightarrow$ (i) By the assumption, $0 \in I$. Let $x, y \in I$. Then by Lemma 2.2.1(4),

$$(x' \to y) \odot x' \odot y' \le y \odot y' = 0,$$

so (iii) implies that $x' \to y \in I$. Now, suppose $x' \odot y \in I$ and $x \in I$. Since $(y \odot x') \odot (x' \odot y)' = 0$, and so by (iii), $y \in I$. Hence, $I \in \mathcal{I}(\mathbf{H})$. $\qquad\square$

Proposition 4.4.10 ([1]) *Let I be a non-empty subset of a bounded hoop* **H**. *Then for any* $x, y, z \in H$, *the following conditions are equivalent:*

(i) $I \in \mathcal{I}(\mathbf{H})$ *and if* $x \odot y' \odot y' \in I$, *then* $x \odot y' \in I$.
(ii) $0 \in I$ *and if* $x \odot y' \odot y' \odot z' \in I$ *and* $z \in I$, *then* $x \odot y' \in I$.

Proof (i) $\Rightarrow$ (ii) By (I1), $0 \in I$. If $x \odot y' \odot y' \odot z' \in I$ and $z \in I$, then by Proposition 4.4.5(ii), $x \odot y' \odot y' \in I$. In conclusion, by (i), $x \odot y' \in I$.

(ii) $\Rightarrow$ (i) First, we prove that $I \in \mathcal{I}(\mathbf{H})$. Let $x \in H$ and $y \in I$ such that $x \leq y$. Then by Remark 2.2.5(ii), $x \leq y'' = y' \rightarrow 0$, which implies $x \odot y' = 0$. From (ii) and $x \odot 0' \odot 0' \odot y' = x \odot y' = 0 \in I$, it follows that $x = x \odot 0' \in I$, which means (I3) holds. Now, assume that $x, y \in I$. Then by Proposition 3.1.4(i),

$$(x' \rightarrow y) \odot 0' \odot 0' \odot x' = (x' \rightarrow y) \odot x' = x' \wedge y \leq y \in I.$$

Thus, by (I3) and (ii), $x' \rightarrow y = (x' \rightarrow y) \odot 0' \in I$, that is (I2) holds. Thus, I is an ideal. The other condition holds clearly, by considering $z = 0$ in (ii). $\square$

Bibliographical Remarks and Suggestions for Further Study

This chapter has aimed to gather the essential results pertaining to filters and congruences of hoops. As we defined and examined filters in Sect. 2.4 within the more general context of pocrims, these properties are directly inherited by hoop algebras. This section primarily focused on results specific to hoops. The results concerning hoop algebras were gathered from various papers in the first section. The proofs, and in some instances the statements, have been modified and presented in a more generalized form than in their original sources. The majority of the results were drawn from the papers [1, 31, 37, 39, 46, 197]. We have endeavored to verify all the interactions between various types of filters of hoops.

To conclude this chapter, we provide additional references concerning filters, deductive systems, and ideals in hoops. Various types of filters, such as regular filters, nodal filters, and obstinate filters, have been investigated in [2, 4, 227, 229]. The concept of annihilators in hoops was introduced and studied in [42]. It was demonstrated that every annihilator of a hoop is an ideal, and that the set of all annihilators of a hoop forms a Boolean algebra. Fuzzy filters, different kinds of n-fold filters, soju filters, and the radical of filters have been studied in [41, 44, 47, 195]. In [23], the Zariski topology induced by prime, minimal prime, and maximal L-ideals is studied. Furthermore, the concept of product ideals in hoops was introduced and investigated in [40].

4.5   Exercises

4.5.1 Prove that the converse of Theorem 4.2.26 does not hold.

4.5.2 Consider the hoop $\mathbf{H} := \mathbf{H}_1 \oplus \mathbf{H}_2$, where $\mathbf{H}_1$ and $\mathbf{H}_2$ are hoops such that $H_1 \cap H_2 = \{1\}$. Prove that

    (i) $\mathcal{F}(\mathbf{H}) = \{F \cup H_2 : F \in \mathcal{F}(\mathbf{H}_1)\} \cup \mathcal{F}(\mathbf{H}_2)$.

    (ii) $\mathbf{H}/(F \cup H_2) \cong \mathbf{H}_1/F$ for all $F \in \mathcal{F}(\mathbf{H}_1)$. In addition, $\mathbf{H}/F \cong \mathbf{H}_1 \oplus (\mathbf{H}/F)$ for all $F \in \mathcal{F}(\mathbf{H}_2)$.

    (iii) If $\mathbf{H}_1$ is linear, then $\mathrm{Spec}(\mathbf{H}) = \{F \cup H_2 : F \in \mathrm{Spec}(\mathbf{H}_1)\} \cup \mathrm{Spec}(\mathbf{H}_2)$.

    (iv) If $\mathbf{H}_1$ is not linear, then $\mathrm{Spec}(\mathbf{H}) = \{F \cup H_2 : F \in \mathrm{Spec}(\mathbf{H}_1)\}$.

    (v) If $M \in \mathrm{Max}(\mathbf{H}_1)$, then $M \cup H_2 \in \mathrm{Max}(\mathbf{H})$.

4.5.3 Let $\varphi : \mathbf{H}_1 \to \mathbf{H}_2$ be a hoop homomorphism. Prove the following statements:

    (i) If φ is an epimorphism and $I \in \mathcal{I}(\mathbf{H}_2)$, then $\varphi^{-1}(I) \in \mathcal{I}(\mathbf{H}_1)$.

    (ii) If φ is an isomorphism and $I \in \mathcal{I}(\mathbf{H}_1)$, then $\varphi(I) \in \mathcal{I}(\mathbf{H}_2)$.

    (iii) $\varphi^{-1}(\{0\}) = \{x \in H_1 : \varphi(x) = 0\}$ is an ideal of $\mathbf{H}_1$.

4.5.4 Let $\mathbf{H}$ be a cancellative hoop and F be a filter of $\mathbf{H}$ such that $\mathbf{H}/F$ is bounded. Prove that $F = H$ (Hint: Use Corollary 3.3.6).

4.5.5 Let $\mathbf{H}$ be a basic hoop and $x \in H$. Show that $x^{\perp} = \bigcap\{P \in \mathrm{Spec}(\mathbf{H}) : x \notin P\}$.

4.5.6 A proper filter F of a hoop $\mathbf{H}$ is said to be an **obstinate** filter if $x \to y, y \to x \in F$ for all $x, y \in H \setminus F$ (see [229]). Let F be a filter of a hoop $\mathbf{H}$. Prove the following statement:

    (i) If F is an obstinate filter, then $F \in \mathrm{Max}(\mathbf{H})$ and $F \in \mathrm{Spec}(\mathbf{H})$.

    (ii) If F is an obstinate filter, then $\mathbf{H}/F \cong \mathbf{2}$. What about the converse?

    (iii) If $\mathbf{H}$ is a bounded hoop, then F is an obstinate filter of $\mathbf{H}$ if and only if for each $x \in \mathbf{H}$, $x \in F$ or $x' \in F$.

    (iv) The filter $F = \{1, a, b\}$ in Example 4.1.3 is an obstinate filter.

4.5.7 Let F be a proper filter of a hoop $\mathbf{H}$. Prove that

$$\mathrm{Spec}(\mathbf{H/F}) = \{P/F : F \subseteq P \in \mathrm{Spec}(\mathbf{H})\}.$$

4.5.8 A **multiplier** on a hoop $\mathbf{H}$ is a map $f : H \to H$ such that $f(x \to y) = x \to f(y)$ for all $x, y \in H$ (see [194]). Show that the multiplier f satisfies the following conditions:

(i) $x \leq f(x)$ for all $x \in H$ and so $f(1) = 1$.

(ii) If $\mathrm{Mult}(\mathbf{H})$ is the set of all multipliers on $\mathbf{H}$, then $(\mathrm{Mult}(\mathbf{H}); \circ, \mathrm{Id}_H)$ is a monoid, where $\circ$ is the composition of functions.

(iii) $f(x) \to y \leq x \to f(y)$ and $f(x \to y) \leq f(x) \to f(y)$.

4.5.9 A $\odot$-closure operator on a hoop $\mathbf{H}$ is a unary operation $c : H \to H$ satisfies the following conditions (see [194]): (c1) c is an order-preserving map such that $x \leq c(x)$ for all $x \in H$; (c2) $c \circ c = c$ and $c(x) \odot c(y) \leq c(x \odot y)$ for all $x, y \in H$. Prove that each closure operator c on a hoop $\mathbf{H}$ satisfies the following properties for all $x, y \in H$:

(i) $x \leq y \leq c(x)$ implies that $c(x) = c(y)$. In addition, $c(x \odot y) = c(x) \odot c(y)$.

(ii) $F := c^{-1}(1)$ is a filter of $\mathbf{H}$. Moreover, $c(H) = \{x \in H : c(x) = x\}$ and it is closed under $\wedge$ and $\to$.

(iii) $(c(H); \odot_c, \to_c, 1)$ is a hoop, where $x \odot_c y = c(x \odot y)$ and $x \to_c y = c(x \to y)$ for all $x, y \in H$.

4.5.10 Let F be a filter of a $\vee$-hoop $\mathbf{H}$. The **co-annihilator**[1] of a relative to F is the set

$$(F, a) := \{x \in H : x \vee a \in F\}. \tag{4.14}$$

If F and G are filters of $\mathbf{H}$, show that the following statements hold:

(i) (F, a) is a filter of $\mathbf{H}$ containing F for every $a \in H$.

(ii) $a \leq b$ implies that $(F, a) \subseteq (F, b)$ for all $a, b \in H$.

(iii) $F \subseteq G$ implies that $(F, a) \subseteq (G, a)$.

(iv) $(F, a) = H$ if and only if $a \in F$ for every $a \in H$.

(v) $(F, a \odot b) = (F, a \wedge b) = (F, a) \cap (F, b)$ for every $a, b \in H$.

(iv) $(F, a) \cap (G, a) = (F \cap G, a)$ for every $a \in H$.

(iiv) $((F, a), b) = ((F, b), a) = (F, a \vee b)$ for every $a, b \in H$.

4.5.11 A non-empty subset I of a bounded hoop $\mathbf{H}$ is said to be **implicative ideal** if (i) $0 \in I$; (ii) $x' \to y \in I$ for every $x, y \in I$; and (iii) $x \odot y' \odot z' \in I$ and $y' \odot z' \in I$ imply that $x \odot z' \in I$. Prove that every implicative ideal of $\mathbf{H}$ is an ideal of $\mathbf{H}$.

---

[1] This concept was introduced in [205] for BL-algebras.

4.5.12 Let I be an ideal of a bounded hoop $\mathbf{H}$. Show that the following statements are equivalent:

    (i) I is an implicative ideal.

    (ii) $x \odot y'' \odot y'' \in I$ implies that $x \odot y \in I$ for every $x, y \in H$.

    (iii) $x^2 \in I$ implies that $x \in I$ for every $x \in H$.

Algebraic Structures Related to Hoops $\qquad$ 5

This chapter explores the intricate connections between hoops and a diverse range of related algebraic structures, demonstrating their fundamental role in algebraic logic. We investigate the relationships between hoops and residuated lattices, MTL-algebras, BL-algebras, MV-algebras, product algebras, $R\ell$-monoids, BCK-algebras, Wajsberg algebras, Heyting and Hertz algebras, lattice implication algebras, Hilbert algebras, and L-algebras. Specifically, we establish key equivalences:

- Bounded $\vee$-hoops are equivalent to divisible residuated lattices.
- Bounded basic hoops are equivalent to BL-algebras
- Bounded Wajsberg hoops are equivalent to both MV-algebras and Wajsberg algebras.
- Product algebras are equivalent to bounded product hoops.
- Hoops are equivalent to BCK-$\wedge$-semilattices with the properties (P) and (Div).

These equivalences, being structural and thus categorical, provide powerful tools for understanding hoops by leveraging the established theories of these related algebras. These relationships can be beneficial for individuals interested in any of the aforementioned algebraic structures.

5.1 Subclasses of Residuated Lattices

This section delves into the intriguing relationships between hoops and some well-established ordered algebraic structures, such as residuated lattices, BL-algebras, MV-algebras, etc. Before we proceed, let us refresh our memory on these essential definitions from earlier sections:

© The Author(s), under exclusive license to Springer Nature Switzerland AG 2026

A. Dvurečenskij et al., *Hoop Algebras*, Frontiers in Mathematics,
https://doi.org/10.1007/978-3-032-11736-6_5

Definition 5.1.1 ([148, 258]) A **residuated lattice**[1] is an algebraic structure $(L; \vee, \wedge, \odot, \rightarrow, 0, 1)$ of type $(2, 2, 2, 2, 0, 0)$ if for all $x, y, z \in L$ it satisfies the following axioms:

(RL1) $(L; \vee, \wedge, 0, 1)$ is a bounded lattice.
(RL2) $(L; \odot, 1)$ is a commutative monoid.
(RL3) $x \odot y \leq z$ if and only if $x \leq y \rightarrow z$.

On each bounded residuated lattice, we set $x' := x \rightarrow 0$ for all $x \in L$.

Proposition 5.1.2 ([148, 235]) *Let* $(L; \vee, \wedge, \odot, \rightarrow, 0, 1)$ *be a residuated lattice. Then, for any* $x, y, z \in L$, *the following conditions are equivalent:*

(r_1) $x \rightarrow (y \vee z) = (x \rightarrow y) \vee (x \rightarrow z)$.
(r_2) $(x \wedge y) \rightarrow z = (x \rightarrow z) \vee (y \rightarrow z)$.
(r_3) $(x \rightarrow y) \vee (y \rightarrow x) = 1$.

Recal that (see [70, 136, 167]) a residuated lattice $(L; \vee, \wedge, \odot, \rightarrow, 0, 1)$ is called:

(i) A **divisible residuated lattice** if

$$x \wedge y = x \odot (x \rightarrow y) \qquad \forall x, y \in L. \tag{Div}$$

(ii) An **MTL-algebra** if

$$(x \rightarrow y) \vee (y \rightarrow x) = 1 \qquad \forall x, y \in L. \tag{Pre}$$

(iii) A **BL-algebra** if L is an MTL-algebra satisfying (Div).
(iv) An **MV-algebraic residuated lattice**[2] if L is a BL-algebra and

$$x'' = x \qquad \forall x \in L. \tag{DNP}$$

(v) A **product algebra** if L is a BL-algebra and satisfying the following conditions:

(PA1) $x \wedge x' = 0$ for all $x \in L$.
(PA2) $x'' \leq \big((x \odot z) \rightarrow (x \odot y)\big) \rightarrow (z \rightarrow y)$ for all $x, y, z \in L$.

Remark 5.1.3 It is important to note that not every residuated lattice is distributive (see [148, p. 164]). However, as shown in [136, p. 275], the lattice reduct of every MTL-algebra must be distributive.

---

[1] More specifically, it is known as a bounded commutative residuated lattice; here, in this section, we call it a residuated lattice.
[2] In some literature, MV-algebras are defined as BL-algebras satisfying (DNP).

Theorem 5.1.4 $\mathbf{H} = (H; \odot, \rightarrow, 0, 1)$ *is a bounded* $\vee$*-hoop if and only if* $\mathbf{H}^* = (H; \vee, \wedge, \odot, \rightarrow, 0, 1)$ *is a divisible residuated lattice.*

Proof Let $\mathbf{H} = (H; \odot, \rightarrow, 0, 1)$ be a bounded $\vee$-hoop. By definition of hoops, $(H; \odot, 1)$ is a commutative monoid. In addition, the lattice structure $(L; \vee, \wedge)$ is a bounded distributive lattice (by Lemma 3.2.10) with the least element 0 and top element 1. Now, (2.2) implies that $\mathbf{H}^* = (H; \vee, \wedge, \odot, \rightarrow, 0, 1)$ is a residuated lattice. Moreover, by Theorem 3.1.5, $\mathbf{H}^*$ is a divisible residuated lattice.

The proof of the converse is straightforward by definition. That is, for each divisible residuated lattice $\mathbf{L} = (L; \vee, \wedge, \odot, \rightarrow, 0, 1)$, the reduct $\mathbf{L}^\circ = (L; \odot, \rightarrow, 0, 1)$ is a bounded $\vee$-hoop. Evidently, $(\mathbf{L}^\circ)^* = \mathbf{L}$ and $(\mathbf{H}^*)^\circ = \mathbf{H}$ for each bounded basic hoop $\mathbf{H}$ and each BL-algebra $\mathbf{L}$. $\qquad\square$

Residuated lattices and hoops are incomparable structures. While hoops form $\wedge$-semilattices under the operation $a \wedge b = a \odot (a \rightarrow b)$, they lack the full lattice structure. Conversely, a counterexample will demonstrate that not all residuated lattices qualify as hoops.

Example 5.1.5 Let $L = [0, 1]$ be the real unit interval, and for all $x, y \in L$, define the operations $\wedge$, $\vee$, $\odot$ and $\rightarrow$ on L as follows:

$$x \wedge y = \min\{x, y\}, \qquad x \vee y = \max\{x, y\},$$

$$x \odot y = \begin{cases} 0 & \text{if } x + y \leq 1 \\ x \wedge y & \text{if } x + y > 1, \end{cases} \qquad x \rightarrow y = \begin{cases} 1 & \text{if } x \leq y \\ (1 - x) \vee x & \text{if } x > y. \end{cases}$$

Then $(L; \vee, \wedge, \odot, \rightarrow, 0, 1)$ is a residuated lattice. Now, let $x > y$ for some $x, y \in L$. Then $x \odot (x \rightarrow y) = x \odot ((1 - x) \vee x)$. If $(1 - x) \vee x = 1 - x$, since $x + (1 - x) = 1$, we have $x \odot (1 - x) = 0 \neq y = x \wedge y$. If $(1 - x) \vee x = x$, then $x \odot x = 0$ or x, and so $x \wedge y = y \neq x \odot (x \rightarrow y)$. Therefore, $(L; \odot, \rightarrow, 1)$ is not a hoop.

Theorem 5.1.6 $\mathbf{H} = (H; \odot, \rightarrow, 0, 1)$ *is a bounded basic hoop (equivalently,* $\vee$*-hoop with condition* (Pre)*) if and only if* $\mathbf{H}^* = (H; \wedge, \vee, \odot, \rightarrow, 0, 1)$ *is an MTL-algebra with condition* (Div)*.*

Proof Recall that, by Theorem 3.3.28, every bounded basic hoop is a bounded $\vee$-hoop with condition (Pre), and conversely. The rest of the proof is straightforward by Theorem 5.1.4. $\qquad\square$

Corollary 5.1.7 $\mathbf{A} = (A; \vee, \wedge, \odot, \rightarrow, 0, 1)$ *is a BL-algebra if and only if* $\mathbf{A}^\circ = (A; \odot, \rightarrow, 0, 1)$ *is a bounded basic hoop.*

Proof According to Definition 5.1(iii), every MTL-algebra with condition (Div) is a BL-algebra, then by Theorem 5.1.6, the proof is clear.   □

Proposition 5.1.8 ([11, Theorem 1.7]) *The class of bounded basic hoops is termwise equivalent to the class of BL-algebras.*

Proof If $\mathbf{H} = (H; \odot, \rightarrow, 0, 1)$ is a bounded basic hoop, then for all $x, y \in H$ we can define $x \wedge y$ and $x \vee y$. Then, algebra $\mathbf{H}^* = (H; \vee, \wedge, \odot, \rightarrow, 0, 1)$ is a BL-algebra. Conversely, if $\mathbf{B} = (B; \vee, \wedge, \odot, \rightarrow, 0, 1)$ is a BL-algebra, then by Corollary 5.1.7, its $\{\odot, \rightarrow, 0, 1\}$-reduct, i.e., $\mathbf{B}^\circ = (B; \odot, \rightarrow, 0, 1)$ is a bounded basic hoop. Therefore, by Theorem 5.1.4, these classes are termwise equivalent.   □

Consider the notation was defined in Proposition 3.2.5.

Theorem 5.1.9 *A bounded hoop with* (DNP) *is a* $\vee$*-hoop. In addition,* $\mathbf{H} = (H; \odot, \rightarrow, 0, 1)$ *is a bounded hoop with* (DNP) *and* (Pre) *if and only if* $\mathbf{H}^* = (H; \vee, \wedge, \odot, \rightarrow, 0, 1)$ *is an MV-algebraic residuated lattice (or equivalently an MV-algebra in the sense of Definition* 5.1.2(iv)*).*

Proof First, assume that $\mathbf{H}$ is a bounded hoop with (DNP). By Proposition 3.1.4, $x' \wedge y'$ exists. We claim that $(x' \wedge y')' = x \vee y$. From $x' \wedge y' \le x', y'$ and Lemma 2.2.1(7) it follows that $x'', y'' \le (x' \wedge y')'$, so by (DNP), $(x' \wedge y')'$ is an upper bound for x and y. Now, let $z \in H$ such that $x, y \le z$. Applying Lemma 2.2.1(7) and (DNP), we get that $z' \le x' \wedge y'$, entails that $(x' \wedge y')' \le z'' = z$. Therefore, $x \vee y = (x' \wedge y')'$.

For the second part, we apply Theorem 3.3.13. Let $\mathbf{H} = (H; \odot, \rightarrow, 0, 1)$ be a bounded hoop with conditions (Pre) and (DNP). By the first part, $\mathbf{H}$ is a $\vee$-hoop. Also, by Corollary 5.1.7, $\mathbf{H}^* = (H; \vee, \wedge, \odot, \rightarrow, 0, 1)$ is a BL-algebra. Since H has (DNP), by Definition 5.1(iv), $\mathbf{H}^* = (H; \vee, \wedge, \odot, \rightarrow, 0, 1)$ is an MV-algebra.

Conversely, let $\mathbf{A} = (H; \vee, \wedge, \odot, \rightarrow, 0, 1)$ be an MV-algebra. Since any MV-algebra is a BL-algebra with (DNP), by Corollary 5.1.7, $\mathbf{A}^\circ = (A; \odot, \rightarrow, 0, 1)$ is a bounded $\vee$-hoop with (DNP) and (Pre).   □

Now, we are ready to show that MV-algebra (see Definition 1.4.1) and MV-algebraic residuated lattice (see Definition 5.1) are termwise equivalent.

Corollary 5.1.10 *The class of MV-algebras is termwise equivalent to the class of residuated lattices satisfying* (Div), (Pre), *and* (DNP).

Proof The proof is a direct consequence of Theorems 5.1.9, 3.3.15, and 3.3.13.   □

Proposition 5.1.11 *Let* **H** *be a basic hoop. Then* $([0, a]; \vee, \wedge, \odot^a, \rightarrow, a, 1)$ *is a BL-algebra for each* $a \in H$. *In addition, if* **H** *is a Wajsberg hoop, then* $([0, a]; \vee, \wedge, \odot^a, \rightarrow, a, 1)$ *is an MV-algebra.*

Proof Let $a \in H$. Since **H** is a $\vee$-hoop (by Lemma 3.3.25(iv)), we have $a \in C_\vee(\mathbf{H})$. Hence, by Proposition 3.2.5, $([0, a]; \odot^a, \rightarrow, a, 1)$ is a bounded hoop. Evidently, $([0, a]; \odot^a, \rightarrow, a, 1)$ satisfies (3.9). Hence, $([0, a]; \odot^a, \rightarrow, a, 1)$ is a bounded basic hoop. Therefore, $([0, a]; \vee, \wedge, \odot^a, \rightarrow, a, 1)$ is a BL-algebra by Proposition 5.1.8. The proof of the second part is straightforward by Theorem 5.1.9. We note that if **H** is a Wajsberg hoop, then $(x \rightarrow y) \rightarrow y = (y \rightarrow x) \rightarrow x$ for every $x, y \in H$, which implies $([0, a]; \odot^a, \rightarrow, a, 1)$ is also a Wajsberg hoop. $\qquad\square$

From Theorems 5.1.9 and 3.3.15 it follows that:

$$\boxed{Bounded\,hoop + (DNP)} \Leftrightarrow \boxed{MV - algebra} \Leftrightarrow \boxed{Bounded\,Wajsberg\,hoop} \qquad (5.1)$$

Example 5.1.12 Let $H = \{0, a, b, 1\}$ be a poset such that $0 < a, b < 1$. Define the operations $\odot$ and $\rightarrow$ on H by Table 5.1:
  Then $\mathbf{H} = (H; \odot, \rightarrow, 0, 1)$ is a $\vee$-hoop with (DNP) and (Pre).

Theorem 5.1.13 ([217]) *A BL-algebra* $\mathbf{H} = (H; \vee, \wedge, \odot, \rightarrow, 0, 1)$ *is a product algebra if and only if it satisfies the following identity*

$$x' \vee \big((x \rightarrow (x \odot y)) \rightarrow y\big) = 1 \quad \forall x, y \in H. \qquad (\text{Prod})$$

Proof Let $\mathbf{H} = (H; \vee, \wedge, \odot, \rightarrow, 0, 1)$ be a BL-algebra. Due to Corollary 5.1.7, $(H; \odot, \rightarrow, 0, 1)$ is a bounded basic hoop. First, suppose that **H** is a product algebra and $x, y \in H$.

$$x'' \leq \big((1 \odot x) \rightarrow (y \odot x)\big) \rightarrow (1 \rightarrow y) = \big(x \rightarrow (y \odot x)\big) \rightarrow y, \text{ by (P2)}.$$

Table 5.1 Operations $\odot$ and $\rightarrow$ of Example 5.1.12

Table 5.1.1

$\odot$	0	a	b	1
0	0	0	0	0
a	0	a	0	a
b	0	0	b	b
1	0	a	b	1

Table 5.1.2

$\rightarrow$	0	a	b	1
0	1	1	1	1
a	b	1	b	1
b	a	a	1	1
1	0	a	b	1

It follows from (P1) and Corollary 3.3.35 that

$$1 = (x \wedge x')' = x' \vee x'' \leq x' \vee \big((x \to (y \odot x)) \to y\big).$$

Thus, (Prod) holds. Conversely, assume that $\mathbf{H}$ is a BL-algebra satisfying (Prod). Choose $x, y \in H$. We have

$$1 = x' \vee \big((x \to (x \odot 0)) \to 0\big) = x' \vee x'',$$

so, Corollary 3.3.35 implies that

$$\begin{aligned}
0 = (x' \vee x'')' = x'' \wedge x''', &\quad \text{by Cor 3.3.35} \\
= x'' \wedge x', &\quad \text{by Remark 2.2.5(iii)} \\
\geq x \wedge x', &\quad \text{by Remark 2.2.5(ii)},
\end{aligned}$$

which means (P1) holds. On the other hand, by (P1) and Theorem 3.3.27, for each $z \in H$, $x' \vee z = 1$ implies that $x'' = x'' \wedge (x' \vee z) = (x'' \wedge x') \vee (x'' \wedge z) = x'' \wedge z$, that is $x'' \leq z$. By the assumption, $x' \vee \big((x \to (x \odot y)) \to y\big) = 1$, so by

$$\begin{aligned}
1 &= x' \vee \big((x \to (x \odot y)) \to y\big) \\
&\leq x' \vee \Big(\big(z \to (x \to (x \odot y))\big) \to (z \to y)\Big), \text{ by Lemma 2.2.1(11)} \\
&= x' \vee \Big(\big((x \odot z) \to (x \odot y)\big) \to (z \to y)\Big), \text{ by Lemma 2.2.1(8)}
\end{aligned}$$

we get $x'' \leq \big((x \odot z) \to (x \odot y)\big) \to (z \to y)$, i.e., (P2) holds. $\qquad\square$

We recall that by Remark 3.3.52, a bounded hoop $\mathbf{H}$ is a product hoop if and only if it satisfies (Prod).

Theorem 5.1.14 *The class of product algebras is termwise equivalent to the class of bounded product hoops.*

Proof Let $\mathbf{H} = (H; \vee, \wedge, \odot, \to, 0, 1)$ be a product algebra. According to Theorem 5.1.6 and Corollary 5.1.7, $\mathbf{H}^\circ = (H; \odot, \to, 0, 1)$ is a bounded basic hoop. Choose $x, y, z \in H$. Since $x \to 0 \leq x \to z$, we have

$$(x \to z) \vee \big((x \to (x \odot y)) \to y\big) \geq x' \vee \big((x \to (x \odot y)) \to y\big) = 1, \text{ by Theorem 5.1.12.}$$

Hence, (PH) holds, which means $(\mathbf{H}; \odot, \to, 0, 1)$ is a bounded product hoop. Conversely, for each bounded product hoop $\mathbf{H} = (\mathbf{H}; \odot, \to, 0, 1)$, by Theorem 5.1.6, $\mathbf{H}^* = (H; \vee, \wedge, \odot, \to, 0, 1)$ is a BL-algebra. In addition, considering $z = 0$ in the identity ((PH)), we have

$$(x \to 0) \vee \big((x \to (x \odot y)) \to y\big) = 1, \quad \forall x, y \in H.$$

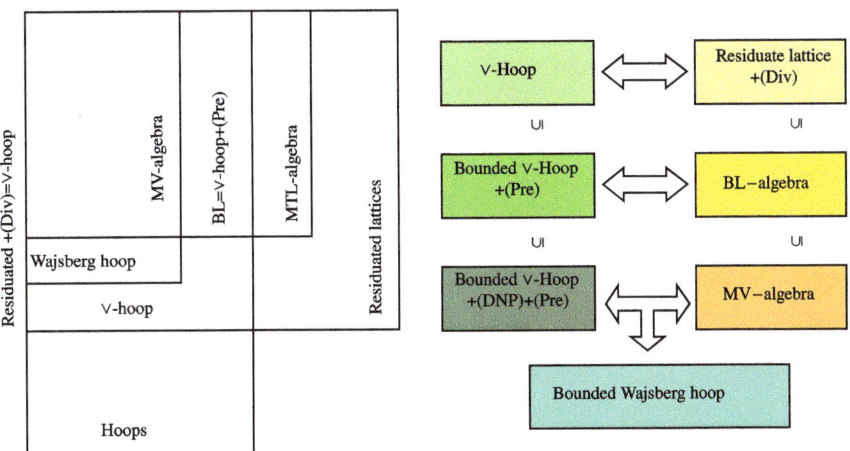

Fig. 5.1 Relationships between hoops, MV, BL, MTL-algebras, and residuated lattices

Thus, (Prod). holds, so due to Theorem 5.1.13, $\mathbf{H}^*$ is a product algebra. Therefore, by Theorem 5.1.4, these classes are termwise equivalent. The proof of the other part is a direct consequence of Theorem 5.1.13 and Proposition 5.1.8. $\qquad\square$

In Fig. 5.1, we present a relation between hoops and the algebraic structures investigated in this section.

The final algebraic structure we will examine in this section is the bounded $R\ell$-monoid. We will demonstrate that every commutative bounded $R\ell$-monoid is a reduct of a bounded ∨-hoop.

According to [124], a **bounded $R\ell$-monoid** is an algebra $\mathbf{M} = (M; \odot, \vee, \wedge, \rightarrow, \rightsquigarrow, 0, 1)$ of type $(2, 2, 2, 2, 2, 0, 0)$ satisfying the following conditions:

$(R\ell1)$ $(M; \odot, 1)$ is a monoid.

$(R\ell2)$ $(M; \vee, \wedge, 0, 1)$ is a bounded lattice.

$(R\ell3)$ $x \odot y \leq z$ if and only if $x \leq y \rightarrow z$ if and only if $y \leq x \rightsquigarrow z$ for all $x, y, z \in M$.

$(R\ell4)$ $\mathbf{M}$ satisfies the identities $(x \rightarrow y) \odot x = x \wedge y = y \odot (y \rightsquigarrow x)$.

An $R\ell$-monoid $(M; \odot, \vee, \wedge, \rightarrow, \rightsquigarrow, 0, 1)$ is commutative if $\odot$ is commutative.

Lemma 5.1.15 ([124]) *An $R\ell$-monoid $(M; \odot, \vee, \wedge, \rightarrow, \rightsquigarrow, 0, 1)$ is commutative if and only if $\rightarrow = \rightsquigarrow$.*

Proof Let M be commutative. For each $x, y, z \in M$ we have $x \leq y \rightarrow z$ iff $x \odot y \leq z$ iff $x \leq y \rightsquigarrow z$. Hence, $x \rightarrow y = x \rightsquigarrow y$ for all $x, y \in M$. Conversely, let $\rightarrow = \rightsquigarrow$ and $x, y \in M$. For every $z \in H$, $x \odot y \leq z$ iff $x \leq y \rightarrow z$ iff $x \leq y \rightsquigarrow z$ iff $y \odot x \leq z$. Thus, $x \odot y = y \odot x$ for all $x, y \in M$. $\qquad\square$

Proposition 5.1.16 *For each commutative bounded $R\ell$-monoid $(M; \odot, \vee, \wedge, \rightarrow, \rightsquigarrow, 0, 1)$, the reduct $(M; \odot, \rightarrow, 0, 1)$ is a bounded $\vee$-hoop.*

Proof The proof is straightforward by Lemma 5.1.15 and the definition of hoops. $\square$

5.1.1 Relation with BCK-Algebras

In Section 2.5, we studied the relations between BCK-algebras and pocrims. By Theorem 2.5.5, every pocrims is a BCK-algebra. In addition, by Theorem 2.5.12, for each BCK(P)-algebra $(X; \rightarrow, 1)$, the structure $(X; \odot, \rightarrow, 1)$ is a pocrim, where $x \odot y = \min\{z \in X : x \leq y \rightarrow z\}$. Applying Theorem 2.5.12 and Proposition 3.1.5, we get the following theorem

Theorem 5.1.17 ([77]) *For every hoop $(H; \odot, \rightarrow, 1)$, the reduct $(H; \rightarrow, 1)$ is a BCK-algebra. In addition, hoops are termwise equivalent to BCK-$\wedge$-semilattices with (P) and (Div).*

Proof If $\mathbf{H} = (H; \odot, \rightarrow, 1)$ is a hoop, then by Theorem 2.5.5, $f(\mathbf{H}) := (H; \rightarrow, 1)$ is a BCK-algebra. Clearly, $f(\mathbf{H})$ is a BCK-$\wedge$-semilattice with (P) and (Div). Now, let $(X; \rightarrow, 1)$ be a BCK-$\wedge$-semilattice with (P) and (Div). Then by Theorem 2.5.12, $g(\mathbf{X}) := (X; \odot, \rightarrow, 1)$ is a pocrim. Given $x, y \in X$ from (Div), we have $y \odot (y \rightarrow x) = y \wedge x = x \wedge y = x \odot (x \rightarrow y)$ so by Proposition 3.1.5, $g(\mathbf{X})$ is a hoop. An easy calculation shows that $g \circ f(\mathbf{H}) = \mathbf{H}$ and $f \circ g(\mathbf{X}) = \mathbf{X}$. Therefore, hoops are termwise equivalent to BCK-$\wedge$-semilattices with (P) and (Div). $\square$

A BCK-algebra $\mathbf{B}$ is called **implicative** if $x = (x \rightarrow y) \rightarrow x$.

Lemma 5.1.18 ([180]) *Let $\mathbf{B} = (B; \rightarrow, 1)$ be a BCK(P)-algebra. Then:*

(i) $(B; \odot, 1)$ *is a commutative monoid with the greatest element* 1.
(ii) *If $\mathbf{B}$ is a BCK(P)-$\wedge$-lattice such that $x^2 = x$, for any $x \in B$, then $x \wedge y = x \odot (x \rightarrow y)$, for any $x, y \in B$.*
(iii) $x \rightarrow (y \rightarrow z) = (x \odot y) \rightarrow z$ *for any $x, y, z \in B$.*

Proof (i) and (iii) follow from Proposition 2.5.11.

(ii) By Proposition 2.5.11, $(B; \odot, \rightarrow, 1)$ is a pocrim. Let $x, y \in B$. Applying Lemma 2.2.3(ii) we have $y \rightarrow (x \wedge y) = (y \rightarrow x) \wedge (y \rightarrow y) = y \rightarrow x \geq x$, so by definition of $x \odot y$, we get $x \odot y \leq x \wedge y$.

Also, by the assumption, $x \wedge y = (x \wedge y) \odot (x \wedge y) \leq x \odot y$, see Lemma 2.2.1(3) (since $(B; \odot, \rightarrow, 1)$ is a pocrim). Hence, $x \odot y = x \wedge y$ for all $x, y \in B$. On the other hand,

$x \odot (x \to y) \leq x, y$, by Lemma 2.2.1(4). Therefore, $y \leq x \to y$ and Lemma 2.2.1(3) imply that $x \odot y \leq x \odot (x \to y) \leq x \wedge y = x \odot y$. $\square$

Theorem 5.1.19 *Let* $\mathbf{B} = (B; \to, 1)$ *be a* $BCK(P)$-$\wedge$-*lattice. If* $x^2 = x$, *for any* $x \in B$, *then* $\mathbf{B}^\circ = (B; \odot, \to, 1)$ *is a hoop.*

Proof Let $\mathbf{B}$ be a BCK(P)-lattice. By Lemma 5.1.18(i), $(B; \odot, 1)$ is a commutative monoid. Since, $x^2 = x$, for any $x \in B$, by Lemma 5.1.18(ii), $x \wedge y = x \odot (x \to y)$, and so Theorem 3.1.2(H3) holds. Moreover, since B is a BCK-algebra, (H1) holds, evidently, and by Lemma 5.1.18(iii), $x \to (y \to z) = (x \odot y) \to z$. Therefore, $\mathbf{B}^\circ = (B; \odot, \to, 1)$ is a hoop, see Theorem 3.1.2. $\square$

Example 5.1.20 Let $B = \{0, a, b, 1\}$ be a chain such that $0 < a < b < 1$. Define the operation $\to$ on B by Table 5.2,

Then $(B; \to, 0, 1)$ is a bounded BCK-algebra. Since B is a chain, it is easy to see that B is a BCK-lattice. The operation $\odot$ on B is given by Table 5.3:

Thus, $(B; \odot, \to, 0, 1)$ is a BCK(P)-$\wedge$-lattice such that for any $x \in B$, $x^2 = x$.

Theorem 5.1.21 *For each bounded hoop* $\mathbf{H} = (H; \odot, \to, 0, 1)$ *with* (DNP), *the reduct* $\mathbf{H}^\circ = (H; \to, 1)$ *is a bounded commutative BCK-algebra.*

Table 5.2 Operation $\to$ of Example 5.1.20

$\to$	0	a	b	1
0	1	1	1	1
a	0	1	1	1
b	0	a	1	1
1	0	a	b	1

Table 5.3 Operation $\odot$ of Example 5.1.20

$\odot$	0	a	b	1
0	0	0	0	0
a	0	a	a	a
b	0	a	b	b
1	0	a	b	1

Proof Let $\mathbf{H} = (H; \odot, \rightarrow, 0, 1)$ be a bounded hoop with (DNP). By Theorem 5.1.17, $\mathbf{H}^\circ = (H; \rightarrow, 1)$ is a BCK-algebra which is bounded, and by Theorem 3.3.15, $(y \rightarrow x) \rightarrow x = (x \rightarrow y) \rightarrow y$ for all $x, y \in H$. Therefore, $\mathbf{H}^\circ = (H; \rightarrow, 1)$ is a bounded commutative BCK-algebra. $\qquad\square$

The following example shows that every $\{\rightarrow, 1\}$-reduct of a bounded hoop is not necessarily a commutative BCK-algebra, in general.

Example 5.1.22 Let $H = \{0, a, b, c, 1\}$ be a poset such that $0 < c < a, b < 1$, but a and b are incomparable. Define the operations $\odot$ and $\rightarrow$ on H by Table 5.4:

Then $\mathbf{H} = (H; \odot, \rightarrow, 0, 1)$ is a bounded hoop. Since $(c \rightarrow 0) \rightarrow 0 = 1 \neq c$, $\mathbf{H}$ does not satisfy (DNP). Moreover, by Theorem 5.1.17, $(H; \rightarrow, 1)$ is a BCK-algebra, but it is not commutative, because,

$$(a \rightarrow 0) \rightarrow 0 = 1 \neq a = (0 \rightarrow a) \rightarrow a.$$

Proposition 5.1.23 *Let* $\mathbf{H} = (H; \odot, \rightarrow, 0, 1)$ *be a bounded hoop with* (DNP) *such that* $x^2 = x$, *for any* $x \in H$. *Then* $\mathbf{H}^\circ = (H; \rightarrow, 1)$ *is an implicative BCK-algebra.*

Proof Let $\mathbf{H} = (H; \odot, \rightarrow, 0, 1)$ be a bounded hoop with (DNP) such that $x^2 = x$ for any $x \in H$. By Theorem 5.1.21, $\mathbf{H}^\circ = (H; \rightarrow, 1)$ is a commutative BCK-algebra. From

$$\begin{aligned}
\big((x \rightarrow y) \rightarrow x\big) \rightarrow x &= \big(x \rightarrow (x \rightarrow y)\big) \rightarrow (x \rightarrow y) \\
&= \big((x \odot x) \rightarrow y\big) \rightarrow (x \rightarrow y) \\
&= (x \rightarrow y) \rightarrow (x \rightarrow y) = 1,
\end{aligned}$$

it follows that $(x \rightarrow y) \rightarrow x \leq x$. Now, Lemma 2.2.1(6) implies that $(x \rightarrow y) \rightarrow x = x$. Therefore, $\mathbf{H}^\circ = (H; \rightarrow, 0)$ is an implicative BCK-algebra. $\qquad\square$

Table 5.4 Operations $\odot$ and $\rightarrow$ of Example 5.1.22

Table 5.4.1

$\odot$	0	c	a	b	1
0	0	0	0	0	0
c	0	c	c	c	c
a	0	c	a	c	a
b	0	c	c	b	b
1	0	c	a	b	1

Table 5.4.2

$\rightarrow$	0	c	a	b	1
0	1	1	1	1	1
c	0	1	1	1	1
a	0	b	1	b	1
b	0	a	a	1	1
1	0	c	a	b	1

Example 5.1.24 (i) If **H** is the hoop as in Example 5.1.12, then **H** is a hoop with (DNP) and for all $x \in H$, $x^2 = x$.

(ii) Let **H** be the hoop as in Example 5.1.22. Then **H** is a BCK-algebra, but is not an implicative BCK-algebra, because, $(a \to 0) \to a = 1 \neq a$.

5.1.2 Relation with Wajsberg Algebras

Now, we show that bounded Wajsberg hoops are termwise equivalent to Wajsberg algebras (see [32, Theorem 1.19]).

Definition 5.1.25 A Wajsberg algebra ([145]) is an algebra $\mathbf{A} = (A; \to, {}^-, 1)$ with a binary operation $\to$, a unary operation $^-$ and one constant 1 satisfying the following axioms:

(i) $1 \to x = x$.
(ii) $(x \to y) \to y = (y \to x) \to x$.
(iii) $(x \to y) \to \big((y \to z) \to (x \to z)\big) = 1$.
(iv) $(x^- \to y^-) \to (y \to x) = 1$.

Proposition 5.1.26 ([145]) *Every Wajsberg algebra* $\mathbf{A} = (A; \to, {}^-, 1)$ *satisfies the following properties for all* $x, y, z \in A$:

(i) $x \to y = 1 = y \to x$ *implies that* $x = y$.
(ii) $x \to x = 1$.
(iii) $x \to 1 = 1$.
(iv) $x \to (y \to x) = 1$.
(v)   *If* $x \to y = 1 = y \to z$, *then* $x \to z = 1$.
(vi)   *If* $x \to (y \to z) = 1$, *then* $y \to (x \to z) = 1$.
(vii) $(x \to y) \to \big((z \to x) \to (z \to y)\big) = 1$.
(viii) $x \to (y \to z) = y \to (x \to z)$.
(ix) $x \to y = 1$ *implies that* $(z \to x) \to (z \to y) = 1 = (y \to z) \to (x \to z)$.
(x) $1^- \to x = 1$.
(xi) $x \to 1^- = x^-$.
(xii) $x^{--} = x$.
(xiii) $x^- \to y^- = y \to x$.

Proof (i) It follows from Definition 5.1.25(i) and (ii).

(ii) Given $x \in A$ by Definition 5.1.25(i) and (iii), we have

$$1 = (1 \to 1) \to \big((1 \to x) \to (1 \to x)\big) = 1 \to (x \to x) = x \to x.$$

(iii) For each $x \in A$ we have $x \to 1 = (1 \to x) \to 1 = (1 \to x) \to (x \to x)$. It follows that

$$x \to 1 = (1 \to x) \to (x \to x) = (1 \to x) \to ((1 \to x) \to x), \text{ by (ii) and Definition 5.1.24(i)}$$
$$= (1 \to x) \to ((x \to 1) \to 1), \text{ Definition 5.1.24(ii)}$$
$$= (1 \to x) \to ((x \to 1) \to (1 \to 1)) = 1, \text{ Definition 5.1.24(iii)}.$$

(iv) Let $x, y \in A$. Then $x \to (y \to x) = (1 \to x) \to (y \to x) = 1 \to ((1 \to x) \to (y \to x)) = (y \to 1) \to ((1 \to x) \to (y \to x)) = 1$.

(v) Suppose that $x \to y = y \to z = 1$ for some $x, y, z \in A$. Then by Definition 5.1.25(iii) and (i), we have $1 = (x \to y) \to ((y \to z) \to (x \to z)) = 1 \to (1 \to (x \to z)) = x \to z$.

(vi) Assume that $x, y, z \in A$ such that $x \to (y \to z) = 1$. Then

$$1 = (x \to (y \to z)) \to (((y \to z) \to z) \to (x \to z))$$
$$= 1 \to (((y \to z) \to z) \to (x \to z))$$
$$= ((z \to y) \to y) \to (x \to z), \text{ Definition 5.1.24(ii)}.$$

Also, by (iv), $y \to ((z \to y) \to y) = 1$, so (v) implies that $y \to (x \to z) = 1$.

(vii) It follows from (vi) and Definition 5.1.25(iii).

(viii) First, we note a remark on the following equalities:

$$1 = y \to ((z \to y) \to y) = y \to ((y \to z) \to z), \text{ by (iv) and Definition 5.1.24(ii)}. \tag{5.2}$$

$$1 = ((y \to z) \to z) \to ((x \to (y \to z)) \to (x \to z)) \text{ by (vii)}. \tag{5.3}$$

Applying (v), from (5.2) and (5.3) we get $y \to ((x \to (y \to z)) \to (x \to z)) = 1$ and so by (vi), $(x \to (y \to z)) \to (y \to (x \to z)) = 1$. Symmetrically, we can show that $(y \to (x \to z)) \to (x \to (y \to z)) = 1$. Therefore, by (i), $x \to (y \to z) = y \to (x \to z)$.

(ix) The proof follows from (vii) and Definition 5.1.25(iii).

(x) By properties (i) and (iv) of Definition 5.1.25, we have

$$(x^- \to 1^-) \to x = (x^- \to 1^-) \to (1 \to x) = 1.$$

Applying (ix), we obtain

$$1 = (1^- \to (x^- \to 1^-)) \to (1^- \to x) = (x^- \to (1^- \to 1^-)) \to (1^- \to x), \text{ by (viii)}$$
$$= (x^- \to 1) \to (1^- \to x) = 1 \to (1^- \to x) = 1^- \to x, \text{ by (ii) and (iii)}.$$

(xi) Let $x \in A$. By (iv), $x^- \to (1^- \to x^-) = 1$. Also, by Definition 5.1.25(iv), $(1^- \to x^-) \to (x \to 1^-) = 1$, so by (v),

$$x^- \to (x \to 1^-) = 1. \tag{5.4}$$

On the other hand, by $1 = (x^- \to 1^-) \to (1 \to x) = (x^- \to 1^-) \to x$ and (ix), we get

$$
\begin{aligned}
1 &= (x \to 1^-) \to \big((x^- \to 1^-) \to 1^-\big) \\
&= (x \to 1^-) \to \big((1^- \to x^-) \to x^-\big) \\
&= (x \to 1^-) \to \big(1 \to x^-\big), \ \text{by (x)} \\
&= (x \to 1^-) \to x^-.
\end{aligned}
$$

Therefore, by (5.4) and (i), $x \to 1^- = x^-$.

(xii) By (xi) and (x), $x^- = (x \to 1^-) \to 1^- = (1^- \to x) \to x = x$.

(xiii) Let $x, y \in A$. By (xii),

$$
(y \to x) \to (x^- \to y^-) = (y^- \to x^-) \to (x^- \to y^-) = 1.
$$

Now, from (i) and Definition 5.1.25(iv) we conclude that $x^- \to y^- = y \to x$. $\qquad\square$

Proposition 5.1.27 *The variety of bounded Wajsberg hoops is termwise equivalent to the variety of Wajsberg algebras.*

Proof Let $\mathbf{A} = (A; \odot, \to, 0, 1)$ be a bounded Wajsberg hoop. For all $x \in A$, we define $x^- = x \to 0$. Then, by Lemma 2.2.1, the algebra $\mathbf{A}^* = (A; \to, ^-, 1)$ is a Wajsberg algebra. Conversely, if $\mathbf{B} = (B; \to, ^-, 1)$ is a Wajsberg algebra, then let $0 = 1^-$ and define

$$
x \odot y = (x \to y^-)^-, \qquad \forall x, y \in B.
$$

(i) $\qquad x \odot y = (x \to y^-)^- = (y^- \to x^-)^- = (y \to x^-)^- = y \odot x$, $\qquad\qquad$ by Proposition 5.1.26(xii) and (xiii). In addition, $1 \odot x = (1 \to x^-)^- = x^{--} = x$.

(ii) $x \to x = 1 = x \to 1$, by Proposition 5.1.26(ii) and (iii).

(iii) Choose $x, y, z \in B$.

$$
\begin{aligned}
(x \odot y) \to z &= (x \to y^-)^- \to z \\
&= z^- \to (x \to y^-), \ \text{by Proposition 5.1.25(xii) and (xiii)} \\
&= x \to (z^- \to y^-), \ \text{by Proposition 5.1.25(viii)} \\
&= x \to (y \to z), \ \text{by Proposition 5.1.25(xiii).}
\end{aligned}
$$

(iv) Finally, given $x, y \in B$, we have

$$
\begin{aligned}
(x \to y) \odot x &= ((x \to y) \to x^-)^- = ((y^- \to x^-) \to x^-)^- \\
&= ((x^- \to y^-) \to y^-)^- = ((y \to x) \to y^-)^- \\
&= (y \to x) \odot y.
\end{aligned}
$$

From (i)–(iv) and Theorem 3.1.3 it follows that $\mathbf{B}^\circ = (B; \odot, \to, 0, 1)$ is a Wajsberg hoop, which is clearly bounded by Proposition 5.1.26(x). It is easy to prove that for any bounded Wajsberg hoop $\mathbf{A}$ and for any Wajsberg algebra $\mathbf{B}$, we have $\mathbf{A}^{*\circ} = \mathbf{A}$ and $\mathbf{B}^{\circ*} = \mathbf{B}$. $\square$

Proposition 5.1.28 *The variety of bounded commutative BCK-algebras is termwise equivalent to the variety of Wajsberg algebras.*

Proof The proof is a direct consequence of Proposition 5.1.27, Theorems 2.5.16, 5.1.9, and 3.3.15. $\square$

For details about Wajsberg algebras, see [32, 145].

5.1.3 Relation with Heyting and Hertz Algebras

This section investigates the relationship between hoops and Hertz algebras as well as Heyting algebras.

Definition 5.1.29 ([236]) A **Hertz algebra** is an algebra $\mathbf{A} = (A; \to, \wedge, 1)$ of type $(2, 2, 0)$ satisfying the following axioms for all $x, y, z \in A$:

(HT1) $x \to x = 1$.
(HT2) $(x \to y) \wedge y = y$.
(HT3) $x \wedge (x \to y) = x \wedge y$.
(HT4) $x \to (y \wedge z) = (x \to y) \wedge (x \to z)$.

Remark 5.1.30 Let $\mathbf{A} = (A; \to, \wedge, 1)$ be a Hertz algebra.
(i) By (HT1) and (HT2), $1 \wedge x = (x \to x) \wedge x = x$ for all $x \in A$.
(ii) Due to (HT1) and (HT3), $x \wedge 1 = x \wedge (x \to x) = x \wedge x$ for all $x \in A$.
(iii) $1 \to x = x$ for all $x \in A$. Indeed,

$$1 \to x = 1 \to (1 \wedge x), \text{ by (i)}$$
$$= (1 \to 1) \wedge (1 \to x), \text{ by (HT4)}$$
$$= 1 \wedge (1 \to x), \text{ by (HT1)}$$
$$= 1 \wedge x = x, \text{ by (HT3) and (i)}.$$

(iv) By (HT2) and (iii), we obtain $x = (1 \to x) \wedge x = x \wedge x$ for all $x \in A$.
(v) Applying (ii) and (iv), we have $1 \wedge x = x = x \wedge x = x \wedge 1$ for all $x \in A$.

(vi) For every $x, y \in A$, $x \wedge y = x$ if and only if $x \to y = 1$. In fact, if $x \wedge y = x$, then by (HT1), (HT4), and (i),

$$1 = x \to (x \wedge y) = (x \to x) \wedge (x \to y) = 1 \wedge (x \to y) = x \to y.$$

Conversely, if $x \to y = 1$, then by (v) and (HT3), $x = x \wedge 1 = x \wedge (x \to y) = x \wedge y$.

Theorem 5.1.31 *If* $\mathbf{H} = (H; \odot, \to, 1)$ *is a hoop satisfying* $x \wedge (x \to y) = x \wedge y$ *for any* $x, y \in H$, *then* $\mathbf{H}^* = (H; \to, \wedge, 1)$ *is a Hertz algebra. In addition, every idempotent hoop is a Hertz algebra.*

Proof Let $\mathbf{H} = (H; \odot, \to, 1)$ be a hoop such that $x \wedge (x \to y) = x \wedge y$ for any $x, y \in H$. By Proposition 3.1.4(i), $(H; \leq)$ is a semilattice, where $x \wedge y = x \odot (x \to y)$. (HT1) and (HT2) follow from Lemma 2.2.1(1) and (10), respectively. Also, by Lemma 2.2.3(ii), (HT4) holds, so by the assumption, $\mathbf{H}^* = (H; \to, \wedge, 1)$ is a Hertz algebra.

Now, assume that $\mathbf{H}$ is an idempotent hoop. Then by Proposition 3.3.40, $\odot = \wedge$ and so by Proposition 3.1.4(i), $\mathbf{H}$ satisfies (HT3), consequently, $(H; \to, \wedge, 1)$ is a Hertz algebra. $\qquad\square$

Example 5.1.32 Let $H = \{0, a, b, 1\}$ be a chain such that $0 < a < b < 1$. Define the binary operations $\to$ and $\odot$ on H by Table 5.5:

Then $\mathbf{H} = (H; \odot, \to, 0, 1)$ is a bounded hoop. Also, clearly $x \wedge (x \to y) = x \wedge y$ for all $x, y \in H$. Then H is a Hertz algebra.

Heyting algebras are the algebraic counterparts of propositional intuitionistic logic, analogous to how Boolean algebras model classical propositional logic. For the definition of Heyting algebra, refer to Definition 1.3.36 and Proposition 1.3.37.

Theorem 5.1.33 *A bounded hoop* $\mathbf{H} = (H; \odot, \to, 1)$ *is a Heyting algebra if and only if it is a* $\vee$*-hoop such that* $\odot = \wedge$.

Table 5.5 Operations $\odot$ and $\to$ of Example 5.1.32

Table 5.5.1

$\odot$	0	a	b	1
0	0	0	0	0
a	0	a	a	a
b	0	a	b	b
1	0	a	b	1

Table 5.5.2

$\to$	0	a	b	1
0	1	1	1	1
a	0	1	1	1
b	0	a	1	1
1	0	a	b	1

Proof Assume that a bounded hoop $\mathbf{H} = (H; \odot, \to, 1)$ is a Heyting algebra. Then clearly, $\mathbf{H}$ is a $\vee$-hoop. Given $x, y \in H$ from $x \odot y \leq x \odot y$ it follows that $x \leq y \to (x \odot y)$, so by (1.1), $x \wedge y \leq x \odot y$. Hence, due to Proposition 3.1.4(i), $x \odot y = x \wedge y$ for all $x, y \in H$. The proof of the converse is evident by definition. $\square$

The following example shows that the condition in the previous theorem is necessary.

Example 5.1.34 Let $H = \{0, a, b, c, d, 1\}$ be a poset with its Hasse diagram given in Fig. 5.2, and Cayley Table 5.6:

For any $x, y \in H$, we define, $x \wedge y = x \odot (x \to y)$ and $x \vee y = ((x \to y) \to y) \wedge ((y \to x) \to x)$. Then $\mathbf{H} = (H; \odot, \to, 0, 1)$ is a bounded $\vee$-hoop. We claim that H is not a Heyting algebra. Since $d \leq a \to b$, but $d \wedge a = d \not\leq b$, because, b and d are incomparable. So, H is not a Heyting algebra.

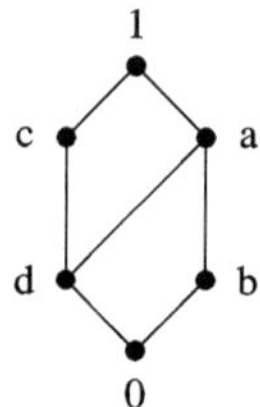

Fig. 5.2 The Hasse diagram of $\mathbf{H}$

Table 5.6 Operations $\odot$ and $\to$ of Example 5.1.34

Table 5.6.1

$\odot$	0	a	b	c	d	1
0	0	0	0	0	0	0
a	0	b	b	d	0	a
b	0	b	b	0	0	b
c	0	d	0	c	d	c
d	d	0	0	d	0	d
1	0	a	b	c	d	1

Table 5.6.2

$\to$	0	a	b	c	d	1
0	1	1	1	1	1	1
a	d	1	a	c	c	1
b	c	1	1	c	c	1
c	b	a	b	1	a	1
d	a	1	a	1	1	1
1	0	a	b	c	d	1

5.1.4 Relation with R_0-Algebras (Lattice Implication Algebras, Hilbert Algebras)

In this section, we investigate the relations among hoops and lattice implication algebras, R_0-algebras, and Hilbert algebras.

Definition 5.1.35 ([261]) Let $\mathbf{A} = (A; \vee, \wedge, 0, 1)$ be a bounded lattice with an order-reversing involution $'$, 1 and 0 be the greatest and the smallest element of A, respectively. The algebraic structure $(A; \vee, \wedge, \rightarrow, ', 0, 1)$ of type $(2, 2, 2, 1, 0, 0)$ is called a **lattice implication algebra** if the following conditions hold for any $x, y, z \in A$:

(I1) $x \rightarrow (y \rightarrow z) = y \rightarrow (x \rightarrow z)$.
(I2) $x \rightarrow x = 1$.
(I3) $x \rightarrow y = y' \rightarrow x'$.
(I4) $x \rightarrow y = y \rightarrow x = 1$ implies $x = y$.
(I5) $(x \rightarrow y) \rightarrow y = (y \rightarrow x) \rightarrow x = x \vee y$.
(I6) $(x \vee y) \rightarrow z = (x \rightarrow z) \wedge (y \rightarrow z)$.
(I7) $(x \wedge y) \rightarrow z = (x \rightarrow z) \vee (y \rightarrow z)$.

Proposition 5.1.36 ([261]) *Each lattice implication algebra $(A; \vee, \wedge, \rightarrow, ', 0, 1)$ satisfies the following assertions for all $x, y, z \in A$:*

(i) $1 \rightarrow x = 1$ *implies that* $x = 1$. *In addition,* $1 \rightarrow x = x$.
(ii) $x \vee y = y$ *if and only if* $x \rightarrow y = 1$. *In addition,* $x \rightarrow 1 = 1$.
(iii) $(x \rightarrow y) \rightarrow ((y \rightarrow z) \rightarrow (x \rightarrow z)) = 1$ *and* $(y \rightarrow z) \rightarrow ((x \rightarrow y) \rightarrow (x \rightarrow z)) = 1$.
(iv) $(A; \otimes, 1)$ *is a commutative monoid, where* $x \otimes y = (x \rightarrow y')'$.
(v) $(x \otimes y) \rightarrow z = x \rightarrow (y \rightarrow z)$.

Proof (i) The proof is straightforward.

(ii) The first part follows from (I6). Additional, since by definition, $(L; \vee, \wedge, 0, 1)$ is a bounded lattice, so $x \leq 1$ implies that $x \rightarrow 1 = 1$.

(iii)

$$
\begin{aligned}
(x \rightarrow y) \rightarrow ((y \rightarrow z) \rightarrow (x \rightarrow z)) &= (x \rightarrow y) \rightarrow (x \rightarrow ((y \rightarrow z) \rightarrow z)), \text{ by (I1)} \\
&= (x \rightarrow y) \rightarrow (x \rightarrow ((z \rightarrow y) \rightarrow y)), \text{ by (I5)} \\
&= (x \rightarrow y) \rightarrow ((z \rightarrow y) \rightarrow (x \rightarrow y)), \text{ by (I1)} \\
&= (z \rightarrow y) \rightarrow ((x \rightarrow y) \rightarrow (x \rightarrow y)), \text{ by (I1)} \\
&= (z \rightarrow y) \rightarrow 1 = 1, \text{ by (ii)}.
\end{aligned}
$$

(iv) Clearly, $\otimes$ is a binary operation on A. By (I3),

$$x \otimes y = (x \to y')' = (y'' \to x')' = (y \to x')' = y \otimes x,$$

which means $\otimes$ is a commutative binary operation. In addition,

$$
\begin{aligned}
(x \otimes y) \otimes z &= \big((x \to y')' \to z'\big)' = \big(z \to (x \to y')\big)', \ \text{ by (I3)} \\
&= \big(x \to (z \to y')\big)', \ \text{ by (I1)} \\
&= \big(x \to (z \to y')''\big)' = \big(x \to (z \otimes y)'\big)' \\
&= x \otimes (z \otimes y) = x \otimes (y \otimes z).
\end{aligned}
$$

(v)

$$
\begin{aligned}
(x \otimes y) \to z &= (x \to y')' \to z = z' \to (x \to y'), \ \text{ by (I3)} \\
&= x \to (z' \to y') = x \to (y \to z), \ \text{ by (I1) and (I3)}.
\end{aligned}
$$

$$\square$$

Theorem 5.1.37 (i) *For each lattice implication algebra* $\mathbf{A} = (A; \vee, \wedge, \to, ', 0, 1)$, *the structure* $\mathbf{A}^* = (A; \otimes, \to, 0, 1)$ *is a bounded Wajsberg hoop, where* $x \otimes y = (x \to y')'$ *and* $x' = x \to 0$ *for every* $x, y \in A$.

(ii) *If* $\mathbf{H} = (H; \odot, \to, 0, 1)$ *is a bounded Wajsberg hoop, then* $\mathbf{H}^\circ = (H; \vee, \wedge, \to, ', 0, 1)$ *is a lattice implication algebra, where* $x' = x \to 0$ *for all* $x \in H$.

Proof (i) Let $\mathbf{A} = (A; \vee, \wedge, \to, ', 0, 1)$ be a lattice implication algebra and $x \otimes y = (x \to y')'$ for all $x, y \in A$. Considering Theorem 2.2.7, by (I4) and Proposition 5.1.36(i)–(v), we conclude that $(A; \otimes, \to, 0, 1)$ is a bounded pocrim. Also, due to (I3), $x \to 0 = 0' \to x' = 1 \to x' = x'$, so $(A; \otimes, \to, 0, 1)$ is a bounded pocrim satisfying (DNP). In addition, for any $x, y \in A$, we have

$$
\begin{aligned}
x \otimes (x \to y) &= (x \to (x \to y)')' = ((x \to y) \to x')' = ((y' \to x') \to x')', \ \text{ by (I3)} \\
&= (x' \vee y')' = x \wedge y, \ \text{ by (I5)},
\end{aligned}
$$

which means the condition (H3) in Theorem 3.1.2 holds. Therefore, by Theorem 3.1.5, $\mathbf{A}^* = (A; \otimes, \to, 0, 1)$ is a bounded hoop. Evidently, due to (I5), this hoop is a Wajsberg hoop.

(ii) Let $\mathbf{H} = (H; \odot, \to, 0, 1)$ be a bounded Wajsberg hoop. Then $\mathbf{H}$ is a bounded $\vee$-hoop satisfying (DNP) and (Pre), see Corollary 3.3.10. Thus, the unary operation $' : H \to H$ defined by $x' := x \to 0$ is an involution. (I1) and (I2) follows from Lemma 2.2.1(8) and (1), respectively. (I3) follows from Remark 2.2.5(viii). Moreover, (I4) is evident, since for each $x, y \in H, x \to y = y \to x = 1$ implies that $x \leq y$ and $y \leq x$, that is $x = y$. Property

(I5) follows from Proposition 3.3.9(iii). Finally, Proposition 3.3.34 leads us to (I6) and (I7). Therefore, $(H; \wedge, \vee, \rightarrow, ', 0, 1)$ is a lattice implication algebra. $\square$

Corollary 5.1.38 *Lattice implication algebras are termwise equivalent to bounded Wajsberg hoops as well as MV-algebras.*

Proof The proof is a direct consequence of Theorems 5.1.37 and 3.3.13. $\square$

For more details on lattice implication algebras, we suggest the monograph [261].

Definition 5.1.39 ([257]) Let $\mathbf{A} = (A; \vee, \wedge, 0, 1)$ be a bounded distributive lattice with an order-reversing involution $'$ and a binary operation $\rightarrow$. Then $\mathbf{A}$ is called an R_0-**algebra** if, for any $x, y, z \in A$, the following conditions hold:

(R1) $x \rightarrow y = y' \rightarrow x'$.
(R2) $1 \rightarrow x = x$.
(R3) $y \rightarrow z \leq (x \rightarrow y) \rightarrow (x \rightarrow z)$.
(R4) $x \rightarrow (y \rightarrow z) = y \rightarrow (x \rightarrow z)$.
(R5) $x \rightarrow (y \vee z) = (x \rightarrow y) \vee (x \rightarrow z)$.
(R6) $(x \rightarrow y) \vee ((x \rightarrow y) \rightarrow (x' \vee y)) = 1$.
(R7) $x \rightarrow x = 1$.
(R8) $x \rightarrow (y \wedge z) = (x \rightarrow y) \wedge (x \rightarrow z)$.

More precisely, an R_0-algebra is an algebra $(A; \vee, \wedge, \rightarrow, ', 0, 1)$ of type $(2, 2, 2, 1, 0, 0)$ satisfies (R1)–(R8). In the sequel, we would prefer to denote R_0-algebras with their full operations. In this way, R_0-algebras and attice implication algebras have the same type.

Theorem 5.1.40 ([261, "Corollary 7.2.2])

(i) *If $(A; \vee, \wedge, \rightarrow, ', 0, 1)$ is an R_0-algebra such that, $(x \rightarrow y) \rightarrow y = (y \rightarrow x) \rightarrow x$ for any $x, y \in A$, then $\mathbf{A} = (A; \wedge, \vee, \rightarrow, ', 0, 1)$ is a lattice implication algebra.*
(ii) *If $(L; \vee, \wedge, \rightarrow, ', 0, 1)$ is a lattice implication algebra such that*

$$(x \rightarrow y) \vee ((x \rightarrow y) \rightarrow (x' \vee y)) = 1 \quad \forall x, y \in L,$$

then $(L; \vee, \wedge, \rightarrow, ', 0, 1)$ is an R_0-algebra.

Proof The proof is straightforward and left to the reader. $\square$

Theorem 5.1.41 (i) *If* $(A; \vee, \wedge, \to, ', 0, 1)$ *is an* R_0-*algebra such that*

$$(x \to y) \to y = (y \to x) \to x \quad \forall x, y \in A,$$

then $\mathbf{A}^* = (A; \otimes, \to, 0, 1)$ *is a bounded Wajsberg hoop, where* $x \otimes y := (x \to y')'$ *for all* $x, y \in A$.

(ii) *If* $(H; \odot, \to, 0, 1)$ *is a bounded Wajsberg hoop satisfying* (R6), *then* $(H; \vee, \wedge, \to, ', 0, 1)$ *is an* R_0-*algebra.*

Proof (i) Let $\mathbf{A} = (A; \vee, \wedge, \to, ', 0, 1)$ be an R_0-algebra satisfying $(x \to y) \to y = (y \to x) \to x$ for all $x, y \in A$, then by Theorem 5.1.40, $\mathbf{A} = (A; \vee, \wedge, \to, ', 0, 1)$ is a lattice implication algebra. Thus, by Theorem 5.1.37, $\mathbf{A}^* = (A; \otimes, \to, 0, 1)$ is a bounded Wajsberg hoop.

(ii) Let $\mathbf{H} = (H; \odot, \to, 0, 1)$ be a bounded Wajsberg $\vee$-hoop. Then by Theorem 5.1.37, $\mathbf{H}^\circ = (H; \vee, \wedge, \to, ', 0, 1)$ is a lattice implication algebra, and if (R6) holds, then by Theorem 5.1.40, $\mathbf{H}^\circ = (H; \vee, \wedge, \to, ', 0, 1)$ is an R_0-algebra. $\qquad\square$

Let us remark that R_0-algebras and hoops are incomparable. Indeed, not all hoops are R_0-algebras. It is noticeable that every hoop is a meet semilattice with respect to the meet operator $a \wedge b = a \odot (a \to b)$, but it does not have a lattice structure, in any case. The following example indicates that not every R_0-algebra is a hoop, in general, and conversely. It shows that the conditions in Theorem 5.1.41 are necessary.

Example 5.1.42 (i) Let $\mathbf{H} = (H; \odot, \to, 1)$ be the hoop as in Example 5.1.22. Clearly, it is a bounded $\vee$-hoop, however, $\mathbf{H}$ does not satisfy (DNP). Furthermore, $b = a \to b \neq a' \to b' = 1$, so (R1) does not hold. Hence, $(H; \vee, \wedge, \to, ', 0, 1)$ is not an R_0-algebra.

(ii) Let $A = \{0, a, b, c, 1\}$ be a chain such that $0 < a < b < c < 1$ and define the operation $\to$ on A as Table 5.7:

Let $\neg a = c, \neg b = b$ and $\neg c = a$. Then $(A; \wedge, \vee, \to, \neg, 0, 1)$ is an R_0-algebra, where $x \wedge y = \min\{x, y\}$ and $x \vee y = \max\{x, y\}$. A routine calculation shows that

Table 5.7 Operation $\to$ of Example 5.1.42(ii)

$\to$	0	a	b	c	1
0	1	1	1	1	1
a	c	1	1	1	1
b	b	b	1	1	1
c	a	a	b	1	1
1	0	a	b	c	1

$$(a \to c) \to c = c \neq 1 = a \to a = (c \to a) \to a.$$

Let define $x \odot y = \neg(x \to \neg y)$ for any $x, y \in A$. We claim that $\mathbf{A}^* = (A; \odot, \to, 0, 1)$ is not a hoop, because

$$c \odot (c \to a) = c \odot a = \neg(c \to \neg a) = \neg(c \to c) = 0 \neq a = c \wedge a.$$

Definition 5.1.43 ([64]) A **Hilbert algebra** is an algebra $\mathbf{A} = (A; \to, 1)$ of type $(2, 0)$ such that the following three axioms are satisfied for all $x, y, z \in A$:

(Hil1) $x \to (y \to x) = 1$.
(Hil2) $x \to (y \to z) = (x \to y) \to (x \to z)$.
(Hil3) $x \to y = y \to x = 1$ implies that $x = y$.

Remark 5.1.44 Let $\mathbf{A} = (A; \to, 1)$ be a Hilbert algebra.
  (i) Due to (Hil1), $1 \to (x \to 1) = 1$ and so

$$(x \to 1) \to 1 = (x \to 1) \to (1 \to (x \to 1)) = 1, \text{ by (Hil1)}.$$

It follows from (Hil3) that $x \to 1 = 1$. Therefore, $x \to 1 = 1$ for every $x \in A$.
  (ii) By (i), $(x \to x) \to 1 = 1$. In addition,

$$1 \to (x \to x) = (x \to 1) \to (x \to x) = x \to (1 \to x) = 1, \text{ by (i), (Hil2) and (Hil1)}.$$

From (Hil3) it follows that $x \to x = 1$ for all $x \in A$.
  (iii) Due to (Hil1), $x \to (1 \to x) = 1$. Moreover, by (ii) and (i) we get

$$1 \to \left(1 \to ((1 \to x) \to x)\right) = \left((1 \to x) \to (1 \to x)\right) \to \left(((1 \to x) \to 1) \to ((1 \to x) \to x)\right)$$
$$= \left((1 \to x) \to (1 \to x)\right) \to \left((1 \to x) \to (1 \to x)\right), \text{ by (Hil2)}$$
$$= 1, \text{ by (ii)}.$$

It follows from (i) and (Hil3) that $(1 \to x) \to x = 1$. Therefore, by (Hil3), $x = 1 \to x$ for all $x \in A$.
  (iv) If $x, y, z \in A$ such that $x \to y = y \to z = 1$, then $x \to z = 1$. Indeed, by (iii) and (Hil2) we get

$$x \to z = 1 \to (x \to z) = (x \to y) \to (x \to z) = x \to (y \to z) = x \to 1 = 1.$$

(v) By (iii), (Hil3), and (iv), the relation $\leq$ defined by $x \leq y$ if and only if $x \to y = 1$ s a partial order on A and due to (i), 1 is the greatest element of the induced poset $(A, \leq)$.

A Hilbert algebra $\mathbf{A}$ is called **bounded** if there exists $0 \in A$ such that $0 \leq x$ for any $x \in A$. On each bounded Hilbert algebra $\mathbf{A}$, we define $x' := x \to 0$ for every $x \in A$.

Proposition 5.1.45 *Every Hilbert algebra* $(A; \to, 1)$ *satisfies the following assertions for all* $x, y, z \in A$:

(i) $x \leq y \to x$.

(ii) $x \leq y$ *implies that* $z \to x \leq z \to y$ *and* $y \to z \leq x \to z$.

(iii) $x \to (y \to z) = y \to (x \to z)$.

Proof (i) It follows from (Hil1).

(ii) Let $x \leq y$. Then by (Hil2) and Remark 5.1.44(i),

$$(z \to x) \to (z \to y) = z \to (x \to y) = z \to 1 = 1.$$

In addition, by (i) and Remark 5.1.44(iii),

$$y \to z \leq x \to (y \to z) = (x \to y) \to (x \to z) = 1 \to (x \to z) = x \to z.$$

(iii) Since $x \leq y \to x$, (ii) and (Hil2) imply that $y \to (x \to z) = (y \to x) \to (y \to z) \leq x \to (y \to z)$. In a similar way, we can show that $x \to (y \to z) \leq y \to (x \to z)$. Therefore, $x \to (y \to z) = y \to (x \to z)$. $\qquad\square$

Theorem 5.1.46 *Let* $(H; \odot, \to, 1)$ *be hoop. Then the reduct* $(H; \to, 0, 1)$ *is a Hilbert algebra if and only if* $x^2 = x$ *for any* $x \in A$.

Proof Let $\mathbf{H} = (H; \odot, \to, 0, 1)$ be a hoop satisfying $x^2 = x$ for any $x \in H$. Then $x \leq y \to x$ implies $x \to (y \to x) = 1$, so (Hil1) holds. Moreover, since $(H; \leq)$ is a poset, if $x \to y = y \to x = 1$, then $x \leq y$ and $y \leq x$ entail that $x = y$, thus, we have (Hil3). By Lemma 2.2.1(10), $y \leq x \to y$, consequently,

$$x \to ((x \to y) \to z) \leq x \to (y \to z), \text{ by Lemma 2.2.1(7)}.$$

Due to Lemma 2.2.1(10) and (11),

$$(x \to y) \to (x \to z) \leq y \to z \leq x \to (y \to z).$$

Furthermore, by Lemma 2.2.1(7),

$$\big(x \to (y \to z)\big) \to \big((x \to y) \to (x \to z)\big) = \big(x \to (y \to z)\big) \to \Big((x \odot (x \to y)) \to z\Big)$$

$$= \Big((x \to y) \odot x \odot (x \to (y \to z))\Big) \to z. \qquad (5.5)$$

Applying (2.2) twice, we obtain

$$(x \to y) \odot x \odot (x \to (y \to z)) = (x \to y) \odot x \odot x \odot (x \to (y \to z)), \text{ since } x^2 = x$$
$$\leq y \odot (y \to z) \leq z.$$

It follows from (5.5) that $x \to (y \to z) \leq (x \to y) \to (x \to z)$. Therefore, $(H; \to, 0, 1)$ is a Hilbert algebra. Conversely, if $\mathbf{H}$ is a hoop such that the reduct $(H; \to, 1)$ is a Hilbert algebra. Choose $x, y, z \in H$. By (Hil2) and Theorem 4.3.12(iii), we obtain $F = \{1\}$ is a implicative filter and so $x \to x^2 \in F$ for every $x \in H$ (see Theorem 4.3.12(v)), equivalently, $x = x^2$ for every $x \in H$.   □

From Theorem 5.1.46 and Proposition 3.3.40 it follows that if $\mathbf{H}$ is a hoop such that the reduct $(H; \to, 1)$ is a Hilbert algebra, then $x \odot y = x \wedge y$ for all $x, y \in H$.

The following example shows that not every hoop is a Hilbert algebra, in general.

Example 5.1.47 If $(H; \odot, \to, 1)$ is the hoop as in Example 5.1.34, then $a \odot a = b$, and so $x^2 \neq x$ for any $x \in H$. Since

$$a \to (d \to b) = a \to a = 1 \neq a = c \to a = (a \to d) \to (a \to b),$$

$(H; \to, 0, 1)$ is not a bounded Hilbert algebra.

Theorem 5.1.48 *An algebra* $(A; \to, 1)$ *is a Hilbert algebra if and only if it is a BCK-algebra satisfying* (Hil2).

Proof Assume that $(A; \to, 1)$ is a Hilbert algebra. (BCK2) follows from (Hil1) and Remark 5.1.44(ii). (BCK3) and (BCK4) follow from Remark 5.1.44(ii) and (i), respectively. (BCK5) is just (Hil3). Now, let $x, y, z \in A$. Then by Proposition 5.1.45(iii) we obtain

$$\begin{aligned}
(x \to y) \to ((y \to z) \to (x \to z)) &= (y \to z) \to ((x \to y) \to (x \to z)), \text{ by Proposition 5.1.44(iii)} \\
&= (y \to z) \to (x \to (y \to z)), \text{ by (Hil2)} \\
&= x \to ((y \to z) \to (y \to z)), \text{ by Proposition 5.1.44(iii)} \\
&= x \to 1 = 1, \text{ by Remark 5.1.43(i) and (ii)}.
\end{aligned}$$

which means (BCK1) holds. Therefore, $(A; \to, 1)$ is a BCK-algebra satisfying (Hil2). The proof of the converse is straightforward by (BCK5) and Lemma 2.5.4.   □

Combining Theorems 5.1.48 and 5.1.17 we conclude the following result:

Theorem 5.1.49 *If* $\mathbf{A}$ *is a Hilbert algebra such* $(A; \to, 1)$ *is a BCK-$\wedge$-semilattices with* (P) *and* (Div), *then* $(A; \odot, \to, 1)$ *is an idempotent bounded Wajsberg hoop, where* $\odot$ *is defined in Definition 2.5.10.*

Proof By Theorems 5.1.17 and 5.1.48, $(A; \odot, \rightarrow, 1)$ is a hoop satisfying (Hil2). Hence, due to Theorem 5.1.46, $x^2 = x$ for all $x \in H$. $\qquad\square$

We conclude this section by presenting a proposition concerning bounded Hilbert algebras, which arises as a special case of Theorem 5.1.49.

Theorem 5.1.50 ([66]) *Let* $(A; \rightarrow, 0, 1)$ *be a bounded Hilbert algebra satisfying* (DNP). *Then* $x \vee y = x' \rightarrow y$ *and* $x \wedge y = (x \rightarrow y')'$ *for all* $x, y \in H$. *In addition,* $(A; \vee, \wedge, ', 0, 1)$ *is a Boolean algebra and* $(A; \odot, \rightarrow, 0, 1)$ *is a bounded idempotent hoop, where* $x \odot y = x \wedge y$ *and* $x \rightarrow y = x' \vee y'$.

Proof Choose $x, y \in A$. By Proposition 5.1.45(iii), $x' \rightarrow y = x' \rightarrow y'' = y' \rightarrow x'' = y' \rightarrow x$, so, $x' \rightarrow y \geq x, y$, see Proposition 5.1.45(i). Now, let $a \in H$ such that $x, y \leq a$. Then Proposition 5.1.45(ii), $a' \leq x'$, similarly, $x' \rightarrow y \leq a' \rightarrow y$. On the other hand, $y \leq a$ implies that $a' \rightarrow y \leq a' \rightarrow a = a' \rightarrow (a' \rightarrow 0) = (a' \rightarrow a') \rightarrow (a' \rightarrow 0) = 1 \rightarrow a'' = a$. In conclusion, $x' \rightarrow y \leq a$. Hence, $x' \rightarrow y = x \vee y$. Evidently, $(x \rightarrow y')' \leq x, y$. If $b \in H$ such that $b \leq x, y$, then $x', y' \leq b'$, so $x \rightarrow y' = x' \vee y' \leq b'$, consequently, $b = b'' \leq (x \rightarrow y')'$. Thus, $x \wedge y = (x \rightarrow y')'$. Now, easy calculations show that $(A; \vee, \wedge, ', 0, 1)$ is a Boolean algebra.

So, by Example 3.3.55, A with the binary operations $x \odot y = x \wedge y$ and $x \rightarrow y := x' \vee y$ forms a bounded idempotent hoop. $\qquad\square$

For more details on Hilbert algebras, we suggest consulting Chapter 8 of [69] and Chap. 5 of [65].

5.1.5 Relation with L-Algebras

L-algebras, defined in the context of the quantum Yang–Baxter equation [138, 243], were first introduced and studied by Rump [244]. He demonstrated that any set X equipped with a binary operation satisfying the equation $(x * y) * (x * z) = (y * x) * (y * z)$ corresponds to a solution of the quantum Yang–Baxter equation, provided that left multiplication is bijective. On the other hand, this equation appears in classical logic (Boolean algebras), intuitionistic logic (Heyting algebras), quantum logic (orthomodular lattices), and Łukasiewicz logic (MV-algebras), see [244].

Definition 5.1.51 ([244]) An algebraic structure $(L; \rightarrow, 1)$ of type $(2, 0)$ is an **L-algebra** if it satisfies the following conditions for all $x, y, z \in L$:

(L1) $x \rightarrow x = x \rightarrow 1 = 1$ and $1 \rightarrow x = x$.
(L2) $(x \rightarrow y) \rightarrow (x \rightarrow z) = (y \rightarrow x) \rightarrow (y \rightarrow z)$.
(L3) $x \rightarrow y = 1 = y \rightarrow x$ implies that $x = y$.

In each L-algebra $(L; \rightarrow, 1)$, the relation $x \leq y$ if and only if $x \rightarrow y = 1$, is a partial order relation with 1 as the greatest element. If $(L, \leq)$ possesses a least element 0, then L is termed **bounded**.

According to [81, 244], an L-algebra $(L; \rightarrow, 1)$ is called a KL-algebra, or CKL-algebra if it satisfies the conditions (K), or (C), respectively:

(K) $x \rightarrow (y \rightarrow x) = 1$ for all $x, y \in L$.
(C) $x \rightarrow (y \rightarrow z) = y \rightarrow (x \rightarrow z)$ for all $x, y, z \in L$.
(D) $x \rightarrow ((x \rightarrow y) \rightarrow y) = 1$ for all $x, y \in L$.

Evidently, each CKL-algebra satisfies both (K) and (D). Since if **L** satisfies (C), then by (C) and (L1), $x \rightarrow (y \rightarrow x) = y \rightarrow (x \rightarrow x) = y \rightarrow 1 = 1$ and $x \rightarrow ((x \rightarrow y) \rightarrow y) = (x \rightarrow y) \rightarrow (x \rightarrow y) = 1$ for every $x, y \in L$. So, every CKL-algebra is a KL-algebra.

Proposition 5.1.52 ([81, 244]) *Every L-algebra* $(L; \rightarrow, 1)$ *satisfies the following statements for all* $x, y, z \in L$:

 (i) *$x \leq y$ implies that $z \rightarrow x \leq z \rightarrow y$.*
 (ii) *If $x, y \leq z$ and $z \rightarrow x = z \rightarrow y$, then $x = y$.*
(iii) *If (K) holds, then $((x \rightarrow y) \rightarrow y) \rightarrow ((x \rightarrow y) \rightarrow z) = y \rightarrow z$.*
(iv) *If (K) holds, then $x \leq y$ implies that $y \rightarrow z \leq x \rightarrow z$.*
 (v) *If (K) and (D) hold, then $x \rightarrow y = ((x \rightarrow y) \rightarrow y) \rightarrow y$.*

Proof (i) Let $x, y, z \in L$ such that $x \leq y$. By (L2), we have $(z \rightarrow x) \rightarrow (z \rightarrow y) = (x \rightarrow z) \rightarrow (x \rightarrow y) = (x \rightarrow z) \rightarrow 1 = 1$, entails that $z \rightarrow x \leq z \rightarrow y$.

(ii) Let $x, y, z \in L$ such that $x, y \leq z$ and $z \rightarrow x = z \rightarrow y$. Then by (L2), $1 = (z \rightarrow x) \rightarrow (z \rightarrow y) = (x \rightarrow z) \rightarrow (x \rightarrow y) = 1 \rightarrow (x \rightarrow y) = x \rightarrow y$. In a similar way, $y \rightarrow x = 1$, so $x = y$.

(iii) Let $x, y, z \in L$. From (L2) and (K), we get $((x \rightarrow y) \rightarrow y) \rightarrow ((x \rightarrow y) \rightarrow z) = (y \rightarrow (x \rightarrow y)) \rightarrow (y \rightarrow z) = 1 \rightarrow (y \rightarrow z) = y \rightarrow z$.

(iv) Let $x, y \in L$ such that $x \leq y$. By (K), we have

$$
\begin{aligned}
1 &= (y \to z) \to ((y \to x) \to (y \to z)), \text{ by (K)} \\
&= (y \to z) \to ((x \to y) \to (x \to z)), \text{ by (L2)} \\
&= (y \to z) \to (1 \to (x \to z)), \text{ since } x \leq y \\
&= (y \to z) \to (x \to z).
\end{aligned}
$$

Hence, $y \to z \leq x \to z$.

(v) By (D), $(x \to y) \to \big(((x \to y) \to y) \to y\big) = 1$, that is $x \to y \leq ((x \to y) \to y) \to y$. Due to (K), $x \leq (x \to y) \to y$, so by (iv), $((x \to y) \to y) \to y \leq x \to y$. Therefore, $x \to y = ((x \to y) \to y) \to y$. $\qquad\square$

Proposition 5.1.53 ([245, Proposition 5.1(c)]) *An L-algebra* $\mathbf{L} = (L; \to, 1)$ *is a CKL-algebra if and only if it satisfies both* (K) *and* (D).

Proof Assume that $\mathbf{L}$ satisfies (K) and (D). Let $x, y, z \in L$ and $w := (x \to (y \to z)) \to (x \to z)$. Then by (L2),

$$
\begin{aligned}
w &= ((y \to z) \to x) \to ((y \to z) \to z), \text{ by (L2)} \\
&\geq (y \to z) \to z, \text{ by (K)} \\
&\geq y, \text{ by (D)}.
\end{aligned}
$$

It follows that

$$
\begin{aligned}
x \to (y \to z) &\leq \big((x \to (y \to z)) \to (x \to z)\big) \to (x \to z), \text{ by (D)} \\
&= w \to (x \to z) \leq y \to (x \to z), \text{ by Proposition 5.1.51(iv)}.
\end{aligned}
$$

The condition (C) now follows by symmetry. The proof of the converse is straightforward. $\qquad\square$

Proposition 5.1.54 ([81, Proposition 2.3 and Remark 3.2(3)]) *Every CKL-algebra* $(L; \to, 1)$ *is a BCK-algebra. In addition, an algebra* $(X; \to, 1)$ *of type* $(2, 0)$ *is a CKL-algebra if and only if* $(X; \to, 1)$ *is a BCK-algebra satisfying* (L2).

Proof Assume that $(L; \to, 1)$ is a CKL-algebra. (BCK3)–(BCK5) follow from (L1) and (L3). In addition, by (C), $x \to ((x \to y) \to y) = (x \to y) \to (x \to y) = 1$ for all $x, y \in L$, i.e., (BCK2) holds. Choose $x, y, z \in H$. By (L1),

$$
\begin{aligned}
1 &= (y \to x) \to 1 \\
&= (y \to x) \to ((y \to z) \to (y \to z)), \text{ by (L1)} \\
&= (y \to z) \to ((y \to x) \to (y \to z)), \text{ by (C)} \\
&= (y \to z) \to ((x \to y) \to (x \to z)), \text{ by (L2)} \\
&= (x \to y) \to ((y \to z) \to (x \to z)), \text{ by (L2).}
\end{aligned}
$$

So, (BCK1) holds. Therefore, $(X; \to, 1)$ is a BCK-algebra.

The proof of the second part follows directly from the definition. Note that, by Lemma 2.5.4(iv), every BCK-algebra satisfies (C). $\square$

Theorem 5.1.55 ([81, 244]) *For each hoop* $(H; \odot, \to, 1)$, *the algebra* $(H; \to, 1)$ *is a CKL-algebra.*

Proof Let $(H; \odot, \to, 1)$ be a hoop. (L1) follows from Lemma 2.2.1(1) and (2). In addition, $x \to y = 1 = y \to x$ implies that $x \le y \le x$, consequently, $x = y$, by (2.3), so (L3) holds. Finally, choose $x, y \in H$. From properties of hoops, we obtain:

$$
\begin{aligned}
(x \to y) \to (x \to z) &= \big(x \odot (x \to y)\big) \to z, \text{ by Lemma 2.2.1(8)} \\
&= \big(y \odot (y \to x)\big) \to z, \text{ by Theorem 3.1.2(H3)} \\
&= (y \to x) \to (y \to z), \text{ by Lemma 2.2.1(8).}
\end{aligned}
$$

Thus, (L2) holds. Therefore, $(H; \to, 1)$ is an L-algebra. Furthermore, due to Lemma 2.2.1(8), $(H; \to, 1)$ is a CKL-algebra. $\square$

For the converse of Theorem 5.1.55, assume that $(L; \to, 1)$ is an L-algebra. If $(L; \to, 1)$ is a reduct of a hoop algebra, then L must satisfy (C), so $(L; \to, 1)$ is a CKL-algebra. Now, applying Theorem 5.1.17, the L-algebra $(L; \to, 1)$ must be a BCK-$\wedge$-semilattice with (P) and (Div).

Bibliographical Remarks and Suggestions for Further Study

This chapter demonstrates the application and significance of hoops by highlighting their extensive connections with related algebraic structures, particularly those algebras relevant to multi-valued logics. The material presented herein has been compiled from various articles, as cited in each section.

In recent times, L-algebras have attracted significant interest from numerous researchers. For further study of L-algebras, we recommend reading paper [241], on the cancellation problem for L-algebras, [244] about L-algebras, self-similarity, and L-groups. The category of L-algebras is studied in [246], the geometry of discrete L-algebras is described in [247], and the relationship between prime L-algebras and right-angled Artin groups is elucidated in [249]. The role of Bosbach's cone algebras for L-algebras is clarified in [250, 253], and pseudo MV-algebras as L-algebras is presented in [264]. We note that pseudo MV-algebras are a non-commutative generalization of MV-algebras presented in [152] and they will be explained in more detail in Chapter 8. In addition, prime ideals of L-algebras were characterized in topological terms in [248, 252]. Main results on BL-algebras, including their classification in the finite case, are reconsidered and extended to a class of L-algebras in [251].

Lattice-Theoretical Properties 6

This chapter is a core component of the book, dedicated to a detailed and in-depth investigation of the structure of hoops. Although the concept of ordinal sum was introduced in the previous chapters, its primary application in the representation of this algebraic structure, particularly irreducible hoops, is demonstrated in this chapter. The chapter delves into the structural analysis of hoops, focusing on specific classes and their representations across seven sections. We begin by exploring ordinal sums and $\oplus$-irreducible hoops, characterizing linear $\oplus$-irreducible hoops as Wajsberg hoops and demonstrating that linear Wajsberg hoops are either cancellative or bounded (bounded case implies that $\mathbf{H}$ is equivalent to an MV-algebra). A key theorem establishes that any linearly ordered hoop is an ordinal sum of Wajsberg hoops, leading to representations for BL-algebras and cancellative hoops.

The second section examines simple and semisimple hoops, proving that simple hoops are linearly ordered Wajsberg hoops and that semisimple hoops are Wajsberg hoops, showing that every finite Wajsberg hoop is semisimple. This provides a complete representation of finite semisimple hoops. Local and perfect hoops are studied in the third section, gathering their basic properties and examples. Representable hoops, explored in the fourth section, are shown to be equivalent to basic hoops, leading to the conclusion that varieties of cancellative, Wajsberg, or product hoops are generated by their linearly ordered elements.

The fifth section investigates subdirectly irreducible hoops, revealing that they contain a minimal non-zero filter that forms a simple, totally ordered Wajsberg subhoop. A crucial representation theorem shows that a hoop is subdirectly irreducible if and only if it is the ordinal sum of another hoop and a non-trivial subdirectly irreducible linear Wajsberg hoop. We then extend the analysis to subdirectly irreducible product hoops as well as basic hoops. We conclude by proving that finite linear hoops are subdirectly irreducible and introducing

© The Author(s), under exclusive license to Springer Nature Switzerland AG 2026

A. Dvurečenskij et al., *Hoop Algebras*, Frontiers in Mathematics,
https://doi.org/10.1007/978-3-032-11736-6_6

k-potent hoops, presenting their main properties and representations for subdirectly irreducible k-potent hoops. This provides a comprehensive understanding of these fundamental structures.

A representation for Wajsberg hoops based on ultrafilters of bounded Wajsberg hoops is studied in the sixth section. First, we establish that for every filter F of a bounded Wajsberg hoop $\mathbf{H}$, $\mathbf{F}_h = (F; \odot, \to, 1)$ is a subhoop of $\mathbf{H}$. Subsequently, it is proven that for every unbounded Wajsberg hoop $\mathbf{H}$, there exists a bounded Wajsberg hoop $\mathbf{A}(\mathbf{H})$ such that $F := H$ is an ultrafilter of $\mathbf{A}(\mathbf{H})$ and $\mathbf{H} \cong \mathbf{F}_h$. Furthermore, the bounded Wajsberg hoop $\mathbf{A}(\mathbf{H})$ is unique up to isomorphism. Hence, every Wajsberg hoop is either bounded or an ultrafilter of a bounded Wajsberg hoop. The strong connection between $\mathbf{H}$ and $\mathbf{A}(\mathbf{H})$ induces a one-to-one correspondence between $\mathrm{Spec}(\mathbf{H})$ and $\mathrm{Spec}(\mathbf{A}(\mathbf{H})) \setminus \{H\}$. The results of this section demonstrate that the variety of Wajsberg hoops is generated by bounded Wajsberg hoops.

The final section of this chapter aims to provide a brief study of the variety of hoops. We begin by examining some fundamental properties, such as the congruence extension property (CEP) and arithmeticity. We then demonstrate that the lattice of varieties of hoops possesses exactly two atoms, namely $\mathrm{V}(\mathbf{L}_1)$ and $\mathrm{V}(\mathbf{L}_\infty)$. Following an investigation of basic results concerning the varieties, we establish that the variety of all hoops has the finite embeddability property (FEP).

6.1 Ordinal Sum of Hoops

In Examples 2.2.8, 3.3.62, and Proposition 3.3.63, we introduced the ordinal sum of two hoops (pocrims). In this section, we extend the concept by defining the ordinal sum for an arbitrary family of hoops indexed by a totally ordered set. We then introduce $\oplus$-irreducible linear hoops and use the concept of cuts in linear hoops to demonstrate that every $\oplus$-irreducible linear hoop is a Wajsberg hoop. Finally, we propose a representation for linear Wajsberg hoops using the negative cone of ℓ-groups.

Example 6.1.1 ([13, 111, 141]) Similar to Example 3.3.62 we can define the ordinal sum of a family $\{\mathbf{H}_i = (H_i; \odot_i, \to_i, 1) : i \in I\}$ of hoops where I is a linearly ordered set and $H_i \cap H_j = \{1\}$ for all distinct elements $i, j \in I$. Set $H := \bigcup_{i \in I} H_i$. Choose $x, y \in H$. There exist $i, j \in I$ such that $x \in H_i$ and $y \in H_j$.

$$x \odot y = \begin{cases} x \odot_i y & \text{if } i = j \\ y & \text{if } x, y \neq 1, \ j < i \\ x & \text{if } x, y \neq 1, \ i < j, \end{cases} \tag{6.1}$$

$$x \to y = \begin{cases} x \to_i y & \text{if } i = j \\ y & \text{if } x, y \neq 1, \ j < i \\ 1 & \text{if } x, y \neq 1, \ i < j. \end{cases} \tag{6.2}$$

We can easily verify that $(H; \odot, \to, 1)$ is a hoop, denoted by $\bigoplus_{i \in I} \mathbf{H}_i$.

In addition, let I be a totally ordered set and $\{\mathbf{H}_i : i \in I\}$ be an arbitrary family of hoops. Consider $\{\mathbf{H}'_i : i \in I\}$ as the disjoint union of $\{\mathbf{H}_i : i \in I\}$, where $H'_i := H_i \times \{i\}$ for all $i \in I$ and

$$(x, i) \odot'_i (y, i) = (x \odot_i y, i), \quad \& \quad (x, i) \to'_i (y, i) = (x \to_i y, i) \quad \forall x, y \in H_i.$$

Then $(H'_i; \odot'_i, \to'_i, (1, i))$ is a hoop, for each $i \in I$. Replacing $(1, i)$ with 1 in each hoop $\mathbf{H}'_i$ we obtain a new hoop $\mathbf{H}''_i$ where 1 plays the role of $(1, i)$. Now, we have a family $\{\mathbf{H}''_i : i \in I\}$ with $H''_i \cap H''_j = \{1\}$ for all distinct elements $i, j \in I$. Clearly, $\mathbf{H}_i \cong \mathbf{H}''_i$ for all $i \in I$. We consider $\bigoplus_{i \in I} \mathbf{H}''_i$ as the ordinal sum of the family $\{\mathbf{H}_i : i \in I\}$.

Since $\mathbf{1} \oplus \mathbf{H} = \mathbf{H} \oplus \mathbf{1}$ for each hoop $\mathbf{H}$, in the sequel, we will only focus on the ordinal sum of a family of non-trivial hoops, i.e., a family $\{\mathbf{H}_i : i \in I\}$ of hoops where $1 < |H_i|$ for all $i \in I$.

Remark 6.1.2 The following properties hold on the ordinal sum $\mathbf{H} = \bigoplus_{i \in I} \mathbf{H}_i$:

(i) $\mathbf{H}$ is linear if and only if $\mathbf{H}_i$ is linear for all $i \in I$.

(ii) It is easy to see that the ordinal sum $\oplus$ of finitely many hoops is associative.

(iii) If $\leq_i$ is the partial order relation on $\mathbf{H}_i$ and $\leq$ is the partial order relation on $\mathbf{H}$, then $x \leq_i y$ if and only if $x \leq y$ for all $x, y \in H_i$ and $i \in I$. In addition, $x < y$ when $x \in H_i \setminus \{1\}$, $y \in H_j$ and $i < j$. Furthermore, if $x, y \in H_i$ such that $x \vee_i y \in H_i \setminus \{1\}$, then $x \vee y$ exists in $\mathbf{H}$, also, $x \vee y = x \vee_i y$ for all $x, y \in H_i$ and $i \in I$. The converse holds true, that is, the existence of $x \vee y \in H_i \setminus \{1\}$ implies the existence of $x \vee_i y \in H_i \setminus \{1\}$ and $x \vee_i y = x \vee y$ for all $x, y \in H_i$ and $i \in I$. However, if $x, y \in H_i \setminus \{1\}$ such that $x \vee_i y = 1$, then for every $i < j$ and $z \in H_j \setminus \{1\}$, we have $x, y \leq z$, which implies $x \vee y \neq 1$. Therefore, $x \vee y$ may differ from $x \vee_i y$.

Finally, if I has a top element $\alpha = \max(I)$, then $x \vee y$ exists in $\mathbf{H}$ if and only if $x \vee_\alpha y$ exists in $\mathbf{H}_\alpha$, additionally, $x \vee y = x \vee_\alpha y$.

(iv) $x \to y = x \to_i y$ and $x \odot_i y = x \odot y$ for all $x, y \in H_i$. Hence, $(H_i; \odot, \to, 1)$ is a subalgebra of the hoop $\bigoplus_{i \in I} \mathbf{H}_i$. Therefore, when there is no ambiguity for the sake of simplicity and consistency, we adopt the same symbols $\to$ and $\odot$ for all hoops $\mathbf{H}_i$ and their direct sum $\bigoplus_{i \in I} \mathbf{H}_i$ throughout this book.

(v) If I has the least element 0 and $\mathbf{H}_0$ is bounded, then so is $\mathbf{H}$. The converse is also true, i.e., if $\mathbf{H}$ is bounded, then I must have a minimum element 0 and $\mathbf{H}_0$ is a bounded hoop.

First, assume that $\min(I) = 0$ and $\mathbf{H}_0$ is a bounded hoop with the least element a. Then $a \leq x$ for all $x \in H_0$. Also, by (iv), $a \leq y$ for each $y \in H_i$ and $i > 0$. Therefore, a is the least element of $\bigoplus_{a \in I} \mathbf{H}_i$. Conversely, suppose that $\bigoplus_{i \in I} \mathbf{H}_i$ is bounded with the least element 0. There exists $i_0 \in I$ such that $0 \in H_{i_0}$. We claim that $\min(I) = i_0$. If there is $j \in I$ such that $j < i_0$, then for each $x \in H_j \setminus \{1\}$ we have $x \to 0 = 1$, which implies $x \leq 0$. Hence, by the assumption, $x = 0 \in H_{i_0} \setminus \{1\}$, a contradiction. Note that every $\mathbf{H}_i$ is a non-trivial hoop. Thus, $i_0 = \min(I)$. We denote i_0 simply by 0.

(vi) Applying (iii), we can easily show that if $\mathbf{H}_1$ is a linear hoop, then $\mathbf{H} := \mathbf{H}_1 \oplus \mathbf{H}_2$ is a $\vee$-hoop if and only if $\mathbf{H}_2$ is a $\vee$-hoop for every hoop $\mathbf{H}_2$. Moreover, if $\vee_1, \vee_2$ and $\vee$ are join operations on $\mathbf{H}_1$, $\mathbf{H}_2$, and $\mathbf{H}$, respectively, then by (iii), we have $x \vee_i y = x \vee y$ for every $x, y \in H_i$ and $i = 1, 2$.

Proposition 6.1.3 ([13, Proposition 3.1]) *Let $\{\mathbf{H}_i : i \in I\}$ be a family of hoops and I be a linearly ordered set. Then the following statements hold:*

(i) *The subhoops $\bigoplus_{i \in I} \mathbf{H}_i$ are exactly the hoops of the form $\bigoplus_{i \in I} \mathbf{B}_i$, where $\mathbf{B}_i$ is a subhoop of $\mathbf{H}_i$ for all $i \in I$.*

(ii) *If $\bigoplus_{i \in I} \mathbf{H}_i$ is bounded, then I has the least element 0, $\mathbf{H}_0$ is a bounded hoop, and bounded subhoops $\bigoplus_{i \in I} \mathbf{H}_i$ are exactly the hoops of the form $\bigoplus_{i \in I} \mathbf{B}_i$, where $\mathbf{B}_0$ is a bounded subhoop of $\mathbf{H}_0$ and $\mathbf{B}_i$ is a subhoop of $\mathbf{H}_i$ for all $i \in I \setminus \{0\}$.*

Proof (i) Let $\mathbf{B} = (B; \odot, \rightarrow, 1)$ be a subhoop of $\bigoplus_{i \in I} \mathbf{H}_i$. Set $B_i := B \cap H_i$ for each $i \in I$. Then $1 \in B_i$, evidently. Choose $x, y \in B_i$. Then $x \rightarrow_i y = x \rightarrow y \in B$ and $x \rightarrow_i y \in H_i$, so $x \rightarrow_i y \in B_i$. Similarly, $x \odot_i y \in B_i$, which means $\mathbf{B}_i = (B_i; \odot_i, \rightarrow_i, 1)$ is a subhoop of $\mathbf{H}_i$ for all $i \in I$. Clearly, $\mathbf{B} = \bigoplus_{i \in I} \mathbf{B}_i$.

(ii) Since $\bigoplus_{a \in I} \mathbf{H}_i$ is bounded, by Remark 6.1.2(v), I has the least element 0 and $\mathbf{H}_0$ is a bounded hoop. Now, let $\mathbf{B}$ be a bounded subhoop of $\bigoplus_{i \in I} \mathbf{H}_i$. By (i), $\mathbf{B} = \bigoplus_{i \in I} \mathbf{B}_i$, where $\mathbf{B}_i$ is a subhoop of $\mathbf{H}_i$ for all $i \in I$. Since $\mathbf{B}$ is bounded, Due to Remark 6.1.2(v), $0 \in \mathbf{B}_0$, which means $(B_0; \odot, \rightarrow, 0, 1)$ is a bounded subhoop of $(\mathbf{H}_0; \odot, \rightarrow, 0, 1)$. $\square$

Example 6.1.4 (i) Let $\mathbf{H}$ be an arbitrary hoop. Then $\mathbf{2} \oplus \mathbf{H}$ is a bounded hoop, since $0 \rightarrow x = 1$ for all $x \in \{0, 1\} \cup H$.

(ii) If $\mathbf{H}_1$ and $\mathbf{H}_2$ are two non-trivial hoops, then $\mathbf{H}_1 \oplus \mathbf{H}_2$ is not a Wajsberg hoop. Indeed, given $a \in H_1 \setminus \{1\}$ and $b \in H_2 \setminus \{1\}$ we have $(a \rightarrow b) \rightarrow b = 1 \rightarrow b = b$ and $(b \rightarrow a) \rightarrow a = a \rightarrow a = 1$.

Proposition 6.1.5 *Let $\mathbf{H}_1$ and $\mathbf{H}_2$ be two hoops. The map $f : \mathbf{H}_1 \oplus \mathbf{H}_2 \rightarrow \mathbf{H}_1$ is defined by $f(x) = x$ for all $x \in H_1$, and $f(x) = 1$ for all $x \in H_2$ is an onto homomorphism. Furthermore, $\mathrm{Ker}(f) = H_2$.*

Proof The proof is straightforward and is left as an exercise for the reader. $\square$

Definition 6.1.6 ([13, 111]) A linear hoop is said to be $\oplus$-**irreducible** if it cannot be written as the ordinal sum of two linearly ordered non-trivial hoops.

Example 6.1.7 Let $\mathbf{H}$ be a linear hoop and $a \in H$ be an idempotent element. By Example 3.3.43(i), $\mathbf{H}_2 := ([a, 1]; \odot, \rightarrow, a, 1)$ is a bounded hoop. On the other hand, $H_1 := \{x \in$

$H : x < a\} \cup \{1\}$ is a universe of a hoop. Indeed, or each $x, y < a$ we have $x \odot y \leq x < a$, so H_1 is closed under $\odot$. We claim that $x \to y < a$ when $y < x$. Otherwise, $a \leq x \to y$ implies that $a \odot x \leq y$. From $a \odot x = a \wedge x = x$ (see Proposition 3.3.40), it follows that $x \leq y$, which is absurd. Therefore, H_1 is closed under $\to$, i.e., $\mathbf{H}_1 := (H_1; \odot, \to, 1)$ is a hoop. For $x \in H_1 \setminus \{1\}$ and $y \in H_2 \setminus \{1\}$ we have $y \to x < a$. Assume the converse, $a \leq y \to x$ implies that $a = a \wedge y = a \odot y \leq x$, contradicting with the assumption. Therefore, $\mathbf{H} = \mathbf{H}_1 \oplus \mathbf{H}_2$.

Proposition 6.1.8 *Let $\mathbf{H}$ be a linear $\oplus$-irreducible hoop. Then $\mathbf{H}$ has only trivial idempotents (1 and 0 (when $\mathbf{H}$ is bounded)).*

Proof It follows from Example 6.1.7. $\square$

In the subsequent section, we delve into the structure of $\oplus$-irreducible linear hoops. This analysis will be instrumental in establishing that all unbounded linear hoops exhibit the cancellative property. First, we need to know the notion of cuts of linear hoops.

Definition 6.1.9 ([13, 111]) A **cut** of a linear hoop $\mathbf{H}$ is a pair (X, Y) of subsets of H if

(i) $X \cup Y = H$.
(ii) $x \leq y$ for all $x \in X$ and $y \in Y$.
(iii) Y is closed under $\odot$.
(iv) $x \odot y = x$ for all $x \in X$ and $y \in Y$.

Remark 6.1.10 Let (X, Y) be a cut of a linear hoop $\mathbf{H}$.

(i) By property (iv) of Definition 6.1.9, either $X \cap Y = \emptyset$ or $X \cap Y$ consists of an idempotent singleton. Indeed, if $x, y \in X \cap Y$, then $x \odot x = x$, and $x = x \odot y = y$, by (iv).

(ii) Choose $x, z \in X$. If $x \odot z \in Y$, then $x \leq x \odot z \leq x$ and so $x \odot z = x \in X$. Hence, X is closed under $\odot$. If $w \in H$ such that $w \leq x \in X$, then $w \in X$, since $w \in H \setminus X$ implies that $w \in Y$, so by Definition 6.1.9(ii), $x \leq w$, in conclusion, $w = x \in X$.

(iii) For each $x \in X \setminus Y$ and $y \in Y \setminus X$, we have $y \to x = x$. Indeed, $y \to x \notin Y$ (otherwise, $x = y \wedge x = y \odot (y \to x) \in Y$, which is absurd), so $y \to x \in X$, which implies $x = y \wedge x = y \odot (y \to x) = y \to x$.

(iv) $z \geq y \in Y$ implies that $z \in Y$. Suppose the converse, i.e., $z \in X$. Then by (ii), $y \in X$. Hence, $z = z \odot y \leq y$, which entails that $y = z$, that is $z \in Y$.

If $\mathbf{H}$ is a linear bounded hoop, a cut (X, Y) is said to be **trivial** if $X = \{0\}$ or $Y = \{1\}$; if $\mathbf{H}$ is an unbounded linear hoop, a cut (X, Y) is said to be trivial if $X = \emptyset$ or $Y = \{1\}$. We note that every bounded linear hoop is a bounded $\vee$-hoop and so it is a linear BL-algebra (see Theorem 3.3.28 and Proposition 5.1.8).

Proposition 6.1.11 ([13, 111]) *Let (X, Y) be a cut of a linear hoop $\mathbf{H}$ such that $X \cap Y = \emptyset$.*

(i) $X \cup \{1\}$ *is a linear subhoop of $\mathbf{H}$ and Y is a linear subhoop of $\mathbf{H}$. Moreover, $\mathbf{H} = (X \cup \{1\}) \oplus Y$. In addition, if (X, Y) is a non-trivial cut, $\mathbf{H}$ is not $\oplus$-irreducible.*

(ii) *If $\mathbf{H} = \mathbf{H}_1 \oplus \mathbf{H}_2$, where $\mathbf{H}_1$ and $\mathbf{H}_2$ are linear hoops, then $(H_1 \setminus \{1\}, H_2)$ is a cut of $\mathbf{H}$. If $\mathbf{H}_1$ and $\mathbf{H}_2$ are non-trivial so is the cut $(H_1 \setminus \{1\}, H_2)$.*

Proof Assume that (X, Y) is a cut of $\mathbf{H}$. By Remark 6.1.10(ii), X is closed under $\odot$. Assume $x_1, x_2 \in X$ and $x_2 < x_1$. Then $x_1 \to x_2 \in X$, otherwise $x_1 \to x_2 \in Y$ implies that $x_2 = x_1 \wedge x_2 = x_1 \odot (x_1 \to x_2) = x_1$, which is a contradiction. Therefore, $H_1 := X \cup \{1\}$ is a linear subhoop $\mathbf{H}$.

On the other hand, by Remark 6.1.10(iv), Y satisfies (F2) and so Y is a filter, by Definition 6.1.9(iii). Choose $x, y \in Y$. From $x \le y \to x$ and $y \le x \to y$ it follows that Y is a subhoop of $\mathbf{H}$. The rest of the statement is evident. $\square$

Proposition 6.1.12 ([13, 111]) *Let $\mathbf{H}$ be a linear hoop, $m \in H \setminus \{1\}$. Set*

$$\overrightarrow{m} := \{x \in H \setminus \{1\} : m \to x = x\}.$$

If $\overrightarrow{m} \ne \emptyset$, then $(\overrightarrow{m}, \overrightarrow{m}^{\,c})$ is a non-trivial cut of $\mathbf{H}$, where $\overrightarrow{m}^{\,c} = H \setminus \overrightarrow{m}$.

Proof Let $\overrightarrow{m} \ne \emptyset$. Choose $a \in \overrightarrow{m}$. Clearly, $H = \overrightarrow{m} \cup \overrightarrow{m}^{\,c}$, $m \in \overrightarrow{m}^{\,c}$ and $a < m$ (otherwise, $m \le a$ implies that $a = m \to a = 1$, which is a contradiction).

(i) $a \odot m = (m \to a) \odot m = a \wedge m = a$.

(ii) Let $y \le x \in \overrightarrow{m}$. To prove that $y \in \overrightarrow{m}$, it suffices to show that $m \to y \le y$. Since $y \le x$, by Lemma 2.2.1(7), $m \to y \le m \to x = x$, so

$$\begin{aligned}
(m \to y) \to y &= (x \wedge (m \to y)) \to y \\
&= \big(x \odot (x \to (m \to y))\big) \to y, \quad \text{by Proposition 3.1.4(i)} \\
&= (x \to (m \to y)) \to (x \to y), \quad \text{by Lemma 2.2.1(8)} \\
&= \big((x \odot m) \to y\big) \to (x \to y) \\
&= (x \to y) \to (x \to y) = 1, \quad \text{by (i).}
\end{aligned}$$

Consequently, $y \in \overrightarrow{m}$, which means $\overrightarrow{m}$ is a down-set that is also closed under $\odot$.

(iii) Choose $x, y \in \overrightarrow{m}$ such that $y < x$. Then $m \to (x \to y) = (m \odot x) \to y = x \to y$, entails that $x \to y \in \overrightarrow{m}$. Note that $x \to y \ne 1$.

(iv) If $x \in \overrightarrow{m}$ and $y \in \overrightarrow{m}^{\,c}$, then $y \to x = x = x \odot y$. Indeed, from $y \le (y \to x) \to x$ and (ii), it follows that $(y \to x) \to x \in \overrightarrow{m}^{\,c}$. By Lemma 2.2.1(8), we have

$$m \to ((y \to x) \to x) = (y \to x) \to (m \to x) = (y \to x) \to x.$$

Thus, either $(y \to x) \to x = 1$ or $(y \to x) \to x \in \overrightarrow{m}$. From the above, we have only the second possibility, i.e., $(y \to x) \to x = 1$, equivalently, $x = y \to x$. In addition, $x \wedge y = x$, otherwise $y \leq x$, so by (ii), $y \in \overrightarrow{m}$, a contradiction. Hence, $x = x \wedge y = (y \to x) \odot y = x \odot y$.

Since $\overrightarrow{m}$ is a down-set (by (ii)), $\overrightarrow{m}^c$ is an up-set, evidently. Hence, $\overrightarrow{m}^c$ is closed under $\to$.

(v) We state that $\overrightarrow{m}^c$ is closed under $\odot$. Otherwise, there are $x, y \in \overrightarrow{m}^c$ such that $x \odot y \in \overrightarrow{m}$. Then by (iv), $y \to (x \odot y) = x \odot y$. But,

$$
\begin{aligned}
x \wedge (y \to (x \odot y)) &= \big(x \to (y \to (x \odot y))\big) \odot x \\
&= \big((x \odot y) \to (x \odot y)\big) \odot x = x,
\end{aligned}
$$

i.e., $x \leq y \to (x \odot y) = x \odot y \in \overrightarrow{m}$ (by (iv)), consequently, due to (ii), $x \in \overrightarrow{m}$, a contradiction.

(vi) If $x \in \overrightarrow{m}$ and $y \in \overrightarrow{m}^c$, then $x \leq y$, otherwise, $y < x$ implies that $x = y \to x = 1$, by (iv), which is a contradiction.

From (i)–(vi) and Definition 6.1.9 it follows that $(\overrightarrow{m}, \overrightarrow{m}^c)$ is a cut of **H**. $\square$

Proposition 6.1.13 ([13, 111]) *Let* **H** *be a linear hoop and a be an element of H which is not idempotent. Set $Y_a := \{z \in H : z \odot a = a\}$. Then (Y_a^c, Y_a) is a cut of* **H**, *where $Y_a^c = H \backslash Y_a$.*

Proof It is clear that $1 \in Y_a$ and $a \in Y_a^c$. We will verify the conditions (i)–(iv) of Definition 6.1.9.

(i) Clear.

(ii) Let $y \in Y_a$ and $z \in Y_a^c$. If $y < z$, then $a = y \odot a \leq z \odot a \leq a$, entails that $z \odot a = a$, that is $z \in Y_a$, which contradict with the assumption. Therefore, $z \leq y$, in particular $z < y$ (since $z \notin Y_a$).

(iii) If $y, z \in Y_a$, then $z \odot y \odot a = z \odot a = a$. In addition, if $y \leq w$, then $a = y \odot a \leq w \odot a \leq a$, consequently, $w \in Y_a$. Therefore, Y_a is a filter of **H**.

(iv) Let $z \in Y_a^c$ and $y \in Y_a$. By (ii), $z < y$. Since $y \odot a = a$, there are two cases:

(Case 1) $z \leq a < y$. This yields $y \odot z = y \odot (z \wedge a) = y \odot (a \odot (a \to z)) = a \odot (a \to z) = a \wedge z = z$.

(Case 2) $a < z < y$. We have $y \odot a = a$, $z \odot a < a$ (since $z \notin Y_a$) and $z \odot y = z$. Otherwise, from $z \odot y \leq z$ we get $z \odot y < z$. Thus, $a < z < y$, $y \odot a = a$, $z \odot y < z$, and $z \odot a < a$ which contradict with Proposition 3.1.6(iii).

(Case 1) and (Case 2) imply that $z \odot y = z$. Hence, (Y_a^c, Y_a) is a cut of **H**. $\square$

Theorem 6.1.14 ([13, 111]) *Let* $\mathbf{H}$ *be a linear hoop. The following statements are equivalent:*

(i) $\mathbf{H}$ *is* $\oplus$-*irreducible.*

(ii) *For all* $x, y \in H$, $y \to x = x$ *implies that* $y = 1$ *or* $x = 1$.

(iii) $\mathbf{H}$ *is a Wajsberg hoop.*

Proof (i) $\Rightarrow$ (ii) Let $y \to x = x$, $x \neq 1$ and $y \neq 1$. Then by Proposition 6.1.12, $(\overrightarrow{y}, \overrightarrow{y}^{\,c})$ is a non-trivial cut of $\mathbf{H}$, so by Proposition 6.1.11, $\mathbf{H}$ is not $\oplus$-irreducible.

(ii) $\Rightarrow$ (i) If $\mathbf{H} = \mathbf{H}_1 \oplus \mathbf{H}_2$ is a non-trivial decomposition, then for all $x \in H_1$ and $y \in H_2$ with $x < y < 1$ we have $y \to x = x$, and $x, y \neq 1$ contradicting (ii). Thus, if (ii) holds, there are no non-trivial decompositions.

Now let z be a fixed element of H. For any $x, y \in [z, 1]$, we define $x^{-z} := x \to z$. Then

(a) $x^{-z} \odot x = (x \to z) \odot x = x \wedge z = z$.

(b) $(x^{-z})^{-z} = (x \to z) \to z \geq x$ (by Lemma 2.2.1(6)).

(c) $x^{-z} = ((x^{-z})^{-z})^{-z}$ (by Lemma 2.2.1(13)).

(d) $(x \odot y)^{-z} = x \to y^{-z}$ (by Lemma 2.2.1(8)).

To establish (ii) $\Rightarrow$ (iii), we first prove the following claim.

Claim A. (ii) implies that $(x^{-z})^{-z} = x$ for any $x \in [z, 1]$ and any $z \in H$.

(e) If $z \leq y \leq x$ and $x^{-z} = y^{-z}$, then $x = y$. Indeed, $y = y \wedge x = x \odot (x \to y)$ which implies $x^{-z} = y^{-z} = ((x \to y) \odot x)^{-z} = (x \to y) \to x^{-z}$ (by (d)). Hence, by (ii), $x^{-z} = 1$ or $x \to y = 1$. The first one implies that $x = y = z$ and the second one implies that $x \leq y$, i.e., $x = y$.

If $x \in H$, then $(x^{-z})^{-z} = x$. In fact, let $z \in H$ and $x \geq z$ be an element of H. From $z \leq x \leq (x^{-z})^{-z}$ and $x^{-z} = ((x^{-z})^{-z})^{-z}$ (by (c)) we get $x = (x^{-z})^{-z}$, see (e). So the claim holds.

(ii) $\Rightarrow$ (iii) Suppose that $x, y \in H$. If $x \leq y$, then $(x \to y) \to y = 1 \to y = y$ and $(y \to x) \to x = y$, by the Claim A. Similarly, $y \leq x$ implies that $(x \to y) \to y = x = (y \to x) \to x$. Hence, (3.6) holds, which proves $\mathbf{H}$ is a Wajsberg hoop.

(iii) $\Rightarrow$ (i) Assume $\mathbf{H} = \mathbf{H}_1 \oplus \mathbf{H}_2$, where $\mathbf{H}_1$ and $\mathbf{H}_2$ are linear hoops which are not trivial. Then there are $y \in H_2 \setminus \{1\}$ and $x \in H_1 \setminus \{1\}$. Check $(x \to y) \to y = 1 \to y = y$ but $(y \to x) \to x = x \to x = 1$ which   entails $\mathbf{H}$ is   not   Wajsberg.   Hence, (iii) implies (i). $\square$

Theorem 6.1.15 ([13, 111]) *Let* $\mathbf{H}$ *be an unbounded linear hoop. Then* $\mathbf{H}$ *is* $\oplus$-*irreducible if and only if* $x = x \odot y$ *implies that* $y = 1$ *for each* $x, y \in H$.

Proof Let $\mathbf{H}$ be $\oplus$-irreducible. Due to Proposition 6.1.8, $\mathbf{H}$ has only trivial idempotents. Assume $x = x \odot y$. By Proposition 6.1.13, (Y_x^c, Y_x) is a cut of $\mathbf{H}$, proving that it has to be only trivial, i.e., $Y_x = \{1\}$ or $Y_x^c = \emptyset$.

Conversely, assume $\mathbf{H} = \mathbf{H}_1 \oplus \mathbf{H}_2$. If there exists $x \in H_1 \setminus \{1\}$, for any $y \in H_2$, we have $x = x \odot y$ giving $y = 1$ and $H_2 = \{1\}$. If there exists $y \in H_2 \setminus \{1\}$, then $H_1 \setminus \{1\}$ has to be empty, indeed if $x \in H_1$, $x < 1$, then $x = x \odot y$, which implies $y = 1$, a contradiction. Consequently, $\mathbf{H}$ is irreducible. $\qquad\square$

Now, we are ready for the last theorem of this section.

Theorem 6.1.16 ([13, 15, 111]) *Let* $\mathbf{H} = (H; \odot, \to, 1)$ *be a linear Wajsberg hoop.*

(i) *If* $\mathbf{H}$ *is unbounded, there is a linearly ordered Abelian ℓ-group* $\mathbf{G} = (G; \vee, \wedge, +, -, 0)$ *such that* $\mathbf{H}$ *is isomorphic to* $\mathbf{N}(\mathbf{G})$.

(ii) *If* $\mathbf{H}$ *is bounded, there is a linearly ordered Abelian ℓ-group* $\mathbf{G} = (G; \vee, \wedge, +, -, 0)$ *with strong unit u such that* $\mathbf{H}$ *is isomorphic* $\mathbf{G}[\mathbf{u}]$.

Proof (i) Suppose that $\mathbf{H}$ is unbounded. Let $x, y, z \in H$ such that $x \odot z = y \odot z$. Since $\mathbf{H}$ is linearly ordered, without loss of generality we assume that $x \leq y$, then $x = x \wedge y = (y \to x) \odot y$ and hence $x \odot z = (y \to x) \odot y \odot z = (y \to x) \odot x \odot z$ which by Theorem 6.1.15 yields $y \to x = 1$, equivalently, $y \leq x$, consequently, $x = y$. Applying Proposition 3.3.19, there is an ℓ-group $(G; \vee, \wedge, +, -, 0)$ such that $\mathbf{H}$ is isomorphic to $\mathbf{N}(\mathbf{G})$. It is clear that $\mathbf{G}$ is linearly ordered.

(ii) If $\mathbf{H}$ is bounded, then by Theorem 3.3.13, $\beta(\mathbf{H}) = (H; \oplus, ', 0, 1)$ is an MV-algebra, and so by Theorem 1.4.15, $\beta(\mathbf{H})$ is isomorphic to $\mathbf{G}[\mathbf{u}]$ for some linearly ordered ℓ-group $(G; \vee, \wedge, +, -, 0)$ with a strong unit u. Now, Corollary 3.3.14 implies that $\mathbf{H} \cong \mathbf{G}[\mathbf{u}]$. $\qquad\square$

Corollary 6.1.17 ([15, Proposition 2.1]) *A linear Wajsberg hoop is either bounded or cancellative. In addition, it is both bounded and cancellative if and only if it is trivial*

Proof It is a straight consequence of Theorem 6.1.16 $\qquad\square$

Now, we proceed to show that any linearly ordered hoop is the ordinal sum of a family of $\oplus$-irreducible hoops. First, we need some notions introduced in [13]. Given a linearly ordered set $(I, \leq)$ a subset $J \subseteq I$ is said to be **connected** if $k \leq i \leq j$ implies that $i \in J$ for all $j, k \in J$ and $i \in I$. A **connected partition** of $(I, \leq)$ is a partition of I into connected subsets.

Definition 6.1.18 ([13]) A **decomposition** of a linear hoop $\mathbf{H}$ is a family $D = \{\mathbf{H}_i : i \in I\}$ of linear hoops such that $\mathbf{H} = \bigoplus_{i \in I} \mathbf{H}_i$.

Let S be the collection of all decompositions of a linear hoop $\mathbf{H}$. We claim that S is a definable class, clearly. Applying the axiom of choice, we can assume without loss of generality that for every decomposition $D = \{\mathbf{H}_i : i \in I\}$ of $\mathbf{H}$, the index set I is a subset of $\mathbf{H}$ (choose $I \subseteq H$ such that $1 \in I$ and $H_i \cap I$ has cardinality 1 for every component $\mathbf{H}_i$). It follows that every element of S is a function from a subset I of H into the powerset of H (since for every $i \in I$ the underlying set H_i of $\mathbf{H}_i$ is a subset of H). Thus, S is a definable subclass of the class Γ of all partial functions from H into the powerset of H. The powerset axiom and the axiom of comprehension guarantee that Γ is a set, and so by the axiom of comprehension, S is a set. Therefore, S can be partially ordered in the following way:

Given $D = \{\mathbf{H}_i : i \in I\}$ and $D' = \{\mathbf{H}'_j : j \in J\}$ we have $D' \leq D$ if and only if there is a connected partition $\{I_j : j \in J\}$ of I such that for $j, j' \in J$:

(i) If $j \leq j'$, then for all $k \in I_j$ and $k' \in I_{j'}$, we have $k < k'$.

(ii) $\mathbf{H}'_j = \bigoplus_{i \in I_j} \mathbf{H}_i$.

Theorem 6.1.19 ([13, Theorem 3.7]) *Every linear hoop (bounded linear hoop) can be uniquely represented as the ordinal sum of a family of linear Wajsberg hoops (whose first component is a bounded linear Wajsberg hoop).*

Proof Let $\mathbf{H}$ be a linear hoop and $(S, \leq)$ be the poset of all decompositions of $\mathbf{H}$. Let C be a chain of decompositions in S. For any $a \in H \setminus \{1\}$ and $D \in C$, let $\mathbf{H}^{D_a}$ be the unique component of D which contains a and let $H_a = \bigcap\{H^{D_a} : D \in C\}$. It is clear that $H_a \cup \{1\}$ is the universe of a subalgebra $\mathbf{H}_a$ of $\mathbf{H}$. For $a, b \in H \setminus \{1\}$, we have $\mathbf{H}_a = \mathbf{H}_b$ if and only if a and b lie in the same component of all the decompositions in C. Let by the axiom of choice $I \subseteq H \setminus \{1\}$ be such that for every $a \in H \setminus \{1\}$, $I \cap H_a$ contains exactly one element. Then $\mathbf{H} = \bigoplus_{a \in I} \mathbf{H}_a$ and the decomposition such obtained is greater than or equal to any element in the chain C. Hence, applying Zorn lemma to the poset $(S, \leq)$, there is a maximal decomposition of $\mathbf{H}$. Each component of the decomposition must be $\oplus$-irreducible and hence by Theorem 6.1.14, they are linear Wajsberg hoops. Now, assume that $\mathbf{H}$ is bounded. By the above argument, there exists a family $\{\mathbf{H}_i : i \in I\}$ of linear Wajsberg hoops such that $\mathbf{H} = \bigoplus_{i \in I} \mathbf{H}_i$. So, by Remark 6.1.2(v), I has the least element 0 and $\mathbf{H}_0$ is bounded with $\min(H) = \min(H_0)$.

Uniqueness. Now, assume that $\bigoplus_{i \in I} \mathbf{H}_i = \mathbf{H} = \bigoplus_{j \in J} \mathbf{K}_j$ for some families $\{\mathbf{H}_i : i \in I\}$ and $\{\mathbf{K}_j : j \in J\}$ of linear Wajsberg hoops. Then for every $x \in H \setminus \{1\}$ there exists unique elements $i \in I$ and $j \in J$ such that $x \in H_i$ and $x \in K_j$. Choose an arbitrary component $\mathbf{H}_i$ of the decomposition and $x \in H_i \setminus \{1\}$. Then $x \in K_j$ for a unique element $j \in J$. Let $y \in H_i \setminus \{1\}$. We claim that $y \in K_j$. If not then either $x < y$ or $y < x$. If $x < y$, then there exists unique $t \in J$ such that $y \in K_t$. We define $X = \bigcup_{r < t}(K_r \setminus \{1\})$ and $Y = \bigcup_{r > t} K_r$. Then (X, Y) is a cut of $\mathbf{H}$ and $(X \cap H_i, Y \cap H_i)$ is a cut of $\mathbf{H}_i$ such that $x \in X \cap H_i$ and $y \in Y \cap H_i$. Thus, by Proposition 6.1.11, $\mathbf{H}$ is not $\oplus$-irreducible, a contradiction (see Theorem 6.1.14). In a similar way, we proceed $y < x$.

Hence, $H_i \subseteq K_j$ and by symmetry $K_j \subseteq H_i$, i.e., $H_i = K_j$. Therefore, every $\mathbf{H}_i$ coincides with some $\mathbf{K}_j$ and vice versa. $\qquad\square$

The following theorem provides a representation for linearly ordered BL-algebras, as established in [13, Theorem 3.7] and [62, Theorem 3.4].

Corollary 6.1.20 ([13, Theorem 3.7]) *Every linearly ordered BL-algebra is the ordinal sum of a family of Wajsberg hoops whose first component is a BL-algebra.*

Proof It follows from Theorem 6.1.19. $\qquad\square$

The upcoming theorem establishes a representation for linearly ordered product algebras in terms of ordinal sums of **2** and cancellative hoops.

Theorem 6.1.21 ([218, Proposition 2.2]) *Let* $\mathbf{H} = (H; \odot, \rightarrow, 1)$ *be a hoop. The following assertions hold:*

(i) *If* $\mathbf{H}$ *is a cancellative hoop, then* $\mathbf{2} \oplus \mathbf{H}$ *is a bounded basic hoop satisfies* (Prod).
(ii) *If* $\mathbf{H}$ *is a bounded linear hoop satisfying* (Prod), *then* $\mathbf{H} = \mathbf{2} \oplus \mathbf{C}$ *for some cancellative hoop* $\mathbf{C}$ (*note that* $\mathbf{C}$ *can be trivial hoop* $\mathbf{1}$).

Proof (i) By Example 3.3.66, $\mathbf{2} \oplus \mathbf{H}$ is a bounded hoop. Recall that the underlying set of $\mathbf{2} \oplus \mathbf{H}$ is the set $\{0, 1\} \cup H$. Choose $x, y \in \{0, 1\} \cup H$. Due to Propositions 3.3.16, and 3.3.9(iii), $\mathbf{H}$ is a Wajsberg hoop. Applying Proposition 3.3.63(iv), we get $\mathbf{2} \oplus \mathbf{H}$ is a basic hoop. In addition, by Remark 6.1.2(vi), the join operation $\vee$ on $\mathbf{2} \oplus \mathbf{H}$ coincides with the join operations on $\mathbf{2}$ and $\mathbf{H}$. Now, we need to show that the identity (Prod) holds, i.e., $x' \vee \big((x \rightarrow (x \odot y)) \rightarrow y\big) = 1$ holds for every $x, y \in \mathbf{2} \oplus \mathbf{H}$. Choose $x, y \in \{0, 1\} \cup H$. If $x = 0$, then $x' = 1$, so $x' \vee \big((x \rightarrow (x \odot y)) \rightarrow y\big) = 1$ for each $y \in \{0, 1\} \cup H$. If $x \neq 0$ and $y = 0$, then $x \rightarrow (x \odot y) = x \rightarrow 0 = 0$, so $x' \vee \big((x \rightarrow (x \odot y)) \rightarrow y\big) = x' \vee (0 \rightarrow 0) = 1$. Finally, if $x, y \neq 0$, then by Proposition 3.3.5, $x \rightarrow (x \odot y) = y$, which implies $x' \vee \big((x \rightarrow (x \odot y)) \rightarrow y\big) = x' \vee (y \rightarrow y) = 1$. Therefore, $\mathbf{2} \oplus \mathbf{H}$ is a product algebra.
   (ii) Let $\mathbf{H} = (H; \odot, \rightarrow, 1)$ be a bounded linear hoop satisfying (Prod). Set $C := H \setminus \{0\}$.
   (1) Clearly, $1 \in C$ and C is closed under $\rightarrow$ (note that $y \leq x \rightarrow y$).
   (2) Given $x \in H \setminus \{0\}$, we have $x' < 1$, so $x \rightarrow (x \odot y) \leq y$ for all $y \in H$, since $x' \vee \big((x \rightarrow (x \odot y)) \rightarrow y\big) = 1$ and $\mathbf{H}$ is linearly ordered.
   (3) Applying (2), if $y = 0$, then we get $x' = x \rightarrow 0 = x \rightarrow (x \odot y) \leq y = 0$, which means $x' = 0$ for all $x \in H \setminus \{0\}$.

(4) If $x, y \in C$, then according to definition of $\oplus$ on $\mathbf{2} \oplus \mathbf{H}$, we have $x \to 0 = 0$ and $y \to 0 = 0$. We claim that $x \odot y \in C$, otherwise, $x \odot y = 0$ implies that

$$1 = (x \odot y) \to 0 = x \to y' = x', \text{ by Lemma 2.2.1(8),}$$

which contradicts part (2).

From (1) and (4) it follows that $\mathbf{C} = (C; \odot, \to, 1)$ is a subhoop of $\mathbf{H}$. In addition, by (2), $x \to (x \odot y) \leq y$ for all $x, y \in C$. On the other hand,

$$y \to (x \to (x \odot y)) = (x \odot y) \to (x \odot y) = 1,$$

which implies $y \leq x \to (x \odot y)$, consequently, $x \to (x \odot y) = y$ for all $x, y \in C$, i.e., $\mathbf{C}$ is a cancellative hoop, see Proposition 3.3.5. Considering the definition of $\mathbf{2} \oplus \mathbf{C}$ we get that $\mathbf{H} = \mathbf{2} \oplus \mathbf{C}$. $\qquad\square$

Applying Theorem 6.1.21 and the termwise equivalent in Proposition 5.1.8, we obtain that for every linearly ordered product algebra $\mathbf{A} = (A; \vee, \wedge, \odot, \to, 0, 1)$ there is a cancellative hoop $\mathbf{H} = (H; \odot, \to, 1)$ such that $\mathbf{A} = (\mathbf{2} \oplus \mathbf{H})^*$. In addition, for every cancellative hoop $\mathbf{H}$, the BL-algebra $(\mathbf{2} \oplus \mathbf{H})^*$ is a product algebra.

At the end of this section, we apply ordinal sum to create another example of a hoop with no maximal filter as follows:

Example 6.1.22 Consider I as the real numbers $\mathbb{R}$ with its natural ordering. Let $\mathbf{H}_i = \mathbf{2}$ for all $i \in I$. By Example 6.1.1, $\mathbf{H} := \bigoplus_{i \in I} \mathbf{H}_i$ is a hoop. We claim that $\mathbf{H}$ has no maximal filter. Let F be a maximal filter of $\mathbf{H}$. There exists $i \in I$ such that H_i is not a subset of F, which entails that $H_j \cap F = \{1\}$ for all $j < i$. We can easily verify that $\bigoplus_{t \geq i-1} \mathbf{H}_t$ is a proper filter of $\mathbf{H}$ properly containing F.

Bibliographical Remarks and Suggestions for Further Study

The ordinal sum is a crucial tool for representing various classes of hoops, particularly subdirectly irreducible hoops, and for studying subvarieties of hoops. Section 6.1 began with fundamental definitions and results, progressing to more advanced topics. The primary results of this section were originally presented in [13, 15, 31, 111, 141].

Due to a crucial result by Aglianò and Montagna [13], we have learned that every linear basic hoop is the ordinal sum of a family of linear Wajsberg hoops. Montagna and Ugolini [218] applied a similar method to find a representation for linear bounded product hoops. More applications of ordinal sums of hoops can be seen in the next sections.

Further applications of ordinal sums can be found in a series of papers related to the variety of hoops, including: varieties of BL-algebras [13, 14], splitting algebras [8, 9], a categorical equivalence for product algebras [218], free product hoops [210], and particularly, the valuable thesis by I. M. A. Ferreirim about varieties and quasivarieties of hoops and their reducts [141], which was very influential for the theory of hoops.

6.2   Simple and Semisimple Hoops

6.2.1   Simple Hoops

According to Definition 1.5.12, a hoop $\mathbf{H}$ is **simple** if $\mathrm{Con}(\mathbf{H}) = \{\Delta, \nabla\}$. Proposition 4.1.13 and Theorem 4.1.14 establish a one-to-one correspondence between congruence relations and filters in hoop algebras. Leveraging this connection, we can propose an alternative definition of simple hoop algebras based on filters as follows:

Definition 6.2.1   A hoop $\mathbf{H}$ is called **simple** if $\mathcal{F}(\mathbf{H}) = \{\{1\}, H\}$.

Let $\mathbf{H}$ be a simple hoop. For each $x \in H \setminus \{1\}$ we have $\{1\} \neq \langle x \rangle \in \mathcal{F}(\mathbf{H})$, which entails that $\langle x \rangle = H$, so by Proposition 4.2.24, $\{1\}$ is the unique maximal filter of $\mathbf{H}$. In a similar way, we can show that $\mathbf{H}$ is simple if $\{1\}$ is a maximal filter of $\mathbf{H}$.

Example 6.2.2   (i) The bounded hoop $([0, 1]; \odot, \rightarrow, 0, 1) = \mathbb{R}[1]$ is simple.

(ii) Consider the cancellative hoop $\mathbf{N}(\mathbb{R}) = (\mathbb{R}^-; \odot, \rightarrow, 0)$ (see Example 3.3.8). Choose $x \in (-\infty, 0)$. For each $y \in (-\infty, 0)$ there exists $n \in \mathbb{N}$ such that $nx \leq y$, equivalently, $\overbrace{x \odot \cdots \odot x}^{n\text{-times}} = nx \leq y$. Thus, $y \in \langle x \rangle$, that is $\langle x \rangle = \mathbb{R}^-$. Therefore, $(\mathbb{R}^-; \odot, \rightarrow, 0)$ is a simple hoop.

(iii) Trivial hoop $\mathbf{1} = (\{1\}; \odot, \rightarrow, 1)$ is simple, clearly.

Lemma 6.2.3   ([31, Lemma 2.1]) *Let $\mathbf{H}$ be a hoop.*

(i) $\mathbf{H}$ *is simple if and only if* $\mathbf{H}$ *satisfies*

$$\text{for all } x \in H \setminus \{1\}, y \in H, \quad \text{there exists } n \in \mathbb{N} \text{ such that } x^n \rightarrow y = 1. \tag{6.3}$$

(ii) *Every simple hoop satisfies*

$$\text{for all } x, y \in H, \ y \rightarrow x = x, \ \text{implies } x = 1 \text{ or } y = 1. \tag{6.4}$$

Proof (i) Observe that $\mathbf{H}$ is simple if and only if for any $x \in H$, if $x \neq 1$, then $\langle x \rangle = H$. Choose $x, y \in H$ such that $x \neq 1$. Then $H = \langle x \rangle$ and Corollary 2.3.9 imply that $x^n \rightarrow y = 1$ for some $n \in \mathbb{N}$. Conversely, let $\mathbf{H}$ satisfies (6.3). Choose $F \in \mathcal{F}(\mathbf{H})$ and $x \in F \setminus \{1\}$. For each $y \in H$ there exists $n \in \mathbb{N}$ such that $x^n \to y = 1 \in F$, so by (F1) and Lemma 2.3.2, $y \in F$, since $x^n \in F$, which implies $F = H$. Hence, $\mathbf{H}$ is simple.

(ii) Assume $\mathbf{H}$ is simple and $x, y \in H$ such that $y \to x = x$. Then

$$y^n \to x = y^{n-1} \to (y \to x) = y^{n-1} \to x = \cdots = y \to (y \to x) = y \to x = x, \quad \forall n \in \mathbb{N}.$$

So, by (i), $x = 1$ or $y = 1$. $\square$

Condition (6.3) is known as the **Archimedean Law**, see [147].

Clearly, any linearly ordered Wajsberg hoop satisfies (6.4). Indeed, let $\mathbf{H}$ be a linearly ordered Wajsberg hoop and $x, y \in H$ such that $y \to x = x$. If $y \le x$, then $y \to x = 1$, which implies $x = 1$. Otherwise, we have $x < y$, and so by Proposition 3.3.9(iii), $y = y \vee x = (y \to x) \to x = x \to x = 1$. Conversely, we have:

Proposition 6.2.4 ([31]) *Let $\mathbf{H}$ be a hoop satisfying (6.4). Then $\mathbf{H}$ is a linearly ordered Wajsberg hoop.*

Proof Choose $x, y \in H \setminus \{1\}$. By Proposition 3.1.4(vi), $(x \to y) \to (y \to x) = y \to x$, so by the assumption, $x \to y = 1$ or $y \to x = 1$, consequently, $\mathbf{H}$ is totally ordered. To see that $\mathbf{H}$ satisfies (3.6), let $x, y \in H$ and assume without loss of generality that $x < y$; since $(x \to y) \to y = 1 \to y = y$, it suffices to show that $y = (y \to x) \to x$. Due to Lemma 2.2.1(13), $y \to x = ((y \to x) \to x) \to x$ entails that

$$\big(((y \to x) \to x) \to y\big) \to (y \to x) = \big(((y \to x) \to x) \to y\big) \to \big(((y \to x) \to x) \to x\big)$$

$$= \Big(\big(((y \to x) \to x) \to y\big) \odot \big((y \to x) \to x\big)\Big) \to x, \ \text{Lemma 2.2.1(8)}$$

$$= \Big(\big((y \to x) \to x\big) \wedge y\Big) \to x, \ \text{by Proposition 3.1.4 (i)}$$

$$= y \to x, \ \text{by Lemma 2.2.1(6)}.$$

Hence, by (6.4) either $y \to x = 1$ or $((y \to x) \to x) \to y = 1$. Since $x < y$, we have $y \to x \ne 1$ and so $((y \to x) \to x) \to y = 1$, which implies $(y \to x) \to x \le y$. Hence, in view of Lemma 2.2.1(6), $(y \to x) \to x = y$. $\square$

Theorem 6.2.5 ([31]) *Every simple hoop is a linearly ordered Wajsberg hoop.*

Proof It follows immediately from Lemma 6.2.3(ii) and Proposition 6.2.4. $\square$

Corollary 6.2.6 ([31]) *Let $\mathbf{H}$ be a non-trivial bounded hoop. Then $\mathbf{H}$ is simple if and only if it is isomorphic to a subalgebra of the bounded hoop $([0, 1]; \odot, \to, 0, 1)$ in Example 3.3.11.*

Proof Suppose that $\mathbf{H} = (H; \odot, \rightarrow, 0, 1)$ is a subalgebra of the bounded hoop $([0, 1]; \odot, \rightarrow, 0, 1)$ in Example 3.3.11. Choose $x \in H \setminus \{1\}$. By Example 3.3.11, $x \odot x = (2x - 1) \vee 0$. It follows that $x \odot x \odot x = (x + ((2x - 1) \vee 0) - 1) \vee 0 = (3x - 2) \vee (x - 1) \vee 0 = (3x - 2) \vee 0$. In a similar way, $\overbrace{x \odot \cdots \odot x}^{n\text{-times}} = (nx - n + 1) \vee 0$ for all $n \in \mathbb{N}$. Since $0 \le x < 1$, there exists $n \in \mathbb{N}$ such that $nx + 1 \le n$, entails that $\overbrace{x \odot \cdots \odot x}^{n\text{-times}} = (nx - n + 1) \vee 0 = 0$, consequently, $0 \in \langle x \rangle$, that is $\langle x \rangle = H$. Therefore, $\mathbf{H}$ is simple.

Conversely, assume that $\mathbf{H} = (H; \odot, \rightarrow, 0, 1)$ is a bounded simple hoop. By Theorem 6.2.5, it is a bounded Wajsberg hoop. Theorem 3.3.13 implies that $\alpha(H; \odot, \rightarrow, 0, 1) = (H; \oplus, ', 0, 1)$ is an MV-algebra. By Example 4.1.4, $\mathcal{I}(\mathbf{M}) = \{F' : F \in \mathcal{F}(\mathbf{H})\}$, where $\mathbf{M} = (M; \oplus, ', 0, 1)$, consequently, $\mathbf{M}$ is simple, since $\mathcal{I}(\mathbf{M}) = \{F' : F \in \mathcal{F}(\mathbf{H})\} = \{\{0\}, H\}$. By [73, Theorem 3.5.1], or Proposition 1.4.6(vi), $\mathbf{M}$ is isomorphic to a subalgebra of the standard MV-algebra $([0, 1]; \oplus, ', 0, 1)$, and so by Remark 3.3.58(ii), $\mathbf{H} = (H; \odot, \rightarrow, 0, 1)$ is isomorphic to a subalgebra of the bounded hoop $([0, 1]; \odot, \rightarrow, 0, 1)$ in Example 3.3.11. $\qquad\square$

Corollary 6.2.7 *Each non-trivial finite simple hoop is isomorphic to the bounded hoop* $\mathbf{L}_n = (L_n; \odot, \rightarrow, 0, 1)$ *for some* $n \in \mathbb{N}$ *where* $L_n = \{0, 1/n, 2/n, \ldots, (n-1)/n, 1\}$ *and* $\odot$ *and* $\rightarrow$ *are defined in Example* 3.3.11.

Proof An easy calculation shows that $\mathbf{L}_n$ is a simple hoop for all $n \in \mathbb{N}$. Consider the map α and β in the proof of Theorem 3.3.13. Let $\mathbf{H} = (H; \odot, \rightarrow, 1)$ be a finite simple hoop with m element. If $m = 1$, then evidently, $\mathbf{H} \cong \mathbf{1}$. Let $|H| = m > 1$. Due to Remark 2.2.2, it is bounded, so by Theorem 6.2.5, $\mathbf{H}$ is a bounded Wajsberg hoop. Similar to the proof of Corollary 6.2.6, we can show that $\mathbf{M} = (H; \oplus, ', 0, 1) = \alpha(\mathbf{H})$ is a finite simple MV-algebra. By Proposition 1.4.6(vii), there exists $n \in \mathbb{N}$ such that $\mathbf{M} \cong (L_n; \oplus, ', 0, 1)$ (see Example 1.4.2(ii)), so by Remark 3.3.58(ii), $\mathbf{H} \cong \beta((L_n; \oplus, ', 0, 1)) = \mathbf{L}_n$. We note that $n + 1 = |L_n| = |H| = m$. $\qquad\square$

Before stating a theorem regarding the presentation of simple unbounded hoops, we recall an important result known as Hölder's theorem:

Theorem 6.2.8 ([197, Theorem 3.2.1]) *An ℓ-group* $\mathbf{G}$ *is simple if and only if* $\mathbf{G}$ *is ℓ-isomorphic to some additive subgroup of* $(\mathbb{R}; +, 0)$ *of the real numbers with the natural order.*

Theorem 6.2.9 ([116, Theorem 2.1]) *Let* $\mathbf{H}$ *be an unbounded simple hoop. Then there exists a subgroup* $\mathbf{G}$ *of* $(\mathbb{R}; +, 0)$ *such as* $\mathbf{H} \cong \mathbf{N}(\mathbf{G})$. *In addition, simple unbounded hoops are cancellative.*

Proof Let $\mathbf{H}$ be an unbounded simple hoop, then by Theorem 6.2.4, it is a linear Wajsberg hoop that is clearly unbounded. From Theorem 6.1.16, it follows that $\mathbf{H}$ is cancellative, so by Proposition 3.3.19, $\mathbf{H} \cong \mathbf{N}(\mathbf{G})$ for some ℓ-group $\mathbf{G}$. Since $\mathbf{H}$ is simple, then by Example 4.1.5, $\mathbf{G}$ is simple (i.e., it has no proper convex ℓ-subgroups), hence by Theorem 6.2.8, $\mathbf{G}$ has to be a subgroup of the group of $(\mathbb{R}; +, 0)$.

The proof of the last part follows from Example 3.3.8. $\qquad\qquad\square$

Example 6.2.10 ([31, Example 2.4]) (i) Take the free monoid on one generator

$$C_\infty = \{1 = a^0, a, a^2, a^3, \ldots\},$$

partially ordered by $1 = a^0 > a = a^1 > a^2 > a^3 > \cdots$, we can define the operations $a^n \odot a^m = a^{n+m}$, and $a^n \to a^m = a^{\max(m-n,0)}$ for all $n, m \in \mathbb{N}$. Then the algebra $\mathbf{C}_\infty = (C_\infty; \odot, \to, 1)$ is a hoop.

On the other hand, by Example 3.3.8, $\mathbf{N}(\mathbb{Z})$ is a cancellative hoop. We can easily verify that $g : C_\infty \to \mathbf{N}(\mathbb{Z})$ defined by $g(a^n) = -n$ is an isomorphism. Therefore, $\mathbf{C}_\infty \cong \mathbf{N}(\mathbb{Z})$.

(ii) Now, consider the Abelian ℓ-group of the integers, $\mathbb{Z} = (\mathbb{Z}; \vee, \wedge, +, -, 0)$ and an arbitrary positive integer m. Let $C_m = \{1 = a^0, a, a^2, \ldots, a^m\}$ and the binary operations $\odot_{a^m}$ and $\to_{a^m}$ on C_m are defined by $a^k \odot_{a^m} a^n = a^{\min(k+n,m)}$, and $a^k \to_{a^m} a^n = a^{\max(n-k,0)}$ for all $k, n \leq m$. Then $\mathbf{C}_m = (C_m; \odot_{a^m}, \to_{a^m}, 1)$ is a hoop. Similarly to (i), $h : \mathbf{C}_m \to \mathbf{Z}[\mathbf{m}]$ (see Example 3.3.57) defined by $h(a^i) = i$ is an isomorphism.

(iii) The finite chains $\mathbf{C}_m$, $m \in \mathbb{N}$, as well as $\mathbf{C}_\infty$ are simple Wajsberg hoops.

It follows from Remark 3.3.58 that $\mathbf{C}_n \cong \mathbf{L}_n$ for all $n \in \mathbb{N}$. While the notation $\mathbf{C}_n$ has been used in some contexts, we prefer to use $\mathbf{L}_n$ throughout this text.

Proposition 6.2.11 ([116, Proposition 2.2]) *Let $\mathbf{H}$ be a hoop. Then $M \in \mathrm{Max}(\mathbf{H})$ if and only if $\mathbf{H}/M$ is simple.*

Proof Let $M \in \mathrm{Max}(\mathbf{H})$ and F be a proper filter of $\mathbf{H}/M$, so by Theorem 4.1.17(i), there exists a $G \in \mathcal{F}(\mathbf{H})$ containing M. From $M \subseteq G \subset H$ and $M \in \mathrm{Max}(\mathbf{H})$ it follows that $M = G$ and so $G/M = M/M = \{1/M\}$. Thus, $\mathcal{F}(\mathbf{H}/F) = \{M, H/F\}$, which entails $\mathbf{H}/M$ is simple.

Conversely, assume that $\mathbf{H}/M$ is simple and there exists $Q \in \mathcal{F}(\mathbf{H})$ such that $M \subseteq Q \subseteq H$. If $M \neq Q$, then $M/M \neq Q/M \in \mathcal{F}(\mathbf{H}/M)$, by Theorem 4.1.17(i). Thus, $Q/M = H/M$ and so $Q = H$. Therefore, $M \in \mathrm{Max}(\mathbf{H})$. $\qquad\qquad\square$

We know that a proper filter F of a bounded hoop $\mathbf{H}$ is an Ultrafilter if and only if $\mathbf{H}/F \cong \mathbf{L}_1$. By combining Proposition 6.2.11, Theorem 6.2.9 and Corollary 6.2.6 we obtain the following result, immediately:

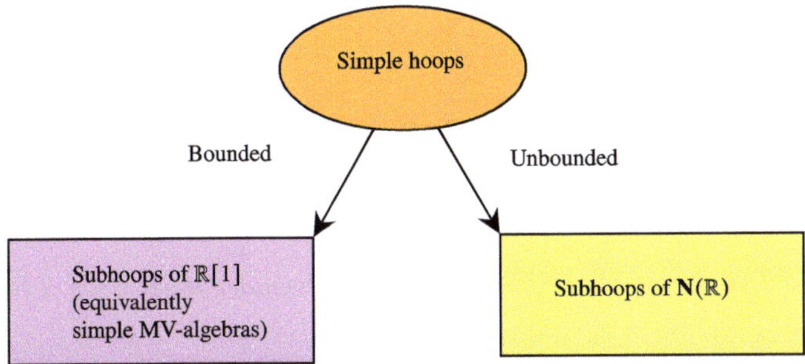

Fig. 6.1 Simple hoops

Corollary 6.2.12 *Let F be a proper filter of a hoop* $\mathbf{H}$. *Then* $F \in \mathrm{Max}(\mathbf{H})$ *if and only if* $\mathbf{H}/F$ *is either isomorphic to* $\mathbf{N}(\mathbf{G})$ *for some subgroup* $\mathbf{G}$ *of* $\mathbb{R}$ *or* $\mathbf{H}/F$ *or isomorphic to a subalgebra of the bounded hoop* $([0, 1]; \odot, \rightarrow, 0, 1)$.

Figure 6.1 summarizes the main results of this section.

6.2.2   Semisimple Hoops

Definition 6.2.13 We define the **radical** of a hoop $\mathbf{H}$, denoted by $\mathrm{Rad}(\mathbf{H})$ as follows:

$$\mathrm{Rad}(\mathbf{H}) = \bigcap \{M : M \in \mathrm{Max}(\mathbf{H})\}. \tag{6.5}$$

Definition 6.2.14 A hoop $\mathbf{H} = (H; \odot, \rightarrow, 1)$ is said to be **semisimple** if $\mathbf{H}$ is non-trivial and $\mathrm{Rad}(\mathbf{H}) = \{1\}$.

Clearly, $\mathrm{Rad}(\mathbf{H}) = \{1\}$ for each simple hoop.

Proposition 6.2.15 *A hoop* $\mathbf{H}$ *is semisimple if and only if it is a subdirect product of simple hoops.*

Proof Assume that $\mathbf{H}$ is semisimple. Then $\mathrm{Rad}(\mathbf{H}) = \{1\}$ implies that $\phi : \mathbf{H} \rightarrow \prod_{M \in \mathrm{Max}(\mathbf{H})} \mathbf{H}/M$ defined by $\phi(x) = (x/M)_{M \in \mathrm{Max}(\mathbf{H})}$ is a one-to-one homomorphism, and $\pi_M \circ \phi(H) = H/M$ for all $M \in \mathrm{Max}(\mathbf{H})$, where $\pi_M : \prod_{M \in \mathrm{Max}(\mathbf{H})} \mathbf{H}/M \rightarrow \mathbf{H}/M$ is the Mth natural projection map. By Proposition 6.2.11, $\mathbf{H}/M$ is simple for all $M \in \mathrm{Max}(\mathbf{H})$.

Conversely, let $\{\mathbf{H}_i = (H_i; \odot, \rightarrow, 1) : i \in I\}$ be a family of simple hoops and $\phi : \mathbf{H} \rightarrow \prod_{i \in I} \mathbf{H}_i$ be a subdirect embedding. For each $i \in I$, set $M_i := \mathrm{Ker}(\pi_i \circ \phi) = (\pi_i \circ \phi)^{-1}(\{1\})$. By Theorem 2.4.8(iii), M_i is a filter of $\mathbf{H}$ for all $i \in I$. In addition, by Theorem 2.4.8(v), $\mathbf{H}/M_i \cong \pi_i \circ \phi(\mathbf{H}) = \mathbf{H}_i$. Since $\mathbf{H}_i$ is simple, Theorem 6.2.11, implies that M_i is a maximal filter of $\mathbf{H}$, that is, $\{M_i : i \in I\} \subseteq \mathrm{Max}(\mathbf{H})$. We have $\bigcap_{i \in I} M_i \subseteq \mathrm{Ker}(\phi) = \{1\}$. Therefore, $\mathrm{Rad}(\mathbf{H}) \subseteq \bigcap_{i \in I} M_i = \{1\}$, which means $\mathbf{H}$ is semisimple. $\quad\square$

The next proposition proposes a representation for $\mathrm{Rad}(\mathbf{H})$, where $\mathbf{H}$ is a bounded basic hoop.

Proposition 6.2.16 *Let* $\mathbf{H} = (H; \odot, \rightarrow, 0, 1)$ *be a bounded basic hoop. Then*

(i) $\mathrm{Rad}(\mathbf{H}) = \{x \in H \setminus \{0\} : (x^n)' \leq x, \ \forall n \in \mathbb{N}\} \cup \{1\}$.

(ii) *If* $\mathbf{H}$ *is a bounded Wajsberg hoop and* $x \in \mathrm{Rad}(\mathbf{H}) \setminus \{1\}$, *then* $x^{n+1} < x^n$ *for all* $n \in \mathbb{N}$.

Proof (i) If $\mathbf{H}$ is trivial, then both sides of the above equality are $\{1\}$ and the proof is evident. Let $\mathbf{H}$ be a non-trivial bounded basic hoop and $x \in H \setminus \mathrm{Rad}(\mathbf{H})$. By Proposition 4.2.28, $\mathrm{Max}(\mathbf{H}) \neq \emptyset$, so there exists a maximal filter M of $\mathbf{H}$ such that $x \notin M$, consequently, $(x^n)' \in M$ for some $n \in \mathbb{N}$, see Corollary 4.2.27. We claim that

$$x \notin \{x \in H \setminus \{0\} : (x^n)' \leq x, \ \forall n \in \mathbb{N}\},$$

otherwise, $(x^n)' \leq x$ implies that $x \in M$, a contradiction. Now, let $x \in H \setminus \{x \in H \setminus \{0\} : (x^n)' \leq x, \ \forall n \in \mathbb{N}\}$ and $x \neq 1$. Then there exists $m \in \mathbb{N}$ such that $(x^m)' \not\leq x$, which implies $(x^m)' \rightarrow x \neq 1$, so by Corollary 4.2.9 and Theorem 4.2.12, there exists a prime filter P such that $(x^m)' \rightarrow x \notin P$, consequently, $x \rightarrow (x^m)' \in P$. On the other hand, due to Proposition 4.2.28, there exists a maximal ideal M of $\mathbf{H}$ such that $P \subseteq M$, which entails $(x^{m+1})' = x \rightarrow (x^m)' \in P \subseteq M$. Hence, $x \notin M$, otherwise, $0 \in \langle M \cup \{x\} \rangle = M$, that is a contradiction. Thus, $x \notin \mathrm{Rad}(\mathbf{H})$. Therefore,

$$\mathrm{Rad}(\mathbf{H}) = \{x \in H \setminus \{0\} : (x^n)' \leq x, \ \forall n \in \mathbb{N}\} \cup \{1\}.$$

(ii) Let $\mathbf{H}$ be a bounded Wajsberg hoop and $x \in \mathrm{Rad}(\mathbf{H}) \setminus \{1\}$. Due to (i), $(x^n)' \leq x$ for all $n \in \mathbb{N}$. Also, by Remark 2.2.5(i), $(x^n)' \leq (x^{n+1})'$ for all $n \in \mathbb{N}$. If there exists $m \in \mathbb{N}$ such that $x^m = x^{m+1}$, then $(x^{m+1})' \rightarrow (x^m)' = 1$. It follows that

$$
\begin{aligned}
1 = (x^{m+1})' \rightarrow (x^m)' &= \big((x \odot x^m) \rightarrow 0\big) \rightarrow (x^m)' \\
&= \big(x \rightarrow (x^m \rightarrow 0)\big) \rightarrow (x^m)', \ \text{by Lemma 2.2.1(8)} \\
&= (x \rightarrow (x^m)') \rightarrow (x^m)' = x \vee (x^m)', \ \text{since } \mathbf{H} \text{ is Wajsberg} \\
&= x, \ \text{by the assumption,}
\end{aligned}
$$

which is a contradiction. Therefore, $(x^n)' < (x^{n+1})'$ for all $n \in \mathbb{N}$, consequently, Theorem 3.3.15 implies that $x^{n+1} < x^n$ for all $n \in \mathbb{N}$ (note that $x^{n+1} = x^n$ entails that $(x^{n+1})' = (x^n)'$, a contradiction). $\quad\square$

Theorem 6.2.17 *Every semisimple hoop is a Wajsberg hoop. The converse holds for finite Wajsberg hoops, i.e., every finite Wajsberg hoop is semisimple.*

Proof Let $\mathbf{H}$ be a semisimple hoop. By Proposition 6.2.15, $\mathbf{H}$ can be embedded in $\prod_{i \in I} \mathbf{H}_i$ with the mapping $\phi : \mathbf{H} \to \prod_{i \in I} \mathbf{H}_i$, where $\mathbf{H}_i$ is a simple hoop for all $i \in I$. By Proposition 6.2.4, each H_i is Wajsberg, that is $(x \to y) \to y = (y \to x) \to x$ for all $x, y \in H_i$. Consequently, $\prod_{i \in I} \mathbf{H}_i$ is a Wajsberg hoop so does $\mathbf{H}$. Indeed, for each $x, y \in H$, we have

$$\phi((x \to y) \to y) = (\phi(x) \to \phi(y)) \to \phi(y) = (\phi(y) \to \phi(x)) \to \phi(x) = \phi((y \to x) \to x).$$

This implies that $(x \to y) \to y = (y \to x) \to x$, since ϕ is one-to-one.

Now, assume that $\mathbf{H}$ is a finite Wajsberg hoop. If $|H| = 1$, then $\mathbf{H}$ is isomorphic to the trivial hoop $\mathbf{1}$, which is clearly simple as well as semisimple. Let $|H| = n > 1$. By Remark 2.2.2, $\mathbf{H}$ is a bounded Wajsberg hoop. If $\mathrm{Rad}(\mathbf{H}) \neq \{1\}$, then by Proposition 6.2.16(ii), $\mathbf{H}$ is infinite, a contradiction. Therefore, $\mathbf{H}$ is semisimple. $\qquad\square$

Corollary 6.2.18 *Finite semisimple hoops are termwise equivalent to finite semisimple MV-algebras.*

Proof The proof follows from Theorem 6.2.17. $\qquad\square$

Corollary 6.2.19 *For every finite semisimple hoop $\mathbf{H}$ there exists $k, n_1, \ldots, n_k \in \mathbb{N}$ such that $\mathbf{H}$ is subdirect product of the family $\{\mathbf{L}_{n_1}, \ldots, \mathbf{L}_{n_k}\}$.*

Proof Let $\mathbf{H}$ be a finite semisimple hoop and $\phi : \mathbf{H} \to \prod_{M \in \mathrm{Max}(\mathbf{H})} \mathbf{H}/M$ be the subdirect embedding in the proof of Proposition 6.2.15. Since $\mathbf{H}$ is a finite hoop, we have $\mathrm{Max}(\mathbf{H})$ and $\mathcal{F}(\mathbf{H})$ are finite sets. In addition, $\mathbf{H}/M$ is a finite hoop for all $M \in \mathrm{Max}(\mathbf{H})$. Suppose that $\mathrm{Max}(\mathbf{H}) = \{M_1, \ldots, M_k\}$. Given a maximal filter M_i of $\mathbf{H}$, the quotient hoop $\mathbf{H}/M_i$ is a finite simple hoop, see Proposition 6.2.11, so by Corollary 6.2.7, there exists $n_i \in \mathbb{N}$ and a hoop isomorphism $g_i : \mathbf{H}/M_i \to \mathbf{L}_{n_i}$, consequently, $g : \prod_{i=1}^{k} \mathbf{H}/M_i \to \prod_{i=1}^{k} \mathbf{L}_{n_i}$ defined by $g(x_1/M_1, \ldots, x_k/M_k) = (g_1(x_1/M_1), \ldots, g_k(x_k/M_k))$ is a hoop isomorphism, too. An easy calculation shows that $g \circ \phi : \mathbf{H} \to \prod_{i=1}^{k} \mathbf{L}_{n_i}$ is a subdirect embedding of hoops. $\quad\square$

Bibliographical Remarks and Suggestions for Further Study

Section 6.2 explored simple and semisimple hoops. Simple hoops play a crucial role in the theory of hoops. The results presented in this section were originally established in [31, 41, 116, 141]. We have demonstrated that every simple hoop is a linear Wajsberg hoop, and consequently, every semisimple hoop is also a Wajsberg hoop. The equivalence between bounded Wajsberg hoops and MV-algebras provided a complete representation for finite semisimple hoops. Considering the notion of the radical of a filter was defined in [41], the radical of the trivial filter $\{1\}$ coincides with $\text{Rad}(\mathbf{H}) = \bigcap\{F : F \in \text{Max}(\mathbf{H})\}$. Consequently, the results of the aforementioned paper can contribute to a more profound understanding of semisimple hoops.

6.3 Local and Perfect Hoops

This section delves into the properties of local and perfect hoops, exploring equivalent conditions that characterize these structures.

6.3.1 Local Hoops

In this section, we characterize locally finite hoops using the notion of the order of elements, establishing their equivalence with simplicity. We also connect element orders, filters, and maximal filters, which provides a structural description of local hoops. These results are supported by several illustrative examples and propositions.

Definition 6.3.1 ([39]) Let $\mathbf{H}$ be a bounded hoop. The order of $x \in H$, in symbols $\text{ord}(x)$, is the smallest $n \in \mathbb{N}$ such that $x^n = 0$. If no such n exists, then $\text{ord}(x) = \infty$.

The hoop $\mathbf{H}$ is called **locally finite** if $\text{ord}(x) < \infty$ for each $x \in H \setminus \{1\}$ (see [39, Definition 5.17]).

Example 6.3.2 (i) In Example 4.3.4, $\text{ord}(0) = 1$, $\text{ord}(a) = 2$ and $\text{ord}(b) = \infty = \text{ord}(1)$. Hence, $\mathbf{H}$ is not locally finite.

(ii) Assume $H = \{0, a, b, c, 1\}$ is a chain where $0 < a < b < c < 1$. Define two operations $\odot$ and $\rightarrow$ on H by Table 6.1.

Then $(H; \odot, \rightarrow, 0, 1)$ is a bounded hoop and $\text{ord}(0) = 1$, $\text{ord}(a) = 2$, $\text{ord}(b) = 2$, $\text{ord}(c) = 3$. So, $\mathbf{H}$ is locally finite.

(iii) Consider the hoop $\mathbf{L}_n = (L_n; \odot, \rightarrow, 0, 1)$ in Example 3.3.11. We have $L_n = \{0, 1/n, \ldots, (n-1)/n, 1\}$. It follows from the definition of $\odot$ that

$$\frac{n-1}{n} \odot \frac{n-1}{n} = \frac{n-1+n-1-n}{n} \vee 0 = \frac{n-2}{n} \vee 0.$$

In a similar way, $((n-2)/n)^k = ((n-2k)/n) \vee 0$ which entails $(m/n)^n = 0$ for each $m \in \{1, \ldots, n-1\}$. Therefore, $\mathbf{L}_n$ is locally finite.

Table 6.1 Operations $\odot$ and $\rightarrow$ of Example 6.3.2(ii)

Table 6.1.1

$\odot$	0	a	b	c	1
0	0	0	0	0	0
a	0	0	0	0	a
b	0	0	0	0	b
c	0	0	a	a	c
1	0	a	b	c	1

Table 6.1.2

$\rightarrow$	0	a	b	c	1
0	1	1	1	1	1
a	c	1	1	1	1
b	b	c	1	1	1
c	b	c	c	1	1
1	0	a	b	c	1

Proposition 6.3.3 *A bounded hoop* **H** *is locally finite if and only if it is simple.*

Proof If **H** is simple, then $0 \in \langle x \rangle$ for all $x \in H \setminus \{1\}$, so by Corollary 2.3.9, there exists $n \in \mathbb{N}$ such that $x^n \leq 0$, or equivalently, $x^n = 0$. Conversely, if **H** is locally finite, then for each $x \in H \setminus \{1\}$ we have $0 = x^n \in \langle x \rangle$, which implies that $\langle x \rangle = H$. Therefore, **H** is simple. $\qquad\square$

Proposition 6.3.4 ([39, Proposition 4.2]) *Let* **H** *be a bounded hoop. The following assertions hold for any* $x, y \in H$:

(i) $\langle x \rangle$ *is a proper filter of* **H** *if and only if* $\mathrm{ord}(x) = \infty$.
(ii) *If* $x \leq y$ *and* $\mathrm{ord}(x) = \infty$, *then* $\mathrm{ord}(y) = \infty$.
(iii) *If* $x \leq y$ *and* $\mathrm{ord}(y) < \infty$, *then* $\mathrm{ord}(x) < \infty$.
(iv) *For every proper filter* F *(as well as* $F \in \mathrm{Max}(\mathbf{H})$), $F \subseteq \{x \in H : \mathrm{ord}(x) = \infty\}$.
(v) $\mathrm{ord}(1) = \infty$ *and* $\mathrm{ord}(0) = 1$.

Proof (i) Let $\langle x \rangle$ be a proper filter of **H**. If there exists $n \in \mathbb{N}$ such that $x^n = 0$, then by Corollary 2.3.9, $0 \in \langle x \rangle$, which is a contradiction.

Conversely, let $\langle x \rangle = H$. Since $0 \in H$, we have $0 \in \langle x \rangle$, and so by Corollary 2.3.9, there exists $n \in \mathbb{N}$ such that $x^n = 0$. Hence, $\mathrm{ord}(x) < \infty$, which is a contradiction.

(ii) Let $x \leq y$ such that $\mathrm{ord}(x) = \infty$ for some $x, y \in H$. From $x^n \leq y^n$ (since $(H; \odot, 1)$ is an ordered commutative monoid) for all $n \in \mathbb{N}$ and $\mathrm{ord}(x) = \infty$, we get $y^n \neq 0$ for all $n \in \mathbb{N}$. Therefore, $\mathrm{ord}(y) = \infty$.

(iii) Let $x \leq y$ such that $\mathrm{ord}(y) < \infty$ for some $x, y \in H$. Then there exists $n \in \mathbb{N}$ such that $y^n = 0$. Since $x^n \leq y^n$, we have $x^n = 0$. Thus, $\mathrm{ord}(x) < \infty$.

The proofs of (iv) and (v) are clear. $\qquad\square$

Recall that if **H** is bounded, due to Proposition 4.2.28, $\mathrm{Max}(\mathbf{H}) \neq \emptyset$.

Definition 6.3.5 ([39, 78]) A hoop $\mathbf{H}$ is called **local** if it has a unique maximal filter.

Theorem 6.3.6 ([39, Theorem 5.11]) *Every bounded linear hoop is local.*

Proof Let $\mathbf{H}$ be a bounded linear hoop. According to Proposition 4.2.28, $\mathbf{H}$ has a maximal filter, M say. Since $\mathbf{H}$ is linearly ordered, $\{1\}$ is a prime filter of $\mathbf{H}$, so by Proposition 4.2.4, $\{F \in \mathcal{F}(\mathbf{H}) : \{1\} \subseteq F\} = \mathcal{F}(\mathbf{H})$ is a chain which implies that M is the only maximal filter of $\mathbf{H}$. Therefore, $\mathbf{H}$ is local. $\square$

Theorem 6.3.7 [39, Theorem 5.3] *A bounded hoop* $\mathbf{H}$ *is local if and only if* $\mathrm{ord}(x) < \infty$ *or* $\mathrm{ord}(x') < \infty$ *for all* $x \in H$.

Proof Let M be the unique maximal filter of a bounded hoop $\mathbf{H}$. If $x \in H$ such that $\mathrm{ord}(x) = \mathrm{ord}(x') = \infty$, then by Proposition 6.3.4(i), $\langle x \rangle$ and $\langle x' \rangle$ are two proper filters of $\mathbf{H}$. Proposition 4.2.28 implies that $\langle x \rangle, \langle x' \rangle \subseteq M$, and so $x, x' \in M$, which entail that $0 = x \odot x' \in M$ (by Remark 2.2.5), a contradiction. Hence, $\mathrm{ord}(x) < \infty$ or $\mathrm{ord}(x') < \infty$.

Conversely, let $\mathbf{H}$ be a bounded hoop such that for each $x \in H$, $\mathrm{ord}(x) < \infty$ or $\mathrm{ord}(x') < \infty$. By Proposition 4.2.28, $\mathbf{H}$ has at least the maximal filter M. We claim that M is the only maximal filter of $\mathbf{H}$. Due to Proposition 6.3.4(iv), $M \subseteq \{x \in H : \mathrm{ord}(x) = \infty\}$. Suppose that there exists $x \in H \setminus M$ such that $\mathrm{ord}(x) = \infty$. Since $M \in \mathrm{Max}(\mathbf{H})$, by Proposition 4.2.25, there exists $n \in \mathbb{N}$ such that $(x^n)' \in M$. Thus, by Proposition 6.3.4(iv), $\mathrm{ord}((x^n)') = \infty$, which entails that $\mathrm{ord}((x^n)'') < \infty$ (since $\mathbf{H}$ is local). Due to Remark 2.2.5(ii), $x \leq x''$ and so by Proposition 6.3.4(iii), $\mathrm{ord}(x^n) < \infty$, consequently, $\mathrm{ord}(x) < \infty$, which is a contradiction. Hence, $M = \{x \in H : \mathrm{ord}(x) = \infty\}$. Then, for any $M \in \mathrm{Max}(\mathbf{H})$, $M = \{x \in H : \mathrm{ord}(x) = \infty\}$. Therefore, $\mathbf{H}$ has a unique maximal filter, see Proposition 6.3.4(iv). $\square$

Example 6.3.8 (i) By Theorem 6.3.6, the hoop $([0, 1]; \odot, \rightarrow, 0, 1)$ in Example 3.3.11 as well as $\mathbf{L}_n$ is local for all $n \in \mathbb{N}$.

(ii) Let $H = \{0, a, b, c, d, 1\}$. Define two operations $\odot$ and $\rightarrow$ on H as Table 6.2. Then $(\mathbf{H}; \odot, \rightarrow, 0, 1)$ is a bounded local hoop and $F := \{a, b, c, 1\}$ is the unique maximal filter of $\mathbf{H}$.

(iii) For every hoop $\mathbf{H}$, the bounded hoop $\mathbf{2} \oplus \mathbf{H}$ is local, possessing the unique maximal filter H. Furthermore, by Exercise 3.4.4, every non-trivial special hoop is local.

Corollary 6.3.9 ([39, Corollary 5.4]) *Let* $\mathbf{H}$ *be a bounded hoop. Then* $\mathbf{H}$ *is local if and only if* $\mathrm{Rad}(\mathbf{H}) = \{x \in H : \mathrm{ord}(x) = \infty\}$.

Proof Let $\mathbf{H}$ be local. Then by the proof of Theorem 6.3.7, it has a unique maximal filter, say M, such that $M = \{x \in H : \mathrm{ord}(x) = \infty\}$. Thus, $\mathrm{Rad}(\mathbf{H}) = M = \{x \in H : \mathrm{ord}(x) = \infty\}$.

Conversely, let $\mathrm{Rad}(\mathbf{H}) = \{x \in H : \mathrm{ord}(x) = \infty\}$ and $M \in \mathrm{Max}(\mathbf{H})$. By Proposition 6.3.4(iv), we have $M \subseteq \{x \in H : \mathrm{ord}(x) = \infty\}$, which implies that $M \subseteq \mathrm{Rad}(\mathbf{H})$. Therefore, $M = \mathrm{Rad}(\mathbf{H})$. $\square$

Table 6.2 Operations $\odot$ and $\rightarrow$ of Example 6.3.8(ii)

Table 6.2.1

$\odot$	0	a	b	c	d	1
0	0	0	0	0	0	0
a	0	a	c	c	d	a
b	0	c	b	c	d	b
c	0	c	c	c	d	c
d	0	d	d	d	0	d
1	0	a	b	c	d	1

Table 6.2.2

$\rightarrow$	0	a	b	c	d	1
0	1	1	1	1	1	1
a	0	1	b	b	d	1
b	0	a	1	a	d	1
c	0	1	1	1	d	1
d	d	1	1	1	1	1
1	0	a	b	c	d	1

Proposition 6.3.10 ([39, Proposition 5.9]) *Let* **H** *be a bounded hoop and* $M = \{x \in H : \mathrm{ord}(x) = \infty\}$. *Then* $M \in \mathrm{Max}(\mathbf{H})$ *if and only if* $\mathrm{ord}(x \odot y) < \infty$ *implies that* $\mathrm{ord}(x) < \infty$ *or* $\mathrm{ord}(y) < \infty$ *for all* $x, y \in H$.

Proof Let $M \in \mathrm{Max}(\mathbf{H})$. Suppose $\mathrm{ord}(x \odot y) < \infty$ for some $x, y \in H$. Then $x \odot y \notin M$, consequently, $x \notin M$ or $y \notin M$ (otherwise by (F1), $x \odot y \in M$), and so $\mathrm{ord}(x) < \infty$ or $\mathrm{ord}(y) < \infty$.

Conversely, suppose that $\mathrm{ord}(x \odot y) < \infty$ implies that $\mathrm{ord}(x) < \infty$ or $\mathrm{ord}(y) < \infty$, for all $x, y \in H$. We prove $M \in \mathrm{Max}(\mathbf{H})$. Since $\mathrm{ord}(0) = 1$, we get $0 \notin M$, and so M is a proper subset of **H**. Let $x \leq y$ and $x \in M$ for some $x, y \in H$. From $x \in M$, $\mathrm{ord}(x) = \infty$, and Proposition 6.3.4(ii) we conclude that $\mathrm{ord}(y) = \infty$, that is $y \in M$. Now, let $x, y \in H$ such that $\mathrm{ord}(x \odot y) < \infty$. Then by assumption, $\mathrm{ord}(x) < \infty$ or $\mathrm{ord}(y) < \infty$, which is a contradiction. Hence, $x \odot y \in M$, and so M is a proper filter of **H**. From Proposition 6.3.4(iv), it follows that M is the greatest proper filter of **H**. Therefore, $M \in \mathrm{Max}(\mathbf{H})$. $\square$

Corollary 6.3.11 *A hoop* **H** *is local if and only if* $\mathrm{ord}(x \odot y) < \infty$ *implies that* $\mathrm{ord}(x) < \infty$ *or* $\mathrm{ord}(y) < \infty$ *for all* $x, y \in H$.

Proof If **H** is local, then by Corollary 6.3.9, $M = \mathrm{Rad}(\mathbf{H}) = \{x \in H : \mathrm{ord}(x) = \infty\}$ is the unique maximal filter of **H**. Now, the proof follows from Proposition 6.3.10.

Conversely, by Proposition 6.3.10, $M = \{x \in H : \mathrm{ord}(x) = \infty\}$ is a maximal filter of **H**. Due to Proposition 6.3.4(iv), it is the greatest proper filter of **H**, which entails that **H** is local. $\square$

Given a bounded hoop $\mathbf{H}$, we define $D(\mathbf{H})$ and $D^*(\mathbf{H})$ as follows:

$$D(\mathbf{H}) = \{x \in H : x^n > 0 \text{ for all } n \in \mathbb{N}\}, \quad D^*(\mathbf{H}) = \{x \in H : x \leq y' \text{ for some } y \in D(\mathbf{H})\}.$$

Note that $D(\mathbf{H}) = \{x \in H : \operatorname{ord}(x) = \infty\}$.

Theorem 6.3.12 *A bounded hoop* $\mathbf{H}$ *is local if and only if* $D(\mathbf{H})$ *is the unique maximal filter of* $\mathbf{H}$. *In addition,* $D(\mathbf{H}) \cap D^*(\mathbf{H}) = \emptyset$ *for every bounded local hoop* $\mathbf{H}$.

Proof The proof of the first part follows from Corollary 6.3.9. Now, assume that $\mathbf{H}$ is a bounded local hoop and $x \in D(\mathbf{H}) \cap D^*(\mathbf{H})$. Since $x \in D^*(\mathbf{H})$, there exists $y \in D(\mathbf{H})$ such that $x \leq y'$. It follows from Remark 2.2.5 that $y \leq y'' \leq x'$, which implies $\operatorname{ord}(x') = \infty$. Consequently, $x, x' \in D(\mathbf{H})$, contradicting Theorem 6.3.7. Therefore, $D(\mathbf{H}) \cap D^*(\mathbf{H}) = \emptyset$. $\qquad\square$

Definition 6.3.13 ([39]) A proper filter F of a bounded hoop $\mathbf{H}$ is called a **primary filter** if for all $x, y \in H$, $(x \odot y)' \in F$ implies that $(x^n)' \in F$ or $(y^n)' \in F$ for some $n \in \mathbb{N}$.

Example 6.3.14 ([39, Examples 3.15 and 3.16]) (i) Assume $H = \{0, a, b, 1\}$ is a chain where $0 < a < b < 1$. Define $\odot$ and $\rightarrow$ on H by Table 6.3.
   Then $(H; \odot, \rightarrow, 0, 1)$ is a bounded hoop and $P = \{b, 1\}$ is a primary filter.
   (ii) Assume $H = \{0, a, b, c, 1\}$ is a chain with $0 < a < b < c < 1$. Define $\odot$ and $\rightarrow$ on H by Table 6.4.
   Then $(H; \odot, \rightarrow, 0, 1)$ is a bounded hoop and $P = \{c, 1\}$ is a prime filter, but it is not a primary filter. Because, $(a \odot b)' = 1 \in P$, while $(a^n)' = b \notin P$ and $(b^n)' = a \notin P$ for any $n \in \mathbb{N}$.

Table 6.3 Operations $\odot$ and $\rightarrow$ of Example 6.3.14(i)

Table 6.3.1

$\odot$	0	a	b	1
0	0	0	0	0
a	0	0	a	a
b	0	a	b	b
1	0	a	b	1

Table 6.3.2

$\rightarrow$	0	a	b	1
0	1	1	1	1
a	a	1	1	1
b	0	a	1	b
1	0	a	b	1

Table 6.4 Operations $\odot$ and $\to$ of Example 6.3.14(ii)

Table 6.4.1

$\odot$	0	a	b	c	1
0	0	0	0	0	0
a	0	a	0	a	a
b	0	0	b	b	b
c	0	a	b	c	c
1	0	a	b	c	1

Table 6.4.2

$\to$	0	a	b	c	1
0	1	1	1	1	1
a	b	1	1	1	1
b	a	a	1	1	1
c	0	a	b	1	1
1	0	a	b	c	1

Theorem 6.3.15 ([39]) *Let $P \in \mathcal{F}(\mathbf{H})$. Then the following statements are equivalent:*

(i)   *P is a primary filter.*
(ii) *For each $x \in H$, there exists $n \in \mathbb{N}$ such that $(x^n)' \in P$ or $((x')^n)' \in P$.*
(iii) *$\mathbf{H}/P$ is a local hoop.*

Proof (i) $\Rightarrow$ (ii) Let P be primary. Given $x \in H$, from $(x \odot x')' = 1$ it follows that $(x^n)' \in P$ or $((x')^n)' \in P$ for some $n \in \mathbb{N}$.

(ii) $\Rightarrow$ (iii) Choose $x \in H$. By (ii), there is $n \in \mathbb{N}$ such that $(x^n)' \in P$ or $((x')^n)' \in P$. The first one implies that $(x^n)'/P = 1/P$, so $(x^n)''/P = 0/P$, consequently, $x^n/P = 0/P$ (by Remark 2.2.5(ii)), and so $\mathrm{ord}(x/P) < \infty$. In a similar way, if $((x')^n)' \in P$, then we can prove that $\mathrm{ord}(x'/P) < \infty$. Therefore, by Theorem 6.3.7, $\mathbf{H}/P$ is local.

(iii) $\Rightarrow$ (i) Let $\mathbf{H}/P$ be local. Then P is a proper filter. Let $(x \odot y)' \in P$ for some $x, y \in H$. Suppose $(x^n)' \notin P$ for all $n \in \mathbb{N}$. Since $(x \odot y)' \in P$, we have $(x \odot y)'/P = 1/P$, equivalently, $x/P \odot y/P = (x \odot y)/P = 0/P$. By (2.2), $x/P \leq y'/P$, so by Remark 2.2.5(iii) and Lemma 2.2.1(7), $((y')^n)'/P \leq (x^n)'/P$. In addition, from $(x^n)' \notin P$ we get $(x^n)'/P \neq 1/P$, which yields that $((y')^n)'/P \neq 1/P$, and $((y')^n)''/P \neq 0/P$, so $((y')^n)/P \neq 0/P$ for all $n \in \mathbb{N}$ (since, $(y')^m/P = 0/P$, implies that $((y')^m)''/P = 0/P$). Thus, $\mathrm{ord}\left(y'/P\right) = \infty$. Since $\mathbf{H}/P$ is local, we have $\mathrm{ord}\left(y''/P\right) < \infty$, and so by Remark 2.2.5(ii) and Proposition 6.3.4(iii), $\mathrm{ord}\left(y/P\right) < \infty$, consequently, there exists $m \in \mathbb{N}$ such that $y^m/P = 0/P$. Therefore, $(y^m)' \in P$. $\qquad\square$

Corollary 6.3.16 ([39]) *If every proper filter of $\mathbf{H}$ is primary, then $\mathbf{H}$ is local.*

Proof Suppose that every proper filter of $\mathbf{H}$ is primary. Since $\{1\}$ is a proper filter of $\mathbf{H}$, we have $\{1\}$ is primary. Thus, by Theorem 6.3.15, $\mathbf{H}/\{1\}$ is local. From $\mathbf{H}/\{1\} \cong \mathbf{H}$, we get $\mathbf{H}$ is local. $\qquad\square$

Theorem 6.3.17 ([39, Theorem 3.17]) *Let* $\mathbf{H}$ *be a bounded* $\vee$*-hoop with* (DNP) *and* $x^2 = x$ *for any* $x \in H$. *Then every primary filter of* $\mathbf{H}$ *is a* $\vee$*-prime filter.*

Proof Let P be a primary filter of $\mathbf{H}$ and $x \vee y \in P$ for some $x, y \in H$. Since $\mathbf{H}$ is bounded satisfying (DNP), by Corollary 3.3.10, we have $(x' \wedge y')' \in P$, so by Proposition 3.1.4(i), $(x' \wedge y')' = (x' \odot (x' \to y'))' \in P$. Now, since P is a primary filter, there exists $n \in \mathbb{N}$ such that $((x')^n)' \in P$ or $((x' \to y')^n)' \in P$. From $x^2 = x$ it follows that $x^n = x$, for any $n \in \mathbb{N}$, and so $(x')' \in P$ or $(x' \to y')' \in P$.

If $x'' \in P$, then $x \in P$, since $\mathbf{H}$ satisfies (DNP).

If $(x' \to y')' \in P$, then $(y \to x)' = (y \to x'')' = (x' \to y')' \in P$. Due to $y' \leq y \to x$ and Lemma 2.2.1(7), we get $(y \to x)' \leq y''$, consequently, $y = y'' \in P$, since $P \in \mathcal{F}(\mathbf{H})$. Therefore, P is a $\vee$-prime filter. $\qquad\square$

Example 6.3.18 Assume $H = \{0, a, b, 1\}$ is a poset where $0 < a, b < 1$. Define $\odot$ and $\to$ on H by Table 6.5.

Then $(H; \odot, \to, 0, 1)$ is a bounded hoop and $\vee$-hoop with (DNP) such that $x^2 = x$ for all $x \in H$.

In the following example, $\mathbf{H}$ is a bounded hoop and $c^2 \neq c$. We show that there exists a primary filter that is not a $\vee$-prime filter.

Example 6.3.19 ([39, Example 3.19]) Assume $H = \{0, a, b, c, d, 1\}$ is a poset where $0 < a, b, c, d < 1$. Define $\odot$ and $\to$ on H by Table 6.6.

Then $(H; \odot, \to, 0, 1)$ is a bounded $\vee$-hoop. Easily, we can see that $P = \{c, d, 1\}$ is a primary filter but is not a $\vee$-prime filter. Because, $a \vee b = 1 \in P$ but $a, b \notin P$.

Table 6.5 Operations $\odot$ and $\to$ of Example 6.3.18

Table 6.5.1

$\odot$	0	a	b	1
0	0	0	0	0
a	0	a	0	a
b	0	0	b	b
1	0	a	b	1

Table 6.5.2

$\to$	0	a	b	1
0	1	1	1	1
a	b	1	b	1
b	a	a	1	1
1	0	a	b	1

Table 6.6 Operations $\odot$ and $\rightarrow$ of Example 6.3.19

Table 6.6.1

$\odot$	0	a	b	c	d	1
0	0	0	0	0	0	0
a	0	a	b	d	d	a
b	0	b	b	0	0	b
c	0	d	0	d	d	c
d	0	d	0	d	d	d
1	0	a	b	c	d	1

Table 6.6.2

$\rightarrow$	0	a	b	c	d	1
0	1	1	1	1	1	1
a	0	1	b	c	c	1
b	c	a	1	c	c	1
c	b	a	b	1	a	1
d	b	a	b	a	1	1
1	0	a	b	c	d	1

6.3.2   Perfect Hoops

In this section, we introduce the concepts of perfect hoops and perfect filters, exploring their connections. To facilitate our investigation, we will assume throughout this section that all hoops under consideration are bounded.

Definition 6.3.20   ([39]) A bounded hoop $\mathbf{H}$ is **perfect** when for any $x \in H$, if $\mathrm{ord}(x) < \infty$, then $\mathrm{ord}(x') = \infty$, and if $\mathrm{ord}(x) = \infty$, then $\mathrm{ord}(x') < \infty$.

According to the definition of the perfect hoops and Theorem 6.3.7, every perfect hoop is local, consequently, $M := \{x \in H : \mathrm{ord}(x) = \infty\}$ is the unique maximal filter of any perfect hoop $\mathbf{H}$.

Example 6.3.21   ([39, Example 4.6]) (i) If $\mathbf{H}$ is the hoop as in Example 6.3.14(ii), then $\mathbf{H}$ is not a perfect bounded hoop.

(ii) Let $H = \{0, a, b, 1\}$ be a chain such that $0 < a < b < 1$. Define the operations $\odot$ and $\rightarrow$ on H as Table 6.7.

Then $(H; \odot, \rightarrow, 0, 1)$ is a bounded perfect hoop.

(iii) The bounded hoop $\mathbf{K[u]}$ in Example 4.2.19(iv) is a perfect hoop. Let $x, y \in \mathbf{G}^+$. By (3.13), we get

$$(0, x) \odot (0, y) = ((0, x) - (1, 0) + (0, y)) \vee (0, 0) = (-1, x + y) \vee (0, 0) = (0, 0)$$

$$(1, -x) \odot (1, -y) = ((1, -x) - (1, 0) + (1, -y)) \vee (0, 0) = (1, -x - y).$$

It follows that, $(0, x)^n = (0, 0)$ for all integer $n \geq 2$ and $(1, -x)^m = (1, -mx)$ for all $m \in \mathbb{N}$, which implies $\mathrm{ord}(0, x) \leq 2$ and $\mathrm{ord}(1, -x) = \infty$ for every $x \in \mathbf{G}^+$. Therefore, by definition, $\mathbf{K[u]}$ is a perfect hoop.

Table 6.7 Operations $\odot$ and $\rightarrow$ of Example 6.3.21(ii)

Table 6.7.1

$\odot$	0	a	b	1
0	0	0	0	0
a	0	a	a	a
b	0	a	a	b
1	0	a	b	1

Table 6.7.2

$\rightarrow$	0	a	b	1
0	1	1	1	1
a	0	1	1	1
b	0	b	1	1
1	0	a	b	1

Based on the definition of perfect MV-algebra in Sect. 1.4.1, we can conclude that

Theorem 6.3.22 *A bounded Wajsberg hoop* $\mathbf{H} = (H; \odot, \rightarrow, 0, 1)$ *is perfect if and only if* $\beta(\mathbf{H}) = (H; \oplus, ', 0, 1)$ *is a perfect MV-algebra, where* β *is the bijection in Theorem 3.3.13.*

Proof The proof is straightforward by definition. Note that on each bounded Wajsberg hoop $\mathbf{H}$ we have $x'' = x$. In addition, $x^n = 0$ if and only if $n.x' = 1$ for all $x \in H$ and $n \in \mathbb{N}$. $\square$

A well-known theorem by Di Nola and Lettieri [100] provides a presentation for perfect MV-algebras using the lexicographic product of Abelian ℓ-groups. Indeed, they proved that each MV-algebra $\mathbf{M}$ is perfect if and only if there exists an Abelian ℓ-group $\mathbf{G}$ such that $\mathbf{M} \cong \Gamma(\mathbb{Z} \overrightarrow{\times} \mathbf{G}, (1, 0))$. In addition, they established a categorically equivalence between the category of perfect MV-algebras and the category of Abelian ℓ-groups. For more details see [24, 25, 100, 104, 154].

Corollary 6.3.23 ([100]) *A bounded Wajsberg hoop* $\mathbf{H} = (H; \odot, \rightarrow, 0, 1)$ *is perfect if and only if there exists an Abelian* ℓ-*group* $\mathbf{G}$ *such that* $\mathbf{H} \cong (\mathbb{Z} \overrightarrow{\times} \mathbf{G})[\mathbf{u}]$, *where* $u = (1, 0)$.

Proof According to [100], an MV-algebra $\mathbf{M} = (M; \oplus, ', 0, 1)$ is perfect if and only if there exists an Abelian ℓ-group $\mathbf{G}$ such that $\mathbf{M} \cong \Gamma(\mathbb{Z} \overrightarrow{\times} \mathbf{G}, (1, 0))$. If α and β are the bijections in Theorem 3.3.13, then $\alpha(\Gamma(\mathbb{Z} \overrightarrow{\times} \mathbf{G}, (1, 0))) = (\mathbb{Z} \overrightarrow{\times} \mathbf{G})[(1, 0)]$ and $\beta((\mathbb{Z} \overrightarrow{\times} \mathbf{G})[(1, 0)]) = \Gamma(\mathbb{Z} \overrightarrow{\times} \mathbf{G}, (1, 0))$, see Corollary 3.3.14. Therefore, Theorem 6.3.22 implies that each bounded Wajsberg hoop $\mathbf{H}$ is perfect if and only if there exists an Abelian ℓ-group $\mathbf{G}$ such that $\mathbf{H} \cong (\mathbb{Z} \overrightarrow{\times} \mathbf{G})[\mathbf{u}]$, where $u = (1, 0)$. $\square$

Definition 6.3.24 ([39]) A bounded hoop $\mathbf{H} = (H; \odot, \rightarrow, 0, 1)$ is called a **bipartite hoop** if there exists $M \in \mathrm{Max}(\mathbf{H})$ such that $\mathbf{H} = M \cup M'$, where $M' := \{x' : x \in M\}$ and $x' = x \rightarrow 0$ for each $x \in H$.

Example 6.3.25 If **H** is the hoop as in Example 3.3.12 and $M = \{a, b, c, 1\}$, then $M \in$ Max(**H**), and $M' = \{0\}$. Hence, **H** is bipartite; more precisely, **H** is a local bipartite hoop.

In the following theorem, we prove that local hoops are perfect under what conditions.

Theorem 6.3.26 ([39, Theorem 5.7]) *Let* **H** *be local bipartite with* (DNP). *Then* **H** *is perfect.*

Proof Let **H** be a local bipartite with (DNP). Since **H** is local, **H** has a unique maximal filter such as M. Since **H** is bipartite, we have $\mathbf{H} = M \cup M'$. Suppose $x \in H$ such that $\mathrm{ord}(x) < \infty$. Then by Proposition 6.3.4(iv), $x \notin M$ and so $x \in M'$. Moreover, since **H** has (DNP), we get $x' \in M$, and by Proposition 6.3.4(iv), $\mathrm{ord}(x') = \infty$. Now, let $x \in H$ and $\mathrm{ord}(x) = \infty$. If $\mathrm{ord}(x') = \infty$, then **H** is not local, which is a contradiction. Hence, $\mathrm{ord}(x') < \infty$. Therefore, **H** is perfect. $\square$

Example 6.3.27 (i) If **H** is the hoop from Example 6.3.14(i), then **H** is local, but it is not perfect. Because $\mathrm{ord}(a) = \mathrm{ord}(a') < \infty$.

(ii) Let $H = \{0, a, b, 1\}$. Define two operations $\odot$ and $\rightarrow$ on H by Table 6.8.

Then $(H; \odot, \rightarrow, 0, 1)$ with these operations is a bounded chain hoop, where $0 < a < b < 1$. Obviously, **H** is local with (DNP). Let $M = \{b, 1\}$. Then $M \in$ Max(**H**), $M' = \{0, a\}$ and $\mathbf{H} = M \cup M'$. Therefore, **H** is local bipartite with (DNP).

Definition 6.3.28 ([39]) A proper filter P of a bounded hoop **H** is called a **perfect filter** of **H** if, for any $x \in H$, $(x^n)' \in P$ for some $n \in \mathbb{N}$ if and only if $((x')^m)' \notin P$ for any $m \in \mathbb{N}$.

Example 6.3.29 (i) If **H** is the hoop as in Example 6.3.14(ii), then $P = \{c, 1\}$ is not a perfect filter of **H**.

(ii) Let $H = \{0, a, b, c, d, 1\}$. Define two operations $\odot$ and $\rightarrow$ on H as Table 6.9.

Then $(H; \odot, \rightarrow, 0, 1)$ is a bounded hoop and $P = \{a, b, 1\}$ is a perfect filter of **H**.

Table 6.8 Operations $\odot$ and $\rightarrow$ of Example 6.3.27(ii)

Table 6.8.1

$\odot$	0	a	b	1
0	0	0	0	0
a	0	0	0	a
b	0	0	b	b
1	0	a	b	1

Table 6.8.2

$\rightarrow$	0	a	b	1
0	1	1	1	1
a	b	1	1	1
b	a	a	1	1
1	0	a	b	1

Table 6.9 Operations $\odot$ and $\to$ of Example 6.3.29(ii)

Table 6.9.1

$\odot$	0	a	b	c	d	1
0	0	0	0	0	0	0
a	0	b	b	d	0	a
b	0	b	b	0	0	b
c	0	d	0	c	d	c
d	0	0	0	d	0	d
1	0	a	b	c	d	1

Table 6.9.2

$\to$	0	a	b	c	d	1
0	1	1	1	1	1	1
a	d	1	a	c	c	1
b	c	1	1	c	c	1
c	b	a	b	1	a	1
d	a	1	a	1	1	1
1	0	a	b	c	d	1

Theorem 6.3.30 ([39, Theorem 4.12]) *Let F be a proper filter of $\mathbf{H}$. Then F is a perfect filter if and only if $\mathbf{H}/F$ is a perfect hoop.*

Proof Let F be a perfect filter and $x \in H$ such that $\mathrm{ord}\,(x/F) < \infty$. Then there exists $n \in \mathbb{N}$ such that $(x/F)^n = 0/F$ and so $(x^n)' \in F$. Since F is a perfect filter, we have $((x')^m)' \notin F$ for all $m \in \mathbb{N}$. Hence, $((x')^m)'/F \neq 1/F$, equivalently, $((x')^m)''/F \neq 0/F$. It follows that $\mathrm{ord}\,(x'/F) = \infty$, otherwise, if there exists $n \in \mathbb{N}$ such that $(x'/F)^n = 0/F$, then $((x')^n)' \in F$, which is a contradiction. Hence, $\mathbf{H}/F$ is perfect. Now, let $\mathrm{ord}\,(x/F) = \infty$. Then $(x/F)^n \neq 0/F$, for all $n \in \mathbb{N}$, equivalently, $(x^n)' \notin F$ for all $n \in \mathbb{N}$. Since F is a perfect filter, there exists $m \in \mathbb{N}$ such that $((x')^m)' \in F$. Hence, $\left((x'/F)^m\right)' = 1/F$ and so $\left((x'/F)^m\right)'' = 0/F$. Thus, $\mathrm{ord}\left(\left((x'/F)^m\right)''\right) < \infty$. Remark 2.2.5(ii) and Proposition 6.3.4(iii) imply that $\mathrm{ord}\left((x'/F)^m\right) < \infty$, and so $\mathrm{ord}\left(x'/F\right) < \infty$. Therefore, $\mathbf{H}/F$ is a perfect hoop.

Conversely, let F be a proper filter of $\mathbf{H}$ such that $(x^n)' \in F$ for some $n \in \mathbb{N}$ and some $x \in H$. Then $(x^n)'/F = 1/F$, so, $(x^n)''/F = 0/F$, i.e., $\mathrm{ord}\left((x^n)''/F\right) < \infty$. Also, by Remark 2.2.5(ii) and Proposition 6.3.4(iii), $\mathrm{ord}\,(x^n/F) < \infty$. Then $\mathrm{ord}\,(x/F) < \infty$. Since $\mathbf{H}/F$ is perfect, we get $\mathrm{ord}\left(x'/F\right) = \infty$, and so $\left(x'/F\right)^m \neq 0/F$ for all $m \in \mathbb{N}$. Thus, $((x')^m)' \notin F$. The proof of the other implication is similar. Therefore, F is a perfect filter. $\square$

Theorem 6.3.31 ([39, Theorem 4.10]) *Every perfect filter of $\mathbf{H}$ is a primary filter.*

Proof Let P be a perfect filter of $\mathbf{H}$. By Theorem 6.3.30, $\mathbf{H}/F$ is a perfect hoop and so it is local (see the note right after Definition 6.3.20). Now, Theorem 6.3.15 implies that P is primary. $\square$

The following example shows that a primary filter may not be a perfect filter, in general.

Table 6.10 Operations $\odot$ and $\rightarrow$ of Example 6.3.32

<table>
<tr><td colspan="5">Table 6.10.1</td><td colspan="5">Table 6.10.2</td></tr>
<tr><td>$\odot$</td><td>0</td><td>a</td><td>b</td><td>1</td><td>$\rightarrow$</td><td>0</td><td>a</td><td>b</td><td>1</td></tr>
<tr><td>0</td><td>0</td><td>0</td><td>0</td><td>0</td><td>0</td><td>1</td><td>1</td><td>1</td><td>1</td></tr>
<tr><td>a</td><td>0</td><td>0</td><td>0</td><td>a</td><td>a</td><td>b</td><td>1</td><td>1</td><td>1</td></tr>
<tr><td>b</td><td>0</td><td>0</td><td>a</td><td>b</td><td>b</td><td>a</td><td>b</td><td>1</td><td>1</td></tr>
<tr><td>1</td><td>0</td><td>a</td><td>b</td><td>1</td><td>1</td><td>0</td><td>a</td><td>b</td><td>1</td></tr>
</table>

Example 6.3.32 Let $H = \{0, a, b, 1\}$ be a chain such that $0 < a < b < 1$. Define the operations $\odot$ and $\rightarrow$ on H as Table 6.10.

Then $(H; \odot, \rightarrow, 0, 1)$ is a bounded perfect hoop. Clearly, $F = \{1\}$ is a primary filter, but it is not a perfect filter. Because, $(a^2)' = (0)' = 1 \in F$, while $((a')^3)' = 1 \in F$.

Theorem 6.3.33 ([39, Theorem 4.13] *A bounded hoop* **H** *is perfect if and only if every proper filter of* **H** *is perfect.*

Proof Let F be a proper filter of **H**, $x \in H$ and there exists $n \in \mathbb{N}$ such that $(x^n)' \in F$. Since F is a proper filter of **H**, by Proposition 6.3.4(i), $\mathrm{ord}((x^n)') = \infty$. Moreover, since **H** is perfect, $\mathrm{ord}((x^n)'') < \infty$. By Remark 2.2.5(ii) and Proposition 6.3.4(iii), $\mathrm{ord}(x^n) < \infty$, and so $\mathrm{ord}(x) < \infty$. Since **H** is a perfect hoop, $\mathrm{ord}(x') = \infty$. Now, suppose $m \in \mathbb{N}$ such that $((x')^m)' \in F$. Then by Proposition 6.3.4(i), $\mathrm{ord}(((x')^m)') = \infty$. Furthermore, **H** is a perfect hoop, so by Remark 2.2.5(ii) and Proposition 6.3.4(iii), $\mathrm{ord}((x')^m) < \infty$. Thus, $\mathrm{ord}(x') < \infty$, which is a contradiction. Hence, $((x')^m)' \notin F$ for all $m \in \mathbb{N}$. For the other implication, let $((x')^m)' \notin F$ for all $m \in \mathbb{N}$. Then $(x')^m \neq 0$. Otherwise, $(x')^m = 0$ implies that $((x')^m)' = 1 \in F$, which is a contradiction. So, $\mathrm{ord}(x') = \infty$. Since **H** is perfect, $\mathrm{ord}(x'') < \infty$ and by Remark 2.2.5(ii) and Proposition 6.3.4(iii), $\mathrm{ord}(x) < \infty$. Hence, there exists $n \in \mathbb{N}$ such that $x^n = 0$, and so $(x^n)' = 1 \in F$. Therefore, every proper filter of **H**, is perfect.

Conversely, suppose that every proper filter of **H** is perfect. Since $\{1\}$ is a proper filter of **H**, it is perfect one. By Theorem 6.3.30, $\mathbf{H} \cong \mathbf{H}/\{1\}$ is perfect. $\square$

Theorem 6.3.34 ([39, Theorem 5.14]) *Let* **H** *be a bounded local hoop. Then* **H** *is perfect if and only if* $\mathbf{H} = D(\mathbf{H}) \cup D^*(\mathbf{H})$.

Proof Let **H** be perfect. Choose $x \in H$ such that $\mathrm{ord}(x) < \infty$. Then $x \notin D(\mathbf{H})$. Since **H** is perfect, we have $\mathrm{ord}(x') = \infty$, and so $x' \in D(\mathbf{H})$. By Remark 2.2.5(ii), $x \leq (x')'$. Since $x' \in D(\mathbf{H})$, by definition of $D^*(\mathbf{H})$, $x \in D^*(\mathbf{H})$. Now, assume $x \in H$ such that $\mathrm{ord}(x) = \infty$. Then $x \in D(\mathbf{H})$. Hence, $H \subseteq D(\mathbf{H}) \cup D^*(\mathbf{H})$. Therefore, $\mathbf{H} = D(\mathbf{H}) \cup D^*(\mathbf{H})$.

Let $\mathbf{H} = D(\mathbf{H}) \cup D^*(\mathbf{H})$ and $\mathrm{ord}(x) < \infty$ for some $x \in H$. Then $x \notin D(\mathbf{H})$, and so $x \in D^*(\mathbf{H})$. Thus, there exists $y \in D(\mathbf{H})$ such that $x \leq y'$. By Remark 2.2.5(i) and (ii),

$y \le y'' \le x'$. Since $y \in D(\mathbf{H})$, by Proposition 6.3.4(ii), $(x')^n > 0$ for all $n \in \mathbb{N}$, and so $\mathrm{ord}(x') = \infty$. Now, suppose $x \in H$ and $\mathrm{ord}(x) = \infty$. Then $x \in D(\mathbf{H})$. If $\mathrm{ord}(x') = \infty$, then by Proposition 6.3.7, $\mathbf{H}$ is not local, a contradiction. Hence, $\mathrm{ord}(x') < \infty$. Therefore, $\mathbf{H}$ is perfect. $\qquad\square$

Corollary 6.3.35 *A bounded local hoop* $\mathbf{H}$ *is perfect if and only if* $D^*(\mathbf{H}) = H \setminus D(\mathbf{H})$.

Proof Let $\mathbf{H}$ be a bounded local hoop. Then by Theorem 6.3.12, $D^*(\mathbf{H}) \cap D(\mathbf{H}) = \emptyset$. Now, the proof of the theorem is a direct consequence of Theorem 6.3.34. $\qquad\square$

Definition 6.3.36 ([39]) A bounded hoop $\mathbf{H}$ is a **peculiar** if it is local and there are $x, y \in H \setminus \{0, 1\}$ such that $\mathrm{ord}(x) = \infty$, $\mathrm{ord}(y) < \infty$ and $\mathrm{ord}(y') < \infty$.

Example 6.3.37 If $\mathbf{H}$ is the hoop from Example 6.3.14(i), then $\mathbf{H}$ is local and $\mathrm{ord}(b) = \infty$, $\mathrm{ord}(a) < \infty$ and $\mathrm{ord}(a') < \infty$, thus, $\mathbf{H}$ is a peculiar.

Theorem 6.3.38 ([39, Theorem 5.23]) *Let* $\mathbf{H} = (H; \odot, \to, 0, 1)$ *be a bounded local hoop such that* $H \neq \{0, 1\}$. *Then* $\mathbf{H}$ *is perfect or locally finite, or peculiar.*

Proof **Case 1:** Suppose that $\mathbf{H}$ is not locally finite and peculiar. We prove that $\mathbf{H}$ is perfect. Since $\mathbf{H}$ is not locally finite, there exists $x \in H \setminus \{1\}$ such that $\mathrm{ord}(x) = \infty$. Since $\mathbf{H}$ is local, we have $\mathrm{ord}(x') < \infty$. Moreover, since $\mathrm{ord}(x) = \infty$, if there exists $y \in H$ such that $\mathrm{ord}(y), \mathrm{ord}(y') < \infty$, then $\mathbf{H}$ is peculiar, which is a contradiction. Hence, for all $y \in H$, $\mathrm{ord}(y) = \infty$ or $\mathrm{ord}(y') = \infty$. If there exists $y \in H$ such that $\mathrm{ord}(y) = \mathrm{ord}(y') = \infty$, then $\mathbf{H}$ is not local, which is a contradiction. Therefore, $\mathbf{H}$ is perfect.

Case 2: Suppose that $\mathbf{H}$ is not perfect and locally finite. We prove that $\mathbf{H}$ is peculiar. By assumption, $\mathbf{H}$ is local. Since $\mathbf{H}$ is not locally finite, there exists $x \in H \setminus \{1\}$ such that $\mathrm{ord}(x) = \infty$. Also, $\mathbf{H}$ is not perfect. Then there exists $y \in H$ such that $\mathrm{ord}(y) = \mathrm{ord}(y') = \infty$ or $\mathrm{ord}(y), \mathrm{ord}(y') < \infty$. If $\mathrm{ord}(y) = \mathrm{ord}(y') = \infty$, then $\mathbf{H}$ is not local, which is a contradiction. Hence, there exists $y \in H$ such that $\mathrm{ord}(y), \mathrm{ord}(y') < \infty$. Therefore, $\mathbf{H}$ is peculiar.

Case 3: Suppose that $\mathbf{H}$ is not perfect and peculiar. We prove that $\mathbf{H}$ is locally finite. Since $\mathbf{H}$ is not perfect, there exists $y \in H$ such that $\mathrm{ord}(y) = \mathrm{ord}(y') = \infty$ or $\mathrm{ord}(y), \mathrm{ord}(y') < \infty$. If $\mathrm{ord}(y) = \mathrm{ord}(y') = \infty$, then $\mathbf{H}$ is not local, which is a contradiction. Thus, there exists $y \in H$ such that $\mathrm{ord}(y), \mathrm{ord}(y') < \infty$. Now, let x be an arbitrary element of $H \setminus \{1\}$. If $x = y$ or $x = y'$, then $\mathrm{ord}(x) < \infty$. Suppose that $x \neq y$. If $\mathrm{ord}(x) = \infty$, then $\mathbf{H}$ is peculiar, which is a contradiction. Thus, $\mathrm{ord}(x) < \infty$. Hence, for all $x \in H \setminus \{1\}$, $\mathrm{ord}(x) < \infty$. Therefore, $\mathbf{H}$ is locally finite. $\qquad\square$

Bibliographical Remarks and Suggestions for Further Study

The majority results of Sect. 6.3 were originally proved in [39]. We attempted to summarize the proof and provide different types of examples. We have learned that every bounded perfect hoop is local. In addition, when a hoop $\mathbf{H}$ satisfies (DNP), then $\mathbf{H}$ is perfect if and only if the MV-algebra corresponding to $\mathbf{H}$ is perfect.

According to Theorem 6.3.34, a bounded hoop $\mathbf{H}$ is perfect if and only if $\mathbf{H} = D(\mathbf{H}) \cup D^*(\mathbf{H})$, where $D(\mathbf{H}) = \{x \in H \colon \mathrm{ord}(x) = \infty\}$ is the unique maximal filter of $\mathbf{H}$, and $D^*(\mathbf{H}) = \{x \in H \colon x \leq y', \exists y \in D(\mathbf{H})\}$. This suggests a natural generalization of perfectness for hoops without least elements. For every local hoop $\mathbf{H}$ with unique maximal filter F, we define $D^*(\mathbf{H}) := \{x \in H \colon \exists y \in F, x \leq y \to a, \forall a \in H\}$. Straightforward calculations demonstrate that this new definition coincides with the previous one for bounded local hoops, since $y \to 0 \leq y \to a$. In addition, if $x \in F \cap D^*(\mathbf{H})$, then $y \leq (y \to a) \to a \leq x \to a$ implies that $x \to a \in F$ for every $a \in H$, consequently, by Lemma 2.3.2(ii), $H \subseteq F$, a contradiction. Thus, $F \cap D^*(\mathbf{H}) = \emptyset$. Now, the concept of perfect hoop can be generalized as: A local hoop $\mathbf{H}$ with unique maximal filter F is perfect if and only if $H = F \cup D^*(\mathbf{H})$.

An application of perfect hoops can be found in [27], where M. Bianchi investigated the smallest variety containing all ordinal sums of linearly ordered perfect MV-algebras (equivalently, perfect Wajsberg hoops). Perfect hoops also find application in the state theory of hoops. Specifically, in Chap. 7, we will demonstrate that every perfect hoop admits a nontrivial state operator.

6.4   Representable Hoops

The class of representable hoops is a prominent subclass of the variety of hoops. This section undertakes a systematic investigation to characterize representable hoops and establish their status as a variety.

Definition 6.4.1 A hoop $\mathbf{H} = (H; \odot, \to, 1)$ is called **representable** if it is a subdirect product of linear hoops, which means there exists a family $\{\mathbf{H}_i \colon i \in I\}$ of linear hoops and an embedding $f : \mathbf{H} \to \prod_{i \in I} \mathbf{H}_i$ such that $\pi_i \circ f(H) = H_i$ for all $i \in I$ where $\pi_i : \prod_{i \in I} H_i \to H_i$ is the ith canonical projection map.

Remark 6.4.2 (i) Every linear hoop is representable, evidently.

(ii) Every representable hoop satisfies the prelinearity condition. Indeed, for each subdirect embedding $f : \mathbf{H} \to \prod_{i \in I} \mathbf{H}_i$ where $\mathbf{H}_i$ is a linear hoop for all $i \in I$, we have $\prod_{i \in I} \mathbf{H}_i$ satisfies (Pre), so does $f(\mathbf{H}) \cong \mathbf{H}$. Therefore, $\mathbf{H}$ satisfies (Pre).

In addition, every linear hoop is a basic hoop, consequently, $\prod_{i \in I} \mathbf{H}_i$ is a basic hoop as well. Since $f(\mathbf{H})$ is a subalgebra of $\prod_{i \in I} \mathbf{H}_i$, we get that $f(\mathbf{H})$ as well as $\mathbf{H}$ is a basic hoop, by Theorem 3.3.28.

(iii) Applying Theorem 4.2.11 with the same method used in Proposition 6.2.15, we can show that $\mathbf{H}$ is representable if and only if $\bigcap\{P : P \in \mathrm{Spec}(\mathbf{H})\} = \{1\}$. Indeed, an easy calculation shows that the map $f : \mathbf{H} \to \prod_{P \in \mathrm{Spec}(\mathbf{H})} \mathbf{H}/P$ is defined by $f(x) = (x/P)_{P \in \mathrm{Spec}(\mathbf{H})}$ is a subdirect embedding of hoops, where $\mathbf{H}/P$ is a linear hoop for every $P \in \mathrm{Spec}(\mathbf{H})$.

(iv) An identity of type $(\vee, \wedge, \odot, \to)$ holds on every representable hoop if and only if it holds on every linear hoop. The proof is clear by definition.

(v) The direct product of any family of linear hoops is a representable hoop.

From now on, we consider the following notations:

$$\mathsf{H} := \text{the variety of all hoops,} \quad \mathsf{RH} := \text{the class of all representable hoops,}$$
$$\mathsf{BH} := \text{the variety of all basic hoops,} \quad \mathsf{WH} := \text{the variety of all Wajsberg hoops,}$$
$$\mathsf{PH} := \text{the variety of all product hoops,} \quad \mathsf{CH} := \text{the variety of all cancellative hoops.}$$

According to Remark 6.4.2(ii), every representable hoop is a basic hoop. Hence, in order to characterize representable hoops, we must focus on the class of basic hoops.

Theorem 6.4.3 ([11, Theorem 1.6]) *Let $\mathbf{H}$ be a hoop. Then $\mathbf{H}$ is representable if and only if it is a basic hoop.*

Proof Let $\mathbf{H}$ be a representable hoop. Then by Remark 6.4.2(ii), it is a basic hoop.

Conversely, suppose that $\mathbf{H}$ is a basic hoop. By Theorem 3.3.28, it is a $\vee$-hoop with (Pre), and so by Corollaries 4.2.9 and 4.2.14, we have

$$\{1\} = \bigcap\{P : P \in \mathrm{Spec}_\vee(\mathbf{H})\} = \bigcap\{P : P \in \mathrm{Spec}(\mathbf{H})\}.$$

Hence, by Remark 6.4.2(iii), $\mathbf{H}$ is representable. $\qquad\square$

Summarizing Theorems 3.3.28, 6.4.3 and Proposition 3.3.26 we have

$$\boxed{\sqcup\text{-hoop+(Pre)}} \Leftrightarrow \boxed{\vee\text{-hoop+(Pre)}} \Leftrightarrow \boxed{\text{Basic hoop}} \Leftrightarrow \boxed{\text{Representable hoop}} \qquad (6.6)$$

Corollary 6.4.4 RH *is a variety.*

Proof The proof is straight consequence of Theorem 6.4.3, since $\mathsf{RH} = \mathsf{BH}$. $\qquad\square$

The next proposition shows that the ordinal sum of representable hoops remains representable.

Proposition 6.4.5 *Let* $\mathbf{H}_1$ *and* $\mathbf{H}_2$ *be two hoops. Then* $\mathbf{H} := \mathbf{H}_1 \oplus \mathbf{H}_2$ *is representable if and only if* $\mathbf{H}_1$ *is linear and* $\mathbf{H}_2$ *are representable.*

Proof Let $\mathbf{H}$ be representable. Then by Theorem 6.4.3, it is a basic hoop. So, by Corollary 3.3.65, $\mathbf{H}_1$ is linear and $\mathbf{H}_2$ is a basic hoop, equivalently, a representable hoop. The proof of the converse follows similarly from Proposition 3.3.63(iv) and Theorem 6.4.3. $\square$

Proposition 6.4.6 *The variety* BH *(respectively,* WH, PH, *and* CH*) is generated by linear basic hoops (respectively, linear Wajsberg hoops, linear product hoops, and linear cancellative hoops).*

Proof Let V be the variety generated by linear Wajsberg hoops. Clearly, $\mathsf{V} \subseteq \mathsf{WH}$. Choose $\mathbf{H} \in \mathsf{WH}$. Due to Theorem 6.4.3, $\mathbf{H}$ is representable, so there exists a family $\{\mathbf{H}_i : i \in I\}$ of linear hoops and a subdirect embedding $f : \mathbf{H} \to \prod_{i \in I} \mathbf{H}_i$. For each $i \in I$, $f(\mathbf{H}) = H_i$, so $H_i \in \mathsf{WH}$, equivalently, $\mathbf{H}_i$ is a Wajsberg hoop. Hence, $\mathbf{H}_i \in \mathsf{V}$ for all $i \in I$, consequently, $\prod_{i \in I} \mathbf{H}_i \in \mathsf{V}$. On the other hand, $\mathbf{H} \cong f(\mathbf{H})$ and $f(\mathbf{H})$ is a subalgebra of $\prod_{i \in I} \mathbf{H}_i$. It follows that $\mathbf{H} \in \mathsf{HSP}(\mathsf{V}) = \mathsf{V}$, see Theorem 1.5.21. Therefore, $\mathsf{V} = \mathsf{WH}$. The proofs of the other parts are similar. $\square$

Corollary 6.4.7 *An identity* $p(x_1, \ldots, x_n) = q(x_1 \ldots, x_n)$ *of type* $\{\odot, \to\}$ *holds on every basic (respectively, Wajsberg, product, or cancellative) hoop if and only if it holds on every linear basic (respectively, Wajsberg, product, or cancellative) hoop.*

Proof The proof is straightforward by Proposition 6.4.6. $\square$

Proposition 6.4.8 *Every bounded basic hoop* $\mathbf{H}$ *satisfies Glivenko property:* $(x'' \to x)' = 0$ *for every* $x \in H$.

Proof By Corollary 6.4.7, it suffices to show that every bounded linear hoop satisfies Glivenko property. Let $\mathbf{H}$ be a bounded linear hoop. Due to Theorem 6.1.19, there exists a family $\{\mathbf{H}_i = (H_i; \odot_i, \to_i, 1)\}_{i \in I}$ of linear Wajsberg hoop such that $\mathbf{H} = \bigoplus_{i \in I} \mathbf{H}_i$, where I is a linearly ordered set such that $\min(I)$ exists. By Remark 6.1.2(v), $0 \in H_{\min(I)}$. Choose $x \in H$. There is $i \in I$ such that $x \in H_i$. If $i = \min(I)$, then $x'' = (x \to_i 0) \to_i 0 = x$ and so $(x'' \to x)' = 0$, since $\mathbf{H}_i$ is a Wajsberg hoop. If $i \neq \min(I)$, then by (6.2), $x' = 0$, which implies $(x'' \to x)' = (1 \to x)' = x' = 0$. Therefore, $\mathbf{H}$ satisfies Glivenko property.

We have established that the concepts of basic hoops and representable hoops are equivalent. As an application of Corollary 6.4.7, consider the following proposition.

Proposition 6.4.9 *If a basic hoop* $\mathbf{H} = (H; \odot, \to, 1)$ *is a product hoop, then it satisfies the following conditions for all* $x, y, z, w \in H$:

(i) $y \to y^2 \le (x \wedge (x \to y)) \to y$.

(ii) $((x \to y) \to y) \odot ((z \odot x) \to (w \odot x)) \odot ((z \odot y) \to (w \odot y)) \le z \to w$.

Proof First, we claim that (i) and (ii) hold on all product hoops. According to Corollary 6.4.7, it suffices to show that all linear product hoops satisfy (i) and (ii). Let **H** be a linear product hoop. By (PH), $(x \to z) \vee ((x \to (x \odot y)) \to y) = 1$ for all $x, y, z \in H$.

(1) Choose $x \in H$. If $x = \min(\mathbf{H})$, then $(x \wedge (x \to y)) \to y = x \to y = 1 \ge y \to y^2$. Hence, **H** satisfies inequality (i). In addition, $(z \odot x) = (w \odot x)$, since $x = \min(\mathbf{H})$, so

$$
\begin{aligned}
((x \to y) \to y) \odot ((z \odot x) \to (w \odot x)) \odot ((z \odot y) \to (w \odot y)) &= y \odot 1 \odot ((z \odot y) \to (w \odot y)) \\
&= y \odot (y \to (z \to (w \odot y))), \\
&\qquad \text{by Lemma 2.2.1(8)} \\
&\le z \to (w \odot y) \\
&\le z \to w, \text{ by Lemma 2.2.1(7).}
\end{aligned}
$$

Thus, (ii) holds. Clearly, if **H** is not bounded, then case (1) is impossible.

(2) If there exists $z \in H$ such that $z < x$, then $x \to z \ne 1$, so by (PH), $(x \to (x \odot y)) \to y = 1$ (since **H** is linearly ordered), equivalently,

$$
x \to (x \odot y) \le y, \qquad \forall y \in H. \tag{6.7}
$$

Specially, $x = y$ implies that $y \to y^2 \le y$, consequently, $y \to y^2 \le y \le (x \wedge (x \to y)) \to y$, see Lemma 2.2.1(10). Hence, **H** satisfies inequality (i).

Now, let $x, y, w, z \in H$ such that x is not the least element of **H**. By Lemma 2.2.1(7) and (6.7) we get:

$$
\begin{aligned}
((x \to y) \to y) \odot ((z \odot x) \to (w \odot x)) \odot ((z \odot y) \to (w \odot y)) &\le (z \odot x) \to (w \odot x) \\
&= z \to (x \to (w \odot x)) \\
&\le z \to w, \text{ by (6.7).}
\end{aligned}
$$

From (1) and (2), we conclude that **H** satisfies (i) and (ii). $\qquad \square$

We note that in [153], a product hoop is defined as a basic hoop satisfying conditions (i) and (ii) of Proposition 6.4.9.

Bibliographical Remarks and Suggestions for Further Study

This section briefly explored the equivalence between basic hoops and representable hoops. This equivalence provides a more straightforward method for verifying identities within subvarieties of basic hoops. Furthermore, we have learned that the class of representable hoops forms a variety. For those interested in the non-commutative case, i.e., pseudo hoops (see Appendix A), the class of representable pseudo hoops may be of particular interest, as it differs from the variety of basic pseudo hoops.

6.5 Subdirectly Irreducible Hoops

In light of Chap. 3, the class of (basic, Wajsberg, product, and cancellative) hoops forms a variety. Consequently, Theorem 1.5.24 implies that subdirectly irreducible hoops generate this variety. This underscores the importance of investigating subdirectly irreducible hoops.

The study of subdirectly irreducible hoops was first undertaken by J. R. Büchi and T. M. Owen. They completely characterized the structure of subdirectly irreducible hoops in general. Then, Blok and Ferreirim [31], provided a precise description of the subdirectly irreducible hoops. This section is devoted to the structural analysis of subdirectly irreducible hoops. We will commence by demonstrating that every subdirectly irreducible basic hoop is linearly ordered. Thereafter, we will employ the ordinal sum construction to provide a representation for arbitrary subdirectly irreducible hoops.

By Definition 1.5.12, a hoop $\mathbf{H}$ is **subdirectly irreducible** if for every subdirect embedding $f : \mathbf{H} \to \prod_{i \in I} \mathbf{H}_i$ there exists $i \in I$ such that $\pi_i \circ f : \mathbf{H} \to \mathbf{H}_i$ is an isomorphism. The trivial hoop $\mathbf{1}$ is subdirectly irreducible. In addition, due to Theorem 1.5.14, a non-trivial hoop $\mathbf{H}$ is subdirectly irreducible if and only if $\mathbf{H}$ has a congruence $\mu \neq \Delta$ that is contained in every non-zero congruence of $\mathbf{H}$, that is $\bigcap \{\theta \in \mathrm{Con}(\mathbf{H}) : \theta \neq \Delta\} \neq \Delta$. On the other hand, by Theorem 4.1.14, a non-trivial hoop $\mathbf{H}$ is subdirectly irreducible if and only if $\mathcal{F}(\mathbf{H}) \setminus \{1\}$ has the least element. Indeed, $\mathrm{U}_* = 1/\mu$ is the unique minimal filter of $\mathbf{H}$ distinct from $\{1\}$. So, U_* is contained in every filter of $\mathbf{H}$ not equal to $\{1\}$. In the sequel, we will always denote this filter by U_*.

Proposition 6.5.1 ([31, Proposition 2.5]) *Let* $\mathbf{H} = (H; \odot, \to, 1)$ *be a subdirectly irreducible hoop.*

(i) U_* *is a subuniverse of* $\mathbf{H}$ *and* $(\mathrm{U}_*; \odot, \to, 1)$ *is a simple totally ordered Wajsberg hoop.*
(ii) *If* $\mathbf{H}$ *is a basic hoop, then* $\mathbf{H}$ *is linear.*

Proof (i) By Corollary 2.3.4, U_* is a subalgebra of $\mathbf{H}$. Moreover, since U_* is the least filter of $\mathbf{H}$ distinct from $\{1\}$, U_* is generated, as a filter, by any $x \in \mathrm{U}_* \setminus \{1\}$, i.e.,

$$\mathrm{U}_* = \{y \in H : x^n \leq y \text{ for some } n \in \mathbb{N}\} = \{y \in H : x \overset{n}{\to} y = 1 \text{ for some } n \in \mathbb{N}\}.$$

Hence, U_* satisfies (6.3). By Lemma 6.2.3 and Theorem 6.2.5, $(\mathrm{U}_*; \odot, \to, 1)$ is a simple totally ordered Wajsberg hoop.

(ii) The proof is straightforward by Theorem 6.4.3, since every basic hoop is representable. $\square$

The following example demonstrates that the converse of Proposition 6.5.1(ii) does not hold, in general.

Example 6.5.2 ([116, p. 1530]) Let $H = (0, 1] \subseteq \mathbb{R}$. Define $x \odot y := x \wedge y$ and

$$x \to y = \begin{cases} 1 & \text{if } x \leq y \\ y & \text{if } y < x. \end{cases}$$

Then $\mathbf{H} = (H; \odot, \to, 1)$ is a hoop. In addition, for each $a \in H$ we have $F_a := [a, 1]$ is a filter of $\mathbf{H}$. From $\bigcap_{a \in H} F_a = \{1\}$ it follows that $\mathbf{H}$ is not subdirectly irreducible.

Definition 6.5.3 Let $\mathbf{H}$ be a subdirectly irreducible hoop with the least filter $U_* \neq \{1\}$. An element $x \in H$ is said to be **fixed** with respect to U_* or simply fix if for all $u \in U$, $u \to x = x$. The set of elements of H which is fixed with respect to U_* is denoted by $\mathrm{Fix}_{U_*}(\mathbf{H})$. Furthermore, the set $S = \big(H \setminus \mathrm{Fix}_{U_*}(\mathbf{H})\big) \cup \{1\}$ is called the **support** of U_*.

Note. From now on, in this section, $\mathbf{H} = (H; \odot, \to, 1)$ is a non-trivial subdirectly irreducible hoop with the least filter $U_* \neq \{1\}$, and S is the support of U_*, unless otherwise stated. We have

(i) $1 \in \mathrm{Fix}_{U_*}(\mathbf{H})$.

(ii) Since there is $u \in U_* \setminus \{1\}$ and $u \to u = 1 \neq u$, so $u \notin \mathrm{Fix}_{U_*}(\mathbf{H})$, which means $\mathrm{Fix}_{U_*}(\mathbf{H})$ is proper subset of H.

(iii) In addition, by (ii), $S \neq \{1\}$.

Proposition 6.5.4 ([31, Proposition 2.7]) *Let $\mathbf{H}$ be a subdirectly irreducible hoop.*

(i) *For all $x \in H \setminus \{1\}$, there exists $u \in U$ such that $u \neq 1$ and $x \leq u$.*
(ii) $U_* \cap \mathrm{Fix}_{U_*}(\mathbf{H}) = \{1\}$.
(iii) $x \in \mathrm{Fix}_{U_*}(\mathbf{H})$ *if and only if $u \to x = x$ for some $u \in U_* \setminus \{1\}$.*

Proof (i) Let $x \in H \setminus \{1\}$. Consider the filter $\langle x \rangle$. By Corollary 2.3.9,

$$\langle x \rangle = \{y \in H : x^n \to y = 1, \text{ for some } n \in \mathbb{N}\}.$$

Since $\langle x \rangle \neq \{1\}$, it follows that $U_* \subseteq \langle x \rangle$. Given $y \in U_* \setminus \{1\}$, there exists $m \in \mathbb{N}$ such that $x^m \to y = 1$. Choose m minimal with respect to the condition $x^m \to y = 1$ and let $u = x^{m-1} \to y$. By Lemmas 2.2.1(10), $y \leq u < 1$, so $u \in U_* \setminus \{1\}$. Moreover, $x \to u = x \to (x^{m-1} \to y) = x^m \to y = 1$, which implies $x \leq u$.

(ii) Let $x \in U_* \cap \mathrm{Fix}_{U_*}(\mathbf{H})$ and $u \in U_* \setminus \{1\}$. Then $u \to x = x$. By Proposition 6.5.1, $(U_*; \odot, \to, 1)$ is a simple hoop, so due to Lemma 6.2.3, U_* satisfies (6.4). It follows that $x = 1$, as claimed.

(iii) Assume $x \neq 1$ and $y \to x = x$ for some $y \in U_* \setminus \{1\}$. Choose an arbitrary element $u \in U$. Since $y \to x = x$, we have $y \overset{n}{\to} x = x$ (see (3.1)) for every $n \in \mathbb{N}$. On the other hand, by Lemma 2.2.1(6), $u \leq (u \to x) \to x$ and so $(u \to x) \to x \in U_*$. Since U_* satisfies (6.3) and $y \neq 1$, there exists $m \in \mathbb{N}$ such that $y^m \to ((u \to x) \to x) = 1$. Hence, by Lemma 2.2.1(8), we have

$$1 = y^m \to ((u \to x) \to x) = (u \to x) \to (y^m \to x) = (u \to x) \to x,$$

consequently, $u \to x = x$. The converse is immediate, since $U_* \neq \{1\}$. $\qquad\square$

Lemma 6.5.5 ([31, Lemma 2.8]) *Let $\mathbf{H}$ be a subdirectly irreducible hoop, S the support of U_* and $x \in \mathrm{Fix}_{U_*}(\mathbf{H}) \setminus \{1\}$. Then:*

(i) $x \leq u$ *for all* $u \in U_*$.
(ii) $u \odot x = x$ *for all* $u \in U_*$.
(iii) *If* $y \leq x$, *then* $y \in \mathrm{Fix}_{U_*}(\mathbf{H})$ *for all* $y \in H$.
(iv) $x \to y, y \to x \in \mathrm{Fix}_{U_*}(\mathbf{H})$ *for all* $y \in H$.
(v) $x \leq y$, $y \to x = x$ *and* $x \odot y = x$ *for all* $y \in S$.

Proof (i) Since $x \neq 1$, and U_* is the least filter distinct from $\{1\}$, $U_* \subseteq \langle x \rangle$. Thus, given $u \in U_*$, $x^n \to u = 1$ for some $n \in \mathbb{N}$. On the other hand,

$$x \to u = (u \to x) \to (x \to u), \ \text{by Proposition 3.1.4(vi)}$$
$$= x \to (x \to u), \ \text{since } x \text{ is fixed}$$
$$= x^2 \to u, \ \text{by Lemma 2.2.1(8)}.$$

By induction, one derives $x^m \to u = x \to u$ for every $m \in \mathbb{N}$. In particular, $1 = x^n \to u = x \to u$, and so $x \leq u$.

(ii) If $u \in U_*$, then from $x \in \mathrm{Fix}_{U_*}(\mathbf{H})$, we get $u \odot x = u \odot (u \to x) = u \wedge x = x$, by (i).

(iii) Let $y \leq x$ and u be an arbitrary element of U_*. Then by Lemma 2.2.1(7), $u \to y \leq u \to x = x$ and so $(u \to y) \to x = 1$. Now,

$$(u \to y) \to y = 1 \to ((u \to y) \to y), \text{ by Lemma 2.2.1(2)}$$
$$= ((u \to y) \to x) \to ((u \to y) \to y)$$
$$= \big(((u \to y) \to x) \odot (u \to y)\big) \to y, \text{ by Lemma 2.2.1(8)}$$
$$= \big(x \odot (x \to (u \to y))\big) \to y, \text{ by Theorem 3.1.2(H3)}$$
$$= \big(x \odot ((x \odot u) \to y)\big) \to y, \text{ by Lemma 2.2.1(8)}$$
$$= \big((u \odot x) \odot ((x \odot u) \to y)\big) \to y, \text{ by (ii)}$$
$$= 1, \text{ by Lemma 2.2.1(2).}$$

So, $u \to y \le y \le u \to y$, consequently, $u \to y = y$, i.e., $y \in \mathrm{Fix}_{U_*}(\mathbf{H})$.

(iv) Let $y \in H$. We will to show that both $x \to y$ and $y \to x$ are in $\mathrm{Fix}_{U_*}(\mathbf{H})$. Consider $u \in U_*$ be arbitrary, then by (ii) and Lemma 2.2.1(8) we have

$$u \to (x \to y) = (u \odot x) \to y = x \to y.$$

Hence, $x \to y \in \mathrm{Fix}_{U_*}(\mathbf{H})$. In addition, by Lemma 2.2.1(8),

$$u \to (y \to x) = y \to (u \to x) = y \to x,$$

and so $y \to x \in \mathrm{Fix}_{U_*}(\mathbf{H})$.

(v) Let $y \in S$. If $y = 1$, all statements are trivially true. Hence, we may assume without loss of generality that $y \ne 1$, i.e., $y \in H \setminus \mathrm{Fix}_{U_*}(\mathbf{H})$. By (iv), $x \to y \in \mathrm{Fix}_{U_*}(\mathbf{H})$. Since, by Lemmas 2.2.1(10), $y \le x \to y$ and $y \notin \mathrm{Fix}_{U_*}(\mathbf{H})$, it follows from (iii) that $x \to y = 1$. Thus, $x \le y$. On the other hand, to show $y \to x = x$ it suffices to show that $(y \to x) \to x = 1$. Let $u \in U_* \setminus \{1\}$. Then $u \to x = x$. Since $x \in \mathrm{Fix}_{U_*}(\mathbf{H})$, we have

$$u \to ((y \to x) \to x) = (y \to x) \to (u \to x) = (y \to x) \to x, \text{ by Lemma 2.2.1(8),}$$

which entails that $(y \to x) \to x \in \mathrm{Fix}_{U_*}(\mathbf{H})$. By (iii) and Lemma 2.2.1(6), the fact that $y \le (y \to x) \to x$ and $y \notin \mathrm{Fix}_{U_*}(\mathbf{H})$, one gets $(y \to x) \to x = 1$ and so $y \to x = x$. Now, using $y \to x = x$ and $x \le y$, we get $x \odot y = (y \to x) \odot y = x \wedge y = x$. $\qquad\square$

In order to present the structure of subdirectly irreducible hoops, we recall the construction of the ordinal sum of two hoops (see Example 3.3.62). W. Cornish first studied this construction in [88], where he showed that for each hoop $\mathbf{H}_1$ and $\mathbf{H}_2$, $\mathbf{H}_1 \oplus \mathbf{H}_2$ is a hoop and $\mathbf{H}_1$ and $\mathbf{H}_2$ are among its subalgebras. Also, observe that $\mathbf{H}_2$ is a filter of the hoop $\mathbf{H}_1 \oplus \mathbf{H}_2$.

By Exercise 4.5, one can easily check that $\mathbf{H}_1 \oplus \mathbf{H}_2$ is subdirectly irreducible if and only if $\mathbf{H}_2$ is subdirectly irreducible. We are now ready to describe subdirectly irreducible hoops.

Theorem 6.5.6 ([31, Theorem 2.9]) *Let* $\mathbf{H}$ *be a subdirectly irreducible hoop.*

(i) $(\mathrm{Fix}_{U_*}(\mathbf{H}); \odot, \to, 1)$ *and* $(S; \odot, \to, 1)$ *are subalgebras of* $\mathbf{H}$*. Moreover,* $S \in \mathcal{F}(\mathbf{H})$*.*

(ii) $U_* \subseteq S$ *and* $\mathbf{S} = (S; \odot, \to, 1)$ *is a subdirectly irreducible Wajsberg hoop. In particular,* S *is totally ordered.*

(iii) $\mathbf{H} = \mathbf{Fix}_{U_*}(\mathbf{H}) \oplus \mathbf{S}$.

Proof (i) $1 \in \mathrm{Fix}_{U_*}(\mathbf{H})$, clearly. To see that $\mathrm{Fix}_{U_*}(\mathbf{H})$ is closed under multiplication and implication, let $x, y \in \mathrm{Fix}_{U_*}(\mathbf{H})$. If $x = 1$, then $x \odot y = 1 \odot y = y \in \mathrm{Fix}_{U_*}(\mathbf{H})$ and $x \to y = 1 \to y = y \in \mathrm{Fix}_{U_*}(\mathbf{H})$. If $x \neq 1$, then by Lemma 2.2.1(4), $x \odot y \leq y$ and so $x \odot y \in \mathrm{Fix}_{U_*}(\mathbf{H})$, by Lemma 6.5.5(iii). In addition, $x \to y$, $y \to x \in \mathrm{Fix}_{U_*}(\mathbf{H})$, by Lemma 6.5.5(iv). This concludes the proof that $\mathrm{Fix}_{U_*}(\mathbf{H})$ is a subuniverse of $\mathbf{H}$.

Now, let $x, y \in S$. Since $y \leq x \to y$, we get $x \to y \in S$, by Lemma 6.5.5(iii). Clearly, $x = 1$ or $y = 1$ implies that $x \odot y \in S$, so suppose that $x, y \neq 1$. We proceed by way of contradiction that $x \odot y \notin S$. Then $x \odot y \neq 1$, and $x, y \notin \mathrm{Fix}_{U_*}(\mathbf{H})$ (by Lemma 6.5.5(iii)), also $x \odot y \in \mathrm{Fix}_{U_*}(\mathbf{H})$, so $y \to (x \odot y) \in \mathrm{Fix}_{U_*}(\mathbf{H})$, see Lemma 6.5.5(iv). On the other hand, by Lemma 2.2.1(15), $x \leq y \to (x \odot y)$ and $x \notin \mathrm{Fix}_{U_*}(\mathbf{H})$, so we have $y \to (x \odot y) = 1$ by Lemma 6.5.5(iii). However, $y \to (x \odot y) = 1$ implies $y = x \odot y \in S$, which contradicts our assumption that $x \odot y \notin S$. Hence, $x \odot y \in S$ and $\mathbf{S}$ is a subalgebra of $\mathbf{H}$ as claimed.

Since $1 \in S$ by definition of S, to show S is also a filter of $\mathbf{H}$, it remains to observe that S satisfies (F2). This follows from Lemma 6.5.5(iii).

(ii) By Proposition 6.5.4(ii) and the definition of S, clearly $U_* \subseteq S$. Since U_* is the least filter of $\mathbf{H}$ not equal to $\{1\}$ we can easily verify that (see Proposition 2.3.5) U_* is also the least filter of S not equal to $\{1\}$. Hence, $\mathbf{S}$ is subdirectly irreducible.

To prove that $\mathbf{S}$ is a totally ordered Wajsberg hoop, it suffices to verify that it satisfies (6.4), see Proposition 6.2.4. Let $x, y \in S$ and assume that $y \to x = x$. If $y \neq 1$, then by Proposition 6.5.4(i), there exists $u \in U_*$ such that $y \leq u$ and $u \neq 1$. Then by Lemmas 2.2.1(10) and (7), $x \leq u \to x \leq y \to x = x$, so $x \in \mathrm{Fix}_{U_*}(\mathbf{H})$, see Proposition 6.5.4(iii). Thus, $x \in \mathrm{Fix}_{U_*}(\mathbf{H}) \cap S$ and hence $x = 1$ due to Proposition 6.5.4(ii).

(iii) This follows immediately from (i) together with Lemma 6.5.5(v). $\qquad\square$

Theorem 6.5.7 ([31]) *A hoop* $\mathbf{H}$ *is subdirectly irreducible if and only if it is the ordinal sum of a hoop* $\mathbf{H}_1$ *and a non-trivial subdirectly irreducible linear Wajsberg hoop* $\mathbf{H}_2$*, i.e.,* $\mathbf{H} = \mathbf{H}_1 \oplus \mathbf{H}_2$*.*

Proof Suppose that $\mathbf{H}$ is subdirectly irreducible. It suffices to set $\mathbf{H}_1 := (\mathrm{Fix}_{U_*}(\mathbf{H}); \odot, \to, 1)$ and $\mathbf{H}_2 := \mathbf{S}$. Due to the note right after Definition 6.5.3, $\mathbf{H}_2$ is non-trivial. In addition, by Theorem 6.5.6(iii), the proof is evident. Conversely, assume that $\mathbf{H} = \mathbf{H}_1 \oplus \mathbf{H}_2$ where $\mathbf{H}_1$ is a hoop and $\mathbf{H}_2$ is a non-trivial subdirectly irreducible linear Wajsberg hoop. Let F be the least element of the set $\mathcal{F}(\mathbf{H}_2) \setminus \{1\}$. Due to Exercise 4.5, F is the least element of $\mathcal{F}(\mathbf{H})$. $\qquad\square$

Proposition 6.5.8 ([11, Proposition 5.2]) *Let* **H** *be a subdirectly irreducible product hoop.*

(i) **H** *is bounded if and only if* **H** $= 2$ *or* **H** $= 2 \oplus$ **C** *where* **C** *is a subdirectly irreducible cancellative hoop.*

(ii) **H** *is unbounded if and only if* **H** *is a subdirectly irreducible cancellative hoop.*

Proof (i) Due to the definition of product hoops, **H** is a basic hoop, so by Proposition 6.5.1(ii), **H** is a linear hoop. Let **H** be bounded. Remark 3.3.52 implies that **H** satisfies (Prod), so by Theorem 6.1.21(ii), **C** $= (H \setminus \{0\}; \odot, \rightarrow, 1)$ is a cancellative hoop and **H** $= 2 \oplus$ **C**. If $C = \{1\}$, then **H** $= 2$. Furthermore, if $C \neq \{1\}$, then easy calculations show that (see Exercise 4.5.2)

$$\min\{F \in \mathcal{F}(\mathbf{H}) : F \neq \{1\}\} = \min\{F \in \mathcal{F}(\mathbf{C}) : F \neq \{1\}\}, \tag{6.8}$$

which means **C** is subdirect irreducible.

Conversely, **2** is a subdirectly irreducible bounded product hoop, evidently. In addition, if **C** is a subdirectly irreducible cancellative hoop, then by Theorem 6.1.21(i), $2 \oplus$ **C** is a bounded basic hoop satisfying (Prod), consequently, it is a product hoop, see Remark 3.3.52. Note that due to Proposition 3.3.63(i), $2 \oplus$ **C** in a linear hoop. Similarly, from (6.8) we get that $2 \oplus$ **C** is subdirectly irreducible. Therefore, $2 \oplus$ **C** is a bounded subdirectly irreducible product hoop.

(ii) The proof is straightforward by Propositions 6.5.1(i) and 3.3.54(ii). It is left as an exercise for the reader. $\qquad\qquad\square$

Proposition 6.5.9 ([11, Lemma 5.5]) *Let* **H** *be a subdirectly irreducible basic hoop satisfying the following inequality:*

$$(x \rightarrow y) \rightarrow y \leq \big((y \rightarrow z) \rightarrow ((y \rightarrow x) \rightarrow x)\big) \rightarrow ((y \rightarrow x) \rightarrow x). \tag{6.9}$$

Then either **H** *is a Wajsberg hoop or* **H** $= 2 \oplus$ **A** *for some non-trivial Wajsberg hoop* **A**.

Proof Due to Proposition 6.5.1(ii), **H** is a linear hoop, since **H** is a subdirectly irreducible basic hoop. In addition, by Theorem 6.5.7, **H** $=$ **H**$_1 \oplus$ **H**$_2$ where **H**$_1$ is a hoop and **H**$_2$ is a non-trivial subdirectly irreducible linearly ordered Wajsberg hoop. Furthermore, **H**$_1$ is a linear hoop, clearly. It suffices to show that **H**$_1 = 1$ or **H**$_1 = 2$. Choose $x \in H_1$ such that $x \neq 1$. If there exists $z \in H_1$ such that $z < x$, then we have $z < x < y < 1$, where y is an element of $H_2 \setminus \{1\}$, consequently, by definition of $\rightarrow$ and $\odot$ on the hoop **H**$_1 \oplus$ **H**$_2$ we have

$$x = y \rightarrow x = y \odot (y \rightarrow x) = y \odot x.$$

In addition, since $x \rightarrow z \in H_1 \setminus \{1\}$ and $y \in H_2$, we have $x \rightarrow z < y$ and so

$$y = 1 \to y = ((x \to z) \to y) \to y, \text{ since } x \to z < y$$
$$= ((x \to z) \to ((x \to y) \to y)) \to ((x \to y) \to y), \text{ since } x \leq y$$
$$\geq (y \to x) \to x = x \to x = 1, \text{ by (6.9)},$$

which is a contradiction. It follows that $x \leq z$. A similar argument for the case $x < z$ leads to a contradiction. Therefore, $H_1 = \{1\}$ or $H_1 = \{0, 1\}$ and the proof is completed. $\qquad\square$

Aglianò et al. [11] leveraged the finite hoops $\mathbf{L}_n$ (see Example 3.3.11) to derive an intriguing representation for finite subdirectly irreducible basic hoops as follows:

Lemma 6.5.10 *Every finite linear hoop is subdirectly irreducible.*

Proof Let $\mathbf{H}$ be a finite linear hoop and $X = \{F \in \mathcal{F}(\mathbf{H}) \colon F \neq \{1\}\}$. Since H is finite, X is a finite set, suppose $X = \{F_1, \ldots, F_m\}$ for some $m \in \mathbb{N}$. Choose $a_i \in F_i \setminus \{1\}$ for all $i \in \{1, \ldots, m\}$. Since $\mathbf{H}$ is linearly ordered, $a = \bigvee_{i=1}^{m} a_i \neq 1$ and $a \in \bigcap_{i=1}^{n} F_i$, which implies $\bigcap_{i=1}^{n} F_i \in X$. Thus, $\mathbf{H}$ is subdirectly irreducible. $\qquad\square$

Theorem 6.5.11 ([11, Corollary 1.9]) *A finite hoop* $\mathbf{H}$ *is subdirectly irreducible basic hoop if and only if there exist* $k, n_1, \ldots n_k \in \mathbb{N}$ *such that* $\mathbf{H} \cong \bigoplus_{i=1}^{k} \mathbf{L}_{n_i}$.

Proof Let $k, n_1, \ldots n_k \in \mathbb{N}$ and $\mathbf{A} = \bigoplus_{i=1}^{k} \mathbf{L}_{n_i}$. By Remark 6.1.2(i), $\mathbf{A}$ is a linear hoop, so it is a basic hoop, see Proposition 3.3.24. Also, $\mathbf{L}_{n_k}$ is the least element of the set $\{F \in \mathcal{F}(\mathbf{A}) \colon F \neq \{1\}\}$, which means $\mathbf{A}$ is subdirectly irreducible. We note that $\mathbf{L}_{n_k}$ is a simple hoop.

Conversely, assume that $\mathbf{H}$ is a finite subdirectly irreducible basic hoop. We prove the claim by induction on the cardinality of H. Suppose $|H| = m > 1$ (since $\mathbf{H}$ is non-trivial). By Proposition 6.5.1(ii), $\mathbf{H}$ is linear hoop and by Theorem 6.5.6, $\mathbf{H} = \mathbf{H}_1 \oplus \mathbf{H}_2$ where $\mathbf{H}_1 = (\mathrm{Fix}_{\mathsf{U}_*}(\mathbf{H}); \odot, \to, 1)$ is a hoop and $\mathbf{H}_2 = (S; \odot, \to, 1)$ is a non-trivial subdirectly irreducible linearly ordered Wajsberg hoop. Evidently, $\mathbf{H}_1$ is a finite linearly ordered basic hoop, so Lemma 6.5.10 entails that $\mathbf{H}_1$ is a finite subdirectly irreducible basic hoop with $|H_1| < m$. It follows from the induction hypothesis there exist $t, n_1, \ldots, n_t \in \mathbb{N}$ such that $\mathbf{H} \cong \bigoplus_{i=1}^{t} \mathbf{L}_{n_i}$. Also, $\mathbf{H}_2$ is simple, see Proposition 6.5.1, so due to Corollary 6.2.7, there exists $m \in \mathbb{N}$ such that $\mathbf{H}_2 \cong \mathbf{L}_m$, consequently,

$$\mathbf{H} = \mathbf{H}_1 \oplus \mathbf{H}_2 \cong \left(\bigoplus_{i=1}^{t} \mathbf{L}_{n_i} \right) \oplus \mathbf{H}_2 \cong \mathbf{L}_{n_1} \oplus \cdots \oplus \mathbf{L}_{n_t} \oplus \mathbf{L}_m, \text{ by Remark 6.1.2(ii)}.$$

It suffices to set $k = t + 1$ and $n_k = m$, then we have $\mathbf{H} \cong \bigoplus_{i=1}^{k} \mathbf{L}_{n_i}$. $\qquad\square$

Recall that an algebraic structure $\mathbf{A}$ is termed **finitely subdirectly irreducible** if any finite family of non-trivial congruence relations on $\mathbf{A}$ possesses a non-trivial intersection.

It is evident that an algebra is finitely subdirectly irreducible if and only if any two non-trivial congruence relations have a non-trivial intersection. Clearly, for finite algebras, being subdirectly irreducible is equivalent to being finitely subdirectly irreducible.

Applying Theorem 4.1.14, we conclude that a hoop $\mathbf{H}$ is finitely subdirectly irreducible if and only if the intersection of any non-trivial filters of $\mathbf{H}$ is not equal to $\{1\}$.

Proposition 6.5.12 ([11, Proposition 1.8]) (i) *A $\vee$-hoop $\mathbf{H}$ is finitely subdirectly irreducible if and only if $(x \to y) \vee (y \to x) = 1$ implies that $x \to y = 1$ or $y \to x = 1$ for every $x, y \in H$.*

(ii) *A basic hoop is finitely subdirectly irreducible if and only if it is linearly ordered.*

Proof (i) Assume that $\mathbf{H}$ is a finitely subdirectly irreducible $\vee$-hoop. If $x, y \in H$ such that $(x \to y) \vee (y \to x) = 1$, then by Proposition 4.1.8, $\langle x \to y \rangle \cap \langle y \to x \rangle = \{1\}$, so by the assumption, $\langle x \to y \rangle = \{1\}$ or $\langle y \to x \rangle = \{1\}$, equivalently, $x \to y = 1$ or $y \to x = 1$. Conversely, let F_1 and F_2 be two non-trivial filters of $\mathbf{H}$ such that $F_1 \cap F_2 = \{1\}$. Then the mapping $f : \mathbf{H} \to \mathbf{H}/F_1 \times \mathbf{H}/F_2$ defined by $f(x) = (x/F_1, x/F_2)$ is an embedding of hoops. Since F_1 and F_2 are non-trivial filters and $F_1 \cap F_2 = \{1\}$ there exist $x_1 \in F_1 \setminus F_2$ and $x_2 \in F_2 \setminus F_1$.

$$
\begin{aligned}
f(x_1 \to x_2) \vee f(x_2 \to x_1) &= \left(\left(\frac{x_1}{F_1}, \frac{x_1}{F_2}\right) \to \left(\frac{x_2}{F_1}, \frac{x_2}{F_2}\right)\right) \vee \left(\left(\frac{x_2}{F_1}, \frac{x_2}{F_2}\right) \to \left(\frac{x_1}{F_1}, \frac{x_1}{F_2}\right)\right) \\
&= \left(\left(\frac{1}{F_1}, \frac{x_1}{F_2}\right) \to \left(\frac{x_2}{F_1}, \frac{1}{F_2}\right)\right) \vee \left(\left(\frac{x_2}{F_1}, \frac{1}{F_2}\right) \to \left(\frac{1}{F_1}, \frac{x_1}{F_2}\right)\right) \\
&= \left(\frac{x_2}{F_1}, \frac{1}{F_2}\right) \vee \left(\frac{1}{F_1}, \frac{x_1}{F_2}\right) = \left(\frac{1}{F_1}, \frac{1}{F_2}\right).
\end{aligned}
$$

It follows that $(x_1 \to x_2) \vee (x_2 \to x_1) = 1$, since f is one-to-one. Thus, due to the assumption, $x_1 \to x_2 = 1$ or $x_2 \to x_1 = 1$, consequently, $x_2 \in F_1$ or $x_1 \in F_2$, which contradicts the choice of x_1 and x_2. Therefore, $\mathbf{H}$ is finitely subdirectly irreducible.

(ii) If $\mathbf{H}$ is a basic hoop, then by Theorem 3.3.28, it is a $\vee$-hoop and $(x \to y) \vee (y \to x) = 1$ for every $x, y \in H$. Now, the proof follows from (i) directly. $\qquad\square$

6.5.1 k-Potent Hoops

The final part of this section will be devoted to the characterization of a specific subclass of hoops, namely subdirectly irreducible k-potent hoops. Recall that, as defined in (3.1), for each hoop $(H; \odot, \to, 1)$, we denote $\overbrace{x \odot \cdots \odot x}^{n\text{-times}}$ by x^n.

Definition 6.5.13 ([31, Definition 1.9]) Let $k \in \mathbb{N}$. A hoop (pocrim) $\mathbf{H}$ is called k-**potent** if it satisfies the identity $x^k = x^{k+1}$.

It is evident that the class of k-potent hoops forms a variety. So, every subhoop of a k-potent hoop $\mathbf{H}$ is k-potent. In addition, if a hoop is k-potent, then it is m-potent for every integer $m \geq k$.

Example 6.5.14 Easy calculations show that

(i)   1-potent hoops coincide with Gödel hoops.
(ii) $\mathbf{L}_n$ is a n-potent hoop for each $n \in \mathbb{N}$.
(iii) $\mathbf{N}(\mathbb{Z})$ as well as $\mathbf{N}(\mathbb{R})$ is not a k-potent hoop for all $k \in \mathbb{N}$.
(iv) No non-trivial cancellative hoop $\mathbf{H}$ is a k-potent hoop for all $k \in \mathbb{N}$. Indeed, $x^n \odot x = x^{n+1} = x^n \odot 1$ implies that $x = 1$.
(v)   The hoop $\prod_{j=1}^{m} \mathbf{L}_{n_j}$ is a k-potent hoop, where $k = \max\{n_1, \ldots, n_m\}$.
(vi) For every finite Wajsberg hoop $\mathbf{H}$, there exists an element $k \in \mathbb{N}$ such that $\mathbf{H}$ is k-potent (see Corollary 6.2.19).
(vii) If $\mathbf{H}_i$ is a k_i-potent hoop for $i = 1, 2$, then $\mathbf{H}_1 \oplus \mathbf{H}_2$ is a k-potent hoop, where $k = \max\{k_1, k_2\}$. Conversely, for every $n \in \mathbb{N}$, if $\mathbf{H}_1 \oplus \mathbf{H}_2$ are n-potent, then so is $\mathbf{H}_1$ and $\mathbf{H}_2$, since $\mathbf{H}_1$ and $\mathbf{H}_2$ are subhoops of $\mathbf{H}_1 \oplus \mathbf{H}_2$ ([26, p. 158]).

Proposition 6.5.15 ([31, Proposition 1.10]) *A hoop $\mathbf{H}$ is k-potent if and only if $\mathbf{H}$ satisfies the identity*

$$x \xrightarrow{\ k\ } y = x \xrightarrow{\ k+1\ } y. \tag{6.10}$$

Proof Assume $\mathbf{H}$ is k-potent. Then by Proposition 3.1.4(ii),

$$x \xrightarrow{\ k\ } y = x^k \to y = x^{k+1} \to y = x \xrightarrow{\ k+1\ } y.$$

Conversely, assume (6.10) holds. Choose $x \in K$.

$$x^k \to x^{k+1} = x \xrightarrow{\ k\ } x^{k+1} = x \xrightarrow{\ k+1\ } x^{k+1} = x^{k+1} \to x^{k+1} = 1.$$

Therefore, $x^k \leq x^{k+1}$. On the other hand, $x^{k+1} = x^k \odot x \leq x^k \odot 1 = x^k$. Therefore, $x^{k+1} = x^k$. $\qquad\square$

In the next theorem, while we are working on the simple bounded hoop $([0, 1]; \odot, \to, 0, 1)$ (see Example 3.3.11), to avoid confusion with the standard product of real numbers, we will use the notation $\odot_{i=1}^{n} x$ instead of the former notation x^n.

Theorem 6.5.16 ([31, Corollary 2.10]) *Let $\mathbf{H}$ be a k-potent hoop. Then $\mathbf{H}$ is subdirectly irreducible if and only if $\mathbf{H} = \mathbf{A} \oplus \mathbf{L}_m$, for some integer $m \leq k$, and some k-potent hoop $\mathbf{A}$.*

Proof Suppose that $\mathbf{H}$ is a k-potent subdirectly irreducible hoop. By Theorem 6.5.6 and Proposition 6.5.1(i), $\mathbf{H} = \mathbf{H}_1 \oplus \mathbf{H}_2$ where $\mathbf{H}_1 = (\mathrm{Fix}_{U_*}(\mathbf{H}); \odot, \to, 1)$ is a hoop and $\mathbf{H}_2 = (S; \odot, \to, 1)$ is a non-trivial subdirectly irreducible simple hoop. Evidently, $\mathbf{H}_1$ and $\mathbf{H}_2$ are k-potent, since $\mathbf{H}_1, \mathbf{H}_2 \leq \mathbf{H}$. By Corollary 6.2.6, $\mathbf{H}_2$ is isomorphic to a subalgebra $\mathbf{B}$ of the bounded Wajsberg hoop $([0, 1]; \odot, \to, 0, 1)$. Evidently, $\mathbf{B}$ is a k-potent. The rest of the proof is followed by the following claims.

Claim 1. If $\mathbf{C}$ is a k-potent simple hoop, then $\mathbf{C}$ is bounded and $x^k = 0$ for all $x \in C \setminus \{1\}$:

Choose $x \in C \setminus \{1\}$. Then $C = \langle x \rangle$. By Corollary 2.3.9, for each $y \in C$, there exists $n \in \mathbb{N}$ such that $x^n \leq y$. If $k \geq n$, then $x^k \leq x^n \leq y$. Also, $k < n$ implies that $x^k = x^{k+1} = \cdots = x^n \leq y$, thus, $x^k = \min(\mathbf{C})$. Therefore, $x^k = 0$ for all $x \in C \setminus \{1\}$.

Claim 2. If $b = \bigvee \{x \in B : x \neq 1\}$, then $b \neq 1$:

Suppose that $b = 1$. Then for each $m > k$ there exists $x \in B \setminus \{1\}$ such that $x > m/(m+1)$, so by the proof of Corollary 6.2.6,

$$\odot_{i=1}^{m} x = (mx - m + 1) \vee 0 \geq (m \frac{m}{m+1} - m + 1) \vee 0$$

$$= (m - \frac{m}{m+1} - m + 1) \vee 0$$

$$= \frac{1}{m+1} > 0.$$

Since $\mathbf{B}$ is a k-potent simple hoop, $\odot_{i=1}^{m} x \neq 0$ contradicts with Claim 1. Hence, $b < 1$.

Claim 3. $\mathbf{B} \cong \mathbf{L}_m$ for some $m \in \mathbb{N}$:

Consider the element b in Claim 2. If $b = 0$, then $B = \{0, 1\}$ and $\mathbf{A} \cong \mathbf{B} \cong \mathbf{L}_1$. Otherwise, $0 < b < 1$ and there exists $x \in B \setminus \{0, 1\}$, consequently, $x, x' \in B \setminus \{0, 1\}$ implies that $b \geq x, 1 - x$, i.e., $b \geq 1/2$. Clearly, $2b - 1 < b$, see Claim 2. We show that $b \in B$, otherwise, we can choose $x, y \in B$ such that $2b - 1 < y < x < b$. We have

$$b < 2b - x < 1 + y - x = (y - x + 1) \wedge 1 = x \to y \in B \setminus \{1\}.$$

Hence, $b \in B \setminus \{1\}$. Let $m \in \mathbb{N}$ be the least integer such that $mb \leq m - 1$, equivalently, $b \leq (m-1)/m$. By Claim 1, $\odot_{i=1}^{k} b = 0$, which entails $kb \leq k - 1$, that is, $m \leq k$. Set

$$S := \{1, b, b \odot b, \ldots, \odot_{i=1}^{m-1} b, \odot_{i=1}^{m} b = 0\}.$$

Note that, according to the choice of m, we have $b \odot b = (2b - 1) \vee 0 = 2b - 1$, $b \odot b \odot b = (3b - 2) \vee 0 = 3b - 2$ and similarly, $\odot_{i=1}^{m-1} b = ((m-1)b - m + 2) \vee 0 = (m-1)b - m + 2$. We will show that $B = S$. Suppose that $y \in B \setminus S$. Clearly, $y > b$ is impossible. In addition, $y < (m-1)b - m + 2$ entails that $1 - y > (m-1) - (m-1)b \geq b$, since $mb \leq m - 1$. So, there exists $n \in \{1, \ldots, m-2\}$ such that $\odot_{i=1}^{n} b > y > \odot_{i=1}^{n+1} b$. Evidently, $(\odot_{i=1}^{n} b) \to y \in B \setminus \{1\}$. Furthermore,

$$\left(\odot_{i=1}^{n} b\right) \to y = \left(1 - (nb - n + 1) + y\right) \wedge 1$$
$$= y - nb + n, \ \text{ since } \left(\odot_{i=1}^{n} b\right) \to y \neq 1$$
$$> (n+1)b - n - nb + n = b, \ \text{ since } y > \odot_{i=1}^{n+1} b$$

which is a contradiction. Thus, $B = S$. By Theorem 3.3.15, every bounded Wajsberg hoop satisfies (DNP), so Remark 2.2.5(i) implies that $1 - b = b' = \odot_{i=1}^{m-1} b = (m-1)b - m + 2$, equivalently, $b = (m-1)/m$. In addition,

$$b \odot b = \frac{m-2}{m}, \quad b \odot b \odot b = \frac{m-3}{m}, \quad \ldots, \quad \odot_{i=1}^{m-1} b = \frac{1}{m}.$$

Therefore, $\mathbf{B} = \mathbf{L}_m$.

From Claim 3, we conclude that $\mathbf{H}_2 \cong \mathbf{B} \cong \mathbf{L}_m$. Conversely, let $\mathbf{H} = \mathbf{A} \oplus \mathbf{L}_m$, for some integer $m \le k$, and some k-potent hoop $\mathbf{A}$. Due to Example 6.5.14(i) and (vii), $\mathbf{H}$ is k-potent. Also, by Theorem 6.5.7, $\mathbf{H}$ is subdirectly irreducible. $\qquad\square$

Bibliographical Remarks and Suggestions for Further Study

Section 6.5 explored subdirectly irreducible hoops, a significant object within the variety of hoops. This section presented fundamental results established in [11, 30, 31, 141], which contribute to a deeper understanding of hoops. A representation for subdirectly irreducible algebras within the variety generated by $\bigoplus_{i=1}^{n} \mathbf{H}_i$, where $\{\mathbf{H}_1, \ldots, \mathbf{H}_n\}$ is a finite set of linearly ordered Wajsberg hoops, was presented in [13, Theorem 7.9]. The final part of this section focused on k-potent hoops; further results on this topic can be found in [26].

6.6   Ultrafilters and Wajsberg Hoops

Section 4.2.2 provides an introduction to ultrafilters and their main features. This type of filter plays a crucial role in bounded Wajsberg hoops. In this section, we will investigate Wajsberg hoops induced from filters of a bounded Wajsberg hoop. A well-known result by Conrad and Darnel [87] demonstrates that every generalized Boolean algebra is either a Boolean algebra or a maximal ideal of a Boolean algebra. Applying the properties of ordinal sums allows us to readily demonstrate that every basic hoop is either bounded or a maximal filter of a bounded basic hoop (equivalently, a BL-algebra). This is because for every non-bounded basic hoop $\mathbf{H}$, $\mathbf{2} \oplus \mathbf{H}$ is a bounded basic hoop (see Proposition 3.3.63(iv)), and H is a maximal filter of $\mathbf{2} \oplus \mathbf{H}$ satisfying $(\mathbf{2} \oplus \mathbf{H})/H \cong \mathbf{2}$. However, establishing this fact for Wajsberg hoops presents a challenge, as $\mathbf{2} \oplus \mathbf{H}$ is not necessarily a Wajsberg hoop (see Example 6.1.4(ii)).

Recently, Abad et al. [6] extended this result to the class of Wajsberg hoops. The primary goal of this section is to explore this generalization. As previously demonstrated in

Theorem 3.3.15, every bounded Wajsberg hoop satisfies the (DNP). In addition, each Wajsberg hoop is a basic hoop (see Proposition 3.3.32), which implies it is representable, see Theorem 6.4.3.

Lemma 6.6.1 ([6]) *Let* $\mathbf{H}$ *be a Wajsberg hoop. Define a binary operation* $\circledast$ *on* $\mathbf{H}$ *by*

$$x \circledast y = (x \to (x \odot y)) \to y, \quad \forall x, y \in H. \tag{6.11}$$

The following assertions hold:

(i) *If* $\mathbf{H}$ *is cancellative, then* $x \circledast y = 1$ *for all* $x, y \in H$.
(ii) *If* $\mathbf{H}$ *is bounded, then* $x \circledast y = x \oplus y$, *where* $x \oplus y = (x' \odot y')'$ *(see Proposition 2.2.6(i)).*
(iii) $(H; \circledast)$ *is a partially ordered commutative monoid and* $x \vee y \leq x \circledast y$ *for all* $x, y \in H$.

Proof (i) Let $\mathbf{H}$ be cancellative and $x, y \in H$. By Proposition 3.3.5, we have

$$\big(x \to (x \odot y)\big) \to y = y \to y = 1.$$

(ii) Let $\mathbf{H}$ be bounded and $x, y \in H$. Then $\mathbf{H}$ is a $\vee$-hoop satisfying (DNP) and

$$
\begin{aligned}
\big(x \to (x \odot y)\big) \to y &= \big(x \to (x \odot y)''\big) \to y, \ \text{ by Theorem 3.3.15}\\
&= \big(x \to (x \to y')'\big) \to y, \ \text{ by Lemma 2.2.1(9)}\\
&= \big(x \odot (x \to y')\big)' \to y, \ \text{ by Lemma 2.2.1(9)}\\
&= (x \wedge y')' \to y, \ \text{ by Proposition 3.1.4(i)}\\
&= (x' \vee y) \to y, \ \text{ by Corollary 3.3.35}\\
&= (x' \to y) \wedge (y \to y), \ \text{ by Proposition 3.3.34}\\
&= x' \to y'' = (x' \odot y')', \ \text{ by Lemma 2.2.1(9)}\\
&= x \oplus y.
\end{aligned}
$$

(iii) Clearly, $(H; \circledast)$ is an algebra of type (2). Choose $x, y, z \in H$.
(1) By Lemma 2.2.1(10), $y \leq x \circledast y$. Lemma 2.2.1(9) and Proposition 3.1.4(i) implies

$$
\begin{aligned}
x \to (x \circledast y) = x \to \big((x \to (x \odot y)) \to y\big) &= \big(x \odot (x \to (x \odot y))\big) \to y\\
&= (x \wedge (x \odot y)) \to y = (x \odot y) \to y = 1.
\end{aligned}
$$

Thus, $x \vee y \leq x \circledast y$.
(2) By Theorem 6.4.3, $\mathbf{H}$ is representable. Let $f : \mathbf{H} \to \prod_{i \in I} \mathbf{H}_i$ be a subdirect embedding, where $\{\mathbf{H}_i : i \in I\}$ is a family of linearly ordered hoops. Evidently, every $\mathbf{H}_i$ is a

Wajsberg hoop, since $\pi_i \circ f(H) = H_i$ for all $i \in I$. Choose $i \in I$. If $\mathbf{H}_i$ is unbounded, then by Theorem 6.1.16, $\mathbf{H}$ is a cancellative hoop, consequently, due to (i), $\pi_i(f(x)) \circledast \pi_i(f(y)) = 1 = \pi_i(f(y)) \circledast \pi_i(f(x))$. Also, if $\mathbf{H}$ is bounded, then by (ii), $\oplus = \circledast$ and by Theorem 3.3.13, $(H_i; \oplus,',0,1)$ is an MV-algebra, which implies $\pi_i(f(x)) \circledast \pi_i(f(y)) = \pi_i(f(y)) \circledast \pi_i(f(x))$. Thus, $f(x \circledast y) = f(x) \circledast f(y) = f(y) \circledast f(x) = f(y \circledast x)$, which implies $x \circledast y = y \circledast x$.

(3) Similar to (2), we can show that $x \circledast (y \circledast z) = (x \circledast y) \circledast z$ for every $x, y, z \in H$. In addition, $x \leq y$ implies that $x \circledast z \leq y \circledast z$.

(1)–(3) entail that $(H; \circledast)$ is a partially ordered commutative monoid. $\qquad\square$

Proposition 6.6.2 *Every Wajsberg hoop can be embedded in a bounded Wajsberg hoop.*

Proof Let $\mathbf{H}$ be a Wajsberg hoop. If $\mathbf{H}$ is bounded, then the proof is evident. Assume that $\mathbf{H}$ is unbounded. Set $S_1 := \{P \in \mathrm{Spec}(\mathbf{H}) : \mathbf{H}/P \text{ is bounded}\}$ and $S_2 := \mathrm{Spec}(\mathbf{H}) \setminus S_1$. Due to Theorem 4.2.11, given $P \in \mathrm{Spec}(\mathbf{H})$, the quotient structure $\mathbf{H}/P$ is a linear Wajsberg hoop, so by Corollary 6.1.17, $\mathbf{H}/P$ is either bounded or cancellative, which implies $S_2 = \{P \in \mathrm{Spec}(\mathbf{H}) : \mathbf{H}/P \text{ is cancellative}\}$. Set $\mathbf{H}_i := \prod_{P \in S_i} \mathbf{H}/P$ for $i = 1, 2$. Define $f : H \to H_1 \times H_2$ by $f(x) = \big((x/P)_{P \in S_1}, (x/P)_{P \in S_2}\big)$. Clearly, f is a homomorphism of hoops. Since every Wajsberg hoop is a basic hoop, by Theorem 6.4.3 and Remark 6.4.2(ii),

$$\mathrm{Ker}(f) = \{x \in H : x \in P, \ \forall P \in S_1 \cup S_2\} = \bigcap \{P : P \in \mathrm{Spec}(\mathbf{H})\} = \{1\}.$$

Hence, f is a one-to-one homomorphism. In addition, $\mathbf{H}_1$ is a bounded Wajsberg hoop. Also, $\mathbf{H}_2$ is a cancellative hoop, see Corollary 3.3.6, so by Proposition 3.3.19, there exists an Abelian ℓ-group $\mathbf{G}_2$ such that $\mathbf{H}_2 = \mathbf{N}(\mathbf{G}_2)$. If $\mathbf{G}$ is the Abelian ℓ-group $\mathbb{Z} \overrightarrow{\times} \mathbf{G}_2$, then by Example 3.3.57, $\mathbf{G}[\mathbf{u}]$ is a bounded Wajsberg hoop, where $u = (1, 0)$. Define $g : H_2 \to (\{0\} \times G_2^+) \cup (\{1\} \times G_2^-)$ by $g(x) = (1, x)$ for all $x \in H_2$. Easy calculations show that $g : \mathbf{H}_2 \to \mathbf{G}[\mathbf{u}]$ is a one-to-one homomorphism of hoops (note that the underlying set of the hoop $\mathbf{G}[\mathbf{u}] = (\{0\} \times G_2^+) \cup (\{1\} \times G_2^-)$). It follows that $k : \mathbf{H} \to \mathbf{H}_1 \times \mathbf{G}[\mathbf{u}]$ defined by

$$k(x) := \big(\pi_1(f(x)), g(\pi_2(f(x)))\big) = \big((x/P)_{P \in S_1}, g((x/P)_{P \in S_2})\big) \quad \forall x \in H,$$

is a one-to-one homomorphism of hoops. Clearly, $\mathbf{H}_1 \times \mathbf{G}[\mathbf{u}]$ is a bounded Wajsberg hoop. Therefore, every Wajsberg hoop can be embedded in a bounded Wajsberg hoop. $\qquad\square$

Recall that by Corollary 2.3.4, for each filter F of a hoop $\mathbf{H} = (H; \odot, \to, 1)$, F is a subuniverse of $\mathbf{H}$, and so $\mathbf{F}_h := (F; \odot, \to, 1)$ is subhoop of $\mathbf{H}$. The following theorem offers a representation for Wajsberg hoops, based on Wajsberg hoops induced from filters of bounded Wajsberg hoops. Note that Theorem 6.6.3 was originally proved in [6] using free MV-algebras. However, we use a different approach based on Proposition 6.6.2 (see also [134, Sect. 3]).

Theorem 6.6.3 *For each unbounded Wajsberg hoop* **H**, *there exist a bounded Wajsberg hoop* **A** *and an ultrafilter F of* **A** *such that* $\mathbf{H} \cong \mathbf{F}_h$.

Proof Let **H** be an unbounded Wajsberg hoop. Due to Proposition 6.6.2, there exist a bounded Wajsberg hoop $\mathbf{M} = (M; \odot, \to, 0, 1)$ and a one-to-one homomorphism $f : \mathbf{H} \to (M; \odot, \to, 1)$ of hoops. Set $A = f(H) \cup f(H)'$. Clearly, $f(H) \cap f(H)' = \emptyset$. Suppose, for contradiction, that there exist $x, y \in H$ such that $a = f(x) = f(y)'$. Then, $0 = f(y) \odot f(y)' = f(y) \odot f(x) = f(y \odot x) \in f(H)$, consequently, for all $z \in H$, we have

$$f((y \odot x) \to z) = f(y \odot x) \to f(z) = 0 \to f(z) = f(1),$$

implying that $y \odot x$ is the least element of H, which contradicts with the assumption. We claim that $\mathbf{A} = (A; \odot, \to, 0, 1)$ is a bounded subhoop of **M**. According to Theorem 3.3.13, it suffices to show that $(A; \oplus, ', 0, 1)$ is a subalgebra of the MV-algebra $(\mathbf{M}; \oplus, ', 0, 1) = \beta(\mathbf{M})$, where β is the bijection in the proof of Theorem 3.3.13. Let $x, y \in H$. Since **M** is a bounded Wajsberg hoop, by Lemma 6.6.1(ii), we obtain

(i) $f(x) \oplus f(y) = f(x) \circledast f(y) = f(x \circledast y) \in f(H) \subseteq A$.

(ii) $f(x) \oplus f(y)' = f(y) \to f(x) = f(y \to x) \in f(H) \subseteq A$. Similarly, $f(x)' \oplus f(y) = f(x \to y) \in f(H) \subseteq A$.

(iii) $f(x)' \oplus f(y)' = (f(x) \odot f(y))' = f(x \odot y)' \in f(H)' \subseteq A$.

(i)–(ii) implies that A is closed under $\oplus$. Clearly, A is closed under $'$, so $(A; \oplus, ', 0, 1)$ is a subalgebra of the MV-algebra $(\mathbf{M}; \oplus, ', 0, 1)$. Note that bounded Wajsberg hoops satisfy (DNP), see Theorem 3.3.15. Hence, $\mathbf{A} = (A; \odot, \to, 0, 1)$ is a bounded subhoop of $(M; \odot, \to, 0, 1)$. Now, we show that $f(H)$ is an ultrafilter of **A**. Clearly, $f(H)$ is closed under $\odot$. Let $a, b \in A$ such that $b \geq a = f(x)$ for some $x \in H$. We claim that $b \in f(H)$, otherwise, there exists $y \in H$ such that $b = f(y)'$, which entails $f(x \odot y) = f(x) \odot f(y) \leq f(y)' \odot f(y) = 0$. Similar to the above argument, we can see that $x \odot y$ is the least element of **H**, a contradiction. Thus, $f(H)$ is a proper filter of **A**. In addition, by definition of A it is evident that $A/f(H)$ equals to $\{0/f(H), 1/f(H)\}$, consequently, $\mathbf{A}/f(H) \cong \mathbf{L}_1$, see Remark 4.2.17(ii). Therefore, $f(H)$ is an ultrafilter of **A**. We note that $0/f(H) = f(H)'$ and $1/f(H) = f(H)$, clearly. $\square$

Consider the assumption of Theorem 6.6.3 and the bounded Wajsberg hoop $(A; \odot, \to, 0, 1)$ in the proof of this theorem, where $A = f(H) \cup f(H)'$. Let $a, b \in A$. Since $f : \mathbf{H} \to \mathbf{A}$ is a hoop embedding, we have

If $a, b \in f(H)$, then there exist $x, y \in H$ such that $a = f(x)$ and $b = f(y)$. Hence, $a \to b = f(x) \to f(y) = f(x \to y)$.

If $a \in f(H)$ and $b \in f(H)'$, then there exist $x, y \in H$ such that $a = f(x)$ and $b = f(y)'$. Hence, $a \to b = f(x) \to f(y)' = (f(x) \odot f(y))' = f(x \odot y)'$.

If $a \in f(H)'$ and $b \in f(H)$, then there exist $x, y \in H$ such that $a = f(x)'$ and $b = f(y)$. Since $\mathbf{A}$ is a bounded Wajsberg hoop by Lemma 6.6.1(ii), $a \to b = f(x)' \to f(y) = f(x)' \to f(y)'' = (f(x)' \odot f(y)')' = f(x) \circledast f(y) = f(x \circledast y)$.

If $a, b \in f(H)'$, then there exist $x, y \in H$ such that $a = f(x)'$ and $b = f(y)'$, consequently, $a \to b = f(x)' \to f(y)' = f(y) \to f(x)'' = f(y) \to f(x) = f(y \to x)$. By summarizing the above arguments, we obtain

$$
a \to b = \begin{cases}
f(x \to y) & \text{if } a = f(x),\ b = f(y),\ x, y \in H \\
f(x \circledast y) & \text{if } a = f(x)',\ b = f(y),\ x, y \in H \\
f(x \odot y)' & \text{if } a = f(x),\ b = f(y)',\ x, y \in H \\
f(y \to x) & \text{if } a = f(x)',\ b = f(y)',\ x, y \in H.
\end{cases}
$$

In a similar way, we can show that

$$
a \odot b = \begin{cases}
f(x \odot y) & \text{if } a = f(x),\ b = f(y),\ x, y \in H \\
f(y \to x)' & \text{if } a = f(x)',\ b = f(y),\ x, y \in H \\
f(x \to y)' & \text{if } a = f(x),\ b = f(y)',\ x, y \in H \\
f(x \circledast y)' & \text{if } a = f(x)',\ b = f(y)',\ x, y \in H.
\end{cases}
$$

Theorem 6.6.4 ([6]) *Every Wajsberg hoop is either bounded or isomorphic to $\mathbf{F}_h$, where F is an ultrafilter of a bounded Wajsberg hoop.*

Proof The proof follows directly from Theorem 6.6.3. $\square$

Since every ultrafilter is a maximal filter, by Theorem 4.2.26, it follows that each Wajsberg hoop is bounded or isomorphic to $\mathbf{F}_h$, where F is a maximal of a bounded Wajsberg hoop.

Theorem 6.6.5 *Let $\mathbf{H}_1$ and $\mathbf{H}_2$ be bounded Wajsberg hoops with ultrafilters F_1 and F_2, respectively such that $\mathbf{F}_{1h} \cong \mathbf{F}_{2h}$. Then $\mathbf{H}_1 \cong \mathbf{H}_2$.*

Proof Let $g : \mathbf{F}_{1h} \to \mathbf{F}_{2h}$ be an isomorphism of hoops. Since $\mathbf{H}_i$ is a bounded Wajsberg hoop and F_i is an ultrafilter of $\mathbf{H}_i$, by Lemma 4.2.20, $H_i = F_i \cup F_i'$ and $F_i \cap F_i' = \emptyset$ for $i = 1, 2$. Define $\bar{g} : H_1 \to H_2$ as follows:

$$
\bar{g}(x) = \begin{cases}
g(x) & \text{if } x \in F_1 \\
g(x')' & \text{if } x \in F_1'.
\end{cases}
$$

Clearly, $\bar{g}$ is a bijection map, $\bar{g}(1) = g(1) = 1$ and $\bar{g}(0) = g(1)' = 0$. Choose $x, y \in H_1$.

(i) $\bar{g}(x \odot y) = g(x \odot y) = g(x) \odot g(y) = \bar{g}(x) \odot \bar{g}(y)$.

(ii) By Lemma 2.2.1(8), $x \odot y' = (x \odot y')'' = (x \to y'')' = (x \to y)' \in F_1'$, so $\bar{g}(x \odot y') = \bar{g}((x \to y)') = g(x \to y)' = (g(x) \to g(y))' = (g(x) \odot g(y)')'' = g(x) \odot g(y)' = \bar{g}(x) \odot \bar{g}(y')$. Similarly, $\bar{g}(x' \odot y) = \bar{g}(x')' \odot \bar{g}(y)$.

(iii) Furthermore, due to Lemma 6.6.1(ii), $x' \odot y' = (x \circledast y)' \in F_1'$, which entails, $\bar{g}(x' \odot y') = g(x \circledast y)' = (g(x) \circledast g(y))' = g(x)' \odot g(y)' = \bar{g}(x') \odot \bar{g}(y')$.

From (i)–(iii) it follows that $\bar{g}$ preserves $\odot$. In a similar way, we can verify that $\bar{g}$ preserves $\to$. Therefore, $\bar{g} : \mathbf{H}_1 \to \mathbf{H}_2$ is an isomorphism of bounded hoops. $\square$

Remark 6.6.6 Let $\mathbf{H} = (H; \odot, \to, 1)$ be an unbounded Wajsberg hoop. By applying Theorems 6.6.3 and 6.6.5, there exists, up to isomorphism, a unique bounded Wajsberg hoop $\mathbf{A} = (A; \odot, \to, 0, 1)$ with an ultrafilter F such that $\mathbf{H} \cong \mathbf{F}_h$. We denote this unique bounded Wajsberg hoop by $\mathbf{A}(\mathbf{H})$ and identify H with its image $f(H)$, where f is the hoop embedding in the proof of Theorem 6.6.3. Considering this identification, we have $\mathbf{A}(\mathbf{H}) = (H \cup H'; \odot, \to, 0, 1)$, where $\odot$ and $\to$ are as follows (see the note right before Theorem 6.6.3):

$$x \odot y = \begin{cases} x \odot y & \text{if } x, y \in H \\ (y \to x')' & \text{if } x \in H', y \in H \\ (x \to y')' & \text{if } x \in H, y \in H' \\ (x' \circledast y')' & \text{if } x, y \in H', \end{cases} \qquad x \to y = \begin{cases} x \to y & \text{if } x, y \in H \\ x' \circledast y & \text{if } x \in H', y \in H \\ (x \odot y')' & \text{if } x \in H, y \in H' \\ y' \to x' & \text{if } x, y \in H'. \end{cases}$$

Furthermore, H is an ultrafilter of $\mathbf{A}(\mathbf{H})$.

As an application of Theorem 6.6.5, let $\mathbf{H}$ be a non-trivial cancellative hoop. Due to Theorem 3.3.19, there exists an Abelian ℓ-group $\mathbf{G}$ such that $\mathbf{H} \cong \mathbf{N}(\mathbf{G})$. Consider the bounded Wajsberg hoop $\mathbf{B} := (\mathbb{Z} \overrightarrow{\times} \mathbf{G})[(1, 0)]$. Clearly, $F = \{1\} \times \mathbf{G}^-$ is an ultrafilter of $\mathbf{B}$ (see Example 4.2.19(iv)) and $\mathbf{F}_h = (\{1\} \times \mathbf{G}^-; \odot, \to, (1, 0))$ is a subhoop of $\mathbf{B}$. In addition, $\mathbf{F}_h \cong \mathbf{H}$, so by Theorem 6.6.6, $\mathbf{A}(\mathbf{H}) \cong \mathbf{B} = (\mathbb{Z} \overrightarrow{\times} \mathbf{G})[(1, 0)]$.

Proposition 6.6.7 *Let $k : \mathbf{H}_1 \to \mathbf{H}_2$ be a homomorphism between two unbounded Wajsberg hoops. Then k can be uniquely extended to a homomorphism $\bar{k} : \mathbf{A}(\mathbf{H}_1) \to \mathbf{A}(\mathbf{H}_2)$ between bounded hoops. In addition, k is one-to-one (onto) if and only if $\bar{k}$ is one-to-one (onto).*

Proof Considering Remark 6.6.6, $A(H_1) = H_1 \cup H_1'$ and $A(H_2) = H_2 \cup H_2'$. Define $\bar{k} : A(H_1) \to A(H_2)$ by

$$\bar{k}(x) = \begin{cases} k(x) & \text{if } x \in H_1 \\ k(x')' & \text{if } x \in H_1'. \end{cases}$$

Let $x, y \in A(H_1)$. Due to Remark 6.6.6, we have

(i) If $x, y \in H_1$, then $\bar{k}(x \odot y) = k(x \odot y) = k(x) \odot k(y) = \bar{k}(x) \odot \bar{k}(y)$.

(ii) If $x \in H_1'$ and $y \in H_1$, then $x \odot y = (y \to x')'$. Since $y \to x' \in H_1$, by definition of $\bar{k}$, we get $\bar{k}(x \odot y) = k(y \to x')' = (k(y) \to k(x'))' = (\bar{k}(y) \to \bar{k}(x)')' = \bar{k}(x) \odot \bar{k}(y)$.

(iii) For $x \in H_1$ and $y \in H_1'$, a proof analogous to that of (ii) demonstrates that $\bar{k}(x \odot y) = \bar{k}(x) \odot \bar{k}(y)$.

(iv) If $x, y \in H_1'$, then $x \odot y = (x' \circledast y')'$ and $x' \circledast y' \in H_1$, which imply

$$\bar{k}(x \odot y) = k(x' \circledast y')' = (k(x') \circledast k(y'))' = (\bar{k}(x)' \circledast \bar{k}(y)')' = \bar{k}(x) \odot \bar{k}(y).$$

(i)–(iv) imply that $\bar{k}$ preserves the binary operation $\odot$. In a similar way, we can deduce that $\bar{k}$ preserves $\to$. In addition, $\bar{k}(0) = k(1)' = 0$ and $\bar{k}(1) = k(1) = 1$, which means $\bar{k} : A(H_1) \to A(H_2)$ is a homomorphism of bounded hoops.

Furthermore, if k is onto, then by definition, $\bar{k}$ is onto. Now, let k be one-to-one and $x, y \in A(H_1)$ such that $\bar{k}(x) = \bar{k}(y)$. If $x, y \in H_1$, then $k(x) = \bar{k}(x) = \bar{k}(y) = k(y)$, which implies $x = y$. For $x, y \in H_1'$, we get $k(x')' = \bar{k}(x) = \bar{k}(y) = k(y')'$, and so, $x' = y'$, consequently, $x = y$. If $x \in H_1$ and $y \in H_1'$, then $\bar{k}(x) = k(x) \in H_2$ and $\bar{k}(y) = k(y')' \in H_2'$ which lead us to a contradiction, since $H_2 \cap H_2' = \emptyset$. Similarly, $x \in H_1'$ and $y \in H_1$ is impossible. Therefore, $\bar{k}$ is one-to-one. On the other hand, if $\bar{k}$ is one-to-one (onto), then so is k, evidently, since $\bar{k}|_{H_1} = k$.

Now, we show that $\bar{k}$ is the unique extension of k. Assume that $g : A(H_1) \to A(H_2)$ be a homomorphism of bounded hoops such that $g(x) = k(x)$ for all $x \in H_1$. If $x \in A(H_1) \setminus H_1$, then $x \in H_1'$, so $g(x) = g(x'') = g(x')' = k(x')' = \bar{k}(x)$, consequently, $g = \bar{k}$. $\square$

Theorem 6.6.8 *Let $g : H_1 \to H_2$ be a homomorphism of hoops, where H_1 is an unbounded Wajsberg hoop and H_2 is a bounded Wajsberg hoop. Then there exists a unique homomorphism $\bar{g} : A(H) \to H_2$ of bounded hoops such that $\bar{g}(x) = g(x)$ for all $x \in H_1$.*

Proof Consider the map $\bar{g} : A(H_1) \to H_2$ is defined by

$$\bar{g}(x) = \begin{cases} g(x) & \text{if } x \in H_1 \\ g(x')' & \text{if } x \in H_1'. \end{cases}$$

The remainder of the proof follows similarly to the proof of Proposition 6.6.7. $\square$

In the sequel, we will investigate the relation between prime filters of $A(H)$ and H, where H is an unbounded Wajsberg hoop. Note that, by Proposition 2.4.7, $\mathcal{F}(H) = \{F \cap H : F \in \mathcal{F}(A(H))\}$.

Lemma 6.6.9 *Let H_1 be a subhoop of H_2 and $P \in \mathrm{Spec}(H_2)$. Then either $H_1 \subseteq P$ or $P \cap H_1 \in \mathrm{Spec}(H_1)$.*

Proof There are two possibilities: $H_1 \cap P = H_1$ or $H_1 \cap P \neq H_1$. The first one implies $H_1 \subseteq P$. The second one entails $P \cap H_1$ is a proper filter of H_1, see Proposition 2.4.7(ii). An easy calculation shows that $f : H_1/(P \cap H_1) \to H_2/P$ defined by $f(x/(P \cap H_1)) = $

x/P is a one-to-one homomorphism of hoops. On the other hand, $\mathbf{H}_2/P$ is a linear hoop, see Theorem 4.2.11. Therefore, $\mathbf{H}_1/(P \cap H_1)$ is linearly ordered, consequently, by Theorem 4.2.11, $P \cap H_1$ is a prime filter of $\mathbf{H}_1$. $\qquad\square$

Proposition 6.6.10 ([6]) *Let* $\mathbf{H}$ *be an unbounded Wajsberg hoop and* $F \in \mathrm{Spec}(\mathbf{H})$.

(i) $\mathcal{F}(\mathbf{H}) \subseteq \mathcal{F}(\mathbf{A}(\mathbf{H}))$.
(ii) *If* $F \in \mathcal{F}(\mathbf{H})$ *such that* $\mathbf{H}/F$ *is unbounded, then* $\mathbf{A}(\mathbf{H}/F) = \mathbf{A}(\mathbf{H})/F$.
(iii) *If* $F \in \mathrm{Spec}(\mathbf{H})$ *such that* $\mathbf{H}/F$ *is unbounded, then* $\mathbf{A}(\mathbf{H})/F$ *is linearly ordered.*
(iv) $\mathbf{H}/F$ *is bounded if and only if* $F \notin \mathrm{Spec}(\mathbf{A}(\mathbf{H}))$.

Proof (i) The proof is straightforward, since H is a filter of $\mathbf{A}(\mathbf{H})$.

(ii) Let F be a filter of $\mathbf{H}$ such that $\mathbf{H}/F$ is unbounded. By (i), F is a filter of $\mathbf{A}(\mathbf{H})$, so $\mathbf{A}(\mathbf{H})/F$ is a bounded Wajsberg hoop. In addition, H/F is a maximal filter of $\mathbf{A}(\mathbf{H})/F$, see Theorem 4.1.17(i). Also, due to part (iii) of the same theorem, $(\mathbf{A}(\mathbf{H})/F)/(H/F) \cong \mathbf{A}(\mathbf{H})/H \cong \mathbf{L}_1$, which implies H/F is an ultrafilter of $\mathbf{A}(\mathbf{H})$. Also, $A(H)/H = (H/F) \cup (H'/F) = (H/F) \cup (H/F)'$, so by Remark 6.6.6, $\mathbf{A}(\mathbf{H}/F) \cong \mathbf{A}(\mathbf{H})/F$.

(iii) Let F be a prime filter of $\mathbf{H}$ such that $\mathbf{H}/F$ is unbounded. Then by Theorem 6.1.16, $\mathbf{H}/F$ is a cancellative linear hoop, so, due to the note right before Proposition 6.6.7, $\mathbf{A}(\mathbf{H}/F) \cong (\mathbb{Z} \overset{\rightarrow}{\times} \mathbf{G})[(1,0)]$, where $\mathbf{G}$ is an Abelian ℓ-group such that $\mathbf{H}/F \cong \mathbf{N}(\mathbf{G})$. Evidently, $\mathbf{G}$ is linearly ordered, (since $\mathbf{H}$ is linearly ordered), so as the ℓ-group $\mathbb{Z} \overset{\rightarrow}{\times} \mathbf{G}$. Hence, $\mathbf{A}(\mathbf{H}/F)$ is a linear hoop. Now, (ii) entails that $\mathbf{A}(\mathbf{H})/F$ is linearly ordered.

(iv) Let F be a prime filter of $\mathbf{H}$ such that $\mathbf{H}/F$ is bounded. There exists $a \in H$ such that $a/F = \min(\mathbf{H}/F)$. By (i), F is a filter of $\mathbf{A}(\mathbf{H})$. We claim that F is not a prime filter of $\mathbf{A}(\mathbf{H})$. Suppose, for contradiction, $\mathbf{A}(\mathbf{H})/F$ is a linearly ordered bounded hoop. Clearly, $\mathbf{H}/F$ is a filter of $\mathbf{A}(\mathbf{H})/F$ and $a/F \odot a/F = a/F$, since $a/F = \min(\mathbf{H}/F)$. It follows that $0/F \leq a/F \leq 1/F$ are a chain of Boolean elements of the bounded linear Wajsberg hoop $\mathbf{A}(\mathbf{H})/F$, see Proposition 3.3.45 (note that since $\mathbf{A}(\mathbf{H})/F$ is a bounded Wajsberg hoop, we have $\mathrm{Reg}(\mathbf{A}(\mathbf{H})/F) = \mathbf{A}(\mathbf{H})/F$). It follows that $a/F = 1/F$ or $a/F = 0/F$. The first one implies that $a \in F$, consequently, $H/F = 1/F$, that is $F = H$, a contradiction. Since H/F is a filter of $\mathbf{A}(\mathbf{H})/F$, the second one entails that $A(H)/F = H/F$, which leads us to a contradiction. Therefore, $F \notin \mathrm{Spec}(\mathbf{A}(\mathbf{H}))$.

Conversely, let $F \in \mathrm{Spec}(\mathbf{H})$ such that $F \notin \mathrm{Spec}(\mathbf{A}(\mathbf{H}))$. We claim that $\mathbf{H}/F$ is bounded. Suppose, to the contrary, that $\mathbf{H}/F$ is not bounded, then by (iii), $\mathbf{A}(\mathbf{H})/F$ is linearly ordered, consequently, by Theorem 4.2.11, F is a prime filter of $\mathbf{A}(\mathbf{H})$. This, however, contradicts the assumption. $\qquad\square$

Theorem 6.6.11 ([6]) *Let* $\mathbf{H}$ *be an unbounded Wajsberg hoop. Then the map*

$$\varphi : \mathrm{Spec}(\mathbf{A}(\mathbf{H})) \setminus \{H\} \to \mathrm{Spec}(\mathbf{H})$$

defined by $\varphi(P) = P \cap H$ is a bijection.

Proof Let $P \in \mathrm{Spec}(\mathbf{A}(\mathbf{H})) \setminus \{H\}$. Since H is a maximal filter of $\mathbf{A}(\mathbf{H})$, we have H is not a subset of P, so by Lemma 6.6.9, $P \cap H \in \mathrm{Spec}(\mathbf{H})$, which means φ is well-defined. First, we show that φ is onto. Choose a prime filter F of $\mathbf{H}$. There are two cases:

(i) If $\mathbf{H}/F$ is unbounded, then by Proposition 6.6.10(iii), F is a prime filter of $\mathbf{A}(\mathbf{H})$, consequently, $F \in \mathrm{Spec}(\mathbf{A}(\mathbf{H})) \setminus \{H\}$. Clearly, $\varphi(F) = F$.

(ii) If $\mathbf{H}/F$ is bounded, then there exists $a \in H$ such that $a/F = \min(\mathbf{H}/F)$. Clearly, $a' \notin H$, so $\langle H \cup \{a'\}\rangle = A(H)$. Let $P := \langle F \cup \{a'\}\rangle$ be the filter of $\mathbf{A}(\mathbf{H})$ generated by $F \cup \{a'\}$. For each $x \in A(H)$, by Corollary 2.3.10, there exist $n \in \mathbb{N}$ such that $(a')^n \to x \in H$. Since $a/F = \min(\mathbf{H}/F)$, we have $a/F \odot a/F = a/F$ which implies $a'/F \odot a'/F = a'/F$, see Proposition 3.3.41. It follows that $(a')^k/F = a'/F$ for every $k \in \mathbb{N}$, equivalently, $a' \to (a')^k \in F$ for each $k \in \mathbb{N}$. From $a' \to (a')^n \in F \subseteq H$, $(a')^n \to x \in H$ and Lemma 2.2.1(15) it follows that

$$a' \to x \in H, \quad \forall x \in A(H). \tag{6.12}$$

Analogously, since $P := \langle F \cup \{a'\}\rangle$, we can demonstrate that, $x \in P$ implies that $a' \to x \in F$ for each $x \in A(H)$.

Choose $x, y \in A(H)$ such that $x \vee y \in P$. By the above argument, $a' \to x \in H$ and $a' \to y \in H$. Furthermore, $a' \to (x \vee y) \in F$ and so by Proposition 3.3.37, $(a' \to x) \vee (a' \to y) \in F$. Since $F \in \mathrm{Spec}(\mathbf{H})$ and $a' \to x, a' \to y \in H$, by Corollary 4.2.9, $a' \to x \in F$ or $a' \to y \in F$, consequently, $x \in P$ or $y \in P$. Therefore, P is a prime ideal of $\mathbf{A}(\mathbf{H})$. Easy calculations show that $P \cap H = F$. Indeed, for every $x \in P \cap H, a' \to x \in F$. Also, due to $a/F \leq x/F$, we get $a \to x \in F$, consequently, $a/F \in \mathrm{Idm}(\mathbf{H}/F) \subseteq \mathrm{Idm}(\mathbf{A}(\mathbf{H})/F)$ entails that

$$\begin{aligned}
x/F = 1/F &\to x/F = (a/F \vee a'/F) \to x/F, \ \text{by Proposition 3.3.45}\\
&= (a/F \to x/F) \wedge (a'/F \to x/F), \ \text{by Lemma 2.2.3(iii)}\\
&= (a \to x)/F \wedge (a' \to x)/F = 1/F \wedge 1/F = 1/F.
\end{aligned}$$

Thus, $x \in F$, i.e., $\varphi(P) = P \cap H = F$.

It remains to show that φ is one-to-one. Let P and Q be elements of $\mathrm{Spec}(\mathbf{A}(\mathbf{H})) \setminus \{H\}$ such that $P \cap H = Q \cap H$. Set $F = P \cap H$. Clearly, F is a filter of $\mathbf{A}(\mathbf{H})$.

If $\mathbf{H}/F$ is unbounded, then by Proposition 6.6.10(iv), F is a prime filter of $\mathbf{A}(\mathbf{H})$. On the other hand, there exists a unique maximal filter M_Q of $\mathbf{A}(\mathbf{H})$ such that $Q \subseteq M_Q$, which implies both H and M_Q are maximal filters of $\mathbf{A}(\mathbf{H})$ containing prime filter F, consequently, $M_Q = H$ and so $Q = P$, see Corollary 4.2.29.

Now, assume that $\mathbf{H}/F$ is bounded. Similar to (ii) above, there exists $a \in A(H)$ such that $a/F = \min(\mathbf{H}/F)$ and a/F is an idempotent element of $\mathbf{A}(\mathbf{H})$. We show that $P = \langle F \cup \{a'\}\rangle = Q$. Since a/F is an idempotent element of the bounded Wajsberg hoop $\mathbf{A}(\mathbf{H})/F$, Proposition 3.3.41 implies that $a/F \vee a'/F = 1/F$, equivalently,

$a \vee a' \in F \subseteq P$. Corollary 4.2.9 entails that $a \in P$ or $a' \in P$. The first one is impossible, since $a/F = \min(\mathbf{H}/F)$ and from $a \in P$ we get $H/F \subseteq P/F$, consequently, $H \subseteq P$, which implies $H = P$, a contradiction. In conclusion, $a' \in P$, and so $\langle F \cup \{a'\} \rangle \subseteq P$. Now, choose $x \in P$. Then by (6.12), $a' \to x \in H$ and so $a' \to x \in P \cap H = F$, hence $a' \to x \in \langle F \cup \{a'\} \rangle$, see Corollary 2.3.10(ii). It follows that $P = \langle F \cup \{a'\} \rangle$. A similar argument shows that $Q = \langle F \cup \{a'\} \rangle$. Therefore, φ is one-to-one. $\square$

Corollary 6.6.12 *Let $\mathbf{H}$ be an unbounded Wajsberg hoop. If F is a maximal filter of $\mathbf{H}$, then F is either a prime filter of $\mathbf{H}$ or $F = P \cap H$ for some $P \in \mathrm{Max}(\mathbf{A}(\mathbf{H})) \setminus \{H\}$.*

Proof Assume that F is a maximal filter of $\mathbf{H}$. Then by Proposition 4.2.32 and Corollary 4.2.9, F is a prime filter. Due to Theorem 6.6.11, there exists $P \in \mathrm{Spec}(\mathbf{A}(\mathbf{H})) \setminus \{H\}$ such that $P \cap H = F$. By Proposition 4.2.28, there is a maximal filter Q of $\mathbf{A}(\mathbf{H})$ containing P. It follows that $F = P \cap H \subseteq Q \cap H \subseteq H$. So, by the assumption, $F = Q \cap H$ or $H = Q \cap H$. The first one implies that $P = Q$ (since φ is a bijection) and $F \in P \cap H$ for some $P \in \mathrm{Max}(\mathbf{A}(\mathbf{H})) \setminus \{H\}$. The second one entails $Q = H$, consequently, $P \subseteq H$ and $F = P \cap H = P$. Thus, F is a prime filter of $\mathbf{A}(\mathbf{H})$. $\square$

Bibliographical Remarks and Suggestions for Further Study

The results presented in Sect. 6.6 were originally established by Abad et al. [6]. They utilized free MV-algebras to demonstrate that $\mathbf{A}(\mathbf{H})$ is a bounded Wajsberg hoop. In contrast, we have employed an alternative approach based on the representation of Wajsberg hoops. Applying Theorem 3.3.13, $\mathbf{H}$ is either bounded (equivalent to an MV-algebra) or unbounded (isomorphic to an ultrafilter of an MV-algebra). Moreover, a non-trivial hoop is cancellative if it is isomorphic to an ultrafilter of a perfect MV-algebra. The results of this section reveal that every Wajsberg hoop $\mathbf{H}$ can be interpreted as a filter of an MV-algebra. While the category of MV-algebras is isomorphic to the category of bounded Wajsberg hoops, and the variety of MV-algebras (considered as bounded Wajsberg hoops) is strictly contained within WH, the subvariety of WH generated by MV-algebras is, in fact, WH, i.e., $\mathrm{HSP}(X) = \mathrm{WH}$. Recently, the relationship between hoops and Bézout domains has been investigated in [133]. A Bézout domain is an integral domain[1] where the sum of any two principal ideals is also a principal ideal. Applying the results from Sect. 6.6, a representation for Wajsberg hoops based on saturated multiplicative systems of Bézout domains was provided. For more detail see [132, 133, 262, 263]. The relationship between Wajsberg hoops and MV-algebras is profoundly deep. By leveraging this strong connection, Aglianò and Panti [15] characterized all free Wajsberg hoops in terms of McNaughton functions over the n-cube.

Let $n \in \mathbb{N}$ and $[0, 1]$ be the real unit interval. A **McNaughton function over the n-cube**, $[0, 1]^n$, [212, 222] is a functions $f : [0, 1]^n \to [0, 1]$ satisfying the following conditions:

---

[1] A commutative ring with a multiplicative identity 1 possesses no non-trivial zero divisors.

(i) f is continuous with respect to the natural topology of $[0, 1]^n$.

(ii) There are linear polynomials $p_1, \ldots, p_k$ with integer coefficients

$$p_i(x_0 \ldots, x_{n-1}) = b_i + m_{i0}x_0 + \cdots + m_{i(n-1)}x_{n-1}, \quad b_i, m_{it} \in \mathbb{Z},$$

such that for each point $y = (y_0, \ldots, y_{n-1}) \in [0, 1]^n$ there is an index $j \in \{1, \ldots k\}$ with $f(y) = p_j(y)$.

In addition, if λ is an infinite cardinal. A function $g : [0, 1]^\lambda \to [0, 1]$ is a **McNaughton function over** λ**-cube**, $[0, 1]^\lambda$, if there are ordinals $\alpha(0) < \alpha(1) < \cdots < \alpha(m-1) < \lambda$ and a McNaughton function f over $[0, 1]^m$ such that for each $x \in [0, 1]^\lambda$, $g(x) = f(x_{\alpha(0)}, \ldots, x_{\alpha(m-1)})$. It is well known that the free MV-algebra over λ generators is the algebra $\mathbf{F}_\lambda(\mathrm{MV})$ of all McNaughton functions over the λ-cube (see [73]). If $\mathbf{0}$ (respectively, $\mathbf{1}$) denotes the element of $[0, 1]^\lambda$ all of whose coordinates are 0 (respectively, 1), then we set

$$\mathbf{F}_\lambda(\mathrm{WH}) := \{f \in \mathbf{F}_\lambda(\mathrm{MV}) : f(\mathbf{1}) = 1\}.$$

Then $\mathbf{F}_\lambda(\mathrm{WH})$ forms a Wajsberg hoop which is a subhoop of the bounded Wajsberg hoop corresponding to the MV-algebra $\mathbf{F}_\lambda(\mathrm{MV})$ (see Theorem 3.3.13). In addition, we have

Theorem 6.6.13 ([15, Theorem 3.1]) $\mathbf{F}_\lambda(\mathrm{WH})$ *is the free Wajsberg hoop over* λ *generators.*

6.7 Variety of Hoops

We begin by recalling some fundamental notions from previous chapters. According to Theorem 1.6.4, and in conjunction with Theorem 3.1.2, the class of hoops, denoted H, forms a variety. Furthermore, by Proposition 3.3.5, Eqs. (3.6), (3.10), Definition 3.3.39, and equation (PH), the class of cancellative hoops CH, the class of Wajsberg hoops WH, the class of basic hoops BH, the class of idempotent hoops IH, and the class of product hoops PH are also variety, respectively. In this section, we will study the class of hoops applying the operators such as I, H, S, P, and P_u and investigate some properties of H.

Let us commence with the statement of two basic properties of the variety of hoops. In light of Theorem 4.1.14, which establishes a one-to-one correspondence between congruence relations and filters, we will show that the variety H satisfies the congruence extension property.

Proposition 6.7.1 ([31, Theorem 1.8]) *The variety of hoops has the congruence extension property.*

Proof Let $\mathbf{H}$ be a hoop and $\mathbf{K}$ be a subhoop of $\mathbf{H}$. By Theorem 4.1.14, it suffices to show that for every filter F of $\mathbf{K}$ there exists a filter $\bar{F}$ of $\mathbf{H}$ such that $\bar{F} \cap K = F$. Let F be a filter of $\mathbf{K}$ and $\bar{F}$ be the filter of $\mathbf{H}$ generated by F, i.e., $\bar{F} = \langle F \rangle$. Clearly, $F \subseteq \bar{F} \cap K$. To see the converse, let $x \in \bar{F} \cap K$. Due to (4.2), there exist $a_1, a_2, \ldots, a_n \in F$ such that $a_1 \odot a_2 \odot \cdots \odot a_n \leq x$. Since $x \in K$ and F is a filter of $\mathbf{K}$, it follows that $x \in F$. $\qquad \square$

Proposition 6.7.2 ([153, Proposition 3.16]) *The variety of hoops is arithmetical.*

Proof Let us consider the following ternary terms:

$$p(a, b, c) = \big((a \to b) \to c\big) \wedge \big((c \to b) \to a\big),$$
$$m(a, b, c) = \big((b \to a) \to a\big) \wedge \big((c \to b) \to b\big) \wedge \big((a \to c) \to c\big).$$

Let $\mathbf{H}$ be a hoop and $x, y \in H$.

$$\begin{aligned} p(x, x, y) &= \big((x \to x) \to y\big) \wedge \big((y \to x) \to x\big) \\ &= (1 \to y) \wedge \big((y \to x) \to x\big) \\ &= y \wedge \big((y \to x) \to x\big) = y, \end{aligned}$$

and

$$p(x, y, y) = \big((x \to y) \to y\big) \wedge \big((y \to y) \to x\big) = \big((x \to y) \to y\big) \wedge x = x,$$

since by Lemma 2.2.1(6), $y \leq (y \to x) \to x$ and $x \leq (x \to y) \to y$. We also have

$$\begin{aligned} m(x, x, y) &= \big((x \to x) \to x\big) \wedge \big((y \to x) \to x\big) \wedge \big((x \to y) \to y\big) \\ &= x \wedge \big((y \to x) \to x\big) \wedge \big((x \to y) \to y\big) = x, \end{aligned}$$

since $x \leq (y \to x) \to x$, and $x \leq (x \to y) \to y$. We prove similarly that $m(x, y, x) = m(y, x, x) = x$. Now, Theorem 1.6.7 implies that the variety of hoops is arithmetical. $\quad \square$

Proposition 6.7.3 ([31, Proposition 1.14]) *Let $\mathbf{H}$ be a Wajsberg hoop and $\mathbf{H}_a$ be the hoop was defined in Proposition 3.2.5 for every $a \in H$. Then $\mathbf{H} \in \mathrm{ISP}_\mathrm{u}\,(\{\mathbf{H}_a : a \in H\})$.*

Proof Let $J_a := \,\downarrow a$ for any $a \in H$. Clearly, for $a_1, \ldots, a_n \in H, \bigcap_{i=1}^{n} J_{a_i} = J_a$, where $a = \bigwedge_{i=1}^{n} a_i$, which means $\{J_a : a \in H\}$ has finite intersection property. Let U be an ultrafilter of $(\mathcal{P}(H), \leq)$ containing $\{J_a : a \in H\}$ and $\mathbf{B} = (\prod_{a \in H} \mathbf{H}_a)/U$ be the ultraproduct of the family $\{\mathbf{H}_a : a \in H\}$ (see Definition 1.5.25). Define $f : \mathbf{H} \to \mathbf{B}$ by

$$f(x) = \frac{(x \vee a)_{a \in H}}{U}, \quad \forall x \in H.$$

Let $x, y \in H$. For every $a \leq x \wedge y$ we have $(x \to y) \vee a = x \to y = (x \vee a) \to (y \vee a)$, which implies

$$\downarrow a \subseteq [[\big((x \to y) \wedge a\big)_{a \in H} = \big((x \vee a) \to (y \vee a)\big)_{a \in H}]].$$

Since $\downarrow a \in U$ and U is a ultrafilter of $(\mathcal{P}(\mathbf{H}), \subseteq)$, we get

$$[[\big((x \to y) \vee a\big)_{a \in H} = \big((x \vee a) \to (y \vee a)\big)_{a \in H}]] \in U,$$

which entails $f(x \to y) = f(x) \to f(y)$. In a similar way, we can show that

$$\downarrow a \subseteq [[\big((x \odot y) \vee a\big)_{a \in H} = \big((x \vee a) \odot (y \vee a)\big)_{a \in H}]],$$

where $a = x \odot y$. In conclusion, $f(x \odot y) = f(x) \odot f(y)$. Thus, f is a homomorphism of hoops. In addition, f is one-to-one. Indeed, if $x \neq y$ such that $f(x) = f(y)$, then either $x < x \vee y$ or $y < x \vee y$. Without loss of generality, assume that $x < x \vee y$. We have $X := [[(x \vee a)_{a \in H} = (y \vee a)_{a \in H}]] = \{a \in H : x \vee a = y \vee a\} \in U$. Clearly, X is an upper set of H containing $a = x \vee y$, so $\uparrow a \subseteq X$, that is $\uparrow a \in U$. On the other hand, by the assumption, $\downarrow x \in U$, consequently, $\emptyset = (\uparrow a) \cap (\downarrow x) \in U$, a contradiction. This leads us to $x = y$. Therefore, $\mathbf{H} \in \mathrm{ISP_u}\,(\{\mathbf{H}_a : a \in H\})$. $\qquad\square$

The next corollary follows from Proposition 6.7.3, directly.

Corollary 6.7.4 ([31, Corollary 1.14]) $\mathbf{L}_\infty \in \mathrm{ISP_u}\,(\{\mathbf{L}_n : n \in \mathbb{N}\})$.

Proof Let $\mathbf{H} := \mathbf{L}_\infty$. For every $n \in \mathbb{N}$, we have $\mathbf{L}_n = \mathbf{H}_n$. Hence, by Proposition 6.7.3, the proof is obvious. $\qquad\square$

Given a class K of algebras, K_{SI} and K_F denote the classes of subdirectly irreducible and finite members of K, respectively. In addition, for every hoop $\mathbf{H}$, $\mathrm{V}(\mathbf{H})$ and $\mathrm{Q}(\mathbf{H})$ denote the variety generated by $\mathbf{H}$ and the quasivariety generated by $\mathbf{H}$, respectively, i.e., $\mathrm{V}(\mathbf{H}) = \mathrm{HSP}(\{\mathbf{H}\})$ and $\mathrm{Q}(\mathbf{H}) = \mathrm{ISPP_u}(\{\mathbf{H}\})$.

Remark 6.7.5 (i) Combining Corollary 1.5.23 and Theorem 6.6.4, we conclude that $\mathrm{V}(\mathbf{H}) = \mathrm{WH}$, where $\mathbf{H}$ is the bounded Wajsberg hoop $\mathbb{R}[1]$, see [15, p. 361].

(ii) A easy application of Proposition 1.5.22 and Theorem 3.3.20 shows that $\mathrm{V}(\mathbf{C}) = \mathrm{V}(\mathbf{N}(\mathbb{Z}))$ for every non-trivial cancellative hoop $\mathbf{C}$. In addition, the variety of cancellative hoops, CH, is generated, as a variety by $\mathbf{N}(\mathbb{Z})$ (see [20] or [141, Remark 4.11]). Therefore, $\mathrm{CH} = \mathrm{V}(\mathbf{C})$ for every non-trivial cancellative hoop $\mathbf{C}$.

Lemma 6.7.6 ([73, Proposition 3.5.3]) *Let $\mathbf{H}$ be a subhoop of $\mathbb{R}[1]$, $H^+ := \{x \in H : 0 < x\}$, and $a := \inf(H^+)$.*

(i) *If $a = 0$, then H is a dense subset of $[0, 1]$.*
(ii) *If $0 < a$, then $\mathbf{H} \cong \mathbf{L}_n$ for some $n \in \mathbb{N} \cup \{0\}$.*

Proof First, we note that due to the proof of Corollary 6.2.6, every non-trivial subhoop of $\mathbb{R}[1]$ contains 0 and so it is a bounded subhoop of $\mathbb{R}[1]$. Indeed, for every $x \in H \setminus \{1\}$ there exists $n \in \mathbb{N}$ such that $0 = \overbrace{x \odot \cdots \odot x}^{n\text{-times}} \in H$.

(i) Assume that $a = 0$. Let $x \in (0, 1]$ and $0 < \epsilon < x/2$. Then there exists $y \in H^+$ such that $0 < y < \epsilon$. Suppose that n is the smallest integer such that $x \leq ny$, noting that $n \geq 2$. It follows that $x - \epsilon \leq (ny) - \epsilon < (ny) - y = (n - 1)y < x$, and so H is dense on $[0, 1]$. Note that by definition of $\odot$ on $\mathbb{R}[1]$, we have $(n - 1)y \in H$.

(ii) Assume that $0 < a$. If $a = 1$, then $H^+ = \{1\}$. There exist two cases, if $0 \in H$, then $H = \{0, 1\}$, which implies $\mathbf{H} = \mathbf{L}_2$. If $0 \notin H$, then $H = H^+ = \{1\}$ and so $\mathbf{H} = \mathbf{L}_0$.

Now, assume that $a < 1$. Then there exists $x \in H^+$ such that $x < 1$, which implies $0 \in H$. It follows that $1 - x = x \to 0 \in H^+$, which entails, $a \leq 1/2$, consequently, $a + a \in [0, 1]$. We claim that $a \in H^+$. For otherwise, there are $x, y \in H^+$ such that $a < x < y < a + a$ (since $a = \inf(H^+)$). We have $a = 2a - a > y - x = y \odot (1 - x) = y \odot x' \in H^+$, a contradiction. Thus, $a \in H^+$. Let m be the unique integer such that $(m - 1)a < 1 \leq ma$. Note that $m \geq 2$. By the definition of $\odot$ on $\mathbb{R}[1]$, we get

$$2a = \left(a - (1 - a) + 1\right) \wedge 1 = a' \to a \in H$$
$$3a = \left(2a - (1 - a) + 1\right) \wedge 1 = a' \to (2a) \in H$$
$$\vdots$$
$$(m - 1)a = \left((m - 2)a - (1 - a) + 1\right) \wedge 1 = a' \to ((m - 2)a) \in H.$$

Set $K := \{0, a, 2a, \ldots, (m - 1)a, 1\}$. We show that $H = K$. Otherwise, choose $x \in H \setminus K$ (absurdum hypothesis).

If $(m - 1)a < x < 1$, then $x' = 1 - x < 1 - (m - 1)a \leq a$, which is impossible, since $x' \in H^+$.

If $ja < x < (j + 1)a$ for some $j \in \{0, 1, \ldots, m - 2\}$, then $a = (j + 1)a - ja > (j + 1)a - x = ((j + 1)a) \odot x' \in H^+$, another contradiction. Hence, $H = K$. Since $\mathbf{H}$ is a bounded subhoop of $\mathbb{R}[1]$, we have $1 - ((m - 1)a) = ((m - 1)a)' \in H$, consequently, $a = 1 - ((m - 1)a)$, whence $a = 1/m$. Therefore, $\mathbf{H} = \mathbf{L}_m$, as required. $\square$

Theorem 6.7.7 ([15, 73]) *If* $\mathbf{H}$ *is an infinite subhoop of* $\mathbb{R}[1]$, *then* $V(\mathbf{H}) = \mathsf{WH}$.

Proof Let $\mathbf{H} = (H; \odot, \to, 0, 1)$ be an infinite subhoop of $\mathbb{R}[1]$ and $V(\mathbf{H})$ be the variety generated by $\mathbf{H}$. We have

$$x \odot y = (x + y - 1) \vee 0 \quad \& \quad x \to y = (y - x + 1) \wedge 1 \quad \& \quad x' = 1 - x, \quad \forall x, y \in [0, 1].$$

The operations $\odot$ and $\to$ as well as $'$ are continuous over $[0, 1]$. Thus, for every term $p(x_1, \ldots, x_n)$ of type $\{\odot, \to\}$, $p : [0, 1]^n \to [0, 1]$ is a continuous function. By Lemma 6.7.6(i), H is dense in $[0, 1]$, so the identity $p(x_1, \ldots, x_n) = 0$ holds in $\mathbb{R}[1]$ if and only if it holds on $\mathbf{H}$. Hence, by Birkhoff's Theorem 1.6.4, $V(\mathbb{R}[1]) = V(\mathbf{H})$. Now, Remark 6.7.5(i) implies that $WH = V(\mathbf{H})$.

We note that, since $(x \to y) \wedge (y \to x) = 1$ if and only if $x = y$, every term $p(x_1, \ldots, x_n) = q(x_1, \ldots, x_n)$ of type $\{\odot, \to\}$ on WH can be consider as a term $t(x_1, \ldots, x_n) = 0$. $\qquad\square$

Theorem 6.7.8 ([31, Theorem 3.1]) *The minimal varieties of hoops are precisely the minimal varieties of Wajsberg hoops, namely* $V(\mathbf{L}_1)$ *and* $V(\mathbf{L}_\infty)$.

Proof Let $V \subseteq H$ be a non-trivial variety. Then V contains a subdirectly irreducible hoop $\mathbf{H}$ and, by Theorem 6.5.6, $\mathbf{H} = \mathbf{Fix}_{U_*}(\mathbf{H}) \oplus \mathbf{S}$, with $\mathbf{S}$ as a subdirectly irreducible Wajsberg hoop. Thus, V contains $V(\mathbf{S})$ (since $\mathbf{S} \leq \mathbf{H}$), and hence a non-trivial variety of Wajsberg hoops. Therefore, V contains a minimal variety of Wajsberg hoops.

It remains to show that the only two minimal varieties of Wajsberg hoops are $V(\mathbf{L}_1)$ and $V(\mathbf{L}_\infty)$. Let V be a non-trivial variety of Wajsberg hoops and $\mathbf{H}$ be a subdirectly irreducible member of V with the least non-trivial filter U_*. Then U_* is a subuniverse of $\mathbf{H}$, which is a simple totally ordered Wajsberg hoop, see Proposition 6.5.1. Hence, V contains a simple totally ordered hoop $\mathbf{U}_* = (U_*; \odot, \to, 1)$. If U_* has a least element, say 0, then it is easy to check that $\{0, 1\}$ is the universe of a subhoop of $\mathbf{U}_*$ isomorphic to $\mathbf{L}_1$. If U_* does not contain a least element, i.e., it is unbounded, choose $x \in U_* \setminus \{1\}$. We show that the one-generated monoid $X := \{1, x, x^2, x^3, \cdots\}$ is the universe of a subhoop of $\mathbf{U}_*$, isomorphic to $\mathbf{L}_\infty$. Since $\mathbf{U}_*$ is simple, for every element $y \in U_*$ there exists $n \in \mathbb{N}$ such that $x^n \leq y$. This fact, together with the unboundedness of $\mathbf{U}_*$ implies $1 > x > x^2 > \cdots > x^n > \cdots$.

To see that X is closed under implication, it suffices to show that for any natural number n and k, if $n < k$, then $x^n \to x^k = x^{k-n}$; indeed, for $n \geq k$, $x^n \leq x^k$ and so $x^n \to x^k = 1$. Given any $n, k \in \mathbb{N}$ with $n < k$, we have $x^n \odot x^{k-n} = x^k$, and so $x^{k-n} \leq x^n \to x^k$, by Lemma 2.2.1(9). Since $x^k > x^{k+1}$, we have $x^k = x^n \odot x^{k-n} \not\leq x^{k+1}$ and hence, $x^{k-n} \not\leq x^n \to x^{k+1}$. The total order in $\mathbf{H}$ implies that $x^n \to x^{k+1} < x^{k-n}$. But $\mathbf{U}_*$ is a Wajsberg hoop, i.e., it satisfies (3.6), and so for all $a, b \in U_*$, if $a \leq b$, then $b = (b \to a) \to a$. In particular, $x^n \to x^{k+1} < x^{k-n}$ implies that

$$
\begin{aligned}
x^{k-n} &= ((x^{k-n} \to (x^n \to x^{k+1})) \to (x^n \to x^{k+1}) \\
&= ((x^{k-n} \odot x^n) \to x^{k+1}) \to (x^n \to x^{k+1}), \text{ by Theorem 3.1.2(H2)} \\
&= (x^k \to x^{k+1}) \to (x^n \to x^{k+1}) \\
&= x^n \to ((x^k \to x^{k+1}) \to x^{k+1}), \text{ by Theorem 3.1.2(H2)} \\
&= x^n \to x^k, \text{ by (3.6) and since } x^{k+1} \leq x^k.
\end{aligned}
$$

This conclude the proof that X is a countable subuniverse of $\mathbf{U}_*$, isomorphic to $\mathbf{L}_\infty$. $\qquad\square$

Let $V \subseteq H$ be a variety. A variant of Mal'sev's varietal product construction [209] allows us to construct V^{n+1} from V^n. Define a sequence of varieties of hoops, $V^1 \subseteq V^2 \subseteq \cdots \subseteq V^n \subseteq \cdots \subseteq V^+$, where $V^+ = \bigvee_{n \in \mathbb{N}} V^n$ and the join is here computed in the lattice of sub-quasivarieties of H.

Definition 6.7.9 ([31, Definition 3.2]) Let V be a variety of Wajsberg hoops. For any $n \in \mathbb{N}$, define the varietal power of V, V^n, as follows:

(i) V^0 is the trivial variety of hoops.
(ii) $V^{n+1} = \mathrm{HSP}((V^n)^*)$, where $(V^n)^*$ is the class of algebras $\mathbf{B} \oplus \mathbf{C}$ such that $\mathbf{B} \in V^n$ and $\mathbf{C} \in V$, $\mathbf{C}$ totally ordered. Finally $V^+ = \mathrm{ISPP_u}(\bigvee_{n \in \mathbb{N}} V^n)$.

In what follows, an n-**generated** hoop refers to a hoop generated by n elements, i.e., $\mathbf{H} = \mathrm{Alg}(X)$ for some n-element subset X of H.

Proposition 6.7.10 ([31, Proposition 3.3]) *Let* $K \subseteq H$ *be any variety and* $\mathbf{H} \in K$ *be an* n-*generated hoop for some* $n \in \mathbb{N}$. *Then* $\mathbf{H} \in V^n$, *where* $V = K \cap WH$.

Proof We proceed by induction on n. If $n = 0$, $\mathbf{H}$ is the 1-element hoop, that is $\mathbf{H} \cong \mathbf{1}$ and the claim holds. Next, assume $n \in \mathbb{N}$ and the claim holds for all $k < n$. To show every n-generated hoop in W belongs to V^n, it suffices to show every n-generated subdirectly irreducible algebra in K belongs to V^n. So, let $\mathbf{H}$ be such an algebra, and U_*, $\mathbf{Fix}_{U_*}(\mathbf{H})$ and $\mathbf{S}$ as in Theorem 6.5.6. Then $\mathbf{H} = \mathbf{Fix}_{U_*}(\mathbf{H}) \oplus \mathbf{S}$ and $\mathbf{S} \in (K \cap WH)_{SI} = V_{SI}$. Let X be a set of generators of $\mathbf{H}$, $|X| = n$. Since $\{1\} \neq U_* \subseteq \mathbf{S}$, and $\mathrm{Fix}_{U_*}(\mathbf{H})$ is a subuniverse of $\mathbf{H}$, we have $X \cap (S \setminus \{1\}) \neq \emptyset$. Let $f : \mathbf{H} \to \mathbf{Fix}_{U_*}(\mathbf{H})$ be the map is defined by

$$f(x) = \begin{cases} 1 & \text{if } x \in S \\ x & \text{if } x \in \mathrm{Fix}_{U_*}(\mathbf{H}). \end{cases}$$

By Proposition 6.1.5, f is an onto homomorphism and by the first isomorphism theorem we obtain $\mathbf{Fix}_{U_*}(\mathbf{H}) \cong \mathbf{H}/\mathbf{S}$. Thus, $\mathbf{Fix}_{U_*}(\mathbf{H}) \in K$, and $\mathbf{Fix}_{U_*}(\mathbf{H})$ is generated by $f(X)$. Since 1 does not contribute to the generation, $\mathbf{Fix}_{U_*}(\mathbf{H})$ is generated by $\{f(x) \mid x \in X, f(x) \neq 1\}$ as well, which is contained in $\{f(x) \mid x \in X, x \notin S\}$. This last set has fewer than n elements and, by the induction hypothesis, it follows that $\mathbf{Fix}_{U_*}(\mathbf{H}) \in V^{n-1}$. From $\mathbf{S} \in V_{SI}$ it follows that $\mathbf{H} \in \left(V^{n-1}\right)^* \subseteq V^n$. $\qquad\square$

Theorem 6.7.11 ([31, Theorem 3.4]) *Let* $k \in \mathbb{N}$ *and* $H(k)$ *(respectively,* $WH(k)$*) denotes the subvariety of* H *(WH) is generated by* k-*generated hoops (Wajsberg hoops). Then*

(i) $H = (WH)^+$.
(ii) $H(k) = (WH(k))^+$, *for all* $k \in \mathbb{N}$.

Proof Note that if $K = H$ or $K = H(k)$, for some $k \in \mathbb{N}$, and $V = K \cap WH$, then $V^+ \subseteq V$. Since any quasivariety is generated by its finitely generated members, the inclusion $K \subseteq V^+$ follows immediately from Proposition 6.7.10. $\qquad\square$

We will now demonstrate that for every $k \in \mathbb{N}$, the varieties $H(k)$ as well as H, are generated by their finite algebras. We will utilize the following lemma. First, we recall Jónsson's Lemma. Essentially, Jónsson's Lemma states that if a variety is congruence-distributive, then its subdirectly irreducible members are closely related to ultraproducts of the algebras that generate the variety. More precisely, they are within HSP_u of the generating class of algebras.

Lemma 6.7.12 ([31, Lemma 3.5]) *Let* $V \subseteq KH$ *be any variety and* $n \in \mathbb{N}$. *Then* $\left(V^{n+1}\right)_{SI} \subseteq (V^n)^*$.

Proof We argue by induction. The claim clearly holds for $n = 0$. Now, assume it holds for $k \leq n$ and let $\mathbf{H} \in \left(V^{n+1}\right)_{SI}$. Since by Proposition 6.7.2, $\mathbf{H}$ is congruence distributive, we may apply Jónsson's Lemma and so $\mathbf{H} \in HS(\mathbf{D})$, where $\mathbf{D} \in P_U\left((V^n)^*\right)$. Let

$$\mathbf{D} = \frac{\prod_{i \in I}(\mathbf{B}_i \oplus \mathbf{C}_i)}{F},$$

where $\mathbf{B}_i \in V^n$, $\mathbf{C}_i \in V$ is totally ordered, and F is a non-principal ultrafilter of $(\mathcal{P}(I), \subseteq)$. By properties of ultraproducts, we see $\mathbf{D} \cong \mathbf{B} \oplus \mathbf{C}$, where $\mathbf{B} \cong (\prod_{i \in I} \mathbf{B}_i)/F$, and $\mathbf{C} \cong (\prod_{i \in I} \mathbf{C}_i)/F$. Indeed, $B = \{(x_i)_{i \in I} \in D \colon \{i \in I \mid x_i \in B_i\} \in F\}$ and $C = \{(x_i)_{i \in I} \in D \colon \{i \in I \mid x(i) \in C_i\} \in F\}$. Thus, $\mathbf{B} \in V^n$, $\mathbf{C} \in V$ and $\mathbf{C}$ is linearly ordered. Now, $\mathbf{H} \in HS(\mathbf{D})$ and it is easy to see that $\mathbf{H} \cong \mathbf{B}' \oplus \mathbf{C}'$, where $\mathbf{B}' \in HS(\mathbf{B})$, $\mathbf{C}' \in HS(\mathbf{C})$. Then $\mathbf{B}' \in V^n$, $\mathbf{C}' \in V$, $\mathbf{C}'$ is linearly ordered and hence $\mathbf{H} \in (V^n)^*$. $\qquad\square$

It is evident from the lemma that in the definition of V^{n+1} in Definition 6.7.9 the operator HSP can be replaced by ISP.

Recall that a variety V of algebras is said to be **locally finite** if every finitely generated member of V is finite.

Theorem 6.7.13 ([31, Theorem 3.6]) $H(k)$ *is locally finite, for all* $k \in \mathbb{N}$. *In particular,* $H(k)$ *is generated by its finite members.*

Proof By Proposition 6.7.10, every finitely generated algebra in $H(k)$ belongs to $(WH(k))^n$, for some $n \in \mathbb{N}$, thus, it suffices to show $(WH(k))^n$ is locally finite, for each $n < \omega$. The case $n = 0$ is trivial. Now, assume that $(WH(k))^n$ is locally finite, for some $n \in \mathbb{N}$. If $\mathbf{H} \in (WH(k))^{n+1}$ is an m-generated subdirectly irreducible hoop, light of the preceding lamma, $\mathbf{H} \cong \mathbf{B} \oplus \mathbf{C}$, with $\mathbf{B} \in (WH(k))^n$ and $\mathbf{C} \in WH(k)$ totally ordered. Moreover, since

$\mathbf{H}$ is subdirectly irreducible, so is $\mathbf{C}$ and hence $\mathbf{C} \in (\mathrm{WH}(k))_{SI}$. Since $\mathbf{B} \in H(\mathbf{H})$, $\mathbf{B}$ is m-generated as well, so $|B| \leq |\mathbf{F}_{(\mathrm{WH}(k))^n}(m)| =: N$. Furthermore, $\mathbf{C} \cong \mathbf{C}_l$, for some $l \leq k$, so $|B| \leq N + k$. We have thus established a uniform upper bound on the size of the m-generated subdirectly irreducible hoops in $(\mathrm{WH}(k))^{n+1}$, which implies that $(\mathrm{WH}(k))^{n+1}$ is locally finite. $\qquad \square$

Note that the upper bound on the size of the m-generated subdirectly irreducibles can be sharpened by using an argument similar to that used in the proof of Proposition 6.7.10.

Clearly, the variety H of all hoops is not locally finite. In fact, the variety of Wajsberg hoops is not locally finite, as $\mathbf{L}_\infty$ is 1-generated. However, H is generated by its finite algebras, as we will see now.

Let $\mathbf{P} = \langle P, \odot^{\mathbf{P}}, \to^{\mathbf{P}}, 1^{\mathbf{P}} \rangle$ be a partial algebra of type $(2, 2, 0)$, thus $\odot^{\mathbf{P}}, \to^{\mathbf{P}}$ are partial binary operations on P, and we will assume $1^{\mathbf{P}} \in P$. We call $\mathbf{P}$ a **partial hoop** if there is a hoop $\mathbf{H} = (H; \odot, \to, 1)$ such that $P \subseteq H$, $1^{\mathbf{P}} = 1$, and $\odot^{\mathbf{P}}$ and $\to^{\mathbf{P}}$ are the restrictions of $\odot$ and $\to$, respectively to P whenever those are defined. Then, $\mathbf{P}$ is said to be a **partial subhoop** of $\mathbf{H}$.

Definition 6.7.14 We say that a class K of hoops has the **finite embeddability property**, (FEP), if for every finite partial subhoop $\mathbf{P}$ of some member $\mathbf{H}$ of K there is a finite member $\mathbf{B}$ of K such that $\mathbf{P}$ is a partial subhoop of $\mathbf{B}$. If we denote the operation of "forming partial subhoops" by S_p, then a class K of hoops has (FEP) if $(\mathsf{S}_P(\mathsf{K}))_F = \mathsf{S}_P(\mathsf{K}_F)$.

T. Evans proved that if a variety V possesses (FEP), then $\mathsf{V} = \mathrm{HSP}(\mathsf{V}_F)$ [139, Theorem 4]. His argument can be readily adapted to demonstrate that, in fact, $\mathsf{V} = \mathrm{ISPP_u}(\mathsf{V}_F)$ (see [141, Theorem 5.2]).

Lemma 6.7.15 ([31, Lemma 3.7]) *Let* V *be a variety. If* V_{SI} *has* (FEP), *then so does* V.

Proof Let $\mathbf{H} \in \mathsf{V}$ and $\mathbf{P}$ be a finite partial subalgebra of $\mathbf{H}$. If $f : \mathbf{H} \to \prod_{i \in I} \mathbf{H}_i$ be a subdirect representation of $\mathbf{H}$, where $\mathbf{H}_i \in \mathsf{V}_{SI}, i \in I$, then $(\pi_i \circ f)(P)$ is the universe of a partial subhoop of $\mathbf{H}_i$, and by the assumption also of a finite subdirectly irreducible algebra in V, say $\mathbf{H}'_i$. Now, $f|_P : \mathbf{P} \to \prod_{i \in I} \mathbf{H}'_i$ is an embedding that preserves all existing operations. Since $\mathbf{P}$ is finite, there is a finite subset $I' \subseteq I$ such that $\pi_{I'} \circ f|_P : \mathbf{P} \to \prod_{i \in I} \mathbf{H}'_i$ yields the desired embedding into a finite algebra of V. $\qquad \square$

In the sequel, we will demonstrate that H possesses (FEP). To this end, we state the following lemma without proof. For a detailed proof, we refer the reader to Lemma 2.4.14 of [141].

Lemma 6.7.16 ([141, Lemma 2.4.14]) *Let* $\mathbf{C}$ *be a linearly ordered Wajsberg hoop. For every finite subset* X *of* C, *there exist a natural number* m *and a one-to-one map* $f : X \to L_m$ *satisfying the following conditions for all* $x, y \in X$:

(i) *If* $x \odot y \in X$, *then* $f(x) \odot f(y) = f(x \odot y)$.

(ii) *If* $x \to y \in X$, *then* $f(x) \to f(y) = f(x \to y)$.

In other words, $\mathbf{X}$ *is a partial subalgebra of* $\mathbf{C}$ *and* f *is an embedding of* $\mathbf{X}$ *to* $\mathbf{C}$.

Theorem 6.7.17 ([141, Lemma 2.5.4]) *The variety* H *of all hoops has* (FEP).

Proof Let $\mathbf{H}$ be a hoop and $\mathbf{H}'$ be a finite partial subhoop of $\mathbf{H}$ with $H' = \{a_1, \dots, a_n\}$. We induct on n. If $n = 1$, then $g(a_1) = 1$ defines a monomorphism from $\mathbf{H}'$ into the subhoop of $\mathbf{H}$ with universe $\{1\}$. Now, let $n \geq 2$ and for each distinct elements i and j of $\{1, \dots, n\}$, θ_{ij} be a maximal congruence on $\mathbf{H}$ for which $(a_i, a_j) \notin \theta_{ij}$. Then there is a natural monomorphism $\mu : \mathbf{H}' \to \prod_{1 \leq i < j \leq n} \mathbf{H}/\theta_{ij}$. Each of the hoops $\mathbf{H}/\theta_{ij}$ is subdirectly irreducible. Hence, by Theorem 6.5.7, $\mathbf{H}/\theta_{ij} = \mathbf{B}_{ij} \oplus \mathbf{C}_{ij}$, where $\mathbf{B}_{ij}$ is a subhoop of $\mathbf{H}/\theta_{ij}$ and $\mathbf{C}_{ij}$ is a subdirectly irreducible linear Wajsberg hoop. If $\pi_{ij} : \mathbf{H} \to \mathbf{H}/\theta_{ij}$ denotes the natural projection of $\mathbf{H}$ onto $\mathbf{H}/\theta_{ij}$, then, by the condition on θ_{ij}, $\pi_{ij}(a_i)$ and $\pi_{ij}(a_j)$ are distinct elements of C_{ij}. It follows that $\pi_{ij}(\mathbf{H}') \cap \mathbf{B}_{ij}$ is a finite partial subhoop of $\mathbf{B}_{ij}$ with fewer than n elements. If $\pi_{ij}(H') \cap B_{ij}$ is empty, then $\pi_{ij}(\mathbf{H}') = \mathbf{C}_{ij}$ and this case reduces to Lemma 6.7.16. Otherwise, by hypothesis of induction, there exist a finite hoop $\mathbf{D}_{ij}$ and a one-to-one homomorphism $g_{ij} : \pi_{ij}(\mathbf{H}') \cap \mathbf{B}_{ij} \to \mathbf{D}_{ij}$. Also, $\pi_{ij}(\mathbf{H}') \cap \mathbf{C}_{ij}$ is a finite partial subhoop of $\mathbf{C}_{ij}$ and thus, by Lemma 6.7.16, there exist a finite linearly ordered Wajsberg hoop $\mathbf{C}_{m(i,j)}$ and a monomorphism $h_{ij} : \pi_{ij}(\mathbf{H}') \cap \mathbf{C}_{ij} \to \mathbf{C}_{m(i,j)}$. Define $f_{ij} : \pi_{ij}(\mathbf{H}') \to \mathbf{D}_{ij} \oplus \mathbf{C}_{m(i,j)}$ by

$$f_{ij}(x) = \begin{cases} g_{ij}(x) & \text{if } x \in B_{ij} \setminus \{1\} \\ h_{ij}(x) & \text{if } x \in C_{ij}. \end{cases}$$

Clearly, f_{ij} is a well-defined one-to-one map. In order to verify that it is also a homomorphism, it suffices to consider $x \in (\pi_{ij}(H') \cap B_{ij}) \setminus \{1\}$ and $y \in \pi_{ij}(H') \cap C_{ij}$. Then, by the definition of the operations of $\odot$ and $\to$ in the ordinal sums $\mathbf{B}_{ij} \oplus \mathbf{C}_{ij}$ and $\mathbf{D}_{ij} \oplus \mathbf{C}_{m(i,j)}$ we obtain

$$f_{ij}(x \to y) = 1 = g_{ij}(x) \to h_{ij}(y) = f_{ij}(x) \to f_{ij}(y).$$
$$f_{ij}(y \to x) = f_{ij}(x) = h_{ij}(y) \to g_{ij}(x) = f_{ij}(y) \to g_{ij}(x).$$
$$f_{ij}(x \odot y) = f_{ij}(x) = g_{ij}(x) \odot h_{ij}(y) = f_{ij}(x) \odot f_{ij}(y).$$

Hence, f_{ij} is a monomorphism from the finite partial subhoop $\pi_{ij}(\mathbf{H}')$ of $\mathbf{B}_{ij} \oplus \mathbf{C}_{ij}$ into the finite hoop $\mathbf{D}_{ij} \oplus \mathbf{C}_{m(i,j)}$. By composition of the monomorphism $\mu : \mathbf{H}' \to \prod_{1 \leq i < j \leq n} \mathbf{H}/\theta_{ij}$ with the product monomorphism

$$f : \prod_{i<j} \pi_{ij}(\mathbf{H}') \to \prod_{i<j} (\mathbf{D}_{ij} \oplus \mathbf{C}_{m(i,j)}),$$

(f is defined componentwisely, which means the factors of f are $f_{i,j}$ for $1 \le i < j \le n$), we obtain an embedding of $\mathbf{H}'$ into the finite hoop $\prod_{i<j}(\mathbf{D}_{ij} \oplus \mathbf{C}_{m(i,j)})$.

Therefore, the variety H has (FEP). $\square$

Bibliographical Remarks and Suggestions for Further Study

Readers interested in this topic are strongly encouraged to consult [141] and the series of articles [7–10] for a more advanced understanding of the variety of hoops. As a concluding note to this section, we would like to briefly highlight the following significant results.

There are more subclasses of H that have (FEP). In [141, Lemma 2.59], it was proved that totally ordered hoops have (FEP), also see [11, Theorem 2.1]. In addition, by [31, Theorem 3.9], subdirectly irreducible Wajsberg hoops as well as (WH)^n have (FEP) for every $n \in \mathbb{N}$. In conclusion, we deduce Corollary 6.7.18.

Corollary 6.7.18 ([31, Corollary 3.11]) *For every natural number n, the varieties* (WH)^n *and* H *are generated, as quasivarieties, by their finite members.*

$$(\text{WH})^{n+1} = \text{ISPP}_{\text{u}}\{\mathbf{B} \oplus \mathbf{C}_m \mid \mathbf{B} \text{ is finite}, \ \mathbf{B} \in (\text{WH})^n, \ m \in \mathbb{N}\}.$$

We now present the following fundamental theorems concerning quasivarieties with the finite embeddability property (FEP):

Theorem 6.7.19 ([141, Theorem 2.5.2]) *A quasivariety* K *has the finite embeddability property if and only if* K $= \text{SPP}_{\text{u}}(\text{K}_{\text{fin}})$, *where* K_{fin} *is the class of finite members of* K.

The following theorem is the quasi-equational version of a well-known result on logical systems due to Harrop.

Theorem 6.7.20 ([170]) *Let* K *be a quasivariety of algebras over a finite language. If* K *is finitely axiomatizable and is generated (as a quasivariety) by its finite algebras, then the quasi-equational theory of* K *is decidable.*

Considering Theorems 6.7.17 and 6.7.20 results, we have

Corollary 6.7.21 ([31, Corollary 3.14]) *The quasi-equational theory of the variety of all hoops is decidable.*

Variety and quasivariety of subclasses of hoops, such as basic hoops and Wajsberg hoops, splitting algebras, were also studied in [8, 10–14, 16–18]. Applying the following notions, Aglianò [10] identified conditions under which a linear Wajsberg hoop $Q(\mathbf{H})$ forms a variety.

Definition 6.7.22 ([13]) Let $\mathbf{H}$ be a bounded Wajsberg hoop.

(i) $\mathbf{H}$ has **rank** n if $\mathbf{H}/\mathrm{Rad}(\mathbf{H}) \cong \mathbf{L}_n$.

(ii) The divisibility index of $\mathbf{H}$, denoted by $d_{\mathbf{H}}$, is the maximum k such that $\mathbf{L}_k$ is embeddable in $\mathbf{H}$ if any, otherwise $d_{\mathbf{H}} = \infty$.

Theorem 6.7.23 ([10, Theorem 5.14, 5.16]) *Let $\mathbf{H}$ be a linear Wajsberg hoop. Then $Q(\mathbf{H})$ is a variety if and only if $\mathbf{H}$ satisfies on of the following conditions:*

(i) *$\mathbf{L}_n$ is embeddable in $\mathbf{H}$ for all $n \in \mathbb{N}$.*

(ii) *$\mathbf{H}$ is finite.*

(iii) *$\mathbf{H}$ is cancellative.*

(iv) *$\mathbf{H}$ is infinite, bounded and the rank of $\mathbf{H}$ is equal to $d_{\mathbf{H}}$.*

In addition, for a finite set $\{\mathbf{H}_1, \ldots \mathbf{H}_m\}$ of linear Wajsberg hoop $Q(\{\mathbf{H}_1, \ldots, \mathbf{H}_m\})$ is a variety if and only if for every $i \in \{1, \ldots, m\}$ $\mathbf{H}_i$ satisfies one of the conditions (i)–(iv).

Let $\{m_1, \ldots, m_n\}$ denotes a finite subset of $\mathbb{N} \cup \{0\}$, where $n = 0$ means $\{m_1, \ldots, m_n\} = \emptyset$, similarly for $\{t_1, \ldots, t_s\}$. Then, based on a result presented in [15], every proper subvariety of Wajsberg hoops takes one of the following forms:

Theorem 6.7.24 ([15, Theorem 2.5]) *Let V be a proper variety of Wajsberg hoops. Then V is either of the form:*

(i) $\mathsf{V}(\mathbf{L}_{m_1}, \ldots, \mathbf{L}_{m_n})$ *for $n \in \mathbb{N}$.*

(ii) $\mathsf{V}(\mathbf{L}_{m_1}, \ldots, \mathbf{L}_{m_n}, \mathbf{L}_\infty)$ *for $n \in \mathbb{N} \cup \{0\}$.*

(iii) $\mathsf{V}(\mathbf{L}_{m_1}, \ldots, \mathbf{L}_{m_n}, \mathbf{L}_\infty^{t_1}, \ldots, \mathbf{L}_\infty^{t_s})$ *for $n \in \mathbb{N} \cup \{0\}$ and $s \in \mathbb{N}$, where $\mathbf{L}_\infty^k$ is the Wajsberg hoop $(\mathbb{Z} \overrightarrow{\times} \mathbb{Z})[(k, 0)]$ for every $k \in \mathbb{N}$.*

In [9], P. Aglianò studied the existence and characterization of splitting algebras within the variety of BL-algebras and its subvarieties.

Aguzzoli and Bianchi [19], focused on varieties of MTL-algebras, BL-algebras, and hoops whose lattices of subvarieties are totally ordered. Further advanced study on the topics related to this section can be found in [157, 169, 214, 218, 256].

It is worth noting that, due to the result in [195], the lattice of subvarieties of the variety of MV-algebras is countable. Furthermore, [101] described finite equational bases for each proper subvariety of MV-algebras.

6.8 Exercises

6.8.1 Let F be a filter of a non-trivial hoop $\mathbf{H}$. Prove that $\mathbf{H}/F$ is semisimple if and only if F is an intersection of maximal filters of $\mathbf{H}$.

6.8.2 Prove the converse of Corollary 6.2.7.

6.8.3 Let $\mathbf{H}$ be a finite hoop. Show that $\mathbf{H}$ is semisimple if and only if $\mathbf{H} \cong \mathbf{L}_{n_1} \times \cdots \times \mathbf{L}_{n_m}$ for some $1 \leq n_1 \leq \cdots \leq n_m \in \mathbb{N}$ (see [73, Proposition 3.6.5]).
Note. A representation for finite BL-algebras was given in [102]. A. Di Nola and A. Lettieri introduced the concept of a BL-comet and demonstrated that every finite BL-algebra is isomorphic to a direct product of BL-comets.

6.8.4 Assume that $\mathbf{H}$ is a bounded local hoop satisfying $x^2 = x$. Prove that $H = \{0, 1\}$.

6.8.5 Let $\mathbf{H}$ be a bounded linear Wajsberg hoop. Prove that $\{x \in H : \mathrm{ord}(x) = \infty\}$ is subuniverse of $\mathbf{H}$ as well as a cancellative hoop.

6.8.6 Let $\mathbf{H}$ be a bounded hoop and P be a proper filter of $\mathbf{H}$. Then P is a primary filter if and only if $x/H \odot y/H = 0/H$ implies that there exists $n \in \mathbb{N}$ such that $(x/H)^n = 0/H$ or $(y/H)^n = 0/H$ for any $x, y \in H$.

6.8.7 Complete the proof of Proposition 6.1.5.

6.8.8 Let P be a proper filter of a bounded hoop $\mathbf{H}$ and $x^2 = x$ for any $x \in H$. Prove that P is a primary filter if and only if $x/H \odot y/H = 0/H$ implies that $x/H = 0/H$ or $y/H = 0/H$ for any $x, y \in H$ (Hint: Apply Exercises 3.4 and 6.8). Therefore, $\mathbf{H}/P \cong \mathbf{2}$. It shows that each primary filter on a hoop with this property is maximal.

6.8.9 Try to prove Theorem 6.3.31 directly.

6.8.10 Let $\mathbf{H}$ be a linearly ordered perfect hoop. Prove that $\mathrm{Rad}(\mathbf{H}) = \{x \in H : x' < x\}$.

6.8.11 Let $\mathbf{H}$ be a linearly ordered hoop. Prove the following statements:

    (i) $\mathrm{Rad}(\mathbf{H}) = \{x \in H : (x^n)' \leq x, \ \forall\, n \in \mathbb{N}\}$.
    (ii) $\mathrm{Rad}(\mathbf{H}) \subseteq \{x \in H : x' \leq x''\}$.
    (iii) If $F = \{x \in H : x' \leq x''\} \in \mathcal{F}(\mathbf{H})$, then $\mathrm{Rad}(\mathbf{H}) = F$.
    (iv) For any $x \in H$, there exists $n \in \mathbb{N}$ such that $(x^n)'' \to x \in \mathrm{Rad}(\mathbf{H})$.

6.8.12 Show that every special hoop is perfect (see Exercise 3.4).

6.8.13 Complete the proof of Proposition 6.5.8(ii).

6.8.14 Let $\mathbf{H} = (H; \odot, \to, 0, 1)$ be a bounded $\vee$-hoop and $\mathrm{Nil}(\mathbf{H}) = \{x \in H : \mathrm{ord}(x) < \infty\}$. Show that $\mathrm{Nil}(\mathbf{H})$ is a lattice ideal of $(H; \vee, \wedge)$ (see Definition 1.3.17).

6.8.15 Let $\mathbf{H}$ be a finite hoop such that $\mathrm{ord}(x) = |H| - 1$ for some $x \in H$. Such an element is called the generator of $\mathbf{H}$. Prove that $\mathbf{H}$ satisfies the following statements:

    (i) $\mathbf{H}$ is a linear Wajsberg hoop, consequently, it satisfies (DNP). In addition, $x = \max(H \setminus \{1\})$.
    (ii) If $y \in H$ such that $\mathrm{ord}(y) = |H| - 1$, then $x = y$.
    (iii) $\mathbf{H} \cong \mathbf{L}_n$, where $n = |H| - 1$.

6.8.16 Complete the proof of Theorem 6.6.5 by showing that $\bar{g}$ preserves the operation $\rightarrow$.

6.8.17 Given a proper filter F of a hoop $\mathbf{H}$, we define $\mathrm{Rad}(F)$ by

$$\mathrm{Rad}(F) = \bigcap \{M \in \mathrm{Max}(\mathbf{H}) : F \subseteq M\}. \tag{6.13}$$

Clearly, $\mathrm{Rad}(F) \in \mathcal{F}(\mathbf{H})$. Let F and G be proper filters of a bounded basic hoop $\mathbf{H} = (H; \odot, \rightarrow, 0, 1)$. Establish the validity of the following statements (see [219]):

  (i)   If F is a prime filter, then $\mathrm{Rad}(F) \in \mathrm{Max}(\mathbf{H})$.
 (ii)   If $\mathbf{H}$ is linearly ordered, then $\mathrm{Rad}(F) = \mathrm{Rad}(G)$.
(iii) $\mathrm{Rad}(F) = \{x \in H : (x^n)' \rightarrow x \in F,\ \forall n \in \mathbb{N}\}$.
(iv) $\mathrm{Rad}(\mathrm{Rad}(F)) = \mathrm{Rad}(F)$.
  (v)   If $\langle F \cup G \rangle \neq H$, then $\mathrm{Rad}(\langle F \cup G \rangle) = \mathrm{Rad}(F) \cap \mathrm{Rad}(G)$.
(vi) $\mathrm{Rad}(\{1/F\}) = \mathrm{Rad}(F)/F$.

6.8.18 Let $\mathbf{H} = (H; \odot, \rightarrow, 0, 1)$ be a bounded product hoop.

  (i) $\mathrm{Reg}(\mathbf{H}) = \{x \in H : x'' = x\}$ is universe of the greatest Boolean subhoop[2] of $\mathbf{H}$.
 (ii) $\mathrm{C}(\mathbf{H}) = \{x \in H : x'' = 1\}$ is the universe of the greatest cancellative subhoop of $(H; \odot, \rightarrow, 1)$.
(iii)   For every $x \in H$ there exists $b \in \mathrm{Reg}(\mathbf{H})$ and $c \in \mathrm{C}(\mathbf{H})$ such that $x = b \odot c$ (Hint: Show that $x = x'' \odot (x'' \rightarrow x)$, $x'' \rightarrow x \in \mathrm{C}(\mathbf{H})$ and $x'' \in \mathrm{Reg}(\mathbf{H})$).

---

[2] $\mathbf{A}$ is a Boolean subhoop of $\mathbf{H}$ if $\mathbf{A}$ is a subhoop of $\mathbf{H}$ and $(A; \vee, \wedge, ', 0, 1)$ forms a Boolean algebra with respect to $\vee$, $\wedge$, and $'$ inherited from $\mathbf{H}$.

State Hoops 7

A state is an analogue of a probability measure. Mundici [224] introduced states as averaging processes for formulas in Łukasiewicz logic. States constitute measures on their associated MV-algebras, which generalize the usual probability measures on Boolean algebras. Dvurečenskij [107, 108] and Georgescu [151] studied states on pseudo MV-algebras and fuzzy logic, respectively. Kroupa [200] showed that the set of all states on every MV-algebra forms a Bauer simplex, and every state on a semisimple MV-algebra arises as an integral with respect to a unique Borel probability measure. See also G. Panti's work [233] as well as the contributions of Marra and Mundici's [212] investigating connections between states and integration with respect to the Lebesgue measure.

Flaminio and Montagna [143, 144] presented another approach to state theory on MV-algebras by adding a new unary operation on the structure of MV-algebras as an internal state. They studied the coherence problem for rational assessments on many-valued events and proposed an algebraic treatment of the Lebesgue integral. They proved that the internal states defined on a subclass of MV-algebras can be represented by means of this more general notion of integral. Di Nola and Dvurečenskij [93] investigated the category of (state-morphism) state MV-algebras and described all subdirectly irreducible state-morphism MV-algebras. In [55], M. Botur and A. Dvurečenskij provided a general approach to state-morphism algebras. These notions for hoops were investigated in [43, 78].

This chapter delves into the concepts of state hoops and state-morphism hoops. The notions of state operator, strong state operator, state-morphism operator, and weak state-morphism operator are introduced, and their properties are investigated.

We will see that every strong state hoop is a state hoop and any state operator on an idempotent hoop is a weak state-morphism operator. Glivenko property is defined, and it is proved that for an idempotent hoop $\mathbf{H}$ having these properties a state operator on $\mathrm{Reg}(\mathbf{H})$ can be extended to a state operator on $\mathbf{H}$. Every perfect hoop admits a non-trivial state operator.

© The Author(s), under exclusive license to Springer Nature Switzerland AG 2026

A. Dvurečenskij et al., *Hoop Algebras*, Frontiers in Mathematics,
https://doi.org/10.1007/978-3-032-11736-6_7

7.1   State Hoop

Definition 7.1.1 ([78]) A **state hoop** is a pair $(\mathbf{H}, \mu)$, where $\mathbf{H}$ is a bounded hoop and $\mu : H \to H$ is a mapping, called a **state operator**, such that for any $x, y \in H$, the following conditions are satisfied:

(S1) $\mu(0) = 0$.
(S2) $\mu(x \to y) = \mu(x) \to \mu(x \wedge y)$.
(S3) $\mu(x \odot y) = \mu(x) \odot \mu(x \to (x \odot y))$.
(S4) $\mu(\mu(x) \odot \mu(y)) = \mu(x) \odot \mu(y)$.
(S5) $\mu(\mu(x) \to \mu(y)) = \mu(x) \to \mu(y)$.

We denotes $\mathrm{Ker}(\mu) = \{x \in H : \mu(x) = 1\}$, which is called the **kernel** of μ. A state operator is called **faithful** if $\mathrm{Ker}(\mu) = \{1\}$.

Proposition 7.1.2 ([76]) *Let $\mathbf{H}$ be a hoop. Define a binary operation $\cup : H \times H \to H$ by $x \cup y = (x \to y) \to y$. Then the following statements hold for all $x, y, z \in H$:*

(i) $1 \cup x = x \cup 1 = 1$.
(ii) $x \leq y$ *implies that* $x \cup y = y$.
(iii) $x \leq y$ *implies that* $x \cup z \leq y \cup z$.
(iv) $(x \cup y) \to y = x \to y$.

Proof (i) $1 \cup x = (1 \to x) \to x = x \to x = 1$ and $x \to 1 = (x \to 1) \to 1 = 1 \to 1 = 1$.
(ii) Let $x \leq y$. Then $(x \to y) \to y = 1 \to y = y$.
(iii) It follows from Lemma 2.2.1(7).
(iv) The proof is straightforward by Lemma 2.2.1(13). $\square$

Proposition 7.1.3 ([76, Proposition 2.6]) *Let $\mathbf{H} = (H; \odot, \to, 0, 1)$ be a bounded hoop. Then $(H; \oplus, 0)$ is a partially ordered monoid, where $x \oplus y = (x' \odot y')'$, satisfying the following conditions:*

(i) $x \oplus 1 = 1 = x \oplus x'$.
(ii) $x, y \leq x \oplus y$.
(iii) $x \oplus y = x' \to y''$.

Proof The proof is left to the reader. $\square$

Proposition 7.1.4   ([78]) *If* $(\mathbf{H}, \mu)$ *is a state hoop, then for all* $x, y \in H$, *the following hold:*

(1) $\mu(1) = 1$.

(2) $\mu(x') = \mu(x)'$.

(3) $x \le y$ *implies* $\mu(x) \le \mu(y)$.

(4) $\mu(x) \odot \mu(y) \le \mu(x \odot y)$. *In addition, either* $x \odot y = 0$ *or* $\mathbf{H}$ *has* (DNP) *and* $y'' \le x$ *implies that* $\mu(x \odot y) = \mu(x) \odot \mu(y)$.

(5) $\mu(x \odot y') \ge \mu(x) \odot \mu(y)'$. *If* $x \le y$, *then* $\mu(x \odot y') = \mu(x) \odot \mu(y)'$.

(6) $\mu(x \wedge y) = \mu(x) \odot \mu(x \to y)$.

(7) $\mu(x \to y) \le \mu(x) \to \mu(y)$. *If* x *and* y *are comparable, then* $\mu(x \to y) = \mu(x) \to \mu(y)$.

(8) $\mu(x \to y) \odot \mu(y \to x) \le (\mu(x) \to \mu(y)) \wedge (\mu(y) \to \mu(x))$.

(9) $\mu^2(x) = \mu(x)$.

(10) *If* $\mathbf{H}$ *has* (DNP), *then* $\mu(x \oplus y) \le \mu(x) \oplus \mu(y)$ *and* $\mu(\mu(x) \oplus \mu(y)) = \mu(x) \oplus \mu(y)$. *If* $x' \le y''$, *then* $\mu(x \oplus y) = \mu(x) \oplus \mu(y)$ *and* $\mu(x \oplus x') = 1$, *where* $x \oplus y = (x' \odot y')'$.

(11) $\mu(H) = \{x \in H : \mu(x) = x\}$.

(12) $\mu(x \to y) = \mu(x) \to \mu(y)$ *if and only if* $\mu(y \to x) = \mu(y) \to \mu(x)$.

(13) $\mu(\mu(x) \wedge \mu(y)) = \mu(x) \wedge \mu(y)$.

(14) $\mu(\mu(x) \cup \mu(y)) = \mu(x) \cup \mu(y)$ *and* $\mu(x \cup y) \le \mu(x) \cup \mu(y)$. *If* x *and* y *are comparable, then* $\mu(x \cup y) = \mu(x) \cup \mu(y)$.

(15) *If* μ *is faithful, then* $x < y$ *implies* $\mu(x) < \mu(y)$.

(16) *If* μ *is faithful, then either* $\mu(x) = x$ *or* $\mu(x)$ *and* x *are not comparable.*

(17) *If* $\mathbf{H}$ *is linearly ordered and* μ *is faithful, then* $\mu(x) = x$ *for all* $x \in H$.

Proof (1) $\mu(1) = \mu(0 \to 0) = \mu(0) \to \mu(0 \wedge 0) = 1$.

(2) $\mu(x') = \mu(x \to 0) = \mu(x) \to \mu(x \wedge 0) = \mu(x) \to \mu(0) = \mu(x) \to 0 = \mu(x)'$.

(3) By Proposition 3.1.4(i), we get $x = y \odot (y \to x)$, so

$$\mu(x) = \mu(y \odot (y \to x)) = \mu(y) \odot \mu\big(y \to (y \odot (y \to x))\big) \le \mu(y).$$

(4) From $x \odot y \le x \odot y$ we get $y \le x \to (x \odot y)$, so by (3), we have $\mu(y) \le \mu(x \to (x \odot y))$. Applying (S3), we get $\mu(x \odot y) = \mu(x) \odot \mu(x \to (x \odot y)) \ge \mu(x) \odot \mu(y)$. If $x \odot y = 0$, then $\mu(x \odot y) = 0$, so that $\mu(x \odot y) = \mu(x) \odot \mu(y) = 0$. If $\mathbf{H}$ has (DNP) and $y'' \le x'$, then $x = x'' \le y''' = y'$ and $x \odot y \le y' \odot y = 0$, yields that $\mu(x \odot y) = \mu(x) \odot \mu(y) = 0$.

(5) The inequality $\mu(x \odot y') \ge \mu(x) \odot \mu(y)'$ follow from (4) and (2). If $x \le y$, then $y' \le x'$, so $x \odot y' \le x \odot x' = 0$. It implies that $\mu(x \odot y') = 0$ and $\mu(y' \odot x) = \mu(y)' \odot \mu(x) = 0$.

(6) We have

$$\mu(x \wedge y) = \mu(x \odot (x \to y)) = \mu(x) \odot \mu\big(x \to (x \odot (x \to y))\big)$$
$$= \mu(x) \odot \mu(x \to (x \wedge y)) = \mu(x) \odot \mu(x \to y).$$

(7) By (S2) and (3), we have $\mu(x \to y) = \mu(x) \to \mu(x \wedge y) \le \mu(x) \to \mu(y)$. If $x \le y$, then $\mu(x) \le \mu(y)$ and $\mu(x \to y) = \mu(x) \to \mu(x \wedge y) = \mu(x) \to \mu(x) = 1$. We also have $\mu(x) \to \mu(y) = 1$, thus $\mu(x \to y) = \mu(x) \to \mu(y)$. If $y \le x$, then $x \wedge y = y$ and the equalities follow from (S2).

(8) By (7), we have $\mu(x \to y) \le \mu(x) \to \mu(y)$ and $\mu(y \to x) \le \mu(y) \to \mu(x)$, hence $\mu(x \to y) \odot \mu(y \to x) \le (\mu(x) \to \mu(y)) \wedge (\mu(y) \to \mu(x))$.

(9) Applying (1) and (S4), we have: $\mu^2(x) = \mu(\mu(x)) = \mu(\mu(x) \odot \mu(1)) = \mu(x) \odot \mu(1) = \mu(x)$.

(10) From $\mu(y' \odot x') \ge \mu(y') \odot \mu(x')$ we get $(\mu(y') \odot \mu(x'))' \ge (\mu(y \odot x'))'$. Applying (2), it follows that $(\mu(y)' \odot \mu(x)')' \ge \mu((y' \odot x')')$. Thus, $\mu(x \oplus y) \le \mu(x) \oplus \mu(y)$. By (2) and (9), we get

$$\mu(\mu(x) \oplus \mu(y)) = \mu((\mu(y)' \odot \mu(x)')') = \big(\mu(\mu(y') \odot \mu(x'))\big)'$$
$$= (\mu(y') \odot \mu(x'))' = (\mu(y)' \odot \mu(x)')' = \mu(x) \oplus \mu(y).$$

Obviously, $\mu(x \oplus 0) = \mu(x) \oplus \mu(0)$. Since $x' \le y''$, by Remark 2.2.5(i) and (iii), $y' \le x''$, which implies $y' \odot x' \le x'' \odot x' = 0$, so by (4) and (2), we have $\mu(y' \odot x') = \mu(y') \odot \mu(x') = \mu(y)' \odot \mu(x)'$. Hence,

$$\mu(x \oplus y) = \mu((y' \odot x')') = (\mu(y' \odot x'))' = (\mu(y)' \odot \mu(x)')' = \mu(x) \oplus \mu(y).$$

For the last assertion, we have $\mu(x \oplus x') = \mu((x'' \odot x')') = (\mu(x'' \odot x'))' = \mu(0)' = 1$.

(11) Consider $y \in \mu(H)$, so there exists $x \in H$ such that $y = \mu(x)$. Hence, $\mu(y) = \mu^2(x) = \mu(x) = y$ and $y \in \{x \in H : \mu(x) = x\}$. Conversely, if $y \in \{x \in H : \mu(x) = x\}$, then $y \in \mu(H)$.

(12) Suppose $\mu(x \to y) = \mu(x) \to \mu(y)$. Applying (S2) and (6), we get

$$\mu(y \to x) = \mu(y) \to \mu(y \wedge x) = \mu(y) \to (\mu(x \to y) \odot \mu(x))$$
$$= \mu(y) \to \big((\mu(x) \to \mu(y)) \odot \mu(x)\big)$$
$$= \mu(y) \to (\mu(x) \wedge \mu(y)) = \mu(y) \to \mu(x).$$

Similarly, if $\mu(y \to x) = \mu(y) \to \mu(x)$, then $\mu(x \to y) = \mu(x) \to \mu(y)$.

(13) Applying (6), (9), and (S5), we get

$$\mu(\mu(x) \wedge \mu(y)) = \mu^2(x) \odot \mu(\mu(x) \to \mu(y)) = \mu(x) \odot (\mu(x) \to \mu(y)) = \mu(x) \wedge \mu(y).$$

(14) Applying (S5) and (9), we get

$$\mu(\mu(x) \cup \mu(y)) = \mu((\mu(x) \to \mu(y)) \to \mu(y)) = \mu(\mu(x) \to \mu(y)) \to \mu(y)$$
$$= (\mu(x) \to \mu(y)) \to \mu(y) = \mu(x) \cup \mu(y).$$

The second part follows by applying (7) twice.

(15) Let μ be faithful. By (3), $x < y$ implies $\mu(x) \le \mu(y)$. Suppose $\mu(x) = \mu(y)$. From (S2) it follows that $\mu(y \to x) = \mu(y) \to \mu(x) = 1$, that is $y \to x \in \mathrm{Ker}(\mu) = \{1\}$. Thus, $y \to x = 1$, hence $y \le x$, a contradiction. It follows that $\mu(x) < \mu(y)$.

(16) Let μ be faithful and $x \in H$ such that $\mu(x) \ne x$. We claim that x and $\mu(x)$ are comparable. Suppose, for contradiction, that they are incomparable. If $x < \mu(x)$, then by (9) and (15), $\mu(x) < \mu^2(x) = \mu(x)$, which is a contradiction. Similarly, $\mu(x) < \mu(x)$ leads to a contradiction. Hence, x and $\mu(x)$ are comparable.

(17) Since $\mathbf{H}$ is linearly ordered, it follows that x and $\mu(x)$ are comparable. Hence, by (16), $\mu(x) = x$ for every $x \in H$. $\qquad\square$

Corollary 7.1.5 *Let $(\mathbf{H}, \mu)$ be a linearly ordered state hoop. Then for all $x, y \in H$ the following hold:*

(i) $\mu(x \to y) = \mu(x) \to \mu(y)$.
(ii) $\mu(x \cup y) = \mu(x) \cup \mu(y)$.

Proof (i) It follows from Proposition 7.1.4(7).

(ii) It follows from Proposition 7.1.4(14). $\qquad\square$

Proposition 7.1.6 *Let $(\mathbf{H}, \mu)$ be a state hoop. Consider the following properties:*

(i) $\mu(x \to y) = \mu(x) \to \mu(y)$ *for all* $x, y \in H$.
(ii) $\mu(x \wedge y) = \mu(x) \wedge \mu(y)$ *for all* $x, y \in H$.
(iii) $\mu(x \odot y) = \mu(x) \odot \mu(y)$ *for all* $x, y \in H$.
(iv) $\mu(x \cup y) = \mu(x) \cup \mu(y)$ *for all* $x, y \in H$.

Then (i) *is equivalent with* (ii) *and* (i) *implies* (iii), (iv).

Proof According to Proposition 7.1.4(12), μ preserves $\to$.

(i) $\Rightarrow$ (ii) By Proposition 7.1.4(6), we have

$$\mu(x \wedge y) = \mu(x) \odot \mu(x \to y) = \mu(x) \odot (\mu(x) \to \mu(y)) = \mu(x) \wedge \mu(y).$$

(ii) $\Rightarrow$ (i) Applying (S2), we get

$$\mu(x \to y) = \mu(x) \to \mu(x \wedge y) = \mu(x) \to (\mu(x) \wedge \mu(y)) = \mu(x) \to \mu(y), \text{ by Lemma 2.2.3(ii)}.$$

(i) $\Rightarrow$ (iii) By (H2), we have

$$\mu(x \odot y) \to \mu(z) = \mu((x \odot y) \to z) = \mu(x \to (y \to z))$$
$$= \mu(x) \to (\mu(y) \to (\mu(z)) = (\mu(x) \odot \mu(y)) \to \mu(z).$$

Taking $z = \mu(x) \odot \mu(y)$ we get

$$\mu(x \odot y) \to \mu(\mu(x) \odot \mu(y)) = (\mu(x) \odot \mu(y)) \to \mu(\mu(x) \odot \mu(y))$$
$$= (\mu(x) \odot \mu(y)) \to (\mu(x) \odot \mu(y)) = 1.$$

Thus, $\mu(x \odot y) \le \mu(\mu(x) \odot \mu(y)) = \mu(x) \odot \mu(y)$. Applying Proposition 7.1.4(4), we get $\mu(x \odot y) = \mu(x) \odot \mu(y)$.

(i) $\Rightarrow$ (iv) This follows from the definition of $\cup$ by applying (i). $\qquad\square$

Let $\mathbf{H}$ be a bounded hoop and $\mu : H \to H$ be a mapping satisfying (S$'$3):
(S$'$3) $\mu(x \odot y) = \mu(y' \cup x) \odot \mu(y)$ for all $x, y \in H$.

Definition 7.1.7 A mapping $\mu : H \to H$ is called a **strong state operator** on a hoop $\mathbf{H}$ if μ satisfies conditions (S1), (S2), (S$'$3), (S4), and (S5).

A pair $(\mathbf{H}, \mu)$ such that $\mathbf{H}$ is a bounded hoop and μ is a strong state operator on $\mathbf{H}$ is called **strong state hoop**.

Theorem 7.1.8 *Every strong state hoop is a state hoop.*

Proof Consider the strong state hoop $(\mathbf{H}, \mu)$ and $x, y \in H$. Taking into consideration that $y' \le y \to x$, we get $y' \cup (y \to x) = y \to x$, since Proposition 7.1.2(ii). Then, we have

$$\mu(x \wedge y) = \mu(x \odot (x \to y)) = \mu(x) \odot \mu(x' \cup (x \to y)) = \mu(x) \odot \mu(x \to y).$$

It follows that

$$\mu(x \odot y) = \mu(x \wedge (x \odot y)) = \mu(x) \odot \mu(x \to (x \odot y)).$$

Thus, condition (S$'$3) implies condition (S3), hence μ is a state operator on $\mathbf{H}$. $\qquad\square$

Proposition 7.1.9 *If μ is a strong state operator on a bounded hoop $\mathbf{H}$ such that $x' \le y$ or $y' \le x$ for some $x, y \in H$, then $\mu(x \odot y) = \mu(x) \odot \mu(y)$.*

Proof Since μ is a strong state operator, it satisfies the condition

$$\mu(x \odot y) = \mu(y' \cup x) \odot \mu(y).$$

Since $y' \le x$ implies $\mu(y' \cup x) = \mu(x)$, we get $\mu(x \odot y) = \mu(x) \odot \mu(y)$. $\qquad\square$

Proposition 7.1.10 *If μ is a state operator on a linearly ordered bounded hoop* **H**, *then μ is a hoop endomorphism such that $\mu^2 = \mu$.*

Proof Since $(\mathbf{H}, \mu)$ is a linearly ordered state hoop, by Corollary 7.1.5, μ preserves $\rightarrow$. Applying Proposition 7.1.6, it follows that μ preserves $\odot$.

Taking into consideration that μ preserves also the constants 0 (by Definition 7.1.1(S1) and 1 (by Proposition 7.1.4(1)), we conclude that μ is an endomorphism. Condition $\mu^2 = \mu$ follows from Proposition 7.1.4(9). $\square$

Proposition 7.1.11 *If $(\mathbf{H}, \mu)$ is a state hoop, then $\mu(H)$ is a subalgebra of* **H**.

Proof By (S4), (S5) and Proposition 7.1.4(1), $\mu(H)$ is closed under the operations $\odot$, $\rightarrow$, and 1. Thus, $\mu(H)$ is a subalgebra of **H**. $\square$

Definition 7.1.12 (i) A state operator μ on a bounded hoop **H** is called a **weak state-morphism operator** on **H** if it satisfies (S6):

(S6) $\mu(x \odot y) = \mu(x) \odot \mu(y)$ for all $x, y \in H$.

In this case, $(\mathbf{H}, \mu)$ is called a **weak state-morphism hoop**.

(ii) A bounded hoop endomorphism $\mu : H \rightarrow H$ is said to be a **state-morphism operator** if $\mu^2 = \mu$.

Obviously, a state-morphism operator is always a weak state-morphism operator.

Example 7.1.13 (i) If **H** is a bounded hoop, then the identity $id_{\mathbf{H}}$ is a state operator on **H**.

(ii) Let **H** be a bounded hoop and $\mathbf{B} = \mathbf{H} \times \mathbf{H}$. Then the mappings $\mu_1, \mu_2 : \mathbf{B} \rightarrow \mathbf{B}$ such that $\mu_1(x_1, x_2) = (x_1, x_1)$ and $\mu_2(x_1, x_2) = (x_2, x_2)$ are state-morphism operators on the bounded hoop **B**.

Remark 7.1.14 From Propositions 7.1.6 and 7.1.9 we have:

(i) If μ is a state operator on **H** preserving $\rightarrow$, then μ is a weak state-morphism operator and a state-morphism operator.

(ii) If μ is a strong state operator on a bounded hoop **H** such that $x' \leq y$ or $y' \leq x$ for all $x, y \in H$, then μ is a weak state-morphism operator.

Proposition 7.1.15 *If* **H** *is a cancellative hoop, then any state operator μ on* **H** *is a weak state-morphism operator.*

Proof According to (S3) and taking into consideration that in a cancellative hoop $y \rightarrow (x \odot y) = x$ for every, we get

$$\mu(x \odot y) = \mu(y \rightarrow (x \odot y)) \odot \mu(y) = \mu(x) \odot \mu(y).$$

Thus, μ is a weak state-morphism operator on $\mathbf{H}$. $\qquad\square$

According to Sect. 3.3.4, an element a of a hoop $\mathbf{H}$ is idempotent if $a^2 = a$. The set of all idempotents of $\mathbf{H}$ is denoted by $\mathrm{Idm}(\mathbf{H})$. By Proposition 3.3.40, if $a \in \mathrm{Idm}(\mathbf{H})$, then for all $x \in H$ we have: $a \odot x = a \wedge x$.

Furthermore, by [189], representable Brouwerian algebras are idempotent basic hoops, and generalized Boolean algebras are idempotent Wajsberg hoops.

Theorem 7.1.16 *If $\mathbf{H}$ is a bounded idempotent hoop, then any state operator μ on $\mathbf{H}$ is a weak state-morphism operator and a state-morphism operator.*

Proof Consider $x, y \in H$. Applying the property of idempotent elements and Proposition 7.1.4(4), we get:

$$\mu(x \wedge y) = \mu(x \odot y) \geq \mu(x) \odot \mu(y) = \mu(x) \wedge \mu(y).$$

On the other hand, $\mu(x \wedge y) \leq \mu(x) \wedge \mu(y) = \mu(x) \odot \mu(y)$. Thus, $\mu(x \wedge y) = \mu(x \odot y)$ and $\mu(x) \odot \mu(y) = \mu(x) \wedge \mu(y)$, so μ is a weak state-morphism operator on $\mathbf{H}$.

Since $\wedge$ is preserved, according to Proposition 7.1.6 ((i) $\Leftrightarrow$ (ii)), the operation $\rightarrow$ is preserved as well.

The constants 0 and 1 are preserved by Definition 7.1.1(S1) and Proposition 7.1.4(1), respectively. Thus, μ is an endomorphism on $\mathbf{H}$.

Since from Proposition 7.1.4(9) we have $\mu^2 = \mu$, it follows that μ is also a state-morphism operator on $\mathbf{H}$. $\qquad\square$

Proposition 7.1.17 *If μ is a state operator on a bounded hoop $\mathbf{H}$, then $\mathrm{Ker}(\mu)$ is a filter of $\mathbf{H}$.*

Proof By Proposition 7.1.4, $\mu(1) = 1$ and so $1 \in \mathrm{Ker}(\mu) \neq \emptyset$. Suppose $x, x \rightarrow y \in \mathrm{Ker}(\mu)$. Then $\mu(x) = \mu(x \rightarrow y) = 1$. By (S2) and Proposition 7.1.4(3), we get

$$1 = \mu(x \rightarrow y) = \mu(x) \rightarrow \mu(x \wedge y) = 1 \rightarrow \mu(x \wedge y) = \mu(x \wedge y) \leq \mu(y),$$

consequently, $\mu(y) = 1$, that is $y \in \mathrm{Ker}(\mu)$. Hence, by Lemma 2.3.2, $\mathrm{Ker}(\mu)$ is a filter of $\mathbf{H}$. $\qquad\square$

7.2 Glivenko and Meet-Negation Properties

We introduce the Glivenko and meet-negation properties, which will be used in the next sections.

Recall that, for a bounded hoop $\mathbf{H} = (H; \odot, \rightarrow, 0, 1)$, we define $\mathrm{Reg}(\mathbf{H}) = \{a \in H : a'' = a\}$. By Remark 2.2.5(vii), it is closed under $\rightarrow$. In addition, if $\mathbf{H}$ is a bounded basic hoop, then $(\mathrm{Reg}(\mathbf{H}); \odot, \rightarrow, 1)$ is a bounded subhoop of $\mathbf{H}$ (see Theorem 3.3.36).

Based on the conditions introduced in [201], we introduce the notion of the Glivenko property for a bounded hoop.

Definition 7.2.1 A bounded hoop $\mathbf{H}$ has the **Glivenko property** if $(x \rightarrow y)'' = x \rightarrow y''$ for all $x, y \in H$.

Remark 7.2.2 ([78]) By Remark 2.2.5, any bounded hoop $\mathbf{H}$ satisfying the Glivenko property, the following identity holds:

$$(x \rightarrow y)'' = x'' \rightarrow y'', \quad \forall x, y \in H. \tag{7.1}$$

Indeed, $x'' \rightarrow y'' = y' \rightarrow x''' = y' \rightarrow x' = x \rightarrow y'' = (x \rightarrow y)''$.

Definition 7.2.3 ([78]) A hoop $\mathbf{H}$ is said to be with the **meet-negation** property (mN for short) if (mN) $(x \wedge y)'' = x'' \wedge y''$ for all $x, y \in H$.

Proposition 7.2.4 ([78]) *Let $(\mathbf{H}, \mu)$ be a state hoop, where $\mathbf{H}$ is an idempotent hoop. Then*

(i) $\mu(x \wedge y) = \mu(x \odot y) = \mu(x) \odot \mu(y) = \mu(x) \wedge \mu(y)$ *for all* $x, y \in H$.
(ii) $\mu(x \rightarrow y) = \mu(x) \rightarrow \mu(y)$ *for all* $x, y \in H$.
(iii) *If* $\mathbf{H}$ *satisfies* (mN) *property, then* $(x \odot y)'' = (x \wedge y)'' = x'' \wedge y'' = x'' \odot y''$ *for all* $x, y \in H$.

Proof (i) It follows from the proof of Theorem 7.1.16 and Proposition 3.3.40.

(ii) By (i) and Proposition 7.1.6, $\mu(x \rightarrow y) = \mu(x) \rightarrow \mu(y)$ for all $x, y \in H$.

(iii) Since $\mathbf{H}$ is idempotent, by Proposition 3.3.40, $x \odot y = x \wedge y$ for all $x, y \in H$, and so

$$(x \odot y)'' = (x \wedge y)'' = x'' \wedge y'' = x'' \odot y''.$$

$\square$

Proposition 7.2.5 *Let* $(\mathbf{H}, \mu)$ *be a state hoop and* $x, y \in \mathrm{Reg}(\mathbf{H})$. *Then:*

(i) $\mu(x) \in \mathrm{Reg}(\mathbf{H})$.

(ii) $x \oplus y \in \mathrm{Reg}(\mathbf{H})$.

(iii) *If* $\mathbf{H}$ *has Glivenko property, then* $x \to y$ *and* $x \cup y \in \mathrm{Reg}(\mathbf{H})$.

(iv) *If* $\mathbf{H}$ *has* (mN) *property, then* $x \wedge y \in \mathrm{Reg}(\mathbf{H})$.

(v) *If* $\mathbf{H}$ *is idempotent with* (mN) *property, then* $x \odot y \in \mathrm{Reg}(\mathbf{H})$.

Proof (i) By Proposition 7.1.4(2), we have $\mu(x)'' = \mu(x'') = \mu(x)$, thus $\mu(x) \in \mathrm{Reg}(\mathbf{H})$.

(ii) Applying Lemma 2.2.1(13), we get:

$$(x \oplus y)'' = ((y' \odot x')')'' = (y' \odot x')' = x \oplus y.$$

Thus, $x \oplus y \in \mathrm{Reg}(\mathbf{H})$.

(iii) By Lemma 2.2.1(8), we have:

$$(x \to y)'' = x \to y'' = x'' \to y'' = x \to y.$$

As a consequence, it follows that

$$((x \to y) \to y)'' = (x'' \to y'') \to y'' = (x \to y) \to y,$$

so $x \cup y \in \mathrm{Reg}(\mathbf{H})$.

(iv) By the (mN) property, we have $(x \wedge y)'' = x'' \wedge y'' = x \wedge y$, thus $x \wedge y \in \mathrm{Reg}(\mathbf{H})$.

(v) From Proposition 7.2.4, we have $(x \odot y)'' = x'' \odot y'' = x \odot y$, so $x \odot y \in \mathrm{Reg}(\mathbf{H})$. $\qquad\square$

7.3　On the Existence of State Operators on Hoops

In this section, we investigate the existence of state operators, proving that every perfect hoop admits a non-trivial state operator. From now on, $\mathbf{H} = (H; \odot, \to, 0, 1)$ will be a bounded hoop unless otherwise stated.

Lemma 7.3.1 ([78, Lemma 5.1]) *Any state operator* μ *on a locally finite hoop* $\mathbf{H}$ *is faithful.*

Proof Assume that there exists $0 < x < 1$ such that $\mu(x) = 1$. Then there is an integer $n \geq 1$ such that $x^n = 0$, hence by Proposition 7.1.4(4), we have $0 = \mu(0) = \mu(x^n) \geq \mu(x)^n = 1$, a contradiction. Thus, μ is faithful. $\qquad\square$

Proposition 7.3.2 *If* $\mathbf{H}$ *is a strongly simple locally finite basic hoop, then the identity map on* H *is the unique state operator on* $\mathbf{H}$.

Proof Let μ be a state operator on **H**. By Lemma 7.3.1, it follows that μ is faithful. Since every strongly simple basic hoop is linearly ordered, applying Proposition 7.1.4(17), we get $\mu(x) = x$ for all $x \in H$. $\square$

Theorem 7.3.3 *Let* **H** *be a bounded basic hoop. If* **H** *is an idempotent hoop satisfying Glivenko and* (mN) *properties and* $\mu : \mathrm{Reg}(\mathbf{H}) \to \mathrm{Reg}(\mathbf{H})$ *be a state operator on* $\mathrm{Reg}(\mathbf{H})$, *then the mapping* $\tilde{\mu} : H \to H$ *defined by* $\tilde{\mu}(x) = \mu(x'')$ *is a state operator on* **H** *such that* $\tilde{\mu}|_{\mathrm{Reg}(\mathbf{H})} = \mu$.

Proof First, we note that, since **H** is idempotent, by Proposition 3.3.45, $\mathrm{Reg}(\mathbf{H}) = \mathrm{Reg}(\mathbf{H}) \cap \mathrm{Idm}(\mathbf{H}) = \mathrm{B}(\mathbf{H})$ is a subhoop of **H**. Obviously, $\tilde{\mu}(0) = \mu(0) = 0$, so the condition (S1) is verified. Applying Proposition 7.2.4, we get

$$\tilde{\mu}(x \to y) = \mu((x \to y)'') = \mu(x'' \to y'') = \mu(x'') \to \mu(x'' \wedge y'')$$
$$= \mu(x'') \to \mu((x \wedge y)'') = \tilde{\mu}(x) \to \tilde{\mu}(x \wedge y).$$

So, $\tilde{\mu}$ satisfies (S2). By Proposition 7.2.4, we also have

$$\tilde{\mu}(x \odot y) = \mu((x \odot y)'') = \mu(x'' \odot y'') = \mu(x'') \odot \mu(x'' \to (x'' \odot y''))$$
$$= \mu(x'') \odot \mu(x'' \to (x \odot y)'') = \mu(x'') \odot \mu((x \to (x \odot y))'')$$
$$= \tilde{\mu}(x) \odot \tilde{\mu}(x \to (x \odot y)).$$

Hence, $\tilde{\mu}$ satisfies (S3). For the condition (S4) we have

$$\tilde{\mu}(\tilde{\mu}(x) \odot \tilde{\mu}(y)) = \mu((\mu(x'') \odot \mu(y''))'') = \mu(\mu(x'')'' \odot \mu(y'')'')$$
$$= \mu(\mu(x'') \odot \mu(y'')) = \mu(x'') \odot \mu(y'') = \tilde{\mu}(x) \odot \tilde{\mu}(y).$$

Finally,

$$\tilde{\mu}(\tilde{\mu}(x) \to \tilde{\mu}(y)) = \mu((\mu(x'') \to \mu(y''))'') = \mu(\mu(x'')'' \to \mu(y'')'')$$
$$= \mu(\mu(x'') \to \mu(y'')) = \mu(x'') \to \mu(y'') = \tilde{\mu}(x) \to \tilde{\mu}(y).$$

It follows that $\tilde{\mu}$ satisfies (S5). Therefore, $\tilde{\mu}$ is a state operator on **H**. In addition, if $x \in \mathrm{Reg}(\mathbf{H})$, then $\tilde{\mu}(x) = \mu(x'') = \mu(x)$, so that $\tilde{\mu}|_{\mathrm{Reg}(\mathbf{H})} = \mu$. $\square$

Theorem 7.3.4 *Any perfect hoop admits a non-trivial state operator.*

Proof Let $\mathbf{H}$ be a perfect hoop. Then $\mathbf{H} = \mathrm{Rad}(\mathbf{H}) \cup \mathrm{Rad}(\mathbf{H})^*$. We will prove that the map $\mu : H \to H$ defined by

$$\mu(x) = \begin{cases} 1 & \text{if } x \in \mathrm{Rad}(\mathbf{H}) \\ 0 & \text{if } x \in \mathrm{Rad}(\mathbf{H})^* \end{cases}$$

is a state operator on $\mathbf{H}$. Obviously, $\mu(0) = 0$, hence (S1) is satisfied.

We consider the following cases:

(1) $a, b \in \mathrm{Rad}(\mathbf{H})$. Obviously, $\mu(a) = \mu(b) = 1$. Since $\mathrm{Rad}(\mathbf{H})$ is a filter of $\mathbf{H}$ and $b \le a \to b$, it follows that $a \to b \in \mathrm{Rad}(\mathbf{H})$. Hence, $\mu(a \to b) = 1$. From the definition of a filter, we have $a \odot b, a \wedge b \in \mathrm{Rad}(\mathbf{H})$. Thus,

$$\mu(a) = \mu(b) = \mu(a \to b) = \mu(a \wedge b) = \mu(a \odot b) = 1.$$

Since $a \odot b \in \mathrm{Rad}(\mathbf{H})$ and $a \odot b \le a \to (a \odot b)$, $a \odot b \le b \to (a \odot b)$ it follows that $a \to (a \odot b)$, $b \to (a \odot b) \in \mathrm{Rad}(\mathbf{H})$, so $\mu(a \to (a \odot b)) = \mu(b \to (a \odot b)) = 1$. One can easily check that the conditions (S2)–(S5) are satisfied.

(2) $a, b \in \mathrm{Rad}(\mathbf{H})^*$. In this case, $\mu(a) = \mu(b) = 0$ and we will prove that $a \to b \in \mathrm{Rad}(\mathbf{H})$. Indeed, suppose that $a \to b \in \mathrm{Rad}(\mathbf{H})^*$. Since $a \le a''$, it follows that $a'' \to b \le a \to b$, so $a'' \to b \in \mathrm{Rad}(\mathbf{H})^*$. But $a' \le a'' \to b$, hence $a' \in \mathrm{Rad}(\mathbf{H})^*$, a contradiction with the condition (ii) in the definition of perfect hoop ($a \in \mathrm{Rad}(\mathbf{H})^* = D(\mathbf{H})^*$ if and only if $a' \in \mathrm{Rad}(\mathbf{H}) = D(\mathbf{H})$). It follows that $a \to b \in \mathrm{Rad}(\mathbf{H})$. Hence, $\mu(a \to b) = 1$. From $a \wedge b \le b, a \odot b \le b$, we get $a \wedge b, a \odot b \in \mathrm{Rad}(\mathbf{H})^*$, thus $\mu(a \wedge b) = \mu(a \odot b) = 0$.

We can see that the conditions (S2)–(S5) are also verified.

(3) $a \in \mathrm{Rad}(\mathbf{H})$, $b \in \mathrm{Rad}(\mathbf{H})^*$. Obviously, $\mu(a) = 1$ and $\mu(b) = 0$. We show that $a \to b \in \mathrm{Rad}(\mathbf{H})^*$. Indeed, suppose that $a \to b \in \mathrm{Rad}(\mathbf{H})$. Because $b \le b''$, we have $a \to b \le a \to b''$, so $a \to b'' \in \mathrm{Rad}(\mathbf{H})$. It means that $(a \odot b')' \in \mathrm{Rad}(\mathbf{H})$, that is, $a \odot b' \in \mathrm{Rad}(\mathbf{H})^*$. On the other hand, since $\mathrm{Rad}(\mathbf{H})$ is a filter of $\mathbf{H}$ and $a, b' \in \mathrm{Rad}(\mathbf{H})$, we have $a \odot b' \in \mathrm{Rad}(\mathbf{H})$, a contradiction. We conclude that $a \to b \in \mathrm{Rad}(\mathbf{H})^*$, so $\mu(a \to b) = 0$. Since $a \wedge b \le b$ and $a \odot b \le b$, we have $a \wedge b, a \odot b \in \mathrm{Rad}(\mathbf{H})^*$, so $\mu(a \wedge b) = \mu(a \odot b) = 0$. Moreover, $a \in \mathrm{Rad}(\mathbf{H})$ and $a \odot b \in \mathrm{Rad}(\mathbf{H})^*$ imply $a \to (a \odot b) \in \mathrm{Rad}(\mathbf{H})^*$, hence $\mu(a \to (a \odot b)) = 0$.

It is easy to see that conditions (S2)–(S5) are satisfied.

(4) $a \in \mathrm{Rad}(\mathbf{H})^*, b \in \mathrm{Rad}(\mathbf{H})$. Taking into consideration that $b \le a \to b$, we have $a \to b \in \mathrm{Rad}(\mathbf{H})$.

From $a \wedge b, a \odot b \le a$, we get $a \wedge b, a \odot b \in \mathrm{Rad}(\mathbf{H})^*$. Hence, $\mu(a) = 0$, $\mu(b) = 1$, $\mu(a \wedge b) = \mu(a \odot b) = 0, \mu(a \to b) = 1$. Applying the Case (3), $b \in \mathrm{Rad}(\mathbf{H})$ and $a \odot b \in \mathrm{Rad}(\mathbf{H})^*$ imply that $b \to (a \odot b) \in \mathrm{Rad}(\mathbf{H})^*$, so $\mu(b \to (a \odot b)) = 0$.

Thus, the conditions (S2)-(S5) are also satisfied.

We conclude that μ is a state operator on $\mathbf{H}$, that is, $(\mathbf{H}, \mu)$ is a state hoop. $\square$

Table 7.1 Operations $\odot$ and $\rightarrow$ of Example 7.4.2

Table 7.1.1

$\odot$	0	a	b	1
0	0	0	0	0
a	0	a	a	a
b	0	a	b	b
1	0	a	b	1

Table 7.1.2

$\rightarrow$	0	a	b	1
0	1	1	1	1
a	0	1	1	1
b	0	a	1	b
1	0	a	b	1

7.4   State Operators and States on Hoops

The notions of states on bounded hoops have been investigated in [43, 76]. In this section, we will study Bosbach and Riečan states on hoops.

Definition 7.4.1 ([76]) Let **H** be a bounded hoop and [0, 1] be the unit interval of real numbers. A **Bosbach state** on **H** is a function $s : H \rightarrow [0, 1]$ such that the following conditions hold:

(BS1) $s(x) + s(x \rightarrow y) = s(y) + s(y \rightarrow x)$ for any $x, y \in H$;
(BS2) $s(0) = 0$ and $s(1) = 1$.

Example 7.4.2 Let $H = \{0, a, b, 1\}$ be a chain such that $0 < a < b < 1$. Consider Table 7.1:

Then $(\mathbf{H}; \odot, \rightarrow, 0, 1)$ is a bounded hoop. Define $s(0) = 0$ and $s(a) = s(b) = s(1) = 1$. Then s is a Bosbach state on **H**.

Proposition 7.4.3 ([76] and [43, Proposition 3.1]) *Let* **H** *be a bounded hoop and s be a Bosbach state on* **H**. *Then the following properties hold for all* $x, y \in H$:

(i) $x \leq y$ *implies* $s(x) \leq s(y)$ *and* $s(y \rightarrow x) = 1 + s(x) - s(y)$.
(ii) $s(x') = 1 - s(x)$.
(iii) $s(x'') = s(x)$.
(iv) $s(x \rightarrow y) = 1 - s(x \cup y) + s(y)$.
(v) $s(x \cup y) = s(y \cup x)$.
(vi) $s(x'' \rightarrow y'') - s(x \rightarrow y) = 1 - s((x \wedge y'') \rightarrow y)$.

Proof (i) Suppose $x, y \in H$ such that $x \leq y$. Then $x \rightarrow y = 1$ and since s is a Bosbach state on **H**, we have $s(x) + s(x \rightarrow y) = s(y) + s(y \rightarrow x)$. Thus, $s(y \rightarrow x) = 1 + s(x) - s(y)$. Also, from $s(y) = 1 + s(x) - s(y \rightarrow x)$, we get that $s(x) \leq s(y)$.

(ii) Since s is a Bosbach state on $\mathbf{H}$, we have $s(x) + s(x \to 0) = s(0) + s(0 \to x) = 0 + 1$. Thus, $s(x') = 1 - s(x)$.

(iii) By (ii), we have $s(x'') = 1 - s(x') = 1 - 1 + s(x) = s(x)$.

(iv) Since $y \le (x \to y) \to y = x \cup y$, (i) implies that

$$s(x \to y) = s((x \cup y) \to y) = 1 - s(x \cup y) + s(y), \text{ by Lemma 2.2.1(13).}$$

(v) By (iv) and (BS1), $s(x \cup y) = 1 - s(x \to y) + s(y) = 1 - s(y \to x) + s(x) = s(y \cup x)$.

(vi) From $y \le y''$ we get $x \to y \le x \to y''$, so by (i),

$$s((x \to y'') \to (x \to y)) = 1 + s(x \to y) - s(x \to y'').$$

Also, $\qquad (x \to y'') \to (x \to y) = (x \odot (x \to y'')) \to y = (x \wedge y'') \to y.$ $\qquad$ By Remark 2.2.5(v), we have $x \to y'' = x'' \to y''$, consequently,

$$
\begin{aligned}
s((x \wedge y'') \to y) &= s((x \to y'') \to (x \to y)) \\
&= 1 + s(x \to y) - s(x \to y'') \\
&= 1 + s(x \to y) - s(x'' \to y'').
\end{aligned}
$$

Equivalently, $s(x'' \to y'') - s(x \to y) = 1 - s((x \wedge y'') \to y)$. $\qquad\square$

Corollary 7.4.4 *Let $\mathbf{H}$ be a bounded Wajsberg hoop with a Bosbach state s. Then $s(x \to y) = 1 - s(x \vee y) + s(y)$.*

Proof Since $\mathbf{H}$ is a Wajsberg hoop, $x \cup y = x \vee y$ for all $x, y \in H$, see Proposition 3.3.9(iii). Hence, by Proposition 7.4.3(iv), the proof is evident. $\qquad\square$

Proposition 7.4.5 ([43, 82]) *Let $\mathbf{H}$ be a bounded hoop and s be a Bosbach state on $\mathbf{H}$. Then $\mathrm{Ker}(s) = \{x \in H : s(x) = 1\}$ is a filter of $\mathbf{H}$.*

Proof According to the definition of Bosbach state, clearly, $s(1) = 1$ and so $1 \in \mathrm{Ker}(s) \ne \emptyset$. Suppose $x, x \to y \in \mathrm{Ker}(s)$. Then $s(x) = s(x \to y) = 1$. By Lemma 2.2.1(6), $x \le (x \to y) \to y$ and by Proposition 7.4.3(i), we have $1 = s(x) \le s((x \to y) \to y)$, and so $s((x \to y) \to y) = 1$. Since by Lemma 2.2.1(10), $y \le x \to y$, we have

$$1 + 1 = s((x \to y) \to y) + s(x \to y) = s(y \to (x \to y)) + s(y) = 1 + s(y).$$

Thus, $s(y) = 1$, and so $y \in \mathrm{Ker}(s)$. Hence, $\mathrm{Ker}(s)$ is a filter of $\mathbf{H}$. $\qquad\square$

Proposition 7.4.6 *Let $\mathbf{H}$ be a bounded hoop and s be a Bosbach state on $\mathbf{H}$. Then $\mathbf{H}/\mathrm{Ker}(s)$ is a bounded hoop with (DNP).*

Proof By Proposition 7.4.5, we know that $\mathbf{H}/\mathrm{Ker}(s)$ is a hoop. On one side, $x \leq x''$. On the other hand, by Proposition 7.4.3(i) and (iii), we have

$$s(x'' \to x) = s(x) + s(x \to x'') - s(x'') = s(x) + 1 - s(x) = 1.$$

Thus, $x'' \to x, x \to x'' \in \mathrm{Ker}(s)$ and so $x/\mathrm{Ker}(s) = x''/\mathrm{Ker}(s) = (x/\mathrm{Ker}(s))''$, consequently, $\mathbf{H}/\mathrm{Ker}(s)$ is a bounded hoop with (DNP). $\qquad\square$

Given a Bosbach state, s of a bounded hoop $\mathbf{H}$, $\mathbf{H}/\mathrm{Ker}(s)$ is a bounded hoop with (DNP), by Proposition 7.4.6. Consequently, $\mathbf{H}/\mathrm{Ker}(s)$ is a bounded Wajsberg hoop, by (5.1). Now, Corollary 4.3.28 implies that $\mathrm{Ker}(s)$ is a fantastic filter of $\mathbf{H}$.

Theorem 7.4.7 *Let $\mathbf{H}$ be a bounded hoop and s be a Bosbach state on $\mathbf{H}$. Then $\mathbf{H}/\mathrm{Ker}(s)$ is termwise equivalent to an MV-algebra.*

Proof Let $\mathbf{H}$ be a bounded hoop and s be a Bosbach state on $\mathbf{H}$. Then by Proposition 7.4.6, $\mathbf{H}/\mathrm{Ker}(s)$ is a bounded hoop with (DNP). Thus, by Theorem 3.3.15, $\mathbf{H}/\mathrm{Ker}(s)$ is bounded Wajsberg hoop. Now, the proof follows from Theorem 3.3.13. $\qquad\square$

Definition 7.4.8 ([43]) Let $\mathbf{H} = (H; \odot, \to, 0, 1)$ be a bounded hoop. We say that two elements $x, y \in H$ are

(i)   **Orthogonal** and we write $x \perp y$ if $x'' \leq y'$ or equivalently, $y'' \leq x'$.
(ii) **n-orthogonal**, denoted by $x \perp_n y$ if $x' \leq y''$.
(iii) $\mathbf{H}$ has the **strong orthogonality property** if $x \perp y$ implies that $x \perp_n y$ for all $x, y \in H \setminus \{0\}$.

For two orthogonal elements x, y, we define a partial binary operation, $+$ on $\mathbf{H}$ by $x + y := y' \to x'' = x \oplus y$.

Proposition 7.4.9 *Let $\mathbf{H}$ be a bounded hoop. For all $x, y \in H$, we have*

(i) $x \perp 0$ *and* $0 \perp x$.
(ii) *If* $x \leq y$, *then* $x \perp y'$.
(iii) *If* $x \perp y$, *then* $x^n \perp y^m$ *for all* $n, m \in \mathbb{N}$.
(iv) *If* $x \perp y$, *then* $y \odot x = 0$.
(v)   *If* $x \perp y$, *then* $x'' \perp y''$.

Proof The proof is left to the reader. $\qquad\square$

By Lemma 2.2.1(8), $y' \to x'' = y' \to (x' \to 0) = x' \to (y' \to 0) = x' \to y''$, which means $x + y = y + x$ for each orthogonal elements x, y of H. It is easy to show that, if $x \perp y$, then $y \le x'$ and $x \odot y = 0$ for all $x, y \in H$. In addition, due to Remark 2.2.5, $x \perp y$ if and only if $y \perp x$ for any $x \in H$.

Proposition 7.4.10 *Let* $(\mathbf{H}, \mu)$ *be a linearly ordered state hoop. If* $\mathbf{H}$ *has the strong orthogonality property, then* $\mu(x \oplus y') = \mu(x) \oplus \mu(y')$ *for all* $x, y \in H$.

Proof (a) If $x = 0$, then applying Proposition 7.1.4(2), we have:

$$\mu(0 \oplus y') = \mu(y''') = \mu(y') = \mu(y)',$$

and

$$\mu(0) \oplus \mu(y') = 0 \oplus \mu(y') = \mu(y')'' = \mu(y)'.$$

Thus, $\mu(x \oplus y') = \mu(x) \oplus \mu(y')$.

  (b) If $y = 0$, then, we have

$$\mu(x \oplus 0') = \mu(x \oplus 1) = \mu(1) = 1 \text{ and } \mu(1) \oplus \mu(y') = 1 \oplus \mu(y') = 1.$$

Thus, $\mu(x \oplus y') = \mu(x) \oplus \mu(y')$.

  (c) Assume $x \ne 0$, $y \ne 0$ and $x \le y$. Since $x \perp y'$, by Proposition 7.1.4(10), we get $\mu(x \oplus y') = \mu(x) \oplus \mu(y')$ and $\mu(y' \oplus x) = \mu(y') \oplus \mu(x)$. $\square$

Definition 7.4.11 Let $\mathbf{H}$ be a bounded hoop. A function $s : H \to [0, 1]$ is called a **Riečan state** if the following conditions hold for all $x, y \in H$:
  (RS1) $s(1) = 1$.
  (RS2) If $x \perp y$, then $s(x + y) = s(x) + s(y)$.

Proposition 7.4.12 ([76, Proposition 5.13]) *Let* $\mathbf{H}$ *be a bounded hoop and* s *be a Riečan state on* $\mathbf{H}$. *The following properties hold for all* $x, y \in H$:

  (i) $s(x') = 1 - s(x)$.
  (ii) $s(0) = 0$ *and* $s(x'') = s(x)$.
  (iii) $x \le y$ *implies that* $s(x) \le s(y)$ *and* $s(y \to x'') = 1 + s(x) - s(y)$.
  (v) $s((x \cup y) \to x'') = 1 - s(x \cup y) + s(x)$.

Proof (i) By Proposition 7.4.9(ii), $x \perp x'$ and $x + x' = x'' \to x'' = 1$, so $s(x) + s(x') = s(x + x') = s(1) = 1$, consequently, $s(x') = 1 - s(x)$.

  (ii) Clearly, $0 \perp 0$ and $0 + 0 = 0' \to 0'' = 0$, so $s(0) + s(0) = s(0 + 0) = s(0)$, which implies $s(0) = 0$. In addition, from $0 \perp x$ it follows that

$$s(x'') = s(0' \to x'') = s(x + 0) = s(x) + s(0) = s(x), \text{ by (ii)}.$$

(iii) Let $x \leq y$. Then by Proposition 7.4.9(ii), $x \perp y'$ and

$$x + y' = y'' \to x'' = x' \to y''' = x' \to y' = y \to x'', \text{ by Remark 2.2.5.}$$

It follows that $s(x) + s(y') = s(x + y') = s(y \to x'')$. Now, due to (i), we get $s(y \to x'') = s(x) + 1 - s(y)$ and $s(x) - s(y) = s(y \to x'') - 1 \leq 0$, equivalently, $s(x) \leq s(y)$.

(iv) Since $x \leq x \cup y$, by Proposition 7.4.9(ii), $x \perp (x \cup y)'$ and so

$$\begin{aligned}
x + (x \cup y)' = (x \cup y)'' \to x'' &= x' \to (x \cup y)''', \text{ by Lemma 2.2.1(8)} \\
&= x' \to (x \cup y)', \text{ by Remark 2.2.5(iii)} \\
&= (x \cup y) \to x'', \text{ by Lemma 2.2.1(8).}
\end{aligned}$$

It follows that

$$s((x \cup y) \to x'') = s(x + (x \cup y)') = s(x) + s((x \cup y)') = s(x) + 1 - s(x \cup y), \text{ by (i).}$$

$\square$

Theorem 7.4.13 *Let* **H** *be a bounded hoop. Then any Bosbach state on* **H** *is a Riečan state.*

Proof Let s be a Bosbach state on **H**. Assume $x \perp y$, i.e., $x'' \leq y'$. By Proposition 7.4.3, $1 + s(x'') = s(x'' \to y') + s(x'') = s(y') + s(y' \to x'')$ and so $1 + s(x) = 1 - s(y) + s(y' \to x'')$. Thus, $s(x + y) = s(y' \to x'') = s(x) + s(y)$. Hence, s is a Riečan state. $\square$

Lemma 7.4.14 *Let* **H** *be a bounded hoop satisfying* (DNP) *and* $s : H \to [0, 1]$ *be a map such that* $s(0) = 0$, $s(1) = 1$ *and* $s((x \to y)'') = 1 - s(x \cup y) + s(y)$. *Then* s *is a Bosbach state.*

Proof By Theorem 3.3.15, **H** is a bounded Wajsberg hoop and due to Proposition 3.3.9, $x \vee y = (x \to y) \to y = (y \to x) \to x$ for all $x, y \in H$. Let $x, y \in H$. Then

$$\begin{aligned}
s(x) + s(x \to y) = s(x) + s((x \to y)'') &= s(x) + 1 - s(x \cup y) + s(y) \\
&= 1 - s(y \cup x) + s(x) + s(y) = s(y \to x) + s(y).
\end{aligned}$$

Hence, s is a Bosbach state. $\square$

Theorem 7.4.15 ([76]) *Let* **H** *be a bounded hoop with* (DNP). *Then a map* $s : H \to [0, 1]$ *is a Bosbach state if and only if* s *is a Riečan state.*

Proof Suppose that s is a Riečan state. By (RS1), $s(1) = 1$ and by Proposition 7.4.12(ii), $s(0) = 0$. In addition, for each $x, y \in H$ we have $s(x \to y) = s((x \to y)'')$. Also, by Theorem 3.3.15, $\mathbf{H}$ is a bounded Wajsberg hoop, so we have

$$
\begin{aligned}
s((x \to y)'') &= s(x \to y) \\
&= s(((x \to y) \to y) \to y), \text{ by Lemma 2.2.1(13)} \\
&= s(((x \to y) \to y) \to y''), \text{ since } \mathbf{H} \text{ satisfies (DNP)} \\
&= s((x \cup y) \to y'') \\
&= s((y \cup x) \to y'') \\
&= 1 - s(x \cup y) + s(y), \text{ by Proposition 7.4.12(v).}
\end{aligned}
$$

Therefore, s is a Bosbach state, see Lemma 7.4.14. The proof of the converse follows from Theorem 7.4.13. $\qquad\square$

Theorem 7.4.16 *Let μ be a state operator on a bounded hoop $\mathbf{H}$ preserving $\to$. If s is a Bosbach state on $\mathbf{H}$, then the mapping $s_\mu : H \to [0, 1]$ defined by $s_\mu(x) = s(\mu(x))$ is a Bosbach state on $\mathbf{H}$.*

Proof Obviously, $s_\mu(0) = 0$ and $s_\mu(1) = 1$, so (SB2) is verified. It follows that

$$
\begin{aligned}
s_\mu(x) + s_\mu(x \to y) &= s(\mu(x)) + s(\mu(x \to y)) \\
&= s(\mu(x)) + s(\mu(x) \to \mu(y)) \\
&= s(\mu(y)) + s(\mu(y) \to \mu(x)) \\
&= s(\mu(y)) + s(\mu(y \to x)) \\
&= s_\mu(y) + s_\mu(y \to x).
\end{aligned}
$$

Thus, s_μ satisfies (SB1). Therefore, s_μ is a Bosbach state on $\mathbf{H}$. $\qquad\square$

Corollary 7.4.17 *Let $(\mathbf{H}, \mu)$ be a linearly ordered state hoop and s be a Bosbach state on $\mathbf{H}$. Then the mapping $s_\mu : H \to [0, 1]$ defined by $s_\mu(x) = s(\mu(x))$ for each $x \in H$, is a Bosbach state on $\mathbf{H}$.*

Proof According to Corollary 7.1.5, μ preserves $\to$, hence by Theorem 7.4.16, s_μ is a Bosbach state on $\mathbf{H}$. $\qquad\square$

Corollary 7.4.18 *Let $(\mathbf{H}, \mu)$ be an idempotent state hoop and s be a Bosbach state on $\mathbf{H}$. Then the mapping $s_\mu : \mathbf{H} \to [0, 1]$ defined by $s_\mu(x) = s(\mu(x))$ is a Bosbach state on $\mathbf{H}$.*

Proof By Proposition 7.2.4, we have $\mu(x \wedge y) = \mu(x) \wedge \mu(y)$. According to Proposition 7.1.6, μ preserves $\to$, hence by Theorem 7.4.16, s_μ is a Bosbach state on $\mathbf{H}$. $\qquad\square$

In Proposition 3.3.36, it was proved that if $\mathbf{H}$ is a bounded basic hoop, then $\mathrm{Reg}(\mathbf{H})$ is a bounded subhoop of $\mathbf{H}$.

Theorem 7.4.19 ([78, Theorem 6.4]) *Let $\mathbf{H}$ be a bounded basic hoop with the Glivenko property and μ be a state operator on $\mathbf{H}$ preserving $\rightarrow$. If s is a Bosbach state on $\mathrm{Reg}(\mathbf{H})$, then the mapping $\tilde{s}_\mu : \mathbf{H} \rightarrow [0, 1]$ defined by $\tilde{s}_\mu(x) = s(\mu(x''))$ is a Bosbach state on $\mathbf{H}$.*

Proof Obviously, $s_\mu(0) = 0$ and $s_\mu(1) = 1$, so (SB2) is verified.

Applying Remark 7.2.2, we have

$$
\begin{aligned}
\tilde{s}_\mu(x) + \tilde{s}_\mu(x \rightarrow y) &= s(\mu(x'')) + s(\mu((x \rightarrow y)'')) \\
&= s(\mu(x'')) + s(\mu(x'' \rightarrow y'')) \\
&= s(\mu(x'')) + s(\mu(x'') \rightarrow \mu(y'')) \\
&= s(\mu(y'')) + s(\mu(y'') \rightarrow \mu(x'')) \\
&= s(\mu(y'')) + s(\mu(y'' \rightarrow x'')) \\
&= s(\mu(y'')) + s(\mu((y \rightarrow x)'')) \\
&= \tilde{s}_\mu(y) + \tilde{s}_\mu(y \rightarrow x).
\end{aligned}
$$

Thus, $\tilde{s}_\mu$ satisfies the condition (SB1). It follows that $\tilde{s}_\mu$ is a Bosbach state on $\mathbf{H}$. $\qquad\square$

Theorem 7.4.20 *Let $\mathbf{H}$ be a hoop satisfying the strong orthogonality property, μ be a state operator, and s be a Riečan state on $\mathbf{H}$. Then the mapping $s_\mu : \mathbf{H} \rightarrow [0, 1]$ defined by $s_\mu(x) = s(\mu(x))$ is a Riečan state on $\mathbf{H}$.*

Proof Obviously, $s_\mu(1) = 1$. It is easy to check that $s_\mu(x \oplus 0) = s_\mu(x) + s_\mu(0)$ and $s_\mu(0 \oplus x) = s_\mu(0) + s_\mu(x)$. Consider $x, y \in H$ such that $x \neq 0$, $y \neq 0$ and $x \perp y$. It follows that $\mu(x) \perp \mu(y)$. By the assumption and Proposition 7.1.4(10), we get $\mu(x \oplus y) = \mu(x) \oplus \mu(y)$. Hence,

$$
\begin{aligned}
s_\mu(x \oplus y) &= s(\mu(x \oplus y)) = s(\mu(x) \oplus \mu(y)) \\
&= s(\mu(x)) \oplus s(\mu(y)) = s_\mu(x) \oplus s_\mu(y).
\end{aligned}
$$

Therefore, s_μ is a Riečan state on $\mathbf{H}$. $\qquad\square$

Theorem 7.4.21 ([78, Theorem 6.7]) *Let $\mathbf{H}$ be a bounded basic hoop with the strong orthogonality property and τ be a state operator on $\mathbf{H}$. If s is a Riečan state on $\mathrm{Reg}(\mathbf{H})$, then the mapping $\tilde{s}_\tau : \mathbf{H} \rightarrow [0, 1]$ defined by $\tilde{s}_\tau(x) = s(\tau(x''))$ is a Riečan state on $\mathbf{H}$.*

Proof Given $x \in H$ we have $x \perp 0$ and $x + 0 = (x' \odot 0')' = x''$, so $s(x'') = s(x + 0) = s(x) + s(0) = s(x) + 0 = s(x)$. Obviously, $\tilde{s}_\tau(1) = 1$, $\tilde{s}_\tau(x \oplus 0) = \tilde{s}_\tau(x) + \tilde{s}_\tau(0)$ and $\tilde{s}_\tau(0 \oplus x) = \tilde{s}_\tau(0) + \tilde{s}_\tau(x)$.

Consider $x, y \in H$ such that $x \neq 0$, $y \neq 0$ and $x \perp y$. Then by Proposition 7.4.9(v), $x'' \perp y''$. Since $(x \oplus y)'' = x'' \oplus y''$ (Lemma 2.2.1(13)), by Proposition 7.1.4(10), we get

$$
\begin{aligned}
\tilde{s}_\tau(x \oplus y) &= s(\tau((x \oplus y)'')) \\
&= s(\tau(x'' \oplus y'')) \\
&= s(\tau(x'') \oplus \tau(y'')) \\
&= s(\tau(x'')) \oplus s(\tau(y'')) \\
&= \tilde{s}_\tau(x) \oplus \tilde{s}_\tau(y)
\end{aligned}
$$

(since s is a Riečan state and from $x'' \perp y''$ it follows that $\tau(x'') \perp \tau(y'')$). Thus, $\tilde{s}_\tau$ is a Riečan state on **H**. $\square$

7.5 States on Intervals of a Hoop

Let **H** be a hoop. Consider $C_c(\mathbf{H})$ and $C_\vee(\mathbf{H})$ which were defined in Chap. 3. In Proposition 3.2.5, we showed that given $a \in C_\vee(\mathbf{H})$, $([a, 1]; \odot^a, \rightarrow, a, 1)$ is a bounded hoop where

$$
x \odot^a y := (x \odot y) \vee a \qquad \forall x, y \in [a, 1].
$$

Now, applying Proposition 3.3.2, we will prove that for each $a \in C_c(\mathbf{H})$ of a bounded hoop **H**, the interval $[0, a]$ forms a bounded hoop.

Proposition 7.5.1 ([76, Theorem 4.2]) *Let* **H** *be a bounded hoop and* $a \in C_c(\mathbf{H})$. *Then* $([0, a]; \otimes_a, \rightarrow_a, 0, a)$ *is a bounded hoop where*

$$
x \otimes y := x \odot (a \rightarrow y) \quad \& \quad x \rightarrow_a y := (x \rightarrow y) \odot a, \qquad \forall x, y \in [0, a].
$$

Proof Choose $x, y, z \in [0, a]$. In order to show $([0, a]; \otimes_a, \rightarrow_a, 0, a)$ is a hoop, we will apply Theorem 3.1.3.

(1) $a \odot (x \otimes y) = a \odot (x \odot (a \rightarrow y)) = x \odot (a \odot (a \rightarrow y)) = x \odot (a \wedge y) = x \odot y$. Similarly, $a \odot (y \otimes x) = x \odot y$. From $a \in C_c(\mathbf{H})$ it follows that $x \otimes y = y \otimes x$.

(2) $x \otimes a = x \odot (a \rightarrow a) = x \odot 1 = x$.

(3) $x \rightarrow_a a = (x \rightarrow a) \odot a = 1 \odot a = a$ and $x \rightarrow_a x = (x \rightarrow x) \odot a = 1 \odot a = a$.

(4) By (1), $(x \otimes y) \to_a z = (y \otimes x) \to_a z = ((y \odot (a \to x)) \to z) \odot a = ((a \to x)$ $\to (y \to z)) \odot a$. Also, $x \to_a (y \to_a z) = x \to_a ((y \to z) \odot a) = (x \to ((y \to z) \odot$ $a)) \odot a$. Since $a \in C_c(\mathbf{H})$, in order to prove that $(x \otimes y) \to_a z = x \to_a (y \to_a z)$ it suffices to show that

$$(a \to x) \to (y \to z) = x \to ((y \to z) \odot a). \tag{7.2}$$

For each $u \in H$ we have

$$u \le (a \to x) \to (y \to z) \Leftrightarrow u \odot (a \to x) \le y \to z, \text{ by (2.2)}$$
$$\Leftrightarrow (u \odot (a \to x)) \odot a \le (y \to z) \odot a, \text{ since } a \in C_c(\mathbf{H})$$
$$\Leftrightarrow u \odot (a \wedge x) \le (y \to z) \odot a, \text{ by Proposition 3.1.4(i)}$$
$$\Leftrightarrow u \odot x \le (y \to z) \odot a, \text{ since } x \le a$$
$$\Leftrightarrow u \le x \to ((y \to z) \odot a), \text{ by (2.2).}$$

Thus, (7.2) holds, consequently, $(x \otimes y) \to_a z = x \to_a (y \to_a z)$.

(5) Given $x, y \in H$ we have

$$(x \to_a y) \otimes x = (y \to_a x) \otimes y \Leftrightarrow ((x \to y) \odot a) \odot (a \to x) = ((y \to x) \odot a) \odot (a \to y)$$
$$\Leftrightarrow (x \to y) \odot (a \odot (a \to x)) = (y \to x) \odot (a \odot (a \to y))$$
$$\Leftrightarrow (x \to y) \odot (a \wedge x) = (y \to x) \odot (a \wedge y)$$
$$\Leftrightarrow (x \to y) \odot x = (y \to x) \odot y.$$

The last equality holds by Proposition 3.1.2(H3). Hence, $(x \to_a y) \otimes x = (y \to_a x) \otimes y$ for all $x, y \in H$.

From (1)–(5) and Theorem 3.1.3 it follows that $([0, a]; \otimes_a, \to_a, a)$ is a hoop. Furthermore, $0 \to x = (0 \to x) \odot a = a$ for all $x \in [0, a]$ which means $([0, a]; \otimes_a, \to_a, 0, a)$ is a bounded hoop. $\qquad\square$

Proposition 7.5.2 ([76]) *Let* $\mathbf{H}$ *be a bounded hoop,* $a \in C_c(\mathbf{H})$ *and* s *be a Bosbach state on* $\mathbf{H}$. *Then the following statements hold*

(i) $s((x \to y) \odot a) = s(a) + s(x \to y) - 1$.
(ii) $s(a^{n+1}) = s(a^n) + s(a) - 1$ *and* $s(a^n) = ns(a) - n + 1$.
(iii) *If* $\mathbf{H}$ *is simple and* $a \ne 1$, *then* $s(a) \ne 1$.

Proof (i) Let $x, y \in H$.

$$s(a) + s(x \to y) - 1 = s(a) + s(a \to ((x \to y) \odot a)) - 1, \text{ by Proposition 3.3.2(i)}$$
$$= s((x \to y) \odot a) + s(((x \to y) \odot a) \to a) - 1, \text{ by (BS1)}$$
$$= s((x \to y) \odot a) + s(1) - 1 = s((x \to y) \odot a), \text{ by (BS2).}$$

(ii) By (BS1), $s(a^n) + s(a^n \to a^{n+1}) = s(a^{n+1}) + s(a^{n+1} \to a^n) = s(a^{n+1}) + s(1) = s(a^{n+1}) + 1$, since $a^{n+1} \leq a^n$. Also, by Proposition 3.3.2(ii), $a^n \to a^{n+1} = a$, so $s(a^n) + s(a) = s(a^{n+1}) + 1$. The proof of the second part follows from mathematical induction on n.

(iii) Since $\mathbf{H}$ is simple and $a \neq 1$, then $0 \in H = \langle a \rangle$ and so $a^n = 0$ for some $n \in \mathbb{N}$. Suppose $s(a) = 1$, then by (ii), $s(a^n) = ns(a) - n + 1$, consequently, $0 = s(a^n) = 1$, a contradiction. Therefore, $s(a) \neq 1$. $\qquad\square$

Theorem 7.5.3 ([76, Theorem 6.4]) *Let $\mathbf{H}$ be a bounded hoop, $a \in C_\vee(\mathbf{H})$ and s be a Bosbach state on $\mathbf{H}$. If $s(a) \neq 1$, then $\lambda : [a, 1] \to [0, 1]$ defined by $\lambda(x) = (s(x) - s(a))/(1 - s(a))$ is a Bosbach state on $([a, 1]; \odot^a, \to, a, 1)$.*

Proof Choose $x, y \in [a, 1]$.

$$
\begin{aligned}
\lambda(x) + \lambda(x \to y) &= \frac{s(x) - s(a)}{1 - s(a)} + \frac{s(x \to y) - s(a)}{1 - s(a)} \\
&= \frac{s(x) + s(x \to y) - 2\,s(a)}{1 - s(a)} \\
&= \frac{s(y) + s(y \to x) - 2\,s(a)}{1 - s(a)} \\
&= \frac{s(y) - s(a)}{1 - s(a)} + \frac{s(y \to x) - s(a)}{1 - s(a)} \\
&= \lambda(y) + \lambda(y \to x).
\end{aligned}
$$

Hence, (BS1) holds. In addition, $\lambda(1) = (s(1) - s(a))/(1 - s(a)) = 1$ and $\lambda(a) = 0$. Therefore, λ is a Bosbach state on $([a, 1]; \odot^a, \to, a, 1)$. $\qquad\square$

From Proposition 7.5.2(iii) and Theorem 7.5.3, it follows that if $a \in C_\vee(\mathbf{H}) \cap C_c(\mathbf{H})$ and $\mathbf{H}$ is a bounded simple hoop, then the map $\lambda : [a, 1] \to [0, 1]$ defined by

$$
\lambda(x) = (s(x) - s(a))/(1 - s(a))
$$

is a Bosbach state on the hoop $([a, 1]; \odot^a, \to, a, 1)$.

Theorem 7.5.4 ([76, Theorem 6.6]) *Let $\mathbf{H}$ be a bounded hoop, $a \in C_c(\mathbf{H})$, and s be a Bosbach state on $\mathbf{H}$. If $s(a) \neq 0$, then $\mu : [0, a] \to [0, 1]$ defined by $\mu(x) = s(x)/s(a)$ is a Bosbach state on $([0, a]; \otimes_a, \to_a, 0, a)$.*

Proof Let $x, y \in [0, a]$.

$$\mu(x) + \mu(x \to_a y) = \mu(x) + \mu((x \to y) \odot a)$$

$$= \frac{s(x)}{s(a)} + \frac{s((x \to y) \odot a)}{s(a)}$$

$$= \frac{s(x)}{s(a)} + \frac{s(a) + s(x \to y) - 1}{s(a)}, \quad \text{by Proposition 7.5.2(i)}$$

$$= \frac{s(x) + s(x \to y) + s(a) - 1}{s(a)}$$

$$= \frac{s(y) + s(y \to x) + s(a) - 1}{s(a)}$$

$$= \frac{s(y)}{s(a)} + \frac{s((y \to x) \odot a)}{s(a)}$$

$$= \mu(y) + \mu(y \to_a x).$$

Also, $\mu(0) = s(0)/s(a) = 0$ and $\mu(a) = s(a)/s(a) = 1$. Therefore, μ is a Bosbach state on $([0, a]; \otimes_a, \to_a, 0, a)$. $\square$

Assume that $\mathbf{H}$ is a bounded hoop such that $a, b \in C_\vee(\mathbf{H}) \cap C_c(\mathbf{H})$ with $a \leq b$. By Proposition 3.2.5, $([a, 1]; \odot^a, \to, a, 1)$ is a bounded hoop. For each $x, y \in [a, 1]$, $x \odot^a b = y \odot^a b$ implies that $x \odot a \odot b = y \odot a \odot b$, consequently, $x = y$, since $a, b \in C_c(\mathbf{H})$, i.e., b belongs to the cancellative center of the bounded hoop $([a, 1]; \odot^a, \to, a, 1)$. From Proposition 7.5.1, it follows that $([a, b]; \otimes_b, \to_b, a, b)$ is a bounded hoop, where

$$x \otimes_b y := x \odot^a (b \to y) = \big(x \odot (b \to y)\big) \vee a, \quad \forall x, y \in [a, b],$$

$$x \to_b y := (x \to y) \odot^a b = \big((x \to y) \odot b\big) \vee a, \quad \forall x, y \in [a, b].$$

On the other hand, since $x \leq b$, we have $b \to y \leq x \to y$, and so

$$a \leq y = b \wedge y = (b \to y) \odot b \leq (x \to y) \odot b,$$

which implies $x \to_b y = (x \to y) \odot b$ for all $x, y \in [a, b]$. Therefore, $([a, b]; \otimes_b, \to_b, a, b)$ with the following operations is a bounded hoop:

$$x \otimes_b y = \big(x \odot (b \to y)\big) \vee a, \quad x \to_b y = (x \to y) \odot b, \quad \forall x, y \in [a, b]. \tag{7.3}$$

Theorem 7.5.5 ([76, Theorem 6.7]) *Let $\mathbf{H}$ be a bounded hoop such that $a, b \in C_\vee(\mathbf{H}) \cap C_c(\mathbf{H})$, $a \leq b$, and s be a Bosbach state on $\mathbf{H}$. If $s(a) \neq s(b)$, then $v : [a, b] \to [a, b]$ defined by $v(x) = (s(x) - s(a))/(s(b) - s(a))$ is a Bosbach state on $([a, b]; \otimes_b, \to_b, a, b)$.*

Proof Given $x, y \in H$ we have

$$
\begin{aligned}
v(x) + v(x \rightarrow_b y) &= \frac{s(x) - s(a)}{s(b) - s(a)} + \frac{s((x \rightarrow y) \odot b) - s(a)}{s(b) - s(a)} \\
&= \frac{s(x) - s(a)}{s(b) - s(a)} + \frac{s(b) + s(x \rightarrow y) - 1 - s(a)}{s(b) - s(a)}, \text{ by Proposition 7.5.2(i)} \\
&= \frac{s(x) + s(x \rightarrow y) - s(a) + s(b) - 1 - s(a)}{s(b) - s(a)} \\
&= \frac{s(y) + s(y \rightarrow x) - s(a) + s(b) - 1 - s(a)}{s(b) - s(a)} \\
&= \frac{s(y) - s(a)}{s(b) - s(a)} + \frac{s(y \rightarrow x) + s(b) - 1 - s(a)}{s(b) - s(a)} \\
&= v(y) + \frac{s((y \rightarrow x) \odot b) - s(a)}{s(b) - s(a)} \\
&= v(y) + v(y \rightarrow x), \text{ by Proposition 7.5.2(i)}.
\end{aligned}
$$

Obviously, $v(a) = 0$ and $v(b) = 1$. Therefore, v is a Bosbach state on $([a, b]; \otimes_b, \rightarrow_b, a, b)$. $\qquad\square$

Bibliographical Remarks and Suggestions for Further Study

Most of the results of this Chap. 7 were originally stated in [45, 78, 79]. Motivated by the observation that the definition of Bosbach states relies on the standard MV-algebra structure of [0, 1], the notion of a state for residuated lattices was generalized to a function with values in a residuated lattice (see [83, 84]). Subsequently, this concept was extended to pseudo BCK-algebras and pseudo hoops [85] (see Appendix A for these concepts). The properties of these generalized states are valuable for developing an algebraic theory of probabilistic models for non-commutative fuzzy logics.

Given a bounded hoop $\mathbf{H}$ and an arbitrary function $s : \mathbf{H} \rightarrow \mathbf{H}$ with $s(0) = 0$, s is a **generalized Bosbach state** of type I (or type I state) if it satisfies one of the following equivalent conditions:

(bsI1) $x \geq y$ implies that $s(x \rightarrow y) = s(x) \rightarrow s(y)$ for all $x, y \in H$.

(bsI2) $s(x \vee y) = s(x \rightarrow y) \rightarrow s(y)$ for all $x, y \in H$.

(bsI3) $s(x \rightarrow y) \rightarrow s(y) = s(y \rightarrow x) \rightarrow s(x)$ and $s(1) = 1$ for all $x, y \in H$.

(bsI4) $x \geq y$ implies that $s(x) = s(x \rightarrow y) \rightarrow s(y)$ for all $x, y \in H$.

(bsI5) $s(x \rightarrow y) = s(x \vee y) \rightarrow s(y)$ for all $x, y \in H$.

(bsI6) $s(x \rightarrow y) = s(x) \rightarrow s(x \wedge y)$ for all $x, y \in H$.

For more details on the generalized Bosbach state, we refer the reader to [78, Sect. 7].

Furthermore, in [147], Y. L. Fu et al. investigated state maps on semihoops. They introduced several types of filters defined by state maps on semihoops, namely SM-filters, state filters, and dual state filters, and discussed the relationships among them. Utilizing SM-filters, they characterized two types of state semihoops.

A comprehensive study of states and state-morphisms on bounded Wajsberg hoops has been conducted in several research papers: [224] discusses the averaging of truth-values in Lukasiewicz logic, [233] is studying measures on MV-algebras, [212] is concerned with the Lebesgue state of unital Abelian ℓ-groups, [144] is investigating the relationship between internal states and probabilistic fuzzy logics, and [93] is about state-morphism MV-algebras. In addition, [122] presents a complete characterization of subdirectly irreducible MV-algebras with internal states. This allows us to classify subdirectly irreducible state-morphism MV-algebras and describe single generators of the variety of state-morphism MV-algebras, and it shows that we have a continuum of varieties of state-morphism MV-algebras. It is in contrast to the result by Komori [196] saying the lattice of subvarieties of MV-algebras is countable. In [54], generators of the varieties of state-morphism algebras, in particular, generators of state-morphism BL-algebras, state-morphism MTL-algebras, state-morphism non-associative BL-algebras, and state-morphism pseudo MV-algebras are presented.

7.6   Exercises

7.6.1 Let $(M; \oplus, ', 0, 1)$ be an MV-algebra. A mapping $\mu : M \to M$ is a state operator if (1) $\mu(1) = 1$; (2) $\mu(x') = \mu(x)'$; (3) $\mu(x \oplus y) = \mu(x) \oplus \mu(y \odot (x \odot y)')$; and (4) $\mu(\mu(x) \oplus \mu(y)) = \mu(x) \oplus \mu(y)$, where $x \odot y = (x' \oplus y')'$. Prove that if s is a state operator of a bounded hoop $\mathbf{H}$ satisfying (DNP), then it is a state operator of the MV-algebra $(H; \oplus, ', 0, 1)$, where $x \oplus y = (x' \odot y')'$.

7.6.2 A homomorphism μ on a hoop $\mathbf{H} = (H; \odot, \to, 1)$ is said to be a **state-morphism** if $\mu^2 := \mu \circ \mu = \mu$. Let μ_1 and μ_2 be two state-morphisms on a hoop $\mathbf{H}$ such that $\mathrm{Ker}(\mu_1) = \mathrm{Ker}(\mu_2)$ and $\mu_1(H) = \mu_2(H)$. Prove that $\mu_1 = \mu_2$.

Appendix A—A Trip into Non-commutative Algebraic Structures—Step into the Realm of Pseudo MV-Algebras

In practice, there are situations where the outcome of the disjunction or conjunction of two events a and b depends on their order; that is, $a \oplus b$ and $b \oplus a$ may differ. In 2001, this fact was considered by Georgescu and Iorgulescu [152], who presented pseudo MV-algebras, which are a non-commutative generalization of MV-algebras. Independently, Rachůnek, [237] has also given another non-commutative generalization of MV-algebras called generalized MV-algebras. Both generalizations are equivalent. The pseudo MV-algebras started a whole scale of non-commutative generalization of known algebras like pseudo BL-algebras, [96, 97], pseudo hoops, [153], pseudo effect algebras, [125, 126], pseudo EMV-algebras, [130, 131], there also exists a non-commutative generalization of EMV-algebras that locally behaves like an MV-algebra [129]. They have a common roof by Bosbach [50, 51] as complemented semigroups and are now also studied as residuated lattices, [149].

In Appendix A, we focus on pseudo MV-algebras, highlighting their considerable complexity. These structures can also be studied in the framework of pseudo hoops or pseudo BL-algebras. This appendix provides an advanced setting for investigating certain non-commutative algebraic systems, with potential applications in non-commutative logic and reasoning.

8.1   Pseudo MV-Algebras

According to [152], a **pseudo MV-algebra** is an algebra $M = (M; \oplus, ^-, ^\sim, 0, 1)$ of type $(2, 1, 1, 0, 0)$ such that the following axioms hold for all $x, y, z \in M$ with an additional binary operation $\odot$ defined via

$$y \odot x = (x^- \oplus y^-)^\sim \tag{8.1}$$

© The Author(s), under exclusive license to Springer Nature Switzerland AG 2026
A. Dvurečenskij et al., *Hoop Algebras*, Frontiers in Mathematics,
https://doi.org/10.1007/978-3-032-11736-6_8

(A1) $x \oplus (y \oplus z) = (x \oplus y) \oplus z$.
(A2) $x \oplus 0 = 0 \oplus x = x$.
(A3) $x \oplus 1 = 1 \oplus x = 1$.
(A4) $1^\sim = 0$; $1^- = 0$.
(A5) $(x^- \oplus y^-)^\sim = (x^\sim \oplus y^\sim)^-$.
(A6) $x \oplus (x^\sim \odot y) = y \oplus (y^\sim \odot x) = (x \odot y^-) \oplus y = (y \odot x^-) \oplus x$.
(A7) $x \odot (x^- \oplus y) = (x \oplus y^\sim) \odot y$.
(A8) $(x^-)^\sim = x$.

If we define $x \le y$ if and only if $x^- \oplus y = 1$, then $\le$ is a partial order on M such that $0 \le x \le 1$, and $(M, \le)$ is a distributive lattice in which $x \vee y$ and $x \wedge y$ exist due to (A6) and (A7), i.e., $x \vee y = x \oplus (x^\sim \odot y)$ and $x \wedge y = x \odot (x^- \oplus y)$.

We show a close connection between pseudo MV-algebras and ℓ-groups not necessarily Abelian. We recall that such an ℓ-group is an algebra $G = (G; \vee, \wedge, +, -, 0)$ (we prefer additively written form) and equipped with a partial order $\le$ such that $g \le h$ implies $u + g + v \le u + h + v$ for all $g, h, u, v \in G$. We define the positive and negative cones by $G^+ = \{g \in G : g \ge 0\}$ and $G^- = \{g \in G : g \le 0\}$. An element $u \in G^+$ is said to be a **strong unit** of G if given $g \in G$, there is an integer $n \ge 1$ such that $g \le nu$, equivalently $G = \bigcup_{n=1}^\infty [-nu, nu]$. A couple (G, u) where u is a fixed strong unit of G is said to be a **unital ℓ-group**. Let G and H be two ℓ-groups. On the direct product $G \times H$ we define a lexicographic order $\le$ defined by $(g_1, h_1) \le (g_2, h_2)$ if and only if $g_1 < g_2$ or $g_1 = g_2$ and $h_1 \le h_2$. Such a group $G \overset{\rightarrow}{\times} H$ is an ℓ-group if and only if G is linearly ordered, [147, p. 26], and in such a case, the **lexicographic product** is denoted by $G \overset{\rightarrow}{\times} H = (G \overset{\rightarrow}{\times} H; \vee, \wedge, +, -, (0, 0))$.

Archetypical examples of pseudo MV-algebras are of the following form:

Example 8.1.1 If (G, u) is a unital ℓ-group (not necessarily Abelian), we define

$$\Gamma(G, u) := [0, u] = \{g \in G : 0 \le g \le u\}$$

and, for all $x, y \in [0, u]$, we put

$$x \oplus y := (x + y) \wedge u, \quad x^- := u - x,$$
$$x \odot y := (x - u + y) \vee 0, \quad x^\sim := -x + u,$$

then $\Gamma(G, u) := ([0, u]; \oplus, ^-, ^\sim, 0, u)$ is a pseudo MV-algebra [152].

A fundamental result on the theory of pseudo MV-algebras, which we show later, says that every pseudo MV-algebra is of this form, see [109].

We recall that if $\oplus$ is commutative, then $x^\sim = x^-$ for each $x \in M$ (M is then said to be **symmetric**) then $(M; \oplus, ^-, 0, 1)$ is an MV-algebra. We note that if M is symmetric, then it does not mean that $\oplus$ is automatically commutative:

Example 8.1.2 Let G be a non-Abelian ℓ-group and $\mathbb{Z}$ the group of integers. The element $(1, 0)$ is a strong unit in the lexicographic product $G \overrightarrow{\times} \mathbb{Z}$, and the pseudo MV-algebra $\Gamma(G \overrightarrow{\times} \mathbb{Z}, (1, 0))$ is symmetric, but $\oplus$ is not commutative.

Basic properties of pseudo MV-algebras can be found in [152]. We note that $\odot$ has higher priority than $\oplus$.

Proposition 8.1.3 ([152]) *Let* $M = (M; \oplus, ^-, ^\sim, 0, 1)$ *be a pseudo MV-algebra. For all* $x, y, z \in M$, *we have*

$$\text{(i)} \quad y \odot x = (x^\sim \oplus y^\sim)^-.$$
$$\text{(ii)} \quad (x^\sim)^- = x.$$
$$\text{(iii)} \quad 0^\sim = 0^- = 1.$$
$$\text{(iv)} \quad x \odot 1 = 1 \odot x = x.$$
$$\text{(v)} \quad x \odot 0 = 0 = 0 \odot x.$$
$$\text{(vi)} \quad x \oplus x^\sim = 1 = x^- \oplus x.$$
$$\text{(vii)} \quad x \odot x^- = 0 = x^\sim \odot x.$$
$$\text{(viii)} \quad (x \oplus y)^- = y^- \odot x^-, \; (x \oplus y)^\sim = y^\sim \odot x^\sim.$$
$$\text{(ix)} \quad (x \odot y)^- = y^- \oplus x^-, \; (x \odot y)^\sim = y^\sim \oplus x^\sim.$$
$$\text{(x)} \quad x \oplus y = (y^- \odot x^-)^\sim = (y^\sim \odot x^\sim)^-.$$
$$\text{(xi)} \quad (x^\sim \odot y) \oplus y^\sim = x^\sim \vee y^\sim = (y^\sim \odot x) \oplus x^\sim.$$
$$\text{(xii)} \quad x \odot (x^- \oplus y) = x \wedge y = y \odot (y^- \oplus x).$$
$$\text{(xiii)} \quad \oplus \text{ is associative.}$$
$$\text{(xiv)} \quad x \leq y \text{ if and only if } y \oplus x^\sim = 1. \text{ In addition, } (M; \oplus, 0) \text{ is a partially ordered monoid.}$$

Proof (i) It follows from (A5) and (8.1). For the proofs of other properties, consult with [152, Proposition 1.7], if necessary. $\qquad\square$

We note that $\oplus$ and $\odot$ are monotone. Moreover, $x \wedge (\bigvee_i y_i) = \bigvee_i (x \wedge y_i)$ and $x \vee (\bigwedge_i y_i) = \bigwedge_i (x \vee y_i)$ whenever corresponding joins and meets are defined (see, e.g., [152, Proposition 1.18]).

Proposition 8.1.4 ([152]) *Let* $M = (M; \oplus, ^-, ^\sim, 0, 1)$ *be a pseudo MV-algebra,* $x \in M$, *and* $\{y_i\}_{i \in I} \subseteq M$. *There hold*

(i) $x \oplus (\bigvee_i y_i) = \bigvee_i (x \oplus y_i), \; (\bigvee_i y_i) \oplus x = \bigvee_i (y_i \oplus x)$ *whenever the arbitrary joins exist.*

(ii) $x \odot (\bigvee_i y_i) = \bigvee_i (x \odot y_i), \; (\bigvee_i y_i) \odot x = \bigvee_i (y_i \odot x)$ *whenever the arbitrary joins exist.*

(iii) $x \oplus (\bigwedge_i y_i) = \bigwedge_i (x \oplus y_i)$, $(\bigwedge_i y_i) \oplus x = \bigwedge_i (y_i \oplus x)$ *whenever the arbitrary meets exist.*

(iv) $x \odot (\bigwedge_i y_i) = \bigwedge_i (x \odot y_i)$, $(\bigwedge_i y_i) \odot x = \bigwedge_i (y_i \odot x)$ *whenever the arbitrary meets exist.*

Proof We prove only (i) and (iii). The other equalities can be proved in similar ways.

First, assume that $y = \bigwedge_i y_i$. By Proposition 8.1.3(xiv), we have $x \oplus y \leq x \oplus y_i$ for each $i \in I$. Let $z \leq x \oplus y_i$ for every $i \in I$. Then $x^\sim \odot z \leq x^\sim \odot (x \oplus y_i) = x^\sim \wedge y_i \leq y_i$, which implies $x^\sim \odot z \leq \bigwedge_i y_i$. It follows that $z \leq x \vee z = x \oplus (x^\sim \odot z) \leq x \oplus y$. Therefore, $x \oplus y = \bigwedge_i (x \oplus y_i)$, which means (iii) holds.

Now, let $y = \bigvee_i y_i$. For each i, we have $x \leq x \oplus y_i \leq x \oplus y$. Assume $x \oplus y_i \leq z$ for each i. Then $x^\sim \odot (x \oplus y_i) \leq x^\sim \odot z$. On the other side, $x^\sim \odot (x \oplus y_i) = x^\sim \odot ((x^\sim)^- \oplus y_i) = x^\sim \wedge y_i$. Therefore, $x^\sim \wedge y_i \leq x^\sim \odot z$.

Hence, $x^\sim \wedge y = \bigvee_i (x^\sim \wedge y_i) \leq x^\sim \odot z$ and $x \oplus (x^\sim \wedge (\bigvee_i y_i)) \leq x \oplus (x^\sim \odot z) = x \vee z = z$. But, by (iii), $x \oplus (x^\sim \wedge (\bigvee_i y_i)) = (x \oplus x^\sim) \wedge (x \oplus y) = x \oplus (\bigvee_i y_i)$, which gives $x \oplus (\bigvee_i y_i) = \bigvee_i (x \oplus y_i)$ as stated. $\qquad\square$

Corollary 8.1.5 ([152]) *Consider the assumptions of Proposition 8.1.4. If I is finite, then conditions (i)–(iv) always hold.*

Proof The proof proceeds in a manner similar to that of Proposition 8.1.4. $\qquad\square$

An element $x \in M$ is said to be **idempotent** or **Boolean** if $x \oplus x = x$. We denote by $B(M)$ the set of Boolean elements of the pseudo MV-algebra M. We have $0, 1 \in B(M)$. Moreover, $B(M)$ is a Boolean algebra. Sometimes, we say that $B(M)$ is a Boolean skeleton of M.

It is possible to show that

Proposition 8.1.6 *The following statements are equivalent:*

(i) x *is Boolean.*
(ii) x^- *is Boolean.*
(iii) $x^\sim$ *is Boolean.*
(iv) $x \odot x = x$.
(v) $x \wedge x^- = 0$.
(vi) $x \wedge x^\sim = 0$.
(vii) $x \vee x^- = 1$.
(viii) $x \vee x^\sim = 1$.

Proof (i) $\Rightarrow$ (v). If $x = x \oplus x$, then $x \wedge x^- = (x \oplus (x^-)^\sim) \odot x^- = (x \oplus x) \odot x^- = x \odot x^- = 0$.

(i) $\Rightarrow$ (vi). Then $x^\sim \wedge x = x^\sim \odot (x \oplus x) = x^\sim \odot x = 0$.

In similar tricks, we can establish all equivalences; for more details, see [152, Proposition 4.2]. $\qquad\qquad\square$

Definition 8.1.7 Let $M = (M; \oplus, ^-, ^\sim, 0, 1)$ be a pseudo MV-algebra. An **ideal** of M is a non-empty subset I of M such that (a) $x, y \in I$ implies, $x \oplus y \in M$, and (b) $x \leq y \in I$ implies $x \in I$. An ideal I is:

(i) **normal** if $x \oplus I = I \oplus x$ for each $x \in M$, where $x \oplus I := \{x \oplus y : y \in I\}$ and $I \oplus x := \{y \oplus x : y \in I\}$;

(ii) **maximal** if I is a proper subset of M and if J is any proper ideal of M such that $I \subseteq J$, then $I = J$. Using the Zorn lemma, every pseudo MV-algebra contains a maximal ideal, as well as there are maximal elements among normal ideals, however, there are pseudo MV-algebras where no maximal ideal is normal, [95];

(iii) **prime** if $x \wedge y \in I$ entails $x \in I$ or $y \in I$.

Every maximal ideal is prime.

If I is a normal ideal of M, then the relation $\cong_I$ on M, defined by $x \cong_I y$ if and only if $x \odot y^-, y \odot x^- \in I$, equivalently, $x^\sim \odot y, y^\sim \odot x \in I$, is an equivalence. Moreover, for each equivalence $\cong$ on M, there is a unique normal ideal I of M such that $\cong = \cong_I$, see [152, Corollary 3.10]. In such a case, given a normal ideal I, we denote by $x/I = \{y \in M : y \cong_I x\}$, $M/I = \{x/I : x \in M\}$, $(x/I)^- = (x^-/I)$, $(x/I)^\sim = (x^\sim/I)$, $x/I \oplus y/I = (x \oplus y)/I$, and $M/I = (M/I; \oplus, ^-, ^\sim, 0/I, 1/I)$ is the **quotient pseudo MV-algebra** corresponding to the normal ideal I. A homomorphism of pseudo MV-algebras means a homomorphism in the standard sense of homomorphisms of algebras.

Define a partial binary operation $+$ on M via: $x + y$ is defined if and only if $x \leq y^-$, and in this case

$$x + y := x \oplus y. \tag{8.2}$$

It is clear that $x + y$ is defined if and only if $x \leq y^-$ if and only if $y \leq x^\sim$.

For each $x \in M$, we put $x^= := (x^-)^-$ and $x^\approx := (x^\sim)^\sim$.

Proposition 8.1.8 *The following properties hold in any pseudo MV-algebra* M.

(i) $x + 0 = x = 0 + x$ *for any* $x \in M$.

(ii) $x + x^\sim = 1 = x^- + x$ *for any* $x \in M$.

(iii) *If* $x^- + y = 1$, *then* $y = x$.

(iv) *If* $x + y = 1$, *then* $y = x^\sim$ *and* $x = y^-$.

  (v) *Let $x + y$ and $(x + y) + z$ be defined in M. Then $y + z$ and $x + (y + z)$ are defined in M, and*

$$(x + y) + z = x + (y + z).$$

    Similarly, if $y + z$ and $x + (y + z)$ are defined in M, so are $x + y$ and $(x + y) + z$.

  (vi) *If $x + b = y$, then $b = x^\sim \odot y$.*

  (vii) *If $x + y_1 = x + y_2$, then $y_1 = y_2$. Cancellation law.*

  (viii) *If $x_1 + y = x_2 + y$, then $x_1 = x_2$. Cancellation law.*

  (ix) *If $y = a + x$, then $a = y \odot x^-$.*

  (x) *If $x \wedge y = 0$, then $x + y$ and $y + x$ are defined in M, and $x + y = x \vee y = y + x$.*

Proof (i) Since $x \leq 1 = 0^-$, then $x + 0 = x \oplus 0 = x$. Similarly, $0 \leq x^-$ implies $0 + x = 0 \oplus x = x$.

  (ii) $(x^\sim)^- = x \geq x$ so that $x + x^\sim = x \oplus x^\sim = 1$, and $x^- \geq x^-$ implies $x^- + x = x^- \oplus x = 1$.

  (iii) Let $x^- + y = 1$, then $x \leq y$. On the other hand, $x^- \leq y^-$ so that $y^{-\sim} \leq x^{-\sim}$, and $y \leq x$. The last two inequalities give $x = y$.

  (iv) Let $x + y = 1$. Then $x \leq y^-$ entails $x^= \leq (y^-)^= = (y^=)^-$ which means that $x^= + y^=$ is defined in M, and $x^= + y^= = 1$. By (iii), we conclude that $(x^-)^- + y^= = 1$ yields $x^- = y^=$ so that $(x^-)^\sim = (y^-)^{-\sim}$ and $x = y^-$, $y = x^\sim$.

  (v) From the existence of $(x + y) + z$ we conclude that $z \leq (x + y)^\sim = y^\sim \odot x^\sim \leq y^\sim$ and $y + z$ is defined in M. In addition, from $z \leq y^\sim \odot x^\sim$ we have $y + z = y \oplus z \leq y \oplus (y^\sim \odot x^\sim) = y \vee x^\sim = x^\sim$ (since $x + y$ exists). Hence, $x + (y + z) \in M$ and $x + (y + z) = x \oplus (y \oplus z) = (x \oplus y) \oplus z = (x + y) + z$.

  (vi) If $x + b = y$, then $y^- + (x + b) = y^- + y = 1$, $(y^- + x) + b = 1$, so that $b = (y^- + x)^\sim = x^\sim \odot y^{-\sim} = x^\sim \odot y$.

  (vii) There exists an element $v \in M$ such that $(x_1 + y) + v = (x_2 + y) + v = 1$, which yields $x_1 = x_2$.

  (viii) It uses ideas from the proof of (vii).

  (ix) Calculate $1 = y + y^\sim = (a + x) + y^\sim = a + (x + y^\sim)$, so that $a = (x + y^\sim)^- = y \odot x^-$.

  (x) We have $x \vee y = x + (x^\sim \odot y) = x + (y \wedge x)^\sim \odot y = x + 0^\sim \odot y$. By symmetry, we get $x \vee y = y + x$. $\qquad\square$

It is worth noting that if $\mathbf{M} = \Gamma(G, u)$ for some unital ℓ-group, then whenever the partial sum $x + y$ exists in $\mathbf{M}$, it coincides with the group addition of x and y in G.

The following lemma will be used to investigate perfect pseudo MV-algebras; consult with [114] for its proof.

Lemma 8.1.9 *Let $(M; \oplus, ^-, ^\sim, 0, 1)$ be a pseudo MV-algebra.*

  (i) *Let $\phi : M \to M$ be a mapping. The following statements are equivalent:*

- *$\phi(x) + y$ exists in M if and only if $y + x$ exists in M, for all $x, y \in M$.*
- *$\phi(x) = x^{--}$ for all $x \in M$.*

  (ii) *The mappings $x \mapsto x^{\sim\sim}$ and $x \mapsto x^{--}$ are automorphisms of M.*

8.2   Representation of Pseudo MV-Algebras and Their Categorical Equivalence

A fundamental result on a representation of pseudo MV-algebras says there is a one-to-one correspondence between pseudo MV-algebras and unital ℓ-groups. Moreover, the category of pseudo MV-algebras and the category of unital ℓ-groups are categorically equivalent; for more details, see [109].

Let G be an ℓ-group. Then it possesses the **Riesz Decomposition Property** (RDP, for a short) [147, Theorem I.1.1], i.e., if $g_1, g_2, h_1, h_2 \in G^+$ such that $g_1 + g_2 = h_1 + h_2$, there are elements $u_{11}, u_{12}, u_{21}, u_{22} \in G^+$ such that $g_1 = u_{11} + u_{12}, g_2 = u_{21} + u_{22}, h_1 = u_{11} + u_{21}$, and $h_2 = u_{12} + u_{22}$.

An analogous property holds in each pseudo MV-algebra (we use the partial addition):

Proposition 8.2.1 *Let M be a pseudo MV-algebra and choose $x_1, x_2, y_1, y_2 \in M$ such that $x_1 + x_2 = y_1 + y_2$. Then there are elements $c_{11}, c_{12}, c_{21}, c_{22} \in M$ such that $x_1 = c_{11} + c_{12}$, $x_2 = c_{21} + c_{22}$, $y_1 = c_{11} + c_{21}$, and $y_2 = c_{12} + c_{22}$.*
  Moreover, if $c_{12} \wedge c_{21} = 0$, the elements $c_{11}, c_{12}, c_{21}, c_{22}$ are uniquely determined.

Proof We set $c_{11} := x_1 \wedge y_1, c_{12} := y_1^\sim \odot x_1 = (y_1 \wedge x_1)^\sim \odot x_1 = c_{11}^\sim \odot x_1, c_{22} := x_2 \wedge y_2$, and $c_{21} := x_2 \odot y_2^- = x_2 \odot (x_2 \wedge y_2)^- = x_2 \odot c_{22}^-$.

Then $x_1 = c_{11} + c_{12}$ and $x_2 = c_{21} + c_{22}$. We now show that $x_2 \odot y_2^- = x_1^- \odot y_1$. Put $y = x_1 + x_2 = y_1 + y_2$. Then $y = y_1 + y_1^\sim \odot y = y \odot y_2^- + y_2 = x_1 + x_1^\sim \odot y = y \odot x_2^- + x_2$. The cancellation property of $+$, see (vii)–(viii) of Proposition 8.1.8, entails $y_2 = y_1^\sim \odot y, y_1 = y \odot y_2^-, x_1 = y \odot x_2^-$ and $x_2 = x_1^\sim \odot y$,
  Check

$$
\begin{aligned}
c_{21} = x_2 \odot y_2^- &= x_1^\sim \odot y \odot (y_1^\sim \odot y)^- = x_1^\sim \odot y \odot (y^- \oplus y_1^{\sim -}) \\
&= x_1^\sim \odot y \odot (y^- \oplus y_1) = x_1^\sim \odot (y \wedge y_1) \\
&= x_1^\sim \odot y_1 = (x_1 \wedge y_1)^\sim \odot y_1 \\
&= c_{11}^\sim \odot y_1.
\end{aligned}
$$

By symmetry, we have

$$y_2 \odot x_2^- = y_1^\sim \odot x_1 = c_{12} = (y_1 \wedge x_1)^\sim \odot x_1 = c_{11}^\sim \odot x_1.$$

Then $c_{12} \wedge c_{21} = c_{11}^\sim \odot y_1 \wedge c_{11}^\sim \odot x_1 = c_{11}^\sim \odot (y_1 \wedge x_1) = 0$. Hence, $c_{12} + c_{21} = c_{21} + c_{12}$, so that $x_1 + x_2 = c_{11} + c_{12} + c_{21} + c_{22} = c_{11} + c_{21} + c_{12} + c_{22} = y_1 + y_2$.

The uniqueness under the condition $c_{12} \wedge c_{21} = 0$ follows from adding the elements c_{11} and c_{21}, respectively to the equality $c_{12} \wedge c_{12} = 0$, we obtain $(c_{11} + c_{12}) \wedge (c_{11} + c_{21}) = c_{11}$ so that $c_{11} = x_1 \wedge y_1$ and $c_{22} = x_2 \wedge y_2$. Using the cancellation property, we see that c_{12} and c_{21} are defined uniquely. $\square$

The latter result can be generalized as follows: If $x_1, \ldots, x_m, y_1, \ldots, y_n \in M$ are such that

$$x_1 + \cdots + x_m = y_1 + \cdots + y_n, \tag{8.3}$$

there exist elements c_{ij} in M for $1 \le i \le m$ and $1 \le j \le n$ such that for all i and j

$$a_i = c_{i1} + \cdots + c_{in},$$
$$b_j = c_{1j} + \cdots + c_{mj}.$$

Moreover, we may assume that

$$(c_{i+1,j} + \cdots + c_{mj}) \wedge (c_{i,j+1} + \cdots + c_{in}) = 0, \tag{8.4}$$

then the elements c_{ij} are unique.

For our aim to present a representation of pseudo MV-algebra, the notion of a semiclan introduced by Bosbach [50] is crucial.

We say that $(C; \wedge, +)$ is a **semiclan** if it is a $\wedge$-semilattice and a partial groupoid with respect to $+$ such that the following axioms are satisfied:

(C1)   If $a \le b$, then there exist $x, y \in C$ such that $b = a + x$ and $b = y + a$.

(C2)   If $a + x$, $a + y \in C$, $a + x = a + y$, then $x = y$, and if $x + a$, $y + a \in C$, $x + a = y + a$, then $x = y$.

(C3)   If $a + x$, $a + y \in C$, then $(a + x) \wedge (a + y) = a + (x \wedge y)$, and if $x + a$, $y + a \in C$, then $(x + a) \wedge (y + a) = (x \wedge y) + a$.

(C4) $a + b$, $(a + b) + c \in C$ if and only if $b + c$, $a + (b + c) \in C$, and in this case we have $(a + b) + c = a + (b + c)$.

(C5)   If $(a \wedge b) + c = c$ and $a \vee b$ exists, then $a + b = a \vee b = b + a$.

It is clear that the positive cone if each ℓ-group $(\mathbf{G}^+; \wedge, +)$ is a semiclan, and by Bosbach [50, p. 321], for any semiclan $(C; \wedge, +)$ there exists an ℓ-group $\mathbf{G}$ with the positive cone

$\mathbf{G}^+$ such that C can be embedded into the semiclan $(\mathbf{G}^+; \wedge, +)$ preserving $+$, $\wedge$, and all existing $\vee$ in C. Therefore, we have

Proposition 8.2.2 *Let* $(M; \oplus, ^-, ^\sim, 0, 1)$ *be a pseudo MV-algebra and let* $+$ *be defined as above. Then* $(M; \wedge, +)$ *is a semiclan, and, in addition, there exist an ℓ-group* $\mathbf{G}_0$ *with the positive cone* $\mathbf{G}_0^+$ *and an injective mapping* $f : M \to \mathbf{G}_0^+$ *such that* f *preserves* $+$, $\wedge$, *and* $\vee$.

Denote by $X_0 := f(M)$ and

$$X := \left\{ \sum_{i=1}^{n} f(a_i) : a_1, \ldots, a_n \in M, \, n \geq 1 \right\}.$$

Then X_0 is a lattice which is a sublattice of $\mathbf{G}_0^+$. It is possible to show that X is also a sublattice of $\mathbf{G}_0^+$; for $x, y \in X$, the expression $x \wedge y \in X$ means that $x \wedge y$ taken in the ℓ-group $\mathbf{G}_0$ is an element of X.

Proposition 8.2.3 *Let* $A, B, U \in X$. *If* $A \wedge B \in X$, $B - (A \wedge B) \in X$, *and* $U \wedge (B - (A \wedge B)) \in X$, *then*

$$(U + A) \wedge B \in X.$$

Proof Since A, B, U are elements of $\mathbf{G}_0^+$, we have

$$U \wedge (B - (A \wedge B)) + (A \wedge B) = (U + A \wedge B) \wedge B = (U + A) \wedge (U + B) \wedge B = (U + A) \wedge B,$$

which proves the assertion in question. $\square$

In [109, Theorem 3.9], we have establish that every pseudo MV-algebra M is isomorphic to some $\Gamma(G, u)$ for some unital ℓ-group (G, u). So, Example 8.1.1 gives a most general form for any pseudo MV-algebra.

Theorem 8.2.4 *Let* $(M; \oplus, ^-, ^\sim, 0, 1)$ *be a pseudo MV-algebra. Then there exists an ℓ-group* G *with strong unit* u *such that* M *and* $\Gamma(G, u)$ *are isomorphic pseudo MV-algebras.*

We say that a pseudo MV-algebra M is **Archimedean** if the existence of $nx := x_1 + \cdots + x_n$, where $x_1 = \cdots = x_n = x$ for any integer $n \geq 1$, entails that $x = 0$. For example, every σ-complete pseudo MV-algebra is Archimedean: Let nx exist in M for each $n \geq 1$. Then $x_\infty = \bigvee_{n=1}^{\infty}(nx)$ exists in M, but $x_\infty = x + \bigvee_{n=2}^{\infty}(nx) = x + x_\infty$ implying $x = 0$. Due to [109, Theorem 4.2], we have

Theorem 8.2.5 ([109]) *In each Archimedean pseudo MV-algebra* $M = (M; \oplus, ^-, ^\sim, 0, 1)$, *the binary operation* $\oplus$ *is commutative, that is,* M *is in fact an MV-algebra.*

We say that a partially ordered group $(G; +, -, \leq, 0)$ with a mapping $f : M \to \mathbf{G}^+$ is a **universal group** for a pseudo MV-algebra $(M; \odot, ^-, ^\sim, 0, 1)$ if:

(i)　The positive cone $\mathbf{G}^+$ is generating for G, i.e., $G = \mathbf{G}^+ - \mathbf{G}^+$.

(ii)　$f(M)$ generates $\mathbf{G}^+$ as a semigroup.

(iii)　$f(x + y) = f(x) + f(y)$ whenever $x + y$ exists in $M, x, y \in M$.

(iv)　For any group K and any $+$ preserving mapping $h : M \to K$, there is a group homomorphism $\phi : G \to K$ such that $h = \phi \circ f$.

The universal group, if it exists, is unique up to isomorphism, and ϕ from (iv) is a unique group homomorphism with that property. We denote the universal group for M by (G, f). In [109, Theorem 5.3], it was shown that if f and G are those from Theorem 8.2.4, then (G, f) is a universal group for M.

In what follows, we establish the basic statement about the categorical equivalence of the category of pseudo MV-algebras (it is also a variety) and the category of unital ℓ-groups (that is not a variety—it is not closed under direct products). It generalizes the categorical equivalence of MV-algebras with the category of unital Abelian ℓ-groups, Theorem 1.4.15, established by Mundici [73, 220].

Let $\mathcal{PMV}$ be the category of pseudo MV-algebras whose objects are pseudo MV-algebras $M = (M; \oplus, ^-, ^\sim, 0, 1)$ and morphisms are homomorphisms of pseudo MV-algebras. Denote by $\mathcal{U}\!\!\vdash\!\mathcal{G}$ the category of unital ℓ-groups whose objects are unital ℓ-groups, i.e., pairs (G, u) where u is a distinguished strong unit of an ℓ-group G. Its morphisms are unital ℓ-group homomorphisms, i.e., any mapping f from a unital ℓ-group (G, u) into a unital ℓ-group (H, v) such that (i) $f(x \pm y) = f(x) \pm f(y)$, (ii) $f(x \vee y) = f(x) \vee f(y)$ and $f(x \wedge y) = f(x) \wedge f(y)$, and (iii) $f(u) = v$.

We define a mapping $\Gamma : \mathcal{U}\!\!\vdash\!\mathcal{G} \to \mathcal{PMV}$ by $\Gamma : (G, u) \mapsto \Gamma(G, u)$, for any object (G, u) in $\mathcal{U}\!\!\vdash\!\mathcal{G}$, where $\Gamma(G, u)$ is defined in Example 8.1.1, and if h is a unital ℓ-homomorphism from (G, u) into (H, v), then $\Gamma(h)(a) := h(a)$, $a \in \Gamma(G, u)$, is a morphism from $\Gamma(G, u)$ into $\Gamma(H, v)$. It is easy to verify that Γ is a functor from $\mathcal{U}\!\!\vdash\!\mathcal{G}$ into $\mathcal{PMV}$. Moreover, Γ is a faithful and full functor:

Proposition 8.2.6 Γ *is a faithful and full functor from the category* $\mathcal{ULG}$ *into the category* $\mathcal{PMV}$ *of pseudo MV-algebras.*

Proof Let h_1 and h_2 be two morphisms from (G, u) into (H, v) such that $\Gamma(h_1) = \Gamma(h_2)$. Then $h_1(x) = h_2(x)$ for any $x \in \Gamma(G, u)$. Since $\Gamma(G, u)$ generates $\mathbf{G}^+$, it is clear that $h_1(x) = h_2(x)$ for any $x \in G$.

To prove that $\mathbf{\Gamma}$ is a full functor, suppose that f is a morphism from $\mathbf{\Gamma}(G, u)$ into $\mathbf{\Gamma}(H, v)$, that is, f is a pseudo MV-homomorphism. Then f preserves the partial addition $+$ and $\wedge$ in $\mathbf{\Gamma}(G, u)$, and $+$ coincides with the group addition taken in G. Since $\mathbf{\Gamma}(G, u)$ generates G, we can show that f can be uniquely extended to a group homomorphism $\hat{f}$ from G into H.

We claim that $\hat{f}$ is a unital ℓ-group homomorphism. The proof will proceed in several steps.

Step 1. Let $a, b, u_0 \in \mathbf{G}^+$. If $\hat{f}(a \wedge b) = \hat{f}(a) \wedge \hat{f}(b)$ and if $\hat{f}(u_0 \wedge (b - (a \wedge b))) = \hat{f}(u_0) \wedge \hat{f}(b - (a \wedge b))$, then

$$\hat{f}((a + u_0) \wedge b) = \hat{f}(a + u_0) \wedge \hat{f}(b).$$

Indeed, we have

$$u_0 \wedge (b - (a \wedge b)) + (a \wedge b) = (u_0 + a \wedge b) \wedge b = (u_0 + a) \wedge (u_0 + b) \wedge b = (u_0 + a) \wedge b,$$

which gives

$$\begin{aligned}
\hat{f}((a + u_0) \wedge b) &= \hat{f}(u_0 \wedge (b - (a \wedge b))) + \hat{f}(a \wedge b) \\
&= [\hat{f}(u_0) \wedge (\hat{f}(b) - (\hat{f}(a) \wedge \hat{f}(b)))] + (\hat{f}(a) \wedge \hat{f}(b)) \\
&= [(\hat{f}(u_0) + (\hat{f}(a) \wedge \hat{f}(b)))] \wedge \hat{f}(b) \\
&= (\hat{f}(u_0) + \hat{f}(a)) \wedge (\hat{f}(u_0) + \hat{f}(b)) \wedge \hat{f}(b) \\
&= \hat{f}(a + u_0) \wedge \hat{f}(b).
\end{aligned}$$

Step 2. $\hat{f}(a \wedge b) = \hat{f}(a) \wedge \hat{f}(b)$ whenever $a \in \mathbf{G}^+$ and $b \in \mathbf{\Gamma}(G, u)$.

Since $\mathbf{\Gamma}(G, u)$ is generating for $\mathbf{G}^+$, a is of the form $a = a_1 + \cdots + a_n$ for some $a_1, \ldots, a_n \in \mathbf{\Gamma}(G, u)$. We will use mathematical induction on n.

If $n = 1$, the statement is trivial. Suppose now that the statement holds for any $a' = a_1 + \cdots + a_i$ with $1 \le i \le n$. Put $a = a_1 + \cdots + a_n$, $u_0 = a_{n+1}$. Then there exist $v_1, \ldots, v_k \in \mathbf{\Gamma}(G, u)$ such that $b = (v_1 + \cdots + v_k) + (a \wedge b)$. Since $v := v_1 + \cdots + v_k \le b \in \mathbf{\Gamma}(G, u)$, $v \in \mathbf{\Gamma}(G, u)$. Hence, $v = b - (a \wedge b)$. Since $\hat{f}$ preserves meets in $\mathbf{\Gamma}(G, u)$, we have $\hat{f}(u_0 \wedge v) = \hat{f}(u_0) \wedge \hat{f}(v)$, so that $\hat{f}(u_0 \wedge (b - (a \wedge b))) = \hat{f}(u_0) \wedge \hat{f}(b - (a \wedge b)) = \hat{f}(u_0) \wedge (\hat{f}(b) - (\hat{f}(a) \wedge \hat{f}(b)))$, where we have used induction hypothesis for a and b. By Step 1, $\hat{f}((a + u_0) \wedge b) = \hat{f}(a + u_0) \wedge \hat{f}(b)$, that is, $\hat{f}((a_1 + \cdots + a_{n+1}) \wedge b) = \hat{f}(a_1 + \cdots + a_{n+1}) \wedge \hat{f}(b)$ for any n.

Step 3. $\hat{f}(a \wedge b) = \hat{f}(a) \wedge \hat{f}(b)$ whenever $a, b \in \mathbf{G}^+$.

Let $a = a_1 + \cdots + a_n$, $b = b_1 + \cdots + b_k$. The proof will follow complete induction on k.

If $k = 1$, we apply Step 2. Suppose now that the assertion holds for any j with $1 \leq j \leq k$. Put $B = a$, $A = b_1 + \cdots + b_k$, $u_0 = b_{k+1}$. By Step 2, $\hat{f}(u_0 \wedge (B - (A \wedge B))) = \hat{f}(u_0) \wedge \hat{f}(B - (A \wedge B))$ and $\hat{f}(A \wedge B) = \hat{f}(A) \wedge \hat{f}(B)$. Therefore, the conditions of Step 1 are satisfied, so that $\hat{f}((A + u_0) \wedge B) = \hat{f}(A + u_0) \wedge \hat{f}(B)$ which proves

$$\hat{f}((a_1 + \cdots + a_n) \wedge (b_1 + \cdots + b_{k+1})) = \hat{f}(a_1 + \cdots + a_n) \wedge \hat{f}(b_1 + \cdots + b_{k+1})$$

for each n and each k.

Step 4. $\hat{f}(a \wedge b) = \hat{f}(a) \wedge \hat{f}(b)$ whenever $a, b \in G$. Then $a = a^+ - a^-$ and $b = b^+ - b^-$, and $a = -a^- + a^+$, $b = -b^- + b^+$. By Step 3, $\hat{f}((a^+ + b^-) \wedge (a^- + b^+)) = \hat{f}(a^+ + b^-) \wedge \hat{f}(a^- + b^+)$. Subtracting $\hat{f}(b^-)$ from the right hand and $\hat{f}(a^-)$ from the left hand, we obtain the assertion in question. $\square$

It is right-adjoint, to show it, we have used the universal group (G, f) for a pseudo MV-algebra M, see [109, Proposition 6.2].

The converting functor $\Psi : \mathcal{PMV} \to \mathcal{U\vdash G}$ is defined by $\Gamma(M) := (G(M), u)$, where $(G(M), u)$ is that from Theorem 8.2.4, and if f is a pseudo MV-homomorphism from M into M_1, then $\Psi(h)$ is a unique ℓ-group homomorphism preserving the fixed strong unit. Therefore, we conclude Ψ is a functor from the category $\mathcal{PMV}$ into the category $\mathcal{U\vdash G}$ which is left-adjoint.

Using results from the theory of categories, [207, (i),(iii) Theorem IV.4.1], and the above results, we present the following final principal result:

Theorem 8.2.7 *The functors Γ and Ψ define a categorical equivalence between the category of pseudo MV-algebras and the category of unital ℓ-groups.*

As a corollary of the last theorem, we have:

Lemma 8.2.8 *Let $h : \Gamma(G, u) \to \Gamma(H, v)$ be a homomorphism of pseudo MV-algebras.*

(i) *There is a unique homomorphism of unital ℓ-groups $f : (G, u) \to (H, v)$ such that $h = \Gamma(f)$.*
(ii) *If h is surjective, so is f.*
(iii) *If is h injective, so is f.*

For another approach to pseudo MV-algebras, consult [149, 255, 264]. To conclude this section, we note that the Riesz Decomposition Property, RDP, plays a significant role in the study of partially ordered groups and pseudo-effect algebras. In [125, 126], it was proven that if a pseudo-effect algebra satisfies a stronger form of RDP, specifically RDP$_1$, then the pseudo-effect algebra is an interval in a po-group with RDP$_1$, which is not necessarily

Abelian. Furthermore, if we define an even stronger variant of RDP, RDP$_2$, then the corresponding pseudo-effect algebra is a pseudo MV-algebra. For further details, we recommend consulting [113, 125–127, 239].

8.2.1   Application of Representation Theorem

The basic representation of pseudo MV-algebras by unital ℓ-groups has many applications. The following result is important to study the ideals of a pseudo MV-algebra using analogous notions in the corresponding unital ℓ-group. Given an ℓ-group G, we denote by $C(G)$ the set of a convex ℓ-subgroups of G.

Theorem 8.2.9 *Let (G, u) be a unital ℓ-group and let $M = \Gamma(G, u)$. For any ideal I of M we assign*

$$\phi(I) := \{x \in G : |x| \wedge u \in I\}. \tag{8.5}$$

(1) The set $\phi(I)$ is a convex ℓ-subgroup of G generated by I.

The mapping $\phi : I \mapsto \phi(I)$ from the set of ideals of M onto $C(G)$ defines a one-to-one mapping preserving the set-theoretical inclusion. The inverse mapping ψ is given by

$$\psi(K) := K \cap [0, u], \quad K \in C(G). \tag{8.6}$$

(2)

$$\phi(I) = \{x \in G : \exists\, x_i, y_j \in I, \ x = x_1 + \cdots + x_n - y_1 - \cdots - y_m\}. \tag{8.7}$$

(3) The restriction of ϕ to prime ideals, or maximal ideals or normal ideals, gives a bijection between corresponding notions in the pseudo MV-algebra M and the unital ℓ-group (G, u) respectively, which preserves the set-theoretical inclusion.

Proof It is clear that $\phi(I)$ contains all elements $x \in G$ of the form $x = x_1 + \cdots + x_n - y_1 - \cdots - y_m$, where $x_i, y_j \in I$. Let now $x \in \phi(I), x \geq 0$. Then $x \leq nu$ for some integer $n \geq 1$. By the Riesz decomposition property, there exist $x_1, \ldots, x_n \in M$ such that $x = x_1 + \cdots + x_n$. Then $x_i \leq x \wedge u \in I$, so that $x_i \in I$. If $x \in \phi(I)$ and $x = x^+ - x^-$, then $x^+, x^- \in \phi(I)$, and we have (8.6).

(3) We now show that if I is a normal ideal of M, then $\phi(I)$ is an ℓ-ideal of G.

Let $h : M \to M/I$ be the canonical mapping defined by $h(a) = a/I, a \in M$, where I is a normal ideal of M. By Lemma 8.2.8, there is a unique ℓ-group-homomorphism f from (G, u) onto (H, v), where $\Gamma(H, v) = M/I$. Then $\phi(I) \subseteq \mathrm{Ker}(f) = \{x \in G : f(x) = 0\}$. Let now $x \in \mathrm{Ker}(f)$. If $x \geq 0$, then $x = x_1 + \cdots + x_n$, where $x_i \in \Gamma(G, u)$, so that $0 \leq f(x_i) \leq f(x) = 0$, which yields $f(x_i) = h(x_i) = 0$, and hence $x_i \in I$ for $i = 1, \ldots, n$.

Therefore, $x \in \phi(I)$. If $x \in \mathrm{Ker}(f)$ is arbitrary, then $x^+ = x \vee 0$, which gives $h(x^+) = h(x) \vee h(0) = 0$, and consequently, $x^+ \in \phi(I)$. In a similar way, $x^- \in \phi(I)$ which gives $x \in \phi(I)$.

We have proved $\mathrm{Ker}(f) = \phi(I)$. Since it is well-known that the kernel of each ℓ-group-homomorphism is an ℓ-ideal, we have $\phi(I)$ is an ℓ-ideal of $\boldsymbol{G}$. $\qquad\square$

The class PMV of pseudo MV-algebras is a variety because equations define it, whereas the class of unital ℓ-groups is not. In the case of MV-algebras, by Komori [195], the lattice of subvarieties of the variety MV of MV-algebras is countable: Each subvariety of MV-algebras is generated by a finite number of MV-algebras of the form $\boldsymbol{\Gamma}(\mathbb{Z}, n)$ or $\boldsymbol{\Gamma}(\mathbb{Z} \overrightarrow{\times} \mathbb{Z}, (m, 0))$ with $n, m \geq 0$. Applying this fact, [101] described an equational base for each proper subvariety of MV-algebras. The situation in the case of all subvarieties of pseudo MV-algebras is more complicated.

An equational class of algebra is the collection of all algebras of the given type, say ℓ-groups, or unital ℓ-groups, or pseudo MV-algebras, which satisfy a given set of universally quantified equations in the language of the algebra. If the type of algebra is itself defined by equations, then an equational class is a variety.

Theorem 8.2.10 ([119, Theorem 3.1]) *Let* V *be an equational class of unital ℓ-groups, and let* $\boldsymbol{\Gamma}(\mathsf{V})$ *be the collection of all pseudo-algebras (isomorphic to)* $\boldsymbol{\Gamma}(\boldsymbol{G}, u))$, *where* $\boldsymbol{G} \in \mathsf{V}$. *Then* $\boldsymbol{\Gamma}$ *is an isomorphism from the lattice of all equational classes of unital ℓ-groups to the lattice of all varieties of pseudo MV-algebras.*

Since the variety of ℓ-groups is uncountable, there are uncountably many subvarieties of pseudo MV-algebras. The minimal non-trivial one is the Boolean subvariety BA of pseudo MV-algebras given by the equation $x = x \oplus x$.

According to [152], a pseudo MV-algebra $\boldsymbol{M}$ is **representable** if it can be represented as a subdirect product of linearly ordered pseudo MV-algebras. It is well-known that every MV-algebra is representable (see, e.g., [73]). We denote by Repr the class of representable pseudo MV-algebras. It is possible to show that $\boldsymbol{M} = \boldsymbol{\Gamma}(\boldsymbol{G}, u)$ is representable if and only if $\boldsymbol{G}$ is a representable ℓ-group, i.e., it is a subdirect product of linearly ordered ℓ-groups. Moreover, Repr is a variety whose equational base is, e.g.,

$$x_1 \wedge (-y + x_2 + y) = 0 \tag{8.8}$$

for all $x_1, x_2, y \in M$ with $x_1 \wedge x_2 = 0$. Or another equation defining Repr is

$$2 \odot (x \wedge y) = (2 \odot x) \wedge (2 \odot y), \quad x, y \in M.$$

According to [152], we say that a **value** of $g \in M \setminus \{0\}$ is an ideal V of M such that $g \notin V$ and V is maximal concerning this property. As it was shown in [152, Theorem 2.15] such an ideal always exists in M and is a prime ideal. For any value V of an element g, we have that there is a unique least ideal V^* containing V and g; it is called a **cover** of V. A pseudo MV-algebra M is said to be **normal-valued** if, for each $g > 0$, any values V of g is normal in its cover V^*. Let NV be the class of normal-valued pseudo MV-algebras. It is possible to show that the pseudo MV-algebra $M = \Gamma(G, u)$ is normal-valued if and only if G is a normal-valued ℓ-group. The class NVLG of normal-valued ℓ-groups forms the largest proper subvariety of the variety LG of ℓ-groups, see [89, p. 419]. An equational base is by [108, Theorem 6.8]

$$(x \oplus y) \wedge (2 \odot y \oplus 2 \odot x) = (x \oplus y), \quad x, y \in M.$$

It is important to say that the variety NV is not the largest proper subvariety of PMV. Indeed, let MN be the class of pseudo MV-algebras whose every maximal ideal is normal. Due to [119, Proposition 6.2], [117], it is a variety. The following example shows MN $\neq$ PMV.

Example 8.2.11 Let Aut($\mathbb{R}$) denote the set of all strictly increasing continuous functions from $\mathbb{R}$ onto $\mathbb{R}$. We write $f \leq g$ for $f, g \in \mathrm{Aut}(\mathbb{R})$ if $\alpha f \leq \alpha g$ for each $\alpha \in \mathbb{R}$. Under the compositions of functions and inversions, Aut($\mathbb{R}$) is an ℓ-group where the identity is a neutral element. Take a translation αu such that $\alpha u = \alpha + 1$. If we define

$$\mathrm{BAut}(\mathbb{R}) := \{g \in \mathrm{Aut}(\mathbb{R}) \colon \exists n \geq 1, \ u^{-n} \leq g \leq u^n\}.$$

Then it is a unital ℓ-group with strong unit u, and $M = \Gamma(\mathrm{BAut}(\mathbb{R}), u)$ is a pseudo MV-algebra where no maximal ideal is normal.

Let O be the variety consisting of the degenerate one-element pseudo MV-algebra, i.e., an MV-algebra with $0 = 1$. Then we have

$$\mathsf{O} \subsetneqq \mathsf{BA} \subsetneqq \mathsf{MV} \subsetneqq \mathsf{Repr} \subsetneqq \mathsf{NV} \subsetneqq \mathsf{MN} \subsetneqq \mathsf{PMV}.$$

For example, the variety of representable pseudo MV-algebras has uncountably many subvarieties, [95]. Suppose we denote by SYM the variety of symmetric pseudo MV-algebras, i.e., satisfying the equation $x^- = x^\sim$. It is easy to verify that the pseudo MV-algebra corresponding to $\Gamma(G, u)$ is symmetric if and only if u commutes with any element $g \in G$. This variety is very interesting; many results known for MV-algebras can be generalized within it. In particular, this is true for its intersection with SYM. On the other hand, an equational base for the variety MN is not yet known.

The well-known Chang Completeness Theorem, see e.g. [73, Theorem 2.5.3], says that an equation holds in the real interval MV-algebra $[0, 1]$ if and only if it holds in every MV-algebra. An analogous role plays the pseudo MV-algebra $\Gamma(\mathrm{BAut}(\mathbb{R}), u)$, see [117, Corollary 4.9]:

Theorem 8.2.12 Completeness Theorem for Pseudo MV-algebras. *An equation holds in every pseudo MV-algebra if it holds in* $\Gamma(\mathrm{BAut}(\mathbb{R}), u)$.

We note that due to [95, p. 1237], the variety of pseudo MV-algebras generated by $\Gamma(\mathrm{BAut}(\mathbb{R}), u)$ is the variety PMV of all pseudo MV-algebras. We note that the ℓ-group $\mathrm{BAut}(\mathbb{R})$ is a doubly transitive ℓ-group. For characterization of doubly transitive ℓ-groups, we recommend consulting, e.g. with [89, 159]. According to [159, Lemma 10.3.1], or [89, Theorem 38.23], we have that if G is a doubly transitive ℓ-group with strong unit u, then the pseudo MV-algebra $\Gamma(G, u)$ generates the variety PMV. Moreover, any equation holds for all pseudo MV-algebras if and only if it holds in $\Gamma(G, u)$. We note that the variety of pseudo MV-algebras generated by the real interval MV-algebra [0, 1] is the variety MV of MV-algebras, see also [73, Proposition 8.1.1].

We conclude this section with a discussion of top varieties, an interesting subclass of varieties.

A **value** of an element g of an ℓ-group G is a maximal convex ℓ-subgroup of G (written in a multiplicative way) which does not contain g. Every convex ℓ-subgroup which does not contain g is contained in a value of g. For any value V of the unit u in (G, u), we set

$$K(V) = \bigcap_{g \in G} g^{-1} V g.$$

Then $K(V)$ is a normal convex ℓ-subgroup of (G, u) contained in V.

For any variety V of pseudo MV-algebras, let $\mathcal{T}(\mathsf{V})$ denote the collection of all $\mathbf{0}(G, u)$ such that for each value V of the unit u in G, $\Gamma(G/K(V), u/K(V)) \in \mathsf{V}$. By [117, Corollary 4.5], $\mathcal{T}(\mathsf{V})$ is a variety; we call it the **top variety** of V. We recall that according to [95, (6.1)] and [117], we have

$$\mathsf{MN} = \mathcal{T}(\mathsf{MV}) = \mathcal{T}(\mathsf{NV}) = \mathcal{T}(\mathsf{MN}).$$

Top varieties were used to describe covers of the variety of pseudo MV-algebras within the frames of symmetric ones and having each maximal ideal normal, that is, any such a variety $\mathsf{V} \subseteq \mathsf{SYM} \cap \mathsf{M}$ that properly contains the variety of MV-algebras and is not properly obtained in any subvariety $\mathsf{W} \supseteq \mathsf{MV}$. It was proved that due to [117, Theorem 3.8], there are continuum many representable covers of the variety of MV-algebras. This fact underlines the richness of the subvariety lattice of pseudo MV-algebras.

For more information about varieties of pseudo MV-algebras, we recommend to consult with [95, 117–119, 172].

8.3   States on Pseudo MV-Algebras

The partial addition $+$ is a natural way to define a state on an MV-algebra or pseudo MV-algebra. We note that a state means an analogy of a finitely additive probability measure in a theory of classical probability systems. Mundici introduced states on MV-algebras in [223] as averaging processes for formulas in Łukasiewicz logic.

A **state** on a pseudo MV-algebra M is a mapping $s : M \to [0, 1]$ such that is additive, i.e., if $x + y$ is defined in M, then $s(x + y) = s(x) + s(y)$, and normalized, i.e., $s(1) = 1$. We denote by $\mathcal{S}(M)$ the system of states on M; it is called the **state space** of M. A σ-**additive state** is any state s such that if $x_n \le x_{n+1}$ for each $n \ge 1$ and $x = \bigvee_n x_n$ exists in M, then $s(x) = \lim_n s(x_n)$. If M is an MV-algebra, then a state always exists. It follows from D. Mundici's representation of MV-algebras by unital Abelian ℓ-groups and by [160, Corollary 4.4]. If M is a pseudo MV-algebra, it can happen $\mathcal{S}(M) = \emptyset$.

Clearly, (i) $s(0) = 0$, (ii) $s(x^-) = 1 - s(x) = s(x^\sim)$, (iii) $s(x) \le s(y)$ whenever $x \le y$.

It is clear that the state space is a convex set, i.e., if s_1, s_2 are states and λ is a real number such $0 \le \lambda \le 1$, then $s = \lambda s_1 + (1 - \lambda)s_2$ is a state. A state s is **extremal** if from $s = \lambda s_1 + (1 - \lambda)s_2$ for $\lambda \in (0, 1)$, we have $s = s_1 = s_2$; we denote by $\partial \mathcal{S}(M)$ the set of extremal states of M.

A mapping s from a pseudo MV-algebra M into the standard MV-algebra of the real unit interval, $([0, 1]; \oplus, ', 0, 1)$ as defined in Example 1.4.2(i), such that, for all $a, b \in M$, $s(a \oplus b) := \min\{s(a) + s(b), 1\}$, (ii) $s(a^-) = s(a^\sim) = 1 - s(a)$, and (iii) $s(1) = 1$ is said to be a **state-morphism**. In other words, a mapping s is a state-morphism if and only if s is a homomorphism from M into the standard MV-algebra of the real interval $[0, 1] = \Gamma(\mathbb{R}, 1)$. We denote by $\mathcal{SM}(M)$ the set of state-morphisms on M.

It is evident that any state-morphism is a state. The converse does not hold, in general. A state s is state-morphism if and only if $s(x \wedge y) = \min\{s(x), s(y)\}$, $x, y \in M$.

We note that a **state** of a unital ℓ-group (G, u) is an additive mapping $f : G \to \mathbb{R}$ such that $f(G^+) \subseteq [0, \infty)$ and $f(u) = 1$. It is clear that if f is a state on (G, u), then its restriction to $[0, u]$ is a state on the pseudo MV-algebra $\Gamma(G, u)$. Conversely, due to the categorical equivalence and Lemma 8.2.8, every state on the pseudo MV-algebra $\Gamma(G, u)$ can be uniquely extended to a state on (G, u).

Proposition 8.3.1 *Let s be a state on M. For all $x, y \in M$, we have*

(i) $s(x \vee y) + s(x \wedge y) = s(x) + s(y)$.
(ii) $s(x \oplus y) + s(x \odot y) = s(x) + s(y)$.
(iii) *The set* $\mathrm{Ker}(s) := \{x \in M : s(x) = 0\}$, *called the kernel of s, is a normal ideal of M.*
(iv) $x/\mathrm{Ker}(s) = y/\mathrm{Ker}(s)$ *if and only if $s(x) = s(x \wedge y) = s(y)$.*
(v) *There is a unique state $\tilde{s}$ on $M/\mathrm{Ker}(s)$ such that $\tilde{s}([x]) = s(x)$, $[x] \in M/\mathrm{Ker}(s)$, where $[x]$ is the coset in $M/\mathrm{Ker}(s)$ determined by $x \in M$.*
(vi) $s(x \oplus y) = s(y \oplus x)$.

Proof We prove only (vi).

Due to (iv), $\tilde{s}([x]) = 0$ if and only if $[x] = [0]$. We claim that $M/\mathrm{Ker}(s)$ is Archimedean. Indeed, let $n[x]$ be defined in $M/\mathrm{Ker}(s)$ for any integer $n \geq 1$. Then $\tilde{s}(n[x]) = n\tilde{s}([x]) = n\,s(x) \leq 1$ for any n. Therefore, $\tilde{s}([a]) = s(a) = 0$. The Archimedeanicity of $M/\mathrm{Ker}(s)$ entails the commutativity of $M/\mathrm{Ker}(s)$ (see Theorem 8.2.5). Therefore, $s(x \oplus y) = \tilde{s}([x] \oplus [y]) = \tilde{s}([y] \oplus [x]) = s(y \oplus x)$. $\qquad\square$

Proposition 8.3.2 *Let s be a state on a pseudo MV-algebra M. Then s is a state-morphism if and only if $\mathrm{Ker}(s)$ is a maximal ideal of M.*

Proof Let s be a state-morphism, and let $x \in M$ be such an element that $s(x) \neq 0$. By (iii) of Proposition 8.3.1, $\mathrm{Ker}(s)$ is a normal ideal of M.

Denote by $\mathrm{Ker}(s)_x$ the ideal of M generated by $\mathrm{Ker}(s)$ and x. By [152, Lemma 3.4], $\mathrm{Ker}(s)_x = \{y \in M : y \leq n \odot x \oplus h \text{ for some } n \geq 1 \text{ and some } h \in \mathrm{Ker}(s)\}$. Let z be an arbitrary element of M. There exists an integer $n \geq 1$ such that $(n-1)s(x) \leq s(z) < ns(x)$. Then $s((n \odot x)^{\sim} \odot z) = 0$. Since $z = (n \odot x) \wedge z \oplus (n \odot x)^{\sim} \odot z \leq n \odot x \oplus (n \odot x)^{\sim} \odot z$, which proves that $z \in \mathrm{Ker}(s)_x$, consequently, $M = \mathrm{Ker}(s)_x$, proving $\mathrm{Ker}(s)$ is a maximal ideal of M.

Conversely, suppose $\mathrm{Ker}(s)$ is a maximal ideal of M. Since by [152, Proposition 1.25], $x \odot y^{-} \wedge y \odot x^{-} = 0$ for all $x, y \in M$, we conclude that either $s(x \odot y^{-}) = 0$ or $s(y \odot x^{-}) = 0$. In the first case we have $0 = s(x \odot y^{-}) = s(x \odot (x \wedge y)^{-}) = s(x) - s(x \wedge y)$. Similarly, $0 = s(y \odot x^{-})$ entails $s(y) - s(x \wedge y) = 0$, i.e., $s(x \wedge y) = \min\{s(x), s(y)\}$, which means that s is a state-morphism. $\qquad\square$

Theorem 8.3.3 *There is a one-to-one relation between state-morphisms and maximal ideals that are normal. Moreover, a state s is a state-morphism if and only if s is an extremal state, i.e., $\mathcal{SM}(M) = \partial \mathcal{S}(M)$.*

In particular, is s_1 and s_2 are state-morphisms such that $\mathrm{Ker}(s_1) = \mathrm{Ker}(s_2)$, then $s_1 = s_2$. For each maximal and normal ideal I, there is a unique state-morphism s such that $\mathrm{Ker}(s) = I$.

Proof See [108, Propositions 4.5–4.7]. Take into account that if I is a maximal ideal and normal of a pseudo MV-algebra M, then the quotient M/I is in fact a state-morphism on M. $\qquad\square$

We define a weak topology of states, saying that a net of states $(s_\alpha)_\alpha$ **converges weakly** to a state s is $s(x) = \lim_\alpha s_\alpha(x)$, $x \in M$. Whence, due to [108, Theorem 4.8], we have:

Theorem 8.3.4 *Let M be a pseudo MV-algebra. If $S(M) \neq \emptyset$, then $S(M)$ is a compact convex Hausdorff space with respect to the weak topology, and the space of all state-morphisms on M is a non-void compact Hausdorff space. Any state is a weak limit of convex linear combinations of the set of extremal points of $S(M)$.*

Due to Theorem 8.3.3, there is a one-to-one correspondence between state-morphisms and maximal ideals that are normal. We introduce the hull-kernel topology on the set $\mathcal{MN}(M)$ of maximal ideals that are normal. We will show that topologically, these spaces are homeomorphic.

For every $a \in M$, we put

$$M(a) := \{I \in \mathcal{NM}(M) : a \notin I\}.$$

Then (i) $M(0) = \emptyset$, $M(a) \subseteq M(b)$ whenever $a \leq b$, $M(a \wedge b) = M(a) \cap M(b)$, $a, b \in M$, $M(a \vee b) = M(a) \cup M(b)$, $a, b \in M$, and $\{M(a) : a \in M\}$ is the base of the so-called **hull-kernel topology** T_{NM} on $\mathcal{NM}(M)$.

Proposition 8.3.5 *Let M be a pseudo MV-algebra. The hull-kernel topology defines a Hausdorff topology such that the closed subspaces of $\mathcal{NM}(M)$ are exactly of the form*

$$C = C(J) := \{I \in \mathcal{NM}(M) : I \supseteq J\}, \tag{8.9}$$

where J is an ideal of M. Similarly, every open set O is of the form

$$O = O(J) := \{I \in \mathcal{NM}(M) : I \not\supseteq J\}. \tag{8.10}$$

Moreover, $T_{NM}(M)$ is a Hausdorff topology on $\mathcal{NM}(M)$.

Proof Let J be any ideal of M. Then $J = \bigvee_{a \in J} I(a)$, where $I(a)$ is an ideal of M generated by the element a. Then $O(J) := \{I \in \mathcal{NM}(M) : I \not\supseteq J\} = \bigcup_{a \in J}\{I : I \not\supseteq I(a)\} = \bigcup_{a \in J} M(a)$ is an open set of $\mathcal{NM}(M)$, and each open set is of the form (8.10). Consequently, every closed subset of $\mathcal{NM}(M)$ is of the form (8.9).

If I_1 and I_2 are two different maximal and normal ideals of M, they are non-comparable, so that there are $x \in I_1 \setminus I_2$ and $y \in I_2 \setminus I_1$. Then $x \wedge y \in I_1 \cap I_2$ and take an idempotent $a \in B(M)$ such that $x, y \leq a$. The elements $x \odot (a \odot y^-)$ and $y \odot (a \odot x^-)$ belong to the pseudo MV-algebra $[0, a]$. Due to $x = (a \odot x^-) \oplus (x \wedge y)$, we have $x \odot (a \odot y^-) \in I_1 \setminus I_2$. In the same way, we have $y \odot (a \odot x^-) \in I_2 \setminus I_1$. We get $I_2 \in M(x \odot y^-)$, $I_1 \in M(y \odot x^-)$, and $(x \odot (a \odot y^-)) \wedge (y \odot (a \odot x^-)) = 0$ which proves that $\mathcal{NM}(M)$ is a Hausdorff space (it can be even empty, see Example 8.3.8 below). $\qquad\square$

The homeomorphism between the weak topology of state-morphisms and the hull-kernel topology of the set of maximal and normal ideals is the following result; for proof details, see [131, Theorem 3.6].

Theorem 8.3.6 *Let M be a pseudo MV-algebra. The mapping $\theta : \mathcal{SM}(M) \to \mathcal{MN}(M)$ given by $\theta(s) = \mathrm{Ker}(s)$, $s \in \mathcal{SM}(M)$, is a homeomorphism. Consequently, both spaces are compact Hausdorff spaces.*

It was already said that every non-degenerated MV-algebra possesses at least one state. From the latter two theorems, we conclude the following basic result about the existence of states on pseudo MV-algebras.

Theorem 8.3.7 *A pseudo MV-algebra has a state if and only if it possesses at least one maximal ideal that is normal.*

Example 8.2.11 gives a pseudo MV-algebra $\boldsymbol{B}\boldsymbol{Aut}(\mathbb{R})$ without any state. Moreover, due to [95, Exercise 3.2], there exists a stateless symmetric pseudo MV-algebra:

Example 8.3.8 Denote by

$$G = \{f \in \mathrm{Aut}(\mathbb{R}) : (\alpha + 1)g = \alpha g + 1, \ \forall \alpha \in \mathbb{R}\}.$$

Then G with compositions of functions form an ℓ-group with a strong unit $\alpha u = \alpha + 1$, $\alpha \in \mathbb{R}$, such that it commutes with every element $g \in G$. Then the pseudo MV-algebra described by $\boldsymbol{\Gamma}(\boldsymbol{G}, u)$ is a symmetric stateless algebra.

We show that every non-degenerate linearly ordered pseudo MV-algebra admits a state.

Theorem 8.3.9 *Each non-degenerate linearly ordered pseudo MV-algebra M possesses a unique state.*

Proof The set of all proper ideals of a linear pseudo MV-algebra $\boldsymbol{M}$ is non-void. Moreover, if I_1 and I_2 are two ideals of $\boldsymbol{M}$, then $I_1 \subseteq I_2$ or $I_2 \subseteq I_1$. If not, then there exist two elements x, y such that $x \in I_1 \setminus I_2$ and $y \in I_2 \setminus I_1$. Since $x \le y$ or $y \le x$, we obtain a contradiction.

Let now I_0 be the set-theoretical union of all proper ideals of $\boldsymbol{M}$. Then I_0 is a maximal ideal. For $x \in M$, $x \in I_0$ if and only if nx is defined in M for any integer $n \ge 1$. Due to Theorem 8.2.4, we can assume that $\boldsymbol{M} = \boldsymbol{\Gamma}(\boldsymbol{G}, u)$ for some unital ℓ-group $(\boldsymbol{G}, u)$. It is possible to show that $\boldsymbol{G}$ is a linearly ordered ℓ-group.

We claim that I_0 is a normal ideal. Let $x \in M$ and $h \in I_0$ be given. If $x \in I_0$, then $x \oplus h = (x \oplus h) \odot x^- + x$ which shows that $h_1 = (x \oplus h) \odot x^- \in I_0$.

Let now $x \notin I_0$. Then, for any $h \in I_0$, $nh < x$ for every $n \geq 1$. Without loss of generality, we can assume that $x + h$ is defined as M. Let $x + h = h_1 + x$ for some $h_1 \in M$. Using the group representation, we have $x + h - x = h_1$. We finish the proof showing that nh_1 is defined in M for any $n \geq 1$. We have $nh_1 = x + nh - x = x - (x - nh)$. But $0 < x - nh \leq x < u$, so that $nh_1 \in M$, and hence $h_1 \in I_0$.

Similarly, we prove that for every $x \in M$ and every $h \in I_0$, there exists an $h_2 \in I_0$ such that $h \oplus x = x + h_2$.

The uniqueness of the maximal ideal is evident.

$\square$

Example 8.3.10 Let G be the set of all pairs (x, y) of real numbers under the decomposition

$$(x_1, y_1) + (x_2, y_2) = (x_1 + x_2, e^{x_2} y_1 + y_2),$$

and let $\mathbf{G}^+$ be the set of all (x, y) such that either $x > 0$ or $x = 0$ and $y \geq 0$. Then $\mathbf{G}$ is a linearly ordered group with a neutral element $(0, 0)$, and with the inverse $-(x, y) = (-x, -e^x y)$ of an element (x, y) and with a strong unit $u = (1, 0)$. Then $\mathbf{M} = \mathbf{\Gamma}(\mathbf{G}, u)$ possesses a unique state-morphism s, namely $s((x, y)) = x$, $(x, y) \in M$.

Theorem 8.3.11 *Every non-degenerate representable pseudo MV-algebra $\mathbf{M}$ possesses at least one state.*

Proof There exists a family of linearly ordered pseudo MV-algebras $\mathbf{M}_t$ for $t \in T$ such that M is a subdirect product of $\prod_{t \in T} \mathbf{M}_t$. We can assume that all are non-degenerated. Every $\mathbf{M}_t$ possesses, due to Theorem 8.3.9, a unique state s_t that is also a state-morphism. If h is the embedding of M into $\prod_{t \in T} \mathbf{M}_t$ such that $h_t := \pi_t \circ h$ is a homomorphism from M onto $\mathbf{M}_t$, then $s(x) = s_t(h_t(x))$, $x \in M$, is a state-morphism on $\mathbf{M}$. $\square$

In what follows, we extend the class of pseudo MV-algebras having a state to normal-valued ones.

Theorem 8.3.12 *Each normal-valued pseudo MV-algebra $\mathbf{M}$ possesses at least one state.*

Proof Every pseudo MV-algebra $\mathbf{M}$ has a maximal ideal. Since $\mathbf{M}$ belongs to NV, it is normal, and by Theorem 8.3.7, its state space is non-void. $\square$

Finding new varieties of non-commutative pseudo MV-algebras with at least one state would be interesting.

8.4   Perfect and n-Perfect Pseudo MV-Algebras

In the paper [100], the authors described perfect MV-algebras. That is, MV-algebra, where each element belongs either to its radical or co-radical (= negations of the radical). It was shown that every perfect MV-algebra M can be represented by an interval in the lexicographic product of the group $\mathbb{Z}$ of integers with an arbitrary Abelian ℓ-group $\mathbb{G}$, that is, $M \cong \Gamma(\mathbb{Z} \overrightarrow{\times} G, (1, 0))$.

8.4.1   Symmetric Perfect Pseudo MV-Algebras

This part presents a generalization of perfect MV-algebras for pseudo MV-algebras. First of all, it is important to redefine the definition from MV-algebraic setup to pseudo MV-algebraic because the radical of an MV-algebra was present as the intersection of maximal ideals of $\mathbb{M}$, whereas, for pseudo MV-algebras, for such an intersection we have maximal ideals which are also normal. A good way to do this is to consider that any perfect MV-algebra can be divided into upper and down parts. This idea is used in this section.

Let $M = (M; \oplus,^- ,^\sim , 0, 1)$ be a pseudo MV-algebra. For any two non-empty subsets A and B of M, we write (i) $A \leq B$ if $a \leq b$ for all $a \in A$ and all $b \in B$, (ii) $A + B = \{a + b : a \in A, b \in B, a + b$ is defined in $M\}$. We denote $A^- = \{a^- : a \in A\}$, $A^\sim = \{a^\sim : a \in A\}$.

A pseudo MV-algebra M is said to be **perfect** if there is a two-valued state s on M such that $M_0 := s^{-1}(\{0\}) \leq M_1 := s^{-1}(\{1\})$; the state s is said to be **strong**.

For example, let G be any ℓ-group. Then

$$M(G) := \Gamma(\mathbb{Z} \overrightarrow{\times} G, (1, 0)) \tag{8.11}$$

is an example of a symmetric perfect pseudo MV-algebra with $M_0 = \{(0, g) : g \in G^+\}$, $M_1 = \{(1, g) : g \in G^-\}$ with a strong state $s(M_0) = \{0\}$ and $s(M_1) = \{1\}$.

Proposition 8.4.1 *Let* $(M; \oplus,^- ,^\sim , 0, 1)$ *be a perfect pseudo MV-algebra. Then*

(i) *M_0 and M_1 are non-empty, mutually disjoint, and they form a decomposition of M such that $M_0^- = M_0^\sim = M_1$ and $M_1^- = M_1^\sim = M_0$.*

(ii) *$M_0 + M_0 = M_0$.*

(iii) *M_0 is a maximal and normal ideal of M and it is a unique maximal ideal.*

(iv) *If s_1 and s_2 are two strong two-valued states, then $\mathrm{Ker}(s_1) = \mathrm{Ker}(s_2)$ and $s_1 = s_2$.*

(v) *A pseudo MV-algebra M is perfect if and only if there is a non-void decomposition M_0, M_1 of M such that (i) and $M_0 \leq M_1$ hold in M. If M_0' and M_1' with $M_0' \leq M_1'$ and (i) form a non-void decomposition, then $M_0 = M_0'$ and $M_1 = M_1'$. We will write $M = (M_0, M_1)$.*

Proof Let s be a strong two-valued state determining the perfectness of M.

(i) It is a simple corollary of the properties of states.

(ii) Let $x, y \in M_0$. Then $x^- \in M_1$ so that $x \le y^-$ and $x + y$ is defined in M and it belongs to M_0. Whence, (ii) holds.

(iii) It is clear that M_0 is an ideal of M. Let $x \in M$ and $y \in M_0$ be such that $x + y$ is defined in M. Then $x + y = ((x + y) \odot y^-) + y$ and $(x + y) \odot y^- \in M_0$ because $s((x + y) \odot x^-) = s(x) + s(y) - s(x) = s(y) = 0$, that is $x + M_0 \subseteq M_0 + x$. Dually, we can show the converse inclusion. Whence, M_0 is a normal ideal of M.

We show that M_0 is maximal. Let $x \in M_1$ and let J be the ideal of M generated by M_0 and x. Then $x^- \in M_0$ and $1 = x^- + x \in J$, so that M_0 is a maximal ideal of M.

Now, let I be any maximal ideal of M. If $I \ne M_0$, there is $x \in I \setminus M_0$. Then $x \in M_1$, and for each $y \in M_0$, we have $y \le x$, that is $M_0 \subseteq I$ which contradicts maximality of I. Whence, $M_0 = I$.

(iv) By (iii), the kernel of a two-valued strong state is a unique maximal ideal of M. Whence $\mathrm{Ker}(s_1) = \mathrm{Ker}(s_2)$, so that $s_1 = s_2$.

(v) One direction is clear. Conversely, let M_0 and M_1 satisfy the conditions. Define a mapping $s : M \to \{0, 1\}$ such that $s(x) = 0$ if and only if $x \in M_0$ and $s(x) = 1$ if and only if $x \in M_1$. Then s is a strong two-valued state, so M is perfect.

If $M = (M_0', M_1')$, then it defines a strong two-valued state s' on M such that $\mathrm{Ker}(s') = M_0'$. By (iv), we conclude $M_0 = M_0'$ and $M_1 = M_1'$. $\qquad\square$

First, we exhibit the situation when a symmetric pseudo MV-algebra M is isomorphic to $\Gamma(\mathbb{Z} \overrightarrow{\times} G, (1, 0))$ for an ℓ-group G.

Let $\mathcal{SPPMV}$ be the category of symmetric perfect pseudo MV-algebras, where the objects are symmetric perfect pseudo MV-algebras, and the morphisms are homomorphisms of pseudo MV-algebras. Let $\mathcal{LG}$ be the category of ℓ-groups, where the objects are ℓ-groups and the morphisms are homomorphisms of ℓ-groups.

We define a mapping $E : \mathcal{LG} \to \mathcal{SPPMV}$ by

$$E(G) := \Gamma(\mathbb{Z} \overrightarrow{\times} G, (1, 0)), \tag{8.12}$$

and if h is a morphism of ℓ-groups, we set

$$E(h)(x) = \begin{cases} (0, h(g)) & \text{if } x = (0, g) \\ (1, -h(g)) & \text{if } x = (1, -g), \end{cases} \tag{8.13}$$

where $g \in G^+$.

Then E is a functor.

Proposition 8.4.2 *E is a faithful and full functor from the category $\mathcal{LG}$ of ℓ-groups into the category $\mathcal{SPPMV}$ of perfect pseudo MV-algebras.*

Proof Let h_1 and h_2 be two morphisms from G into G' such that $E(h_1) = E(h_2)$. Then $(0, h_1(g)) = (0, h_2(g))$ for any $g \in G^+$, consequently $h_1 = h_2$.

To prove that E is a full functor, suppose that $f : \Gamma(\mathbb{Z} \overset{\rightarrow}{\times} G, (1, 0)) \to \Gamma(\mathbb{Z} \overset{\rightarrow}{\times} G', (1, 0))$ is a morphism of pseudo MV-algebras. Then $f(0, g) = (0, g')$ for a unique $g' \in G'^+$. Define a mapping $h : G^+ \to G'^+$ by $h(g) = g'$ if and only if $f(0, g) = (0, g')$. Then $h(g_1 + g_2) = h(g_1) + h(g_2)$ if $g_1, g_2 \in G^+$, and h preserves $\vee$ and $\wedge$. Assume now that $g \in G$ is arbitrary. Then $g = g_1 - g_2 = g'_1 - g'_2$, where $g_1, g_2, g'_1, g'_2 \in G^+$, which gives $g_1 + g'_2 = g'_1 + g_2$, i.e., $h(g) = h(g_1) - h(g_2)$ is a well-defined extension of h from G^+ onto G. We assert that h preserves $\wedge$ in G, i.e., $h(a \wedge b) = h(a) \wedge h(b)$ whenever $a, b \in G$. Let $a = a^+ - a^-$ and $b = b^+ - b^-$, and $a = -a^- + a^+$, $b = -b^- + b^+$. Since , $h((a^+ + b^-) \wedge (a^- + b^+)) = h(a^+ + b^-) \wedge h(a^- + b^+)$. We obtain the assertion in question by subtracting $h(b^-)$ from the right hand and $h(a^-)$ from the left hand.

Therefore, h is a homomorphism of ℓ-groups, and $E(h) = f$ as desired. $\qquad\square$

The following result is important for us.

Proposition 8.4.3 *Let M be a perfect pseudo MV-algebra. Then there is a unique (up to isomorphism) ℓ-group G such that $M \cong \Gamma(\mathbb{Z} \overset{\rightarrow}{\times} G, (1, 0))$.*

Proof Let $M = (M_0, M_1)$ be a symmetric perfect pseudo MV-algebra. By Theorem 8.2.4, we can assume $M = \Gamma(G_0, u)$ for some unital ℓ-group (G_0, u). By Proposition 8.4.1, M_0 is a cancellative semigroup satisfying conditions of [147, Theorem II.4], which guarantees that M_0 is a positive cone of a unique (up to isomorphism) directed ℓ-group G. Since M_0 is a lattice, G is an ℓ-group.

The algebra $\Gamma(\mathbb{Z} \overset{\rightarrow}{\times} G, (1, 0))$ is a symmetric perfect pseudo MV-algebra, and let us define a mapping $f : M \to \Gamma(\mathbb{Z} \overset{\rightarrow}{\times} G, (1, 0))$ by $f(a) = (0, a)$ and $f(a^-) = (1, -a)$ if $a \in M_0$. Then f is injective, onto, and preserves $^-$, $+$, $\vee$ and $\wedge$, and consequently, f can be extended to an injective group homomorphism $\hat{f} : G_0 \to \mathbb{Z} \overset{\rightarrow}{\times} G$.

We show that $\hat{f}$ is an ℓ-group homomorphism. The proof will proceed in several steps.

Step 1. Let $a, b, u_0 \in G_0^+$. If $\hat{f}(a \wedge b) = \hat{f}(a) \wedge \hat{f}(b)$ and if $\hat{f}(u_0 \wedge (b - (a \wedge b))) = \hat{f}(u_0) \wedge \hat{f}(b - (a \wedge b))$, then

$$\hat{f}((a + u_0) \wedge b) = \hat{f}(a + u_0) \wedge \hat{f}(b).$$

Indeed, we have

$$u_0 \wedge (b - (a \wedge b)) + (a \wedge b) = (u_0 + a \wedge b) \wedge b = (u_0 + a) \wedge (u_0 + b) \wedge b = (u_0 + a) \wedge b,$$

which gives

$$\hat{f}((a + u_0) \wedge b) = \hat{f}(u_0 \wedge (b - (a \wedge b))) + \hat{f}(a \wedge b)$$
$$= [\hat{f}(u_0) \wedge (\hat{f}(b) - (\hat{f}(a) \wedge \hat{f}(b)))] + (\hat{f}(a) \wedge \hat{f}(b))$$
$$= [(\hat{f}(u_0) + (\hat{f}(a) \wedge \hat{f}(b)))] \wedge \hat{f}(b)$$
$$= (\hat{f}(u_0) + \hat{f}(a)) \wedge (\hat{f}(u_0) + \hat{f}(b)) \wedge \hat{f}(b) = \hat{f}(a + u_0) \wedge \hat{f}(b).$$

Step 2. $\hat{f}(a \wedge b) = \hat{f}(a) \wedge \hat{f}(b)$ whenever $a \in G_0^+$ and $b \in \Gamma(G_0, u)$.

Since $\Gamma(G_0, u)$ is generating for G_0^+, a is of the form $a = a_1 + \cdots + a_n$ for some $a_1, \ldots, a_n \in \Gamma(G_0, u)$. The proof will follow mathematical induction on n.

If $n = 1$, the statement is trivial. Suppose now that the statement holds for any $a' = a_1 + \cdots + a_i$ with $1 \le i \le n$. Put $a = a_1 + \cdots + a_n$, $u_0 = a_{n+1}$. Then there exist $v_1, \ldots, v_k \in \Gamma(G_0, u)$ such that $b = (v_1 + \cdots + v_k) + (a \wedge b)$. Since $v := v_1 + \cdots + v_k \le b \in \Gamma(G_0, u)$, $v \in \Gamma(G_0, u)$. Hence, $v = b - (a \wedge b)$. Since $\hat{f}$ preserves meets in $\Gamma(G_0, u)$, we have $\hat{f}(u_0 \wedge v) = \hat{f}(u_0) \wedge \hat{f}(v)$, so that $\hat{f}(u_0 \wedge (b - (a \wedge b))) = \hat{f}(u_0) \wedge \hat{f}(b - (a \wedge b)) = \hat{f}(u_0) \wedge (\hat{f}(b) - (\hat{f}(a) \wedge \hat{f}(b)))$, where we have used induction hypothesis for a and b. By Step 1, $\hat{f}((a + u_0) \wedge b) = \hat{f}(a + u_0) \wedge \hat{f}(b)$, that is, $\hat{f}((a_1 + \cdots + a_{n+1}) \wedge b) = \hat{f}(a_1 + \cdots + a_{n+1}) \wedge \hat{f}(b)$ for any n.

Step 3. $\hat{f}(a \wedge b) = \hat{f}(a) \wedge \hat{f}(b)$ whenever $a, b \in G_0^+$.

Let $a = a_1 + \cdots + a_n$, $b = b_1 + \cdots + b_k$. The proof will follow complete induction on k.

If $k = 1$, we apply Step 2. Suppose now that the assertion holds for any j with $1 \le j \le k$. Put $B = a$, $A = b_1 + \cdots + b_k$, $u_0 = b_{k+1}$. By Step 2, $\hat{f}(u_0 \wedge (B - (A \wedge B))) = \hat{f}(u_0) \wedge \hat{f}(B - (A \wedge B))$ and $\hat{f}(A \wedge B) = \hat{f}(A) \wedge \hat{f}(B)$. Therefore, the conditions of Step 1 are satisfied, so that $\hat{f}((A + u_0) \wedge B) = \hat{f}(A + u_0) \wedge \hat{f}(B)$ which proves $\hat{f}((a_1 + \cdots + a_n) \wedge (b_1 + \cdots + b_{k+1})) = \hat{f}(a_1 + \cdots + a_n) \wedge \hat{f}(b_1 + \cdots + b_{k+1})$ for each n and each k.

Step 4. $\hat{f}(a \wedge b) = \hat{f}(a) \wedge \hat{f}(b)$ whenever $a, b \in G_0$. Then $a = a^+ - a^-$ and $b = b^+ - b^-$, and $a = -a^- + a^+$, $b = -b^- + b^+$. By Step 3, $\hat{f}((a^+ + b^-) \wedge (a^- + b^+)) = \hat{f}(a^+ + b^-) \wedge \hat{f}(a^- + b^+)$. Subtracting $\hat{f}(b^-)$ from the right hand and $\hat{f}(a^-)$ from the left hand, we obtain the assertion in question.

Consequently f is an MV-isomorphism from the pseudo MV-algebras M onto $\Gamma(\mathbb{Z} \overrightarrow{\times} G, (1, 0))$. $\qquad\qquad\square$

Using ideas from the general categorical equivalence of pseudo MV-algebras, it is possible to show that $(\mathbb{Z} \overrightarrow{\times} G, f)$ is a universal group for M, which implies that the functor is E right-adjoint.

Define a morphism $P : \mathcal{SPPMV} \to \mathcal{LG}$ via $P(M) := G$ whenever $(\mathbb{Z} \overrightarrow{\times} G, f)$ is a universal group for M. It is a left-invariant functor.

We now present the main result on the categorical equivalence of the category of symmetric perfect pseudo MV-algebras and the category of ℓ-groups. It follows from the last propositions and general ideas of the categorical equivalence of pseudo MV-algebras, see Theorem 8.2.7.

Theorem 8.4.4 *The functor E defines a categorical equivalence of the category $\mathcal{LG}$ of ℓ-groups and the category $\mathcal{SPPMV}$ of symmetric perfect pseudo MV-algebras.*

8.4.2 Perfect Pseudo MV-Algebras Not Necessarily Symmetric

In what follows, we represent perfect pseudo MV-algebras that are not necessarily symmetric. This case was studied in [56, 114]. We need the following notion.

Let $G = (G; \vee, \wedge, +, -, 0)$ be an ℓ-group and $\phi : G \to G$ be an automorphism of ℓ-groups. We define the lexicographic semidirect product $\overset{\rightarrow}{\rtimes}_\phi$ on the direct product $\mathbb{Z} \times G$ with the lexicographic order, where the group operations are given by

$$(n, g) \cdot (m, h) := (n + m, \phi^m(g) + h),$$
$$(n, g)^{-1} := (-n, \phi^{-n}(-g)) = (-n, -\phi^{-n}(g))$$

for all $n \in \mathbb{Z}$ and $g \in G$, with the neutral element $(0, 0)$. Then $\mathbb{Z} \overset{\rightarrow}{\rtimes}_\phi G$ defines an ℓ-group, and $(1, 0)$ is its strong unit. We define a pseudo MV-algebra

$$E(G)_\phi := \Gamma(\mathbb{Z} \overset{\rightarrow}{\rtimes}_\phi G, (1, 0)). \tag{8.14}$$

For each $g \in G^+$, we have $(0, g)^{-1} = (0, -g)$ and $(1, -g)^{-1} = (-1, \phi^{-1}(g))$, so that

$$(0, g)^- = (1, -g), \quad (0, g)^\sim = (1, -\phi(g)),$$
$$(1, -g)^- = (0, \phi^{-1}(g)), \quad (1, -g)^\sim = (0, g),$$
$$(0, g)^{--} = (0, \phi^{-1}(g)), \quad (0, g)^{\sim\sim} = (0, \phi(g)),$$
$$(1, -g)^{--} = (1, -\phi(g)), \quad (1, -g)^{\sim\sim} = (1, -\phi(g)).$$

Then $E(G)_\phi$ gives an example of a symmetric pseudo MV-algebra if and only if $\phi = Id_G$, that is, ϕ is the identity on G. We recall that

$$(E(G)_\phi)_0 = \{(0, g) : g \in G^+\}, \quad (E(G)_\phi)_1 = \{(1, -\phi(g)) : g \in G^+\}.$$

A perfect pseudo MV-algebra $E(G_2)_{\phi_2}$ is an isomorphic (homomorphic) image of a perfect $E(G_1)_{\phi_1}$ if and only if there is an isomorphism (surjective homomorphism) of ℓ-groups $h : G_1 \to G_2$ such that

$$h \circ \phi_1 = \phi_2 \circ h. \tag{8.15}$$

Theorem 8.4.5 *Let M be a perfect pseudo MV-algebra. There is a unique (up to isomorphism) ℓ-group G and an automorphism of ℓ-groups $\phi : G \to G$ such that $M \cong \Gamma(\mathbb{Z} \overrightarrow{\rtimes}_\phi G, (1, 0))$.*

Proof According to Proposition 8.4.1(v), M can be expressed as $M = (M_0, M_1)$, where M_0 and M_1 satisfy (i) of Proposition 8.4.1 and $M_0 \le M_1$, and $(E_0; +)$ is a semigroup. It is cancellative, containing the neutral element 0, $x + y = 0$ implies $x = y = 0$ and $x + E_0 = E_0 + x$ for each $x \in M_0$, so that by the Birkhoff's theorem, [147, Theorem II.4], M_0 is the positive cone of some (unique up to isomorphism) ℓ-group G.

Let ϕ_0 be the restriction of the automorphism $x \mapsto x^{\sim\sim}$, $x \in M$, onto M_0, see Lemma 8.1.9(ii). Then ϕ_0 is an injective mapping preserving $+$ and the order $\le$ in M_0. Moreover, ϕ_0 maps M_0 onto M_0 and M_1 onto itself. Then ϕ_0 can be uniquely extended onto the automorphism ϕ of the ℓ-group G.

Let us define the perfect pseudo MV-algebra $M(G)_\phi$ by (8.14).

Define a mapping $h : M \to M(G)_\phi$ as follows

$$h(x) = \begin{cases} (0, x) & \text{if } x \in E_0, \\ (1, -x^\sim) & \text{if } x \in E_1. \end{cases}$$

We show that h is an isomorphism from M onto $E(G)_\phi$. Clearly, $h(0) = (0, 0)$ and $h(1) = (1, 0)$. Check and use equations just after (8.14).

Let $x \in M_0$, then $x^-, x^\sim \in M_1$ and

$$\begin{aligned} h(x^-) &= (1, -x^{-\sim}) = (1, -x), \\ h(x)^- &= (0, x)^- = (1, -x), \\ h(x^\sim) &= (1, -x^{\sim\sim}), \\ h(x)^\sim &= (0, x)^\sim = (1, -x^{\sim\sim}). \end{aligned}$$

Let $x \in M_1$, then $x^-, x^\sim \in M_0$ and

$$\begin{aligned} h(x^-) &= (0, x^-), \\ h(x)^- &= (1, -x^\sim)^- = (0, \phi^{-1}(x^\sim)) = (0, x^-), \\ h(x^\sim) &= (0, x^\sim), \\ h(x)^\sim &= (1, -x^\sim)^\sim = (0, x^\sim). \end{aligned}$$

We denote the addition in $M(G)_\phi$ by $+_\phi$.

(a) Now, let $x, y \in M_0$. Then $x + y \in M_0$ and $h(x + y) = (0, x + y) = (0, x) +_\phi (0, y) = h(x) +_\phi h(y)$.

(b) Let $x \in M_0$, $y \in M_1$. Then $x + y$ is defined in M if and only if $x \le y^-$ if and only if $y \le x^\sim$ if and only if $x^{\sim\sim} \le y^\sim$. In either case, $x + y \in M_1$. Check

$$h(x + y) = (1, -(x + y)^\sim),$$

$$h(x) +_\phi h(y) = (0, x) +_\phi (1, -y^\sim) = (1, x^{\sim\sim} - y^\sim).$$

We see that $h(x) +_\phi h(y)$ is defined in $M(G)_\phi$. We show $-(x + y)^\sim = x^{\sim\sim} - y^\sim$ in the ℓ-group G.

Calculate in M:

$$(x + y)^\sim + (x + y)^{\sim\sim} = (x + y)^\sim + (x^{\sim\sim} + y^{\sim\sim}) = 1 = y^\sim + y^{\sim\sim},$$

$$(x + y)^\sim + x^{\sim\sim} = y^\sim.$$

Then

$$(x + y)^\sim = y^\sim \setminus x^{\sim\sim} = y^\sim - x^{\sim\sim},$$

where $-$ and $+$ are calculated in the ℓ-group G which is possible because $y^\sim, x^{\sim\sim} \in M_0$ and $y^\sim \ge x^{\sim\sim}$. Therefore, $h(x + y) = h(x) +_\phi h(y)$.

(c) Let $x \in M_1$ and $y \in M_0$ and let $x + y \in M$. Then $x + y \in M_1$ and $x + y$ is defined iff $x \le y^-$ iff $x^{\sim\sim} \le y^\sim$ iff $y \le x^\sim$. Check

$$h(x + y) = (1, -(x + y)^\sim)$$

$$h(x) +_\phi h(y) = (1, -x^\sim) +_\phi (0, y) = (1, -x^\sim + y) \in M(G)_\phi.$$

In the pseudo MV-algebra M, we have

$$(x + y) + (x + y)^\sim = 1 = x + x^\sim$$

$$y + (x + y)^\sim = x^\sim = y + y \diagup x^\sim$$

$$(x + y)^\sim = y \diagup x^\sim$$

$$(x + y)^\sim = -y + x^\sim \in G.$$

Then $h(x + y) = h(x) +_\phi h(y)$.

Summarizing, we have $M \cong E(G)_\phi$.

Uniqueness. Assume G and ℓ-groups G_1 and G_2 with a fixed automorphism $\phi_i : G_i \to G_i$, $i = 1, 2$ such that $M' := M(G_1)_{\phi_1} \cong E \cong M(G_2)_{\phi_2} =: M''$. Then $M'_0 = \{(0, g) : g \in G_1^+\} \cong M''_0 = \{(0, g) : g \in G_2^+\}$. Define a symmetric perfect pseudo MV-algebra $F^i = \Gamma(\mathbb{Z} \overrightarrow{\times} G_i, (1, 0))$ for $i = 1, 2$. Then $\mathbb{Z} \overrightarrow{\times} G_1 \cong \mathbb{Z} \overrightarrow{\times} G_2$, and $F_0^1 = E'_0$, $F_0^1 = E''_0$. By Theorem 8.2.7, we have $\mathbb{Z} \overrightarrow{\times} G_1$ and $\mathbb{Z} \overrightarrow{\times} G_2$ are isomorphic ℓ-groups. Consequently, G_1 and G_2 are isomorphic ℓ-groups. $\square$

Now, we present the categorical equivalence of the category of perfect pseudo MV-algebras (not necessarily symmetric), $\mathcal{PPMV}$, with the category $\mathcal{LG}_{Aut}$ of ℓ-groups G with a fixed automorphism $\phi : G \to G$. That is objects of $\mathcal{LG}_{Aut}$ are pairs (G, ϕ), where $G = (G; \vee, \wedge, +, -, 0)$ is an ℓ-group with a fixed automorphism $\phi : G \to G$. The morphisms are homomorphisms of ℓ-groups commuting with the fixed automorphism. That is, T is a morphism from (G_1, ϕ_1) to (G_2, ϕ_2) if and only if $T : G_1 \to G_2$ is a group homomorphism such that $\phi_1 \circ T = T \circ \phi_2$. It is easy to verify that $\mathcal{LG}_{Aut}$ is indeed a category.

Given an object (G, ϕ) from $\mathcal{LG}_{Aut}$, we define

$$E_A(G, \phi) := \Gamma(\mathbb{Z} \overrightarrow{\rtimes}_\phi G, (1, 0)) \tag{8.16}$$

and, for each morphisms $h : (G, \phi) \to (G', \phi')$, we set

$$E_A(h)(x) = \begin{cases} (0, h(g)) & \text{if } x = (0, g), \\ (1, -\phi'(h(g))) & \text{if } x = (1, -g), \end{cases} \quad g \in G^+. \tag{8.17}$$

Then E_A is a functor that is both faithful and full.

Using a universal group for pseudo MV-algebras, we define an opposite functor $P_A : \mathcal{PPMV} \to \mathcal{LG}_{Aut}$ as follows. If M is a perfect pseudo MV-algebra from $\mathcal{PPMV}$, let $P_A(M) = (G, \phi)$, where $(\mathbb{Z} \overrightarrow{\rtimes}_\phi G, \gamma)$ is the universal group for M. It is clear that if $f_0 : M \to N$ is a morphism of perfect pseudo MV-algebras, then it can be uniquely extended to a homomorphism $P_A(f_0)$ from G into G_1, where $\mathbb{Z} \overrightarrow{\rtimes}_{\phi_1} G_1$ is a universal group for the perfect pseudo MV-algebra N. Therefore, we can establish:

Theorem 8.4.6 *The functors P_A and E_A of functors constitute a categorical equivalence between the category $\mathcal{PPMV}$ of perfect pseudo MV-algebras and the category $\mathcal{LG}_{Aut}$ of ℓ-groups with fixed automorphism.*

8.4.3 n-Perfect Pseudo MV-Algebras

Now, we extend perfect pseudo MV-algebras to n-perfect pseudo MV-algebras. Let $n \geq 1$ be a fixed integer. We say that a pseudo MV-algebra M is n-**perfect** if there is a state s on M such that (i) $s(M) = \{0, 1/n, \ldots, n/n\}$ and (ii) $M_i := s^{-1}(\{i/n\}) \leq M_{i+1} := s^{-1}(\{(i + 1)/n\})$ for each $i = 0, 1, \ldots, n - 1$. Such an $(n + 1)$-valued state s is said to be **strong**. Using ideas of the proof of Proposition 8.3.1, s is state-morphism. If $n = 1$, then we have a perfect MV-algebra.

Proposition 8.4.7 *Let M be an n-perfect pseudo MV-algebra. Then*

(i) $M_i^- = M_{n-i} = M_i^\sim$ *for each* $i = 0, 1, \ldots, n$.

(ii) $M_i + M_j = M_{i+j}$ *whenever* $j + i < n$. *If* $x \in M_i$ *and* $y \in M_j$ *for* $i + j > n$, *then* $x + y$ *is not defined in* M. *The set* M_0 *is a normal and maximal ideal.*

(iii) *A pseudo MV-algebra* M *is n-perfect if and only if there is a decomposition of* M *consisting of non-empty subsets* $M_0, M_1, \ldots, M_n$ *of* M *such that* (i) *and* $M_i \leq M_{i+1}$ *hold in* M *for each* $i = 0, 1, \ldots, n - 1$. *In such a case,* $M_0, M_1, \ldots, M_n$ *are unique, we can write* $M = (M_0, M_1, \ldots, M_n)$, *and* M *has a unique* $(n + 1)$-*valued strong state.*

Proof (i) and (ii) are simple.

(iii) Since M_0 is the kernel of the $(n + 1)$-valued strong state s determining $M_0, M_1, \ldots, M_n$, the set M_0 is a normal ideal. Let $b \notin M_0$. Then $n.b \in M_n$ and $(n.b)^- \in M_0$ which by [152, Proposition 3.5] means that M_0 is a maximal ideal. By Proposition 8.3.2 and Theorem 8.3.3, s is an extremal state. We show that M_0 is a unique maximal ideal. Thus, let I be any maximal ideal of M and let $I \neq M_0$. There is $a \in M_0 \setminus I$ so that 1 belongs to the ideal generated by I and a. From normality of M_0, we have $1 = a_0 + x = a^- + a_0$ for some $a_0 \in M_0$ and $x \in I$. Due to $a_0 \leq a_0^-$, we have $a_0 \leq x$, so that $a_0 \in I$ and $1 = a_0 + x \in I$ which is impossible. Hence, $M_0 \subseteq I$ and maximality of M_0 gives $M_0 = I$. Since there is a one-to-one relationship between maximal ideals that are normal and extremal states, we conclude that M has a unique $(n + 1)$-valued strong state.

Now, let non-empty sets $M_0, M_1, \ldots, M_n$ satisfy the conditions. Define a mapping $s : M \to \{0, 1\}$ such that $s(x) = i/n$ if and only if $x \in M_i$ for $i = 0, 1, \ldots, n$. Then M_0 is a maximal and normal ideal. Then, s is a strong $(n + 1)$-valued state, so that M is n-perfect. $\square$

Let G be an ℓ-group and $\phi : G \to G$ be an automorphism of ℓ-groups. Given an integer $n \geq 1$, we define an ℓ-group $\frac{1}{n}\mathbb{Z} \overset{\rightarrow}{\rtimes}_\phi G$ in an analogous way as we did for $\mathbb{Z} \overset{\rightarrow}{\rtimes}_\phi G$, that is $(i/n, g) \cdot (j/n, h) = ((i + j)/n, \phi^j(g) + h)$ and $(i/n, g)^{-1} = (-i/n, \phi^{-i}(-g)) = (-i/n, -\phi^{-i}(g))$ for all $i, j \in \mathbb{Z}$. In addition, the element $(1, 0)$ is a strong unit for $\frac{1}{n}\mathbb{Z} \overset{\rightarrow}{\times} G$ as well as for $\frac{1}{n}\mathbb{Z} \overset{\rightarrow}{\rtimes}_\phi G$.

We note that subsets $(M_n(G)_\phi)_0 = \{(0, g) : g \in \mathbf{G}^+\}$, $(M_n(G)_\phi)_i = \{(i/n, g) : g \in G\}$ for $i = 1, \ldots, n - 1$, and $(M_n(G)_\phi)_n = \{(n/n, g) : g \in \mathbf{G}^-\}$ show that $M_n(G)_\phi$ is an n-perfect pseudo MV-algebra.

We define

$$M_n(G) = \Gamma(\tfrac{1}{n}\mathbb{Z} \overset{\rightarrow}{\times} G, (1, 0)),$$
$$M_n(G)_\phi = \Gamma(\tfrac{1}{n}\mathbb{Z} \overset{\rightarrow}{\rtimes}_\phi G, (1, 0)).$$

Then $M_n(G)$ and $M_n(G)_\phi$ are a symmetric n-perfect pseudo MV-algebra and an n-perfect pseudo MV-algebra (not necessarily symmetric), respectively. It is clear that $M_n(G)_\phi$ is symmetric if and only if $\phi = Id_G$.

We note that in $M_n(G)_\phi$, we have

$$(i/n, g)^- = ((n-i)/n, -\phi^{-i}(g)), \quad (i/n, g)^\sim = ((n-i)/n, -\phi^{n-i}(g)),$$
$$(i/n, g)^{--} = (i/n, \phi^{-n}(g)), \quad (i/n)^{\sim\sim} = (i/n, \phi^n(g)).$$

Now, we present a representation theorem for pseudo MV-algebras, [114, Theorem 5.2]. Its proof follows ideas of analogous proof for perfect MV-algebras, but if $n > 1$, we need a special element $a \in M_1$ such that na is defined in M and $na = 1$. For $n = 1$, such an element exists automatically, namely $a = 1 \in M_1$ and $1a = 1$.

Theorem 8.4.8 *An n-perfect pseudo MV-algebra $M = (M_0, M_1, \dots, M_n)$ is isomorphic to an n-perfect pseudo MV-algebra $\Gamma(\frac{1}{n}\mathbb{Z} \overset{\rightarrow}{\rtimes}_\phi G, (1, 0))$ for some ℓ-group G and an automorphism ϕ if and only if, there is an element $a \in M_1$ such that na exists in M and $na = 1$.*

In such a case, let G' be an ℓ-group and $\phi' : G' \to G'$ be an isomorphism of ℓ-groups. The perfect pseudo MV-algebra $M(G')_{\phi'}$ is isomorphic to M if and only if there is an isomorphism of ℓ-groups $h : G \to G'$ such that $h \circ \phi = \phi' \circ h$.

Proof If $M \cong M_n(G)_\phi$, and $\psi : M \to M_n(G)_\phi$ is an automorphism, then the element $a = \psi^{-1}((1/n, 0))$ is an element in question. We have $(M_n(G)_\phi)_0 = \{(0, g) : g \in G^+\}$ and it is a normal ideal. Let $x = (i/n, g)$ be an element of $M_n(G)_\phi$ with $i > 0$ and let I be an ideal of $M_n(G)_\phi$ generated by $(M_n(G)_\phi)_0$ and x. If $n = 1$, then clearly $(M_n(G)_\phi)_0$ is a maximal ideal. If $n > 2$, then we can assume that $x = (1/n, g)$, with $g \in G$. Then $(1/n, 0) \leq (n-1)x = ((n-1)/n, \phi^{n-2}(g) + \cdots + \phi^0(g)) \in (M_n(G)_\phi)_{n-1} = (M_n(G)_\phi)_1^-$, so that $1 \in I$. If $n = 2$, then either $x \in (M_n(G)_\phi)_2$, or $x \in (M_n(G)_\phi)_1$. In the first case, every element of $(M_n(G)_\phi)_1$ is less than x, so that $(M_n(G)_\phi)_1^\sim = (M_n(G)_\phi)_1 \subseteq I$, therefore, $1 \in I$. In the second case, g can be expressed as $g = g_1 - g_2$, where $g_1, g_2 \in G^+$. Since $(0, g_2) \in (M_n(G)_\phi)_0$, we have $(1/n, 0) \leq (1/n, g_1) = (1/n, g) + (0, g_2) \in (M_n(G)_\phi)_1$ so that $(1, 0) \in I$ and $1 \in I$.

For the converse statement, M_0 is a maximal ideal of $M = (M_0, M_1, \dots, M_n)$ and let $a \in M_1$ be an element such that $na \in M$ and $na = 1$. It is a normal ideal while it is the kernel of some $(n + 1)$-valued strong state.

First, we show that M_0 is a maximal ideal. Let $x \notin M_0$ and let J be the ideal of M generated by M_0 and x. Since every $M_i \leq M_{i+1}$ for $i = 0, 1, \dots, n$, we can assume that $x \in M_1$. If $n = 1$, then M_0 is maximal by Proposition 8.3.1(v). If $n \geq 2$, then $x \leq na$, so that $x = a_1 + \cdots + a_n$, where $a_i \leq a$. Exactly one of a_i's, say a_1, is from $M_1 \cap J$ and all others are from M_0. Then $na_1 \leq na$, so that $na_1 \in M_n \cap J$ and $(na)^- \in M_0$ which gets $1 = (na)^- + (na) \in J$.

Now, we prove that every M_i is a directed set. If $x, y \in I_0$, then $0 \leq x, y \leq x + y \in M_0$, so that M_0 is directed. Since $M_n = M_0^\sim$, so is directed M_n. Now, take $i = 1, \dots, n-1$ and let $x, y \in M_1$. Since M_0 is a maximal and normal ideal, y is in the ideal generated by M_0 and x, and due to the Riesz Decomposition Property, $y = z + x_1$, where $z \in M_0$ and

$x_1 \leq x$. Since $y \in M_i$, we have $x_1 \in M_i$. Then $x_1 \leq x$, y and M_i is downwards directed and so M_{n-i} is upwards directed. Since $M_i = M_{n-i}^{\sim}$, every M_i is also upwards directed, so that every M_i is directed.

Since $M_0 + M_0 = M_0$, $(M_0; +)$ is a semigroup satisfying the conditions of [147, Theorem II.4], therefore, M_0 can be assumed to be the positive cone of some (unique up to isomorphism) ℓ-group G.

Given the element a, let $\phi_a(g) := -a + g - a$, $g \in G$. The mapping ϕ_a is an automorphism of the ℓ-group G.

Let $M = (M_0, \ldots, M_n)$ and define a mapping $h : M \to M_n(G)_{\phi_a}$ as follows

$$h(x) = (i/n, -ia + x) \quad \text{if } x \in M_i, \ i = 0, 1, \ldots, n. \tag{8.18}$$

We show that $h(x)$ is defined correctly. If $x \in M_0$, then $h(x) = (0, -0a + x) = (0, x)$. If $x \in M_n$, then $h(x) = (n/n, -na + x) = (n/n, -1 + x) = (n/n, -x^{\sim})$. Let $i = 1, \ldots, n - 1$ and let (H, u) be a unital ℓ-group such that $M \cong \Gamma(H, u)$. We can assume that $M = \Gamma(H, u)$. Since M_i is downwards directed, there is an element $z_i \in M_i$ such that $z_i \leq ia, x$. Then in the ℓ-group H, we have

$$-ia + x = -ia + z_i - z_i + x = -(-z_i + ia) + (-z_i + x),$$

and the elements $(-z_i + ia)$ and $(-z_i + x)$ belong to M_0, that is $-ia + x \in G$.

Since $na = a + (n - 1)a = 1 = (n - 1)a + a$, we have $a^- = (n - 1)a = a^{\sim}$. Let $x \in M_i$, $y \in M_j$, and $x + y \in M_{i+j}$. Then $x \leq y^-$. We can show that

$$h(y^-) = ((n - j)/n, -(n - j)a + y^-) = ((n - j)/n, -(n - j)a + na - y)$$
$$= ((n - j)/n, ja - y) = h(y)^-.$$

Thus, if $i + j < n$, then $i < n - j$, so that $h(x) = (i/n, -ia + x) \leq ((n - j)/n, ja - y) = h(y)^-$. If $i = n - j$, then $h(x) = ((n - j)/n, -(n - j)a + x) \leq ((n - j)/n, -(n - j)a + y^-) = h(y)^-$. Therefore, $h(x) +_{\phi_a} h(y)$ is defined in $M_n(G)_{\phi_a}$. We note that in the following, ϕ_a^j denotes the j-many compositions of ϕ_1 with itself. Whence, we have

$$h(x + y) = ((i + j)/n, -(i + j)a + x + y)$$
$$= ((i + j)/n, -ia + (-ja + x + ja) + (-ja + y))$$
$$= ((i + j)/n, -ia + \phi_a^j(x) + (-ja + y))$$
$$= ((i + j)/n, \phi_a^j(-ia + x) + (-ja + y)).$$
$$h(x) +_\phi h(y) = (i/n, -ia + x) +_{\phi_a} (j/n, -ja + x)$$
$$= ((i + j)/n, \phi_a(-ia + x) + (-ja + y)).$$

Clearly $h(0) = (0, 0)$ and $h(1) = (n/n, -na + 1) = (1, 0)$, so that h is an injective homomorphism of pseudo MV-algebras. Moreover, h^{-1} is also an isomorphism.

Let G' be another ℓ-group and let $\phi' : G' \to G'$ be an automorphism of ℓ-groups. Set $\phi := \phi_a$. Using the relations $(i/n, g)^- = ((n - i)/n, -\phi^{-i}(g))$, $(i/n, g')^- = ((n - i)/n, -\phi^{-i}(g'))$ and an analogous approach as in (8.15), we have $M \cong M_n(G)_\phi \cong M_n(G')_{\phi'}$ if and only if there is an isomorphism $h : G \to G'$ such that $h \circ \phi = \phi' \circ h$. $\qquad\qquad\square$

In the following example, we present an $(n - 1)$-perfect pseudo MV-algebra that has no element $a \in M_1$ such that $(n - 1)a = 1$.

Example 8.4.9 Let $n \geq 2$ be a fixed integer and let $G_n = \mathbb{Z}^n \overset{\leftarrow}{\times} \mathbb{Z}$ be ordered antilexicographically, where $\mathbb{Z}^n$ is ordered by coordinates. We convert G_n into the Scrimger n-group, i.e., the addition in G_n is defined by $(a_1, a_1, \ldots, a_n, n_1) + (b_1, b_2, \ldots, b_n, n_2) :=$

$$
\begin{cases}
(a_n + b_1, a_1 + b_2, \ldots, a_{n-1} + b_n, n_1 + n_2) & \text{if } n_2 = n - 1 | \mathrm{mod}\, n \\
(a_{n-1} + b_1, a_n + b_2, \ldots, a_{n-2} + b_n, n_1 + n_2) & \text{if } n_2 = n - 2 | \mathrm{mod}\, n \\
\quad\vdots & \\
(a_3 + b_1, a_4 + b_2, \ldots, a_1 + b_{n-1}, a_2 + b_n, n_1 + n_2) & \text{if } n_2 = 2 | \mathrm{mod}\, n \\
(a_2 + b_1, a_3 + b_2, \ldots, a_n + b_{n-1}, a_1 + b_n, n_1 + n_2) & \text{if } n_2 = 1 | \mathrm{mod}\, n \\
(a_1 + b_1, a_2 + b_2, \ldots, a_{n-1} + b_{n-1}, a_n + b_n, n_1 + n_2) & \text{if } n_2 = 0 | \mathrm{mod}\, n.
\end{cases}
$$

The element $u_n = (1, \ldots, 1, n - 1)$ is a strong unit for G_n. Then $M = \Gamma(G_n, u_n)$ is an $(n - 1)$-perfect pseudo MV-algebra that is not symmetric; if we define $M_0 = (\mathbb{Z}^+)^n \times \{0\}$, $M_i = \mathbb{Z}^n \times \{i\}$ if $0 < i < n - 1$, and $M_{n-1} = \mathbb{Z}^n_{\leq 1} \times \{n - 1\}$, then $M = (M_0, M_1, \ldots, M_{n-1})$, M admits a unique state; this state takes $i/(n - 1)$ on M_i (it is a strong state), $i = 0, 1, \ldots, n - 1$. In M_1, there is no element $(a_1, \ldots, a_n, 1)$ such that $(n - 1)(a_1, \ldots, a_n, 1) = u_n$. So M cannot be represented by any $\Gamma(\frac{1}{n-1}\mathbb{Z} \overset{\rightarrow}{\times}_\phi G, (1, 0))$.

We present a categorical representation for n-perfect pseudo MV-algebras for each integer $n \geq 1$. Let $\mathcal{PPMV}_n$ be the category of n-perfect pseudo MV-algebras whose objects are pairs (M, a), where $M = (M_0, M_1, \ldots, M_n)$ is an n-perfect pseudo MV-algebra with a fixed element $a \in M_1$ such that na exists in M and $na = 1$. Morphisms are homomorphisms of pseudo MV-algebras preserving fixed elements. Then indeed $\mathcal{PPMV}_n$ is a category. Given an object (G, ϕ) from $\mathcal{LG}_{Aut}$, we define

$$
M_n(G, \phi) = (\Gamma(\tfrac{1}{n}\mathbb{Z} \overset{\rightarrow}{\times}_\phi G, (1, 0)), (1/n, 0))
$$

and, for each morphism $h : (G, \phi) \to (G', \phi')$, we define

$$
M_n(h)(i/n, g) = (i/n, \phi'^i(h(g))), \quad (i/n, g) \in M_n(G)_\phi.
$$

We define also a functor $\boldsymbol{P}_n : \mathcal{PPMV}_n \to \mathcal{LG}_{Aut}$ as follows: $\boldsymbol{P}_n(\boldsymbol{M}, a) = (\boldsymbol{G}, \phi_a)$, where $\boldsymbol{G}$ is a unique (up to isomorphism) ℓ-group from Theorem 8.4.8 and $\phi_a(g) = -a + g + a$, $g \in \boldsymbol{G}$, is an automorphism of the po-group $\boldsymbol{G}$.

Using similar ideas from the proof of the categorical equivalence of $\mathcal{PPMV}$ and $\mathcal{LG}_{Aut}$, we can establish that $\boldsymbol{M}_n$ is a faithful and full functor and $\boldsymbol{P}_n$ is a left adjoint of $\boldsymbol{M}_n$. Therefore, the following result holds:

Theorem 8.4.10 *The category of $\mathcal{PPMV}_n$ of n-perfect pseudo MV-algebras and the category of $\mathcal{LG}_{Aut}$ of ℓ-groups are categorically equivalent.*

In addition, the categories PerfPEA *and* PerfPEA$_n$ *are categorically equivalent.*

Summarizing the above established results about perfect and n-perfect pseudo MV-algebras, we conclude:

Corollary 8.4.11 *The category $\mathcal{PPMV}_n$ of n-perfect pseudo MV-algebras, the category $\mathcal{PPMV}$ of perfect pseudo MV-algebras, and the category $\mathcal{LG}_{Aut}$ are categorically equivalent.*

At the end of the section, we can outline a more general kind of perfect pseudo MV-algebras, called $(\boldsymbol{H}, u)$-perfect pseudo MV-algebras. Let $(\boldsymbol{H}, u)$ be a unital linearly ordered ℓ-group and $\boldsymbol{G}$ be an ℓ-group, both groups written additively.

Take a group homomorphism $\phi : H \to \mathrm{Aut}(\boldsymbol{G})$, where $\mathrm{Aut}(\boldsymbol{G})$ is the set of group homomorphisms of $\boldsymbol{G}$. The automorphism $\phi(h) : \boldsymbol{G} \to \boldsymbol{G}$ is denoted simply by ϕ_h for each $h \in H$. Then $\phi_{h_1+h_2} = \phi_{h_1} \circ \phi_{h_2}$ for all $h_1, h_2 \in H$, and ϕ_0 is the identity on $\boldsymbol{G}$. We define a **semidirect product** with respect to ϕ of $\boldsymbol{H}$ and $\boldsymbol{G}$, $H \overrightarrow{\rtimes}_\phi G$, as a special group structure on the direct product $\boldsymbol{H} \times \boldsymbol{G}$ ordered lexicographically, where the group operations on $H \overrightarrow{\rtimes}_\phi G$ are defined by

$$(h_1, g_1) \cdot (h_2, g_2) = (h_1 + h_2, \phi_{h_2}(g_1) + g_2), \quad (h, g)^{-1} = (-h, \phi_{-h}(-g)).$$

The neutral element is $(0_H, 0_G)$ and $(u, 0_G)$ is a strong unit of $H \overrightarrow{\rtimes}_\phi G$. It is clear that $\boldsymbol{H} \overrightarrow{\rtimes}_\phi \boldsymbol{G}$ is an ℓ-group. We define a pseudo MV-algebra $\boldsymbol{\Gamma}(\boldsymbol{H}, u)$ and a pseudo MV-algebra

$$M^\phi_{\boldsymbol{H}, u}(\boldsymbol{G}) := \boldsymbol{\Gamma}(\boldsymbol{H} \overrightarrow{\rtimes}_\phi \boldsymbol{G}, (u, 0)). \tag{8.19}$$

Let $(\boldsymbol{H}, u) = (\frac{1}{n}\mathbb{Z}, 1)$ and let $\phi_0 \in \mathrm{Aut}(\boldsymbol{G})$. We define a group homomorphism $\phi : \frac{1}{n}\mathbb{Z} \to \mathrm{Aut}(\boldsymbol{G})$ defined by $\phi(i/n) = \phi_0^i$, $i \in \mathbb{Z}$. Then $\boldsymbol{\Gamma}(\frac{1}{n}\mathbb{Z} \overrightarrow{\rtimes}_\phi \boldsymbol{G}, (1, 0))$ is an n-perfect pseudo MV-algebra studied above.

We present two examples of this new kind of perfect pseudo MV-algebras.

Example 8.4.12 Let $H_1 = (0, \infty)$ be the group of positive real numbers with the natural product, natural order, and neutral element $0_{H_1} = 1$. The element $u = 2$ is a strong unit for H_1. Take $G_1 = \mathbb{R}$ with the natural addition and with the order of real numbers, $0_{G_1} = 0$. For each $h > 0$, the mapping $\phi_h : \mathbb{R} \to \mathbb{R}$ defined by $\phi_h(g) = hg$, $g \in \mathbb{R}$, is a group automorphism of $\mathbb{R}$. Define the pseudo MV-algebra $M = M^{\phi}_{H_1,2}(G_1)$, its top element is $(2, 0)$. It is totally ordered and not symmetric: We have

$$(h, g)^{-1} = \left(\frac{1}{h}, -\frac{1}{h}g\right), \quad (h, g)^{-} = \left(\frac{2}{h}, -\frac{1}{h}g\right), \quad (h, g)^{\sim} = \left(\frac{2}{h}, -\frac{2}{h}g\right).$$

Example 8.4.13 Let $H_2 = \mathbb{R} = G_2$ be endowed with the natural addition and natural product, then $0_{H_2} = 0 = 0_{G_2}$. We define $\phi : H_2 \to \mathrm{Aut}(G)$ by $\phi(h)(g) = \mathrm{e}^h g$, $h \in H_2$ and $g \in G_2$. Then

$$(h_1, g_1) + (h_2, g_2) = (h_1 + h_2, \mathrm{e}^h g_1 + g_2), \quad -(h, g) = (-h, \mathrm{e}^{-h} g).$$

The semidirect products $H_1 \overrightarrow{\rtimes}_\phi G_1$ from the previous example and $H_2 \overrightarrow{\rtimes}_\phi G_2$ are isomorphic linearly ordered groups, the ℓ-group isomorphism $\psi : H_1 \overrightarrow{\rtimes}_\phi G_1 \to H_2 \overrightarrow{\rtimes}_\phi G_2$ is given by

$$\psi(h, g) = (\ln h, g) \quad , (h, g) \in H_1 \overrightarrow{\rtimes}_\phi G_1. \tag{8.20}$$

(1) If $u_2 = 1$, then $M_2 = M^{\phi}_{H_2,1}(G_2)$ is a non-symmetric pseudo MV-algebra, and ψ is not an isomorphism of unital ℓ-groups.

(2) If $u_2 = \ln 2$, then $\Gamma(H_1 \overrightarrow{\rtimes}_\phi G_1, (2, 0))$ and $\Gamma(H_2 \overrightarrow{\rtimes}_\phi , (\ln 2, 0))$ are isomorphic pseudo MV-algebras; the isomorphism is the restriction of ψ onto $M = M^{\phi}_{H_1,2}(G_1)$.

8.5   Kite Construction of Pseudo MV-Algebras and Pseudo BL-Algebras

In this section, we demonstrate how an ℓ-group $\mathbf{G}$, together with the disjoint sum of powers of its positive and negative cones, can be used to construct pseudo MV-algebras and pseudo BL-algebras, that are special kinds of pseudo hoops. In our construction, we will use ℓ-groups written in a multiplicative way, and therefore, instead of two arrows, we will use the left and right divisions. This is used for residuated structures due to [149]. We present a construction of how, from $\mathbf{G}$, we can obtain pseudo BL-algebras and pseudo MV-algebras. This was the first time used this way in [190] and developed in [121]. Since the Hasse diagram of this pseudo-BL-algebra from [190] has a kite shape, they were named in [121] kite algebras or simply kites.

We recall the definition of a pseudo hoop using divisions.

According to [153], an algebra $\mathbf{A} = (A; \cdot, \backslash, /, 1)$ of type $(2, 2, 2, 0)$ is a **pseudo hoop** if the following holds, for all $a, b \in A$

(i) $a \cdot 1 = 1 \cdot a = a$.

(ii) $a\backslash a = 1 = a/a$.

(iii) $c/(a \cdot b) = (c/b)/a$.

(iv) $(a \cdot b)\backslash c = b\backslash(a\backslash c)$.

(v) $(b/a) \cdot a = (a/b) \cdot a = a \cdot (a\backslash b) = b \cdot (b\backslash a)$.

This means that $(A; \cdot, \backslash, /, 1)$ is a residuated monoid, (i.e., $\cdot$ is associative, with unit element 1, and $x \cdot y \le z$ if and only if $y \le x\backslash z$ if and only if $x \le z/y$ for all $x, y, z \in A$). We will write xy instead of $x \cdot y$. We note that in [153], there are used arrows $\rightarrow$ and $\rightsquigarrow$, which are connected with divisions as follows: $y \rightarrow z = z/y$ and $y \rightsquigarrow z = y\backslash z$. A pseudo hoop $\mathbf{A}$ is non-trivial if $A \ne \{1\}$.

The operations $\backslash$ and $/$ are called *left division* (or *right residuation*) and *right division* (or *left residuation*), respectively. We will assume that $\cdot$ has higher binding priority than $\backslash$ and $/$, which bind stronger than the lattice connectives.

Then the relation $\le$ defined by $a \le b$ if and only if $a\backslash b = 1$ (iff $b/a = 1$) is a partial order on A, in addition $a \le b$ if and only if there is $c \in A$ such that $a = c \cdot b$, and A is a $\wedge$-lattice because $a \wedge b = (b/a) \cdot a = (a/b) \cdot a = a \cdot (a\backslash b) = b \cdot (b\backslash a)$ with the top element 1.

By [153, Remark 2.3], the operation $\cdot$ is commutative if and only if $\backslash = /$, and in such a case, $\mathbf{A}$ is said to be a **hoop**.

For example, let $\mathbf{G} = (G; \vee, \wedge, \cdot, ^{-1}, e)$ be an ℓ-group written in a multiplicative way. If we endow the negative cone $\mathbf{G}^-$ with binary operations $a \cdot b = ab$, $a\backslash b = (a^{-1}b) \wedge e$ and $a/b = (ab^{-1}) \wedge e$, then $(\mathbf{G}^-; \cdot, \backslash, /, e)$ is an example of a pseudo hoop.

Pseudo MV-algebras can also be studied in the realm of pseudo hoops. Indeed, let $\mathbf{M} = (M; \oplus, ^-, ^\sim, 0, 1)$ be a pseudo MV-algebra. Then the algebraic structure $\mathbf{H}(M) := (M; \cdot, \backslash, /, 1)$, where

$$x \cdot y := (y^\sim \oplus x^\sim)^- \quad \& \quad x\backslash y := x^- \oplus y \quad \& \quad x/y = x \oplus y^\sim, \quad \forall x, y \in M,$$

is a bounded pseudo hoop.

A pseudo BL-algebra is a special kind of a pseudo hoop with an additional fixed element 0 introduced in [96, 97]; we present an equivalent definition, see e.g. [121]: An algebra $\mathbf{A} = (A; \cdot, \backslash, /, \wedge, \vee, 0, 1)$ of type $(2, 2, 2, 2, 2, 0, 0)$ is a **pseudo BL-algebra** if

(i) $(A; \wedge, \vee, 0, 1)$ is a bounded lattice.

(ii) $(A; \cdot, \wedge, \vee, 1)$ is a residuated monoid.

(iii) $x(x\backslash(x \wedge y)) = x \wedge y = ((x \wedge y)/x)x$.

(iv) $x(x\backslash y) = x \wedge y = (y/x)x$ (divisibility).

(v) $x\backslash y \vee y\backslash x = 1 = y/x \vee x/y$ (prelinearity).

We define two negations: $x^- = 0/x$ and $x^\sim = x\backslash 0$ of $x \in A$, and we say that a pseudo BL-algebra is **good** if $x^{-\sim} = x^{\sim-}$ for all $x \in A$. A pseudo MV-algebra is a good BL-algebra satisfying $x^{-\sim} = x = x^{\sim-}$ for all $x \in A$. In this section, we will use this definition of pseudo MV-algebras. (Verify that this approach is correct; Exercise 8.8.2).

We define positive and negative cones as usual by $\mathbf{G}^+ := \{g \in G : e \le g\}$ and $\mathbf{G}^- := \{g \in G : g \le e\}$. Let I, J be sets with $|J| \le |I|$. Since only the cardinalities of I and J matter for the construction to follow, it is harmless to think of these sets as ordinals. Let further $\lambda, \rho : J \to I$ be injections. Now, we define an algebra with the universe $(\mathbf{G}^+)^J \uplus (\mathbf{G}^-)^I$, where $\uplus$ denotes a union of disjoint sets. We order its universe by keeping the original co-ordinatewise ordering within $(\mathbf{G}^+)^J$ and $(\mathbf{G}^-)^I$, and setting $x \le y$ for all $x \in (\mathbf{G}^+)^J$, $y \in (\mathbf{G}^-)^I$. It is easy to verify that this is a (bounded) lattice ordering of $(\mathbf{G}^+)^J \uplus (\mathbf{G}^-)^I$. Notice also that the case $I = J$ is not excluded, so the element $e^I := \langle e : i \in I \rangle$ may appear twice: at the bottom of $(\mathbf{G}^+)^J$ and at the top of $(\mathbf{G}^-)^I$. To avoid confusion in the definitions below, we adopt a convention of writing $a_i^{-1}, b_i^{-1}, \ldots$ for co-ordinates of elements of $(\mathbf{G}^-)^I$ and $f_j, g_j, \ldots$ for co-ordinates of elements of $(\mathbf{G}^+)^J$. In particular, we will write e^{-1} for e as an element of $\mathbf{G}^-$. We also put 1 for the constant sequence $(e^{-1})^I := \langle e^{-1} : i \in I \rangle$ and 0 for the constant sequence e^J. With these conventions in place, we are ready to define multiplication, putting:

$$\langle a_i^{-1} : i \in I \rangle \cdot \langle b_i^{-1} : i \in I \rangle = \langle (b_i a_i)^{-1} : i \in I \rangle$$

$$\langle a_i^{-1} : i \in I \rangle \cdot \langle f_j : j \in J \rangle = \langle a_{\lambda(j)}^{-1} f_j \vee e : j \in J \rangle$$

$$\langle f_j : j \in J \rangle \cdot \langle a_i^{-1} : i \in I \rangle = \langle f_j a_{\rho(j)}^{-1} \vee e : j \in J \rangle$$

$$\langle f_j : j \in J \rangle \cdot \langle g_j : j \in J \rangle = \langle e : j \in J \rangle = 0.$$

Definition 8.5.1 Divisions, $/$ and $\backslash$, corresponding to multiplication defined as above on $(\mathbf{G}^+)^J \uplus (\mathbf{G}^-)^I$ are defined by:

$$\langle a_i^{-1} : i \in I \rangle \backslash \langle b_i^{-1} : i \in I \rangle = \langle a_i b_i^{-1} \wedge e^{-1} : i \in I \rangle$$

$$\langle b_i^{-1} : i \in I \rangle / \langle a_i^{-1} : i \in I \rangle = \langle b_i^{-1} a_i \wedge e^{-1} : i \in I \rangle$$

$$\langle a_i^{-1} : i \in I \rangle \backslash \langle f_j : j \in J \rangle = \langle a_{\lambda(j)} f_j : j \in J \rangle$$

$$\langle f_j : j \in J \rangle / \langle a_i^{-1} : i \in I \rangle = \langle f_j a_{\rho(j)} : j \in J \rangle$$

$$\langle f_j : j \in J \rangle \backslash \langle g_j : j \in J \rangle = \langle a_i^{-1} : i \in I \rangle,$$

$$\text{where } a_i^{-1} = \begin{cases} f_{\rho^{-1}(i)}^{-1} g_{\rho^{-1}(i)} \wedge e^{-1} & \text{if } \rho^{-1}(i) \text{ is defined} \\ e^{-1} & \text{otherwise} \end{cases}$$

$$\langle g_j : j \in J \rangle / \langle f_j : j \in J \rangle = \langle b_i^{-1} : i \in I \rangle,$$

$$\text{where } b_i^{-1} = \begin{cases} g_{\lambda^{-1}(i)} f_{\lambda^{-1}(i)}^{-1} \wedge e^{-1} & \text{if } \lambda^{-1}(i) \text{ is defined} \\ e^{-1} & \text{otherwise,} \end{cases}$$

$$\langle a_i^{-1} : i \in I \rangle / \langle f_j : j \in J \rangle = (e^{-1})^I = \langle f_j : j \in J \rangle \backslash \langle a_i^{-1} : i \in I \rangle.$$

Fig. 8.1 Hasse diagram of the kite $K_{2,1}^{\lambda,\rho}(\mathbb{Z})$

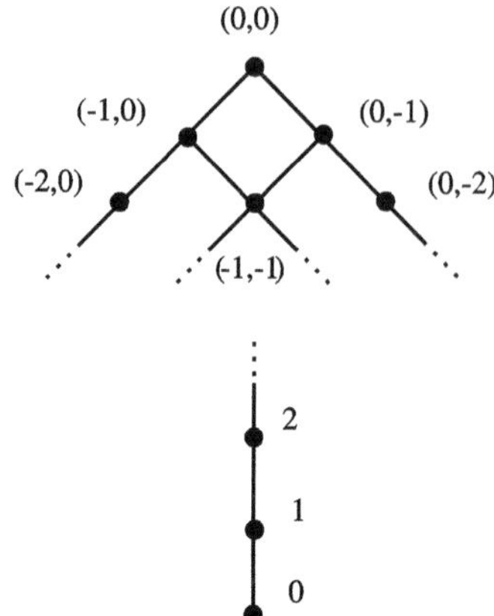

We will call the algebra we have just defined a *kite* of **G**, and write $K_{I,J}^{\lambda,\rho}(\mathbf{G})$ for it. Imagine that if $G = \mathbb{Z}$, and taking $I = 2$, $J = 1$, $\lambda(0) = 0$, and $\rho(0) = 1$, we get an algebra $K_{2,1}^{\lambda,\rho}(\mathbf{G})$ from [190] which is a pseudo BL-algebra that is not good [121] and whose shape resembles a kite form, see Fig. 8.1. Moreover, as this example shows, starting with a commutative ℓ-group G, we can obtain a non-commutative kite algebra.

The following result is from [121]:

Theorem 8.5.2 *For any ℓ-group* **G***, any choice of appropriate sets* I, J, *and maps* λ, ρ, *the algebra* $K_{I,J}^{\lambda,\rho}(\mathbf{G})$ *is a pseudo BL-algebra.*

Proof It is clear that $K_{I,J}^{\lambda,\rho}(\mathbf{G})$ is a lattice with $0 = e^J$, $1 = e^I$, and 1 is a unit of multiplication. To show that multiplication is associative, first observe that triples from $(\mathbf{G}^-)^I$ associate because $(\mathbf{G}^-)^I$ is just the negative cone of $\mathbf{G}^I$. Next, triples involving at least two elements from $(\mathbf{G}^+)^J$ associate because both the products equal 0. The remaining cases all involve one element from $(\mathbf{G}^+)^J$ and two from $(\mathbf{G}^-)^I$. One such case is:

$$
\begin{aligned}
(\langle a_i^{-1}: i \in I \rangle \cdot \langle f_j: j \in J \rangle) \cdot \langle b_i^{-1}: i \in I \rangle &= \langle a_{\lambda(j)}^{-1} f_j \vee e: j \in J \rangle \cdot \langle b_i^{-1}: i \in I \rangle \\
&= \langle (a_{\lambda(j)}^{-1} f_j \vee e) b_{\rho(j)}^{-1} \vee e: j \in J \rangle \\
&= \langle a_{\lambda(j)}^{-1} f_j b_{\rho(j)}^{-1} \vee b_{\rho(j)}^{-1} \vee e: j \in J \rangle \\
&= \langle a_{\lambda(j)}^{-1} f_j b_{\rho(j)}^{-1} \vee e: j \in J \rangle \\
&= \langle a_{\lambda(j)}^{-1} f_j b_{\rho(j)}^{-1} \vee a_{\lambda(j)}^{-1} \vee e: j \in J \rangle \\
&= \langle a_{\lambda(j)}^{-1} (f_j b_{\rho(j)}^{-1} \vee e): j \in J \rangle \\
&= \langle a_i^{-1}: i \in I \rangle \cdot \langle f_j b_{\rho(j)}^{-1} \vee e: j \in J \rangle \\
&= \langle a_i^{-1}: i \in I \rangle \cdot (\langle f_j: j \in J \rangle \cdot \langle b_i^{-1}: i \in I \rangle).
\end{aligned}
$$

Other cases follow by similar calculations. Now, to show that divisibility holds, we also proceed case by case. Let us deal with two cases here. The first is:

$$\langle f_j : j \in J \rangle \cdot (\langle f_j : j \in J \rangle \backslash \langle g_j : j \in J \rangle) = \langle f_j : j \in J \rangle \cdot \langle a_i^{-1} : i \in I \rangle$$
$$= \langle f_j a_{\rho(j)}^{-1} \vee e : j \in J \rangle$$

where

$$a_i^{-1} = \begin{cases} f_{\rho^{-1}(i)}^{-1} g_{\rho^{-1}(i)} \wedge e^{-1} & \text{if } \rho^{-1}(i) \text{ is defined} \\ e^{-1} & \text{otherwise} \end{cases}$$

but, observe that $\rho^{-1}(\rho(j))$ is always defined and equals j, so, calculating further, we obtain $f_j a_{\rho(j)}^{-1} = g_j \wedge f_j$ for every $j \in J$, and therefore

$$\langle f_j a_{\rho(j)}^{-1} \vee e : j \in J \rangle = \langle (f_j \wedge g_j) \vee e : j \in J \rangle = \langle f_j \wedge g_j : j \in J \rangle$$

as required. For the second, take:

$$\langle a_i^{-1} : i \in I \rangle \cdot (\langle a_i^{-1} : i \in I \rangle \backslash \langle f_j : j \in J \rangle) = \langle a_i^{-1} : i \in I \rangle \cdot \langle a_{\lambda(j)} f_j : j \in J \rangle$$
$$= \langle a_{\lambda(j)}^{-1} a_{\lambda(j)} f_j \vee e : j \in J \rangle$$
$$= \langle f_j : j \in J \rangle.$$

All other cases are straightforward.

Residuation property. First, we show that $x \cdot y \leq z$ if and only if $x \leq z/y$. If $x, y \in (\mathbf{G}^-)^I$, then $z \in (\mathbf{G}^-)^I$ and the statement follows from residuation in $\mathbf{G}^-$.

Let

$$\langle a_i^{-1} : i \in I \rangle \cdot \langle f_j : j \in J \rangle \leq \langle g_j : j \in J \rangle.$$

Then

$$\langle a_{\lambda(j)}^{-1} f_j \vee e : j \in J \rangle \leq \langle g_j : j \in J \rangle.$$

For any $j \in J$, $a_{\lambda(j)}^{-1} f_j \vee e \leq g_j$, $a_{\lambda(j)}^{-1} f_j \leq g_j$, $a_{\lambda(j)}^{-1} \leq g_j f_j^{-1}$ and $a_{\lambda(j)}^{-1} \leq g_j f_j^{-1} \wedge e$.

On the other hand,

$$\langle g_j : j \in J \rangle / \langle f_j : j \in J \rangle = \langle b_i^{-1} : i \in I \rangle,$$

where $b_i^{-1} = g_{\lambda^{-1}(i)} f_{\lambda^{-1}(i)}^{-1} \wedge e$ if $\lambda^{-1}(i)$ is defined, otherwise $b_i^{-1} = e$. If we set $i = \lambda(j)$ and $j = \lambda^{-1}(i) = j$, we have

$$\langle a_i^{-1} : i \in I \rangle \leq \langle g_j : j \in J \rangle / \langle f_j : j \in J \rangle = \langle b_i^{-1} : i \in I \rangle,$$

and conversely, the latter identity implies $\langle a_i^{-1} : i \in I \rangle \cdot \langle f_j : j \in J \rangle \leq \langle g_j : j \in J \rangle$.

Now let

$$\langle a_i^{-1} : i \in I \rangle \cdot \langle f_j : j \in J \rangle \leq \langle b_i^{-1} : i \in I \rangle$$
$$\langle a_{\lambda(j)}^{-1} f_j \vee e : j \in J \rangle \leq \langle b_i^{-1} : i \in I \rangle.$$

Then

$$\langle b_i^{-1} : i \in I \rangle / \langle f_j : j \in J \rangle = \langle e^{-1} : i \in I \rangle,$$

which entails trivially

$$\langle a_i^{-1} : i \in I \rangle \leq \langle b_i^{-1} : i \in I \rangle / \langle f_j : j \in J \rangle.$$

Conversely, assume the latter inequality. Then trivially $\langle a_i^{-1} : i \in I \rangle \cdot \langle f_j : j \in J \rangle \leq \langle b_i^{-1} : i \in I \rangle$.

In the same way, we prove all remaining cases of the proof $x \cdot y \leq z$ if and only if $x \leq z/y$, and dually, we can prove that $x \cdot y \leq z$ if and only if $y \leq x \backslash z$.

It remains to show prelinearity. Since multiplication and divisions in $(\mathbf{G}^-)^I$ are defined co-ordinatewise, prelinearity for $x, y \in (\mathbf{G}^-)^I$ is inherited from $\mathbf{G}^-$. If $x \in (\mathbf{G}^-)^I$ and $y \in (\mathbf{G}^+)^J$, or *vice versa* prelinearity holds trivially. For the only remaining case, calculating

$$\langle f_j : j \in J \rangle \backslash \langle g_j : j \in J \rangle \vee \langle g_j : j \in J \rangle \backslash \langle f_j : j \in J \rangle$$

yields two cases: (1) if $\rho^{-1}(i)$ is defined, we have

$$f_{\rho^{-1}(i)}^{-1} g_{\rho^{-1}(i)} \wedge e^{-1} \vee g_{\rho^{-1}(i)}^{-1} f_{\rho^{-1}(i)} \wedge e^{-1} = (f_{\rho^{-1}(i)}^{-1} g_{\rho^{-1}(i)} \vee g_{\rho^{-1}(i)}^{-1} f_{\rho^{-1}(i)}) \wedge e^{-1} = e^{-1}$$

and (2) if $\rho^{-1}(i)$ is not defined, we have

$$a_i^{-1} \vee a_i^{-1} = e^{-1} \vee e^{-1} = e^{-1}$$

as well. Thus, prelinearity holds, and that finishes the proof of all the claims in the theorem. $\square$

Interestingly, many kites turn out to be pseudo MV-algebras.

Theorem 8.5.3 *Let* $\mathbf{G}$ *be an ℓ-group, and suppose* $|I| = |J|$ *and* λ, ρ *are bijections. Then* $K_{I,J}^{\lambda,\rho}(\mathbf{G})$ *is a pseudo MV-algebra.*

Proof By Theorem 8.5.2, we only need to show that under the conditions of the lemma, the identity $x/(y \backslash x) = x \vee y = (x/y) \backslash x$ holds. We have two non-trivial cases to consider:

Case 1. $x \in (\mathbf{G}^+)^J$ and $y \in (\mathbf{G}^-)^I$. Then, $x = \langle f_j : j \in J \rangle \le \langle a_i^{-1} : i \in I \rangle = y$, and so we calculate:

$$
\begin{aligned}
\langle f_j : j \in J \rangle / (\langle a_i^{-1} : i \in I \rangle \backslash \langle f_j : j \in J \rangle) &= \langle f_j : j \in J \rangle / \langle a_{\lambda(j)} f_j : j \in J \rangle \\
&= \langle f_{\lambda^{-1}(i)} (a_{\lambda(j)} f_j)^{-1}_{\lambda^{-1}(i)} \wedge e^{-1} : i \in I \rangle \\
&= \langle f_{\lambda^{-1}(i)} f^{-1}_{\lambda^{-1}(i)} a_i^{-1} : i \in I \rangle \\
&= \langle a_i^{-1} : i \in I \rangle
\end{aligned}
$$

which shows that $x/(y\backslash x) = y = x \vee y$ holds. Notice that we used bijectiveness of λ to pass from the first to the second equality above.

Case 2. $x, y \in (\mathbf{G}^+)^J$. Then, $x = \langle f_j : j \in J \rangle$ and $y = \langle g_j : j \in J \rangle$, and so we calculate:

$$
\begin{aligned}
\langle f_j : j \in J \rangle / (\langle g_j : j \in J \rangle \backslash \langle f_j : j \in J \rangle) &= \langle f_j : j \in J \rangle / \langle g^{-1}_{\rho^{-1}(i)} f_{\rho^{-1}(i)} \wedge e^{-1} : i \in I \rangle \\
&= \langle f_j (g^{-1}_{\rho^{-1}(i)} f_{\rho^{-1}(i)} \wedge e^{-1})^{-1}_{\rho(j)} : j \in J \rangle \\
&= \langle f_j (f_j^{-1} g_j \vee e) : j \in J \rangle \\
&= \langle g_j \vee f_j : j \in J \rangle
\end{aligned}
$$

which again shows that $x/(y\backslash x) = x \vee y$ holds. The proofs for $x \vee y = (x/y)\backslash x$ are symmetric. $\qquad\square$

Now, we present good kites and pseudo MV-algebras:

Theorem 8.5.4 *Let $\mathbf{G}$ be an ℓ-group, and $K^{\lambda,\rho}_{I,J}(\mathbf{G})$ a kite.*

1. $K^{\lambda,\rho}_{I,J}(\mathbf{G})$ *is good if and only if $\lambda(J) = \rho(J)$.*
2. $K^{\lambda,\rho}_{I,J}(\mathbf{G})$ *is a pseudo MV-algebra if and only if $\lambda(J) = I = \rho(J)$.*

Proof (1) Let $x = \langle a_i : i \in I \rangle$. Then $x^- = \langle a_{\rho(j)} : j \in J \rangle$ and $x^\sim = \langle a_{\lambda(j)} : j \in J \rangle$. In addition, $x^{-\sim} = \langle z_i : i \in I \rangle$, where

$$
z_i^{-1} = \begin{cases} a_i^{-1} & \text{if } \lambda^{-1}(i) \text{ is defined} \\ e^{-1} & \text{otherwise,} \end{cases}
$$

and $x^{\sim -} = \langle y_i : i \in I \rangle$, where

$$
y_i^{-1} = \begin{cases} a_i^{-1} & \text{if } \rho^{-1}(i) \text{ is defined} \\ e^{-1} & \text{otherwise.} \end{cases}
$$

Now, if $x = \langle f_j : j \in J \rangle$, we have $x^{-\sim} = \langle g_j : j \in J \rangle$, where

$$g_j = \begin{cases} f_j & \text{if } \rho^{-1}(i) \text{ is defined} \\ e & \text{otherwise,} \end{cases}$$

and $x^{\sim -} = \langle h_j : j \in J \rangle$, where

$$h_j = \begin{cases} f_j & \text{if } \lambda^{-1}(i) \text{ is defined} \\ e & \text{otherwise.} \end{cases}$$

Hence, if $\lambda(J) = \rho(J)$, the kite $K_{I,J}^{\lambda,\rho}(\mathbf{G})$ is good.

Conversely, assume that the kite $K_{I,J}^{\lambda,\rho}(\mathbf{G})$ is good, and let $\lambda(J) \neq \rho(J)$. For our aims we can assume that each $a_i^{-1} \neq e^{-1}$. There is an $i \in I$ such that either $\lambda^{-1}(i)$ or $\rho^{-1}(i)$ is not defined. Equivalently, $x_i^{-1} = e^{-1}$ and $x_i^{-1} = a_i^{-1}$ or $y_i^{-1} = e^{-1}$ and $y_i^{-1} = a_i^{-1}$. Hence, $\lambda(J) = \rho(J)$.

(2) If $\lambda(J) = I = \rho(J)$, then $K_{I,J}^{\lambda,\rho}(\mathbf{G})$ is a pseudo MV-algebra by Theorem 8.5.3. Conversely, let the kite be a pseudo MV-algebra. Since every pseudo MV-algebra is good, by the first part of the present proof, we have $\lambda(J) = \rho(J)$. Now assume, with a view to contradiction, that there is an $i \in I \setminus \lambda(J)$. Then both $\lambda^{-1}(i)$ and $\rho^{-1}(i)$ are not defined, whence $x_i^{-1} = e^{-1} = y_i^{-1} \neq a_i$ which contradicts the property $x^{-\sim} = x = x^{\sim -}$. Therefore, $\lambda(J) = I = \rho(J)$. $\qquad\qquad\square$

Now we present some classes of kite pseudo BL-algebras.

8.5.1   Boolean Algebras

Let $I = 0 = J$. Then, $(\mathbf{G}^-)^I$ and $(\mathbf{G}^+)^J$ are both singletons, and $\lambda = \rho$ can only be the empty function (hence, an injection). Thus, $K_{0,0}^{\emptyset,\emptyset}(\mathbf{G})$ is the two-element Boolean algebra for any ℓ-group $\mathbf{G}$. We also get the two-element Boolean algebra in another way. Namely, let $\mathbf{O}$ be the trivial ℓ-group. Then $\lambda = id = \rho$, and $K_{I,J}^{id,id}(\mathbf{O})$ is the two-element Boolean algebra for any choice of I and J.

8.5.2   Product Logic Algebras

Let $I = 1$ and $J = 0$. As the only function from J to I is the empty function (which is an injection), the kite $K_{1,0}^{\emptyset,\emptyset}(\mathbf{G})$ is well-defined for any ℓ-group $\mathbf{G}$. Let $\mathbb{R}^+$ stand for the ℓ-group of positive reals under multiplication. Then $K_{1,0}^{\emptyset,\emptyset}(\mathbb{R}^+)$ is isomorphic to the *standard product logic algebra,* i.e., the real interval $[0, 1]$ with usual multiplication, and divisions given by $x/y = \frac{x}{y} = y \backslash x$ for $y \neq 0$, $x/0 = 1 = 0 \backslash x$, and $0 \backslash 0 = 1 = 0/0$. Taking $\mathbb{Z}$ for

G, we obtain as $K_{1,0}^{\emptyset,\emptyset}(\mathbb{Z})$ the algebra $\mathbb{Z}_\perp^- = \langle \mathbb{Z}^- \cup \{\perp\}; \min, \max, +, \backslash, /, \perp, 0 \rangle$, where $\perp = -\infty$ and $\perp + \perp = \perp = \perp + x$, $x \backslash y = \min\{(y - x), 0\} = y/x$, $\perp \backslash \perp = 0 = \perp/\perp$, $x/\perp = \perp \backslash x = 0$, $x \backslash \perp = \perp = \perp/x$. This is also a product logic algebra. Both $K_{1,0}^{\emptyset,\emptyset}(\mathbb{R}^+)$ and $K_{1,0}^{\emptyset,\emptyset}(\mathbb{Z})$ generate the whole variety of product logic algebras. This variety covers the variety of Boolean algebras.

8.5.3   Jipsen-Montagna Algebras

Taking $I = 2$, $J = 1$, $\lambda(0) = 0$, and $\rho(0) = 1$, we get an algebra $K_{2,1}^{\lambda,\rho}(\mathbf{G})$ of the kind considered by Jipsen-Montagna [190], for any ℓ-group **G**. In particular, the algebra $K_{2,1}^{\lambda,\rho}(\mathbb{Z})$ is a pseudo BL-algebra which is not good, and it generates another cover of the variety of Boolean algebras. In [121], it was shown that the same holds for $K_{n+1,n}^{\lambda,\rho}(\mathbb{Z})$ with $\lambda(i) = i$ and $\rho(i) = i + 1$, for an arbitrary $n \in \omega$. The variety Kite of kite pseudo BL-algebras is a proper subvariety of the variety PBL of pseudo BL-algebras.

8.5.4   Chang Chain

Now, let $I = 1 = J$. Then $K_{1,1}^{id,id}(\mathbb{Z})$ is the Chang MV-algebra. The variety generated by $K_{1,1}^{id,id}(\mathbb{Z})$ is also a cover of the variety of Boolean algebras in the variety PBL of pseudo BL-algebras.

8.5.5   Intervals in Scrimger Groups

Taking $I = J = n$ for $n \geq 2$, and putting $\lambda(i) = i$ and $\rho(i) = i + 1 \pmod{n}$, we get that $K_{I,I}^{\lambda,\rho}(\mathbb{Z})$ is isomorphic to $\mathbf{\Gamma}(\mathbf{G}_n, (\langle 0 \rangle, 1))$, where $\mathbf{G}_n$ is the subgroup of the wreath product $\mathbb{Z} \wr \mathbb{Z}$ (antilexicographically ordered), consisting of the elements $\langle \langle a_i : i \in \mathbb{Z} \rangle, b \rangle$, such that $i = j \pmod{n}$ implies $a_i = a_j$.

8.5.6   Subdirectly Irreducible Kites

At the end of this section, we present the building stones of kites theory—subdirectly irreducible kites. For more information, consult with [121, Sects. 5 and 6].

Theorem 8.5.5 *If a kite $K_{I,J}^{\lambda,\rho}(\mathbf{G})$ is subdirectly irreducible, so is the ℓ-group* **G**, *and I and J are at most countably infinite.*

Moreover, the following are equivalent:

(i) **G** *is subdirectly irreducible and for all* $i, j \in I$ *there exists* $m \in \omega$ *such that* $(\rho \circ \lambda^{-1})^m(i) = j$ *or* $(\lambda \circ \rho^{-1})^m(i) = j$.
(ii) $K_{I,J}^{\lambda,\rho}(\mathbf{G})$ *is subdirectly irreducible.*

A complete characterization of subdirectly irreducible kites is the following result:

Theorem 8.5.6 *Let* $K_{I,J}^{\lambda,\rho}(\mathbf{G})$ *be a subdirectly irreducible kite. Then,* $K_{I,J}^{\lambda,\rho}(\mathbf{G})$ *is isomorphic to precisely one of:*

(0) $K_{0,0}^{\emptyset,\emptyset}(\mathbf{G})$, $K_{1,1}^{id,id}(\mathbf{G})$, $K_{1,0}^{\emptyset,\emptyset}(\mathbf{G})$,
(1) $K_{n,n}^{\lambda,\rho}(\mathbf{G})$, *with* $\lambda(j) = j$ *and* $\rho(j) = j + 1 \pmod{n}$.
(2) $K_{\mathbb{Z},\mathbb{Z}}^{\lambda,\rho}(\mathbf{G})$, *with* $\lambda(j) = j$ *and* $\rho(j) = j + 1$.
(3) $K_{\omega,\omega}^{\lambda,\rho}(\mathbf{G})$, *with* $\lambda(j) = j$ *and* $\rho(j) = j + 1$.
(4) $K_{\omega,\omega}^{\lambda,\rho}(\mathbf{G})$, *with* $\lambda(j) = j + 1$ *and* $\rho(j) = j$.
(5) $K_{n+1,n}^{\lambda,\rho}(\mathbf{G})$, *with* $\lambda(j) = j$ *and* $\rho(j) = j + 1$.

Moreover, types (1) and (2) consist entirely of pseudo MV-algebras, and the other types contain no pseudo MV-algebras except the two-element Boolean algebra. A kite of type (3) or (4) is good if and only if it is a two-element Boolean algebra. A kite of type (5) is good if and only if $J = \emptyset$.

8.6   Loomis–Sikorski Theorem for σ-Complete MV-Algebras

A famous Loomis–Sikorski theorem says that every σ-complete Boolean algebra is a σ-homomorphic image of a σ-complete algebra of subsets of a non-void set Ω. It was proved independently by Loomis and Sikorski; see a note in [254, Sect. 29]. We recall that a family S of subsets of a set $\Omega \neq \emptyset$ is a σ-**algebra of subsets** if (i) $\Omega \in S$, (ii) if $A \in S$, then $\Omega \setminus A \in S$, and (iii) $A_n \in S$ for each $n \geq 1$, then $\bigcup_n A_n \in S$.

This section presents a generalization of this theorem for σ-complete MV-algebras. (Due to Theorem 8.2.5, every σ-complete pseudo MV-algebra is an MV-algebra.) The Loomis–Sikorski theorem type for pseudo MV-algebras was established independently in [106, 171, 224].

We start with the following basic notions important for the Loomis–Sikorski Theorem, with MV-algebras of fuzzy sets where MV-operations are defined pointwisely. We note that a **fuzzy set** B on $\Omega \neq \emptyset$ is any mapping $f := f_B : \Omega \to [0, 1]$ (f_B is a membership function of B). For example, if $A \subseteq \Omega$, then $f := \chi_A$, the indicator (characteristic function) of A, is a (strict, i.e., $0 - 1$-valued) fuzzy set. A **tribe** of fuzzy sets on a set $\Omega \neq \emptyset$ is a non-void system $\mathcal{T} \subseteq [0, 1]^\Omega$ such that:

(i) $1_\Omega \in \mathcal{T}$, where $1_\Omega(\omega) = 1$, $\omega \in \Omega$.

(ii)   If $f \in \mathcal{T}$, then $1 - f \in \mathcal{T}$.

(iii)   If $(f_n)_{n=1}^\infty$ is a sequence of elements of $\mathcal{T}$, then the fuzzy set $\bigoplus_{n=1}^\infty f_n$ belongs to $\mathcal{T}$, where

$$\bigoplus_{n=1}^\infty f_n(\omega) := \min\{\sum_{n=1}^\infty f_n(\omega), 1\}, \quad \omega \in \Omega. \tag{8.21}$$

Whence, any tribe is a σ-complete MV-algebra where all MV-operations are defined by points. For example, if $f_n = \chi_{A_n}$ for $n \geq 1$, then $\bigoplus_n \chi_{A_n} = \chi_{\bigcup_n A_n}$. Therefore, if S is a σ-algebra of subsets of Ω, then $\mathcal{T}(S) := \{\chi_A : A \in S\}$ is a tribe of (crisp) fuzzy sets, so that any tribe is a generalization of a σ-algebra of subsets.

A **Bold algebra** of fuzzy sets on a set $\Omega \neq \emptyset$ is a non-void system $C \subseteq [0, 1]^\Omega$ such that:

(i) $1_\Omega \in C$.

(ii)   If $f \in C$, then $1 - f \in C$.

(iii)   If $f, g \in C$, then the fuzzy set $f \oplus g$ belongs to C, where

$$(f \oplus g)(\omega) := \min\{f(\omega) + g(\omega), 1\}, \quad \omega \in \Omega. \tag{8.22}$$

Every Bold algebra is a semisimple MV-algebra of fuzzy sets where all MV-operations are defined by points. We use notation $C = C(\Omega)$ and $\mathcal{T} = \mathcal{T}(\Omega)$, respectively.

Let C be a Bold algebra. Given a point $\omega \in \Omega$, the mapping $s_\omega : C \to [0, 1]$ defined by $s_\omega(f) := f(\omega)$, $f \in C$, is a state-morphism, so by Proposition 8.3.2, $\mathrm{Ker}(s_\omega)$ is a maximal ideal of C. We note that not every state-morphism on a Bold algebra is of this form, in general. For example, if $\Omega = \mathbb{N}$ and C is the set of characteristic functions of subsets $A \subseteq \mathbb{N}$ that are finite or co-finite, then $C = C(\mathbb{N})$ possesses a two-valued state that is zero on each χ_A, where A is finite and one when A is co-finite. It is an extremal state and thus a state-morphism not of the form s_ω. If C is a tribe, then s_ω is a σ-additive state that is also a state-morphism.

Proposition 8.6.1 *For every MV-algebra M, there are a Bold-algebra $C(\Omega)$ and a homomorphism from M onto $C(\Omega)$.*

An MV-algebra M is semisimple if and only if it is isomorphic to some Bold algebra $C(\Omega)$.

Proof (1) Given an element $a \in M$, let $\bar{a} : \mathcal{SM}(M) \to [0, 1]$ be defined by

$$\bar{a}(s) := s(a), \quad s \in \mathcal{SM}(M). \tag{8.23}$$

If we set $\Omega = \mathcal{SM}(M)$, then $\Omega \neq \emptyset$ whenever M is not degenerated, and $\bar{a}$ is a fuzzy set on Ω. Let $\overline{M} = \{\bar{a} : a \in M\}$ and put $C := C(\mathcal{SM}(M)) = \overline{M}$. Then C is a Bold algebra and the mapping $a \mapsto \bar{a}$ is a surjective homomorphism in question.

(2) Since every Bold algebra is a semisimple MV-algebra, one direction is evident. Conversely, let M be a semisimple MV-algebra. Then the intersection of all maximal ideals of M is the zero set $\{0\}$. Take the surjective homomorphism $h : M \to \overline{M}$ defined in (1), i.e., $h(a) = \bar{a}$, $a \in M$, and assume $h(a) = \bar{0}$, i.e., $s(a) = 0$ for each state-morphism s. Due to Theorem 8.3.3, for each $s \in \mathcal{SM}(M)$, there exists a unique maximal ideal $I = \mathrm{Ker}(s)$ such that $s(a) = M/I$, and a belongs to each maximal ideal of M. The semisimplicity yields $a = 0$, and h is injective, i.e., an isomorphism. $\qquad\square$

Lemma 8.6.2 *Let* $\mathrm{B}(M)$ *be the set of Boolean elements of an MV-algebra* M*. If* M *is* σ*-complete, then* $\mathrm{B}(M)$ *is a* σ*-complete MV-algebra.*

Proof Let $(a_i)_i$ be a sequence of elements of $\mathrm{B}(M)$ such $a = \bigvee_i a_i$. Then

$$a \wedge a = \bigvee_i (a_i \wedge a') \leq \bigvee_i (a_i \wedge a_i') = 0,$$

implying $a \in \mathrm{B}(M)$. $\qquad\square$

We remind that a topological space $\Omega \neq \emptyset$ is

(i) **extremally disconnected** if every two different points are separated by a clopen subset of Ω.

(ii) **basically disconnected** if the closure of every open F_σ subset of Ω is open.

We note that the weak topology of state-morphisms on a σ-complete MV-algebra, as well as the hull-kernel topology of maximal ideals, Theorem 8.3.6, are basically disconnected, see e.g. [106, Proposition 4.3]:

Proposition 8.6.3 *If* M *is a* σ*-complete or complete MV-algebra, then the space* $\mathcal{M}(M)$ *of maximal ideals of* M *is basically disconnected or extremally disconnected, respectively.*

Proof Since $M = \Gamma(G, u)$ is σ-complete (complete) if and only if its representation ℓ-group G is Dedekind σ-complete (complete), according to [160, Theorem 8.14], the space $\mathcal{SM}(M)$ is homeomorphic with the set $\mathcal{M}(\mathrm{B}(M))$ of all maximal ideals of the Boolean algebra $\mathrm{B}(M)$. Hence, by [254, Theorem 22.4], a Boolean algebra is σ-complete (complete) if and only if $\mathcal{M}(\mathrm{B}(M))$ is basically disconnected (extremally disconnected). $\qquad\square$

For a bounded function $g : X \to \mathbb{R}$ on a topological space X we define

$$\tilde{g}(x) = \inf_{U \in \mathcal{N}(x)} \sup\{g(y) : y \in U\}, \tag{8.24}$$

where $\mathcal{N}(x)$ is the system of open neighborhoods for $x \in X$. Then

(i) $g(x) \le \tilde{g}(x)$ for any $x \in X$.
(ii) $\tilde{g}(x) = g(x)$ if g is continuous in x.

If $D(g)$ is the set of discontinuity points for g, then

$$\{x \in X : g(x) \ne \tilde{g}(x)\} \subseteq D(g) = \bigcup_n \{g^{-1}(R_n) - Int(g^{-1}(R_n))\}, \tag{8.25}$$

where $\{R_n\}$ is an open basis in $\mathbb{R}$.

Let f be a real-valued function on $\Omega \ne \emptyset$. We define

$$N(f) := \{\omega \in \Omega : |f(\omega)| > 0\}.$$

Suppose that $\mathcal{F}$ is a Bold algebra of fuzzy sets on Ω. Then for all $f, g \in \mathcal{F}$ we have

(i) $N(f \oplus g) = N(f) \cup N(g)$.
(ii) $N(f * g) = \{\omega \in \Omega : f(\omega) > g(\omega)\}$.
(iii) $(f * g) \oplus (g * f) = (f * g) + (g * f)$, where $f * g = f - (f \wedge g)$.
(iv) $N((f * g) \oplus (g * f)) = N(f - g)$.
(v) $N(f) \subseteq N(g)$ if $f \le g$.

Theorem 8.6.4 (Loomis-Sikorski Theorem) *For every σ-complete MV-algebra $\mathbf{M}$, there exist a tribe $\mathcal{T} = \mathcal{T}(\Omega)$ of fuzzy sets on a compact Hausdorff topological space Ω and an MV-σ-homomorphism h from $\mathcal{T}$ onto $\mathbf{M}$.*

Proof We set $\Omega = \mathcal{SM}(M)$ endowed with the topology of weak convergence of state-morphisms. Let $\mathcal{T}$ be the tribe of fuzzy sets on $\mathcal{SM}(M)$ generated by the Bold algebra $\{\bar{a} : a \in M\}$ (it always exists). Consider by $\mathcal{T}'$ the class of all functions $f \in \mathcal{T}$ with the property that for some $b \in M$, $N((f * \bar{b}) \oplus (\bar{b} * f))$ is a meager set in Ω.

If b_1 and b_2 are two elements of M such that $N((f * \bar{b}_i) \oplus (\bar{b}_i * f))$ is a meager set for $i = 1, 2$, then

$$N((\bar{b}_1 * \bar{b}_2) \oplus (\bar{b}_2 * \bar{b}_1)) = N(\bar{b}_1 - \bar{b}_2) \subseteq N(\bar{b}_1 - f) \cup N(f - \bar{b}_2)$$

is a meager set. Due to the Baire theorem, saying that any non-empty open set of a compact Hausdorff space cannot be a meager set, [202], we conclude that $\bar{b}_1 = \bar{b}_2$, that is, $b_1 = b_2$.

It is clear that $\mathcal{T}'$ is closed under the formation of complement $f \mapsto 1 - f$ and it is a Bold algebra containing $\{\bar{a} \colon a \in M\}$.

To show that $\mathcal{T}'$ is a tribe is necessary to verify that $\mathcal{T}'$ is closed under limits of non-decreasing sequences from $\mathcal{T}'$.

Let $(f_n)_n$ be a sequence of non-decreasing elements from $\mathcal{T}'$. Choose $b_n \in M$ such that $N(f_n - \bar{b}_n)$ is a meager set. Without loss of generality, we can assume that $b_n \leq b_{n+1}$. Denote $f = \lim_n f_n$, $b = \bigvee_{n=1}^{\infty} b_n$, $b_0 = \lim_n \bar{b}_n$. Then $f, b_0 \in \mathcal{T}$ and $b \in M$. We have

$$N(f - \bar{b}) \subseteq N(f - b_0) \cup N(\bar{b} - b_0)$$

and $N(f - b_0) = \{m \colon f(m) < b_0(m)\} \cup \{m \colon b_0(m) < f(m)\}$.

If $m \in \{m \colon f(m) < b_0(m)\}$, then there is an integer $n \geq 1$ such that $f(m) < \bar{b}_n(m) \leq b_0(m)$. Hence, $f_n(m) \leq f(m) < \bar{b}_n(m) \leq b_0(m)$ so that

$$m \in \{m \colon f_n(m) < \bar{b}_n(m)\}.$$

Similarly we can prove that if $m \in \{m \colon b_0(m) < f(m)\}$, then there is an integer $n \geq 1$ such that $m \in \{m \colon \bar{b}_n(m) < f_n(m)\}$.

The last two cases imply

$$N(f - b_0) \subseteq \bigcup_{n=1}^{\infty} N(\bar{b}_n - f_n)$$

which is a meager set.

Apply now (8.24) to the function b_0 to obtain $\tilde{b}_0$, that is

$$\tilde{b}_0(m) := \inf_{U \in \mathcal{N}(m)} \sup\{b_0(y) \colon y \in U\}.$$

Since $\mathcal{SM}(M))$ is basically disconnected, compact, Hausdorff, and $b_0^{-1}(\alpha, \infty) = \bigcup_n \bar{b}_n^{-1}(\alpha, \infty)$ for any $\alpha \in \mathbb{R}$, $b_0^{-1}(\alpha, \infty)$ is an open F_σ set, by [160, Lemma 9.1], $\tilde{b}_0$ is continuous. Since b_0 is a point limit of a sequence of continuous functions, by [202, pp. 86, 405–406], $D(b_0) \supseteq N(\tilde{b}_0 - b_0)$ is a meager set,

$$b_0 \leq \tilde{b}_0 \leq \bar{b},$$

and $N(\bar{b} - b_0) \subseteq N(\bar{b} - \tilde{b}_0) \cup D(b_0)$.

Finally we show $\tilde{b}_0 = \bar{b}$. Define $C(X)$, where $X = \mathcal{SM}(M))$, as the set of all continuous real valued functions defined on X. Since X is a basically disconnected, compact, Hausdorff space, the space $C(X)$ of continuous functions on X is a Dedekind σ-complete ℓ-group with a strong unit 1_X, [160, Corollary 9.3]. Using D. Mundici's functor Γ, we see that the system of all continuous fuzzy functions on X, $C(X) = \Gamma(C(X), 1_X)$ is a σ-complete MV-algebra.

Applying [160, Lemma 9.12], we conclude the mapping $a \mapsto \bar{a}, a \in M$, preserves countable suprema and infima. Consequently, $\tilde{b}_0 = \bar{b}$, which proves that $\mathcal{T}'$ is a tribe, and whence, $\mathcal{T}' = \mathcal{T}$.

Due to definition of $\mathcal{T}'$, for any $f \in \mathcal{T}$ there is a unique element $h(f) := b \in M$ such that $N(f - \bar{b})$ is meager, which proves that $h : \mathcal{T} \to M$ is a surjective MV-σ-homomorphism in question. $\qquad \square$

The Loomis–Sikorski theorem has many generalizations to different algebraic structures, e.g., monotone σ-complete effect algebras [60, 110] or EMV-algebras [128], etc.

8.7 EMV-Algebras and Pseudo EMV-Algebras

A generalization of Boolean algebras is a generalized Boolean algebra (or a Boolean ring) where the top element is not guaranteed. Such an algebra without top element can always be embedded into a Boolean algebra as its maximal ideal; see [87]. This idea was used in [129] to present EMV-algebras (extended MV-algebras) as a common generalization of MV-algebras and generalized Boolean algebras. Then EMV-algebras were generalized to non-commutative pseudo EMV-algebras as a common area for pseudo MV-algebras and Boolean rings, [130, 131]. The main results concern a representation of pseudo EMV-algebras without top element and the categorical equivalence. We note that ideas of EMV-algebras and pseudo EMV-algebras have been used in [205, 260] in introducing Ehoops and pseudo Ehoops (extended hoops, extended pseudo hoops) as a generalization of hoops and pseudo hoops.

Motivation. Let $M = (M; \oplus, ^-, ^\sim, 0, 1)$ be a pseudo MV-algebra and $x \in M$. From the basic properties of x^- and $x^\sim$, we have

$$x^- = \min\{z \in M : z \oplus x = 1\}, \quad x^\sim = \min\{z \in M : x \oplus z = 1\}. \tag{8.26}$$

Let $a \in M$ be a Boolean element of a pseudo MV-algebra M. The interval $[0, a] := \{x \in M : 0 \le x \le a\}$ can be converted into a pseudo MV-algebra as follows: For $x \in [0, a]$, we set $x^{-a} = a/x$ and $x^{\sim a} = x \backslash a$, and since $a \oplus a = a$, we put $\oplus_a = \oplus|_{[0,a] \times [0,a]}$. Then $M_a = ([0, a]; \oplus_a, ^{-a}, ^{\sim a}, 0, a)$ is a pseudo MV-algebra. We define two unary operators λ_a and ρ_a on $M_a = [0, a]$ by

$$\lambda_a(x) := \min\{z \in [0, a] : z \oplus x = a\}, \quad \rho_a(x) := \min\{z \in [0, a] : x \oplus z = a\}, \tag{8.27}$$

for each $x \in [0, a]$. Using (8.26)–(8.27), we have for each $x \in [0, a]$

$$x^{-a} = \lambda_a(x), \quad x^{\sim a} = \rho_a(x).$$

In other words, if $a \in B(M)$, then $([0, a]; \oplus, \lambda_a, \rho_a, 0, a)$ is a pseudo MV-algebra. These algebras will play an important role in the definition of pseudo EMV-algebras.

Let $M = (M; \oplus, 0)$ be a monoid with a neutral element 0. The monoid M is not assumed to be commutative. An element $a \in M$ is said to be an **idempotent** if $a \oplus a = a$. We denote by $\mathrm{Idm}(M)$ the set of idempotents of M; then (i) $0 \in \mathrm{Idm}(M)$, (ii) if $a, b \in \mathrm{Idm}(M)$ and $a \oplus b = b \oplus a$, then $a \oplus b \in \mathrm{Idm}(M)$. We say that a monoid $(M; \oplus, 0)$ endowed with a partial order $\leq$ is (i) **ordered** if $x \leq y$ implies $z_1 \oplus x \oplus z_2 \leq z_1 \oplus y \oplus z_2$ for all $z_1, z_2 \in M$, (ii) **naturally ordered** if $x \leq y$ if and only if there are $z_1, z_2 \in M$ such that $x \oplus z_1 = y = z_2 \oplus x$.

According to [130, 131], an algebra $(M; \vee, \wedge, \oplus, 0)$ of type $(2, 2, 2, 0)$ is called a **pseudo EMV-algebra** if it satisfies the following conditions:

(E1) $(M; \vee, \wedge, 0)$ is a distributive lattice with the least element 0.

(E2) $(M; \oplus, 0)$ is an ordered monoid with a neutral element 0.

(E3) For each $a \in \mathrm{Idm}(M)$, the elements

$$\lambda_a(x) := \min\{z \in [0, a]: z \oplus x = a\}, \quad \rho_a(x) := \min\{z \in [0, a]: x \oplus z = a\}$$

exist in M for all $x \in [0, a]$, and the algebra $([0, a]; \oplus, \lambda_a, \rho_a, 0, a)$ is a pseudo MV-algebra.

(E4) For each $x \in M$, there is $a \in \mathrm{Idm}(M)$ such that $x \leq a$.

As we see, no top element in M is assumed. If M has no top element M is said to be a **proper pseudo EMV-algebra**.

We note:

(i) If $\oplus$ is commutative, then $\lambda_a = \rho_a$, and we obtain an EMV-algebra in the sense of [129].

(ii) If $(M; \oplus, ^-, ^\sim, 0, 1)$ is a pseudo MV-algebra, then $(M; \vee, \wedge, \oplus, 0)$ is a pseudo EMV-algebra with a top element 1. Conversely, if a pseudo EMV-algebra $(M; \vee, \wedge, \oplus, 0)$ has a top element 1, then M can be converted into a pseudo MV-algebra whose algebraic pseudo MV-structure is compatible with the pseudo EMV-algebraic structure of M. Of course, every finite pseudo EMV-algebra possesses a top element, so M gives an MV-algebra.

(iii) Every generalized Boolean algebra is an EMV-algebra.

(iv) Let $\{M_i : i \in I\}$ be a family of pseudo EMV-algebras. We can easily show that $\sum_{i \in I} M_i := \{(x_i)_{i \in I} \in \prod_{i \in I} M_i : x_i = 0 \text{ for all but a finite number of } i \in I\}$ is a pseudo EMV-algebra with componentwise operations. For example, let $A_1 = M$ be a pseudo EMV-algebra with top element and $A_i = \{0, 1\}$ for all $i \in \mathbb{N}$, then $\sum_{i \in I} A_i$ is a pseudo EMV-algebra.

The reduct $(M; \vee, \wedge)$ is a distributive lattice. Therefore, it yields a partial order $x \leq y$ iff $x \vee y = y$ iff $x \wedge y = x$. If a is a fixed idempotent element of M, $([0, a]; \oplus, \lambda_a, \rho_a, 0, a)$ is a pseudo MV-algebra. Recall that, in each pseudo MV-algebra $(A; \oplus, ^-, ^\sim, 0, 1)$ there is a partial order relation $\preccurlyeq$ (induced by $\oplus, ^-, ^\sim, 1$) defined by $x \preccurlyeq y$ iff $x^- \oplus y = 1$ iff

$y \oplus x^\sim = 1$, see [152, Proposition 1.9(g)]. So, the partial order on the pseudo MV-algebra $([0, a]; \oplus, \lambda_a, \rho_a, 0, a)$ is defined by $x \preccurlyeq y$ iff $\lambda_a(x) \oplus y = a$ iff $y \oplus \rho_a(x) = a$. Therefore, from (8.26)–(8.27) and (E3), we have for each $x \in [0, a]$

$$\min_{\leq}\{z \in [0, a]: z \oplus x = a\} = \lambda_a(x) = \min_{\preccurlyeq}\{z \in [0, a]: z \oplus x = a\},$$

$$\min_{\leq}\{z \in [0, a]: x \oplus z = a\} = \rho_a(x) = \min_{\preccurlyeq}\{z \in [0, a]: x \oplus z = a\},$$

where $\min_{\leq}$ and $\min_{\preccurlyeq}$ are minimal taken with respect to $\leq$ in M and $\preccurlyeq$ taken in $[0, a]$, respectively. It is possible to show that

$$x \leq y \Leftrightarrow x \preccurlyeq y, \tag{8.28}$$

in addition, $(M; \oplus, 0)$ is naturally ordered, see Exercise 8.8.5.

In an analogy with pseudo MV-algebras, we can define a total binary operation $\odot$ on pseudo EMV-algebras.

Proposition 8.7.1 ([130, Proposition 3.3]) *Let* $\mathbf{M} = (M; \vee, \wedge, \oplus, 0)$ *be a pseudo EMV-algebra. For all* $x, y \in M$, *we define*

$$x \odot y = \rho_a(\lambda_a(y) \oplus \lambda_a(x)), \tag{8.29}$$

where $a \in \mathrm{Idm}(M)$ *and* $x, y \in [0, a]$. *Then* $x \odot y$ *is correctly defined and it does not depend on* $a \in \mathrm{Idm}(M)$, *and*

$$x \odot y = \lambda_a(\rho_a(y) \oplus \rho_a(x)). \tag{8.30}$$

In addition, if $x, y \in M$, $x \leq y$, *then*

$$y \odot \lambda_a(x) = y \odot \lambda_b(x), \quad \rho_a(x) \odot y = \rho_b(x) \odot y \tag{8.31}$$

for all idempotents a, b *of* M *with* $x, y \leq a, b$, *and*

$$y = (y \odot \lambda_a(x)) \oplus x = x \oplus (\rho_a(x) \odot y). \tag{8.32}$$

If $x, y \in [0, a]$ *for some idempotent* $a \in M$, *then*

$$x \odot \lambda_a(y) = x \odot \lambda_a(x \wedge y), \quad \rho_a(y) \odot x = \rho_a(x \wedge y) \odot x, \tag{8.33}$$

and

$$(x \odot \lambda_a(y)) \oplus (x \wedge y) = x = (x \wedge y) \oplus (\rho_a(y) \odot x). \tag{8.34}$$

Finally, if $x, y \leq a \in \mathrm{Idm}(M)$, *then*

$$((x \oplus y) \odot \lambda_a(x)) \oplus x = x \oplus y = y \oplus (\rho_a(y) \odot (x \oplus y)) \tag{8.35}$$

and if $x \leq a \in \mathrm{Idm}(M)$, then

$$x \odot \lambda_a(x) = 0 = \rho_a(x) \odot x. \tag{8.36}$$

The notions of an ideal, a maximal ideal, and a normal ideal are defined similarly to pseudo MV-algebras. Let M_1 and M_2 be pseudo EMV-algebras. A mapping $f : M_1 \to M_2$ is said to be a **homomorphism** of pseudo EMV-algebras if f preserves $\vee, \wedge, \oplus, 0$ and, for each idempotent $a \in \mathrm{Idm}(M_1)$ and for each $x \in [0, a]$, $f(\lambda_a(x)) = \lambda_{f(a)}(f(x))$ and $f(\rho_a(x)) = \rho_{f(a)}(f(x))$.

Unfortunately, the class of pseudo EMV-algebras is neither a variety nor a quasivariety since even the class of EMV-algebras is not such a structure because it is not closed under forming subalgebras with respect to the original operations $\vee, \wedge, \oplus$, and 0, see [129, Theorem 3.11]. Indeed, take the Chang MV-algebra $M = \mathbf{\Gamma}(\mathbb{Z} \overrightarrow{\times} \mathbb{Z}, (1, 0))$. It defines an EMV-algebra with top element having a unique maximal ideal I, namely $I = \{(0, n) \colon n \geq 0\}$. The set I is closed under $\vee, \wedge, \oplus, 0$, so $I = (I; \vee, \wedge, \oplus, 0)$ is a subalgebra of M, but I is not an EMV-subalgebra of the EMV-algebra M because it does not have enough idempotent elements. Therefore, the pseudo EMV-algebras form a more general form, called a **q-variety**, which is analogous to a variety as it was studied in [130, Theorem 4.4].

8.7.1   Representation of Pseudo EMV-Algebras

We show that every pseudo EMV-algebra M is either a pseudo EMV-algebra with top element or it can be embedded into a pseudo EMV-algebra N with top element as a maximal and normal ideal of N, [131, Theorem 6.4]. It generalizes an analogous result known for generalized Boolean algebras [87].

We start with the following result.

Theorem 8.7.2 ([130]) *Let $M = (M; \vee, \wedge, \oplus, 0)$ be a proper pseudo EMV-algebra. If there is a pseudo EMV-algebra N with top element such that M can be embedded into N. There is a pseudo EMV-subalgebra N_0 of N, N_0 with top element, such that M can be embedded into N_0 as a maximal and normal ideal of N_0.*

Proof Let N be a pseudo EMV-algebra with top element 1 such that M can be embedded into N. Without loss of generality, we can assume that M is a pseudo EMV-subalgebra of N. We define

$$N_0 = \{x \in N \colon \text{ either } x = x_0 \in M \text{ or } x = \rho_1(x_0) \text{ for some } x_0 \in M\}.$$

Claim. If $x_0 \in M$ and $x_0 \le a \in \mathrm{Idm}(M)$, then $\lambda_1^2(x_0), \rho_1^2(x_0) \in M$. Moreover, $\lambda_1^2(x_0) = \lambda_a^2(x_0) =: \varphi_\lambda(x_0)$, $\rho_1^2(x_0) = \rho_a^2(x_0) =: \varphi_\rho(x_0)$, and $\varphi_\lambda(x_0)$ and $\varphi_\rho(x_0)$ do not depend on $a \ge x_0$.

Let $a \in \mathrm{Idm}(M)$ be an idempotent such that $x_0 \le a$. By [130, Proposition 3.3(iii)], we have $\lambda_1(x_0) = \lambda_a(x_0) \oplus \lambda_1(a)$, so that by (8.33) and (8.31), we have

$$
\begin{aligned}
\lambda_1^2(x_0) &= \lambda_1(\lambda_a(x_0) \oplus \lambda_1(a)) = a \odot \lambda_1(\lambda_a(x_0)) \\
&= a \odot (\lambda_a^2(x_0) \oplus \lambda_1(a)) = a \wedge (\lambda_a^2(x_0) \vee \lambda_1(a)) \\
&= (a \wedge \lambda_a^2(x_0)) \vee (a \odot \lambda_1(a)) = a \wedge \lambda_a^2(x_0) \\
&= \lambda_a^2(x_0) = \varphi_\lambda(x_0) \in M.
\end{aligned}
$$

In the same way we establish $\rho_1^2(x_0) = \varphi_\rho(x_0) \in M$, and $\varphi_\lambda(x_0)$ and $\varphi_\rho(x_0)$ do not depend on $a \in \mathrm{Idm}(M)$ such that $x_0 \le a$.

We show that N_0 defines a pseudo EMV-algebra with top element.

Since $(N; \oplus, \lambda_1, \rho_1, 0, 1)$ is a pseudo MV-algebra that is termwise equivalent to the pseudo EMV-algebra N, we will use $x^- := \lambda_1(x)$ and $x^\sim := \rho_1(x)$, $x \in N$. First we note that according to Claim, we have $x_0^{--} = (x_0^{--})^\sim = \varphi_\lambda(x_0)^\sim \in N_0$ for each $x_0 \in M$, so that, N_0 is closed under $^-$ and $^\sim$.

Clearly, $M \subseteq N_0$ and $1 \in N_0$. Let $x, y \in N_0$. We show that $x \oplus y \in N_0$. There are four cases. Case (i): $x = x_0$, $y = y_0 \in M$. Then $x \vee y, x \wedge y, x \oplus y \in N_0$.

Case (ii): $x = x_0^\sim$, $y = y_0^\sim$ for some $x_0, y_0 \in M$. Then $x \vee y = x_0^\sim \vee y_0^\sim = (x_0 \wedge y_0)^\sim$, $x \wedge y = (x_0 \vee y_0)^\sim$ and $x \oplus y = x_0^\sim \oplus y_0^\sim = (y_0 \odot x_0)^\sim \in N_0$.

Case (iii): $x = x_0$ and $y = y_0^\sim$ for some $x_0, y_0 \in M$. Then

$$
x \oplus y = x_0 \oplus y_0^\sim = (y_0 \odot x_0^-)^\sim = (y_0 \odot (x_0 \wedge y_0)^-)^\sim = (y_0 \odot \lambda_b(x_0 \wedge y_0))^\sim,
$$

where b is an idempotent of M such that $x_0, y_0 \le b$; for the last equality we use equality (8.31) of Proposition 8.7.1. Using again Proposition 8.7.1, we have $y_0 \odot \lambda_b(x_0 \wedge y_0) \in M$ so that $x \oplus y \in N_0$.

Case (iv): $x = x_0^\sim$ and $y = y_0$ for some $x_0, y_0 \in M$. There is $b \in \mathrm{Idm}(M)$ such that $x_0, y_0 \le b$. Check

$$
\begin{aligned}
x \oplus y = x_0^\sim \oplus y_0 &= (y_0^\sim \odot x_0^{\sim\sim})^- = (y_0^\sim \odot \varphi_\rho(x_0))^- = \big((y_0 \wedge \varphi_\rho(x_0))^\sim \odot \varphi_\rho(x_0)\big)^- \\
&= \big((\rho_b(y_0 \wedge \varphi_\rho(x_0)) \vee b^\sim) \odot \varphi_\rho(x_0)\big)^{--\sim} = \big(\rho_b(y_0 \wedge \varphi_\rho(x_0)) \odot \varphi_\rho(x_0)\big)^{--\sim} \\
&= \big(\varphi_\lambda(\rho_b(y_0 \wedge \varphi_\rho(x_0)) \odot \varphi_\rho(x_0))\big)^\sim \in N_0.
\end{aligned}
$$

Now let $x_0 \in M$. Then $x_0^\sim \in N_0 \setminus M$ and if we set $y_0 = x_0^{\sim-} = \varphi_\lambda(x_0) \in M$, then $x_0^- = y_0^\sim \in N_0 \setminus M$. Claim implies $x_0^{\sim\sim} = \varphi_\rho(x_0) \in M$ and $x_0^{--} = \varphi_\lambda(x_0) \in M$. Hence, $(N_0; \oplus, ^-, ^\sim, 0, 1)$ is a pseudo MV-algebra, so that $(N_0; \vee, \wedge, \oplus, 0)$ is its termwise equivalent pseudo EMV-algebra with top element.

The set M is a proper subset of N_0 and M is closed under $\oplus$. Let $y \le x_0 \in M$. Then y cannot be from $N_0 \setminus M$, otherwise $1 \in M$. Hence, $y \in M$ and M is an ideal of N_0. If we take $y \in N_0 \setminus M$, then the ideal I of N_0 generated by $M \cup \{y\}$ contains 1, so that $I = N_0$ which establishes M is a maximal ideal of N_0. To prove normality of M in N_0, let $y \oplus x_0 \in y \oplus M$, $y \in N_0$. It is sufficient to assume $y = y_0^\sim$ for some $y_0 \in M$. Then $y_0^\sim \oplus x_0 = (y_0^\sim \oplus x_0) \vee y_0^\sim = ((y_0^\sim \oplus x_0) \odot y_0) \oplus y_0^\sim$. Since $(y_0^\sim \oplus x_0) \odot y_0 \le y_0 \in M$, we have $(y_0^\sim \oplus x_0) \odot y_0 \in M$, so that $y_0^\sim \oplus M \subseteq M \oplus y_0^\sim$. In a similar way, we prove the opposite inclusion. $\qquad\square$

Theorem 8.7.3 (Basic Representation Theorem) *Every pseudo EMV-algebra M is either a pseudo EMV-algebra with top element or M can be embedded into a pseudo EMV-algebra N_0 with top element as a maximal and normal ideal of N_0. In the second case, every element $x \in N_0$ is either from the image of M or there is a unique x_0 from the image of M such that $x = \rho_1(x_0)$.*

Proof If M has a top element, then the statement holds trivially.

Thus, assume that M is a proper pseudo EMV-algebra. We show that M can be embedded into a pseudo EMV-algebra with top element:

For each idempotent $a \in \mathrm{Idm}(M)$, $([0, a]; \vee, \wedge, \oplus, 0)$ is a pseudo EMV-algebra with top element. Therefore, $N = \prod_{a \in \mathrm{Idm}(M)}[0, a]$ is a pseudo EMV-algebra with top element. The mapping $\varphi : M \to N$ defined by $\varphi(x) = (x \wedge a)_{a \in \mathrm{Idm}(M)}[0, a]$, $x \in M$, is an embedding of M into the pseudo EMV-algebra N with top element and it satisfies the condition of Theorem 8.7.2.

Applying Theorem 8.7.2, we have the assertion question. $\qquad\square$

8.7.2 Category of Proper Pseudo EMV-Algebras

We show that the system of proper pseudo EMV-algebras forms a category. We present one category of pseudo MV-algebras with a fixed maximal and normal ideal and a category of unital ℓ-groups with special properties that are mutually categorically equivalent.

Let $\mathcal{PPEMV}$ be the category of proper pseudo EMV-algebras whose objects are proper pseudo EMV-algebras and morphisms are homomorphisms of pseudo EMV-algebras. On the other hand, let $\mathcal{FPMV}$ be the category whose objects are pairs (M, I), where M is a pseudo MV-algebra with a fixed maximal and normal ideal I of M such that (i) I has enough idempotents, i.e., for each $x \in I$, there is an idempotent $a \in \mathrm{Idm}(I)$ such that $x \le a$, (ii) no $x \in I$ is a top element of I, and (iii) $I \cup I^\sim = M$, where $I^\sim := \{x^\sim : x \in I\}$. Morphisms $\phi : (M_1, I_1) \to (M_2, I_2)$ in the category $\mathcal{FPMV}$ are homomorphisms of pseudo MV-algebras $\phi : M_1 \to M_2$ such that $\phi(I_1) \subseteq I_2$. Then $I \cap I^\sim = \emptyset$: If $x = y^\sim$ for $x, y \in I$, then $1 = x^- \oplus x = y \oplus x \in I$.

For example, if M is the Chang MV-algebra $\mathbf{\Gamma}(\mathbb{Z} \overset{\rightarrow}{\times} \mathbb{Z}, (1, 0))$, then $I = \{(0, n) \colon n \geq 0\}$ is a unique maximal ideal of M but I does not have enough idempotents. So (M, I) does not belong to $\mathcal{FPMV}$.

Define a mapping $\mathbf{\Phi} \colon \mathcal{FPMV} \to \mathcal{PPEMV}$ as follows: For any object $(N, I) \in \mathcal{FPMV}$, let

$$\mathbf{\Phi}(N, I) := (I; \vee, \wedge, \oplus, 0),$$

and if (N_1, I_1) and (N_2, I_2) are objects of $\mathcal{FPMV}$ and $\phi \colon (N_1, I_1) \to (N_2, I_2)$ is a morphism, then

$$\mathbf{\Phi}(\phi)(x) := \phi(x), \quad x \in I_1.$$

Proposition 8.7.4 $\mathbf{\Phi}$ *is a well-defined functor that is faithful and full from the category* $\mathcal{FPMV}$ *into the category* $\mathcal{PPEMV}$.

Proof We start to show that $\mathbf{\Phi}$ is a well-defined functor. That is, we show if $\phi \colon (N_1, I_1) \to (N_2, I_2)$ is a morphism in the category $\mathcal{FPMV}$, then $\mathbf{\Phi}(\phi)$ is a morphism in $\mathcal{PPEMV}$. At any rate, the mapping $\mathbf{\Phi}(\phi)$ is, in fact, a homomorphism from the proper pseudo EMV-algebra determined by I_1 into the pseudo proper EMV-algebra determined by I_2.

Let ϕ_1 and ϕ_2 be two morphisms from (N_1, I_1) into (N_2, I_2) such that $\mathbf{\Phi}(\phi_1) = \mathbf{\Phi}(\phi_2)$. Then $\phi_1(x) = \phi_2(x)$ for each $x \in I_1$. If $x \in N_1 \setminus I_1$, there is a unique element $x_0 \in I_1$ such that $x = x_0^{\sim}$. Then $\phi_1(x) = \phi_1(x_0^{\sim}) = (\phi_1(x_0))^{\sim} = (\phi_2(x_0))^{\sim} = \phi_2(x)$ which entails $\phi_1 = \phi_2$, i.e., $\mathbf{\Phi}$ is a faithful functor.

Now, we show $\mathbf{\Phi}$ is a full functor: Let $h \colon I_1 \to I_2$ be a morphism from $\mathcal{PPEMV}$, i.e., h is a homomorphism. By Theorem 8.7.3, there are pseudo EMV-algebras N_1 and N_2 with top elements such that I_1 and I_2 can be embedded into N_1 and N_2, respectively, as their maximal ideals. We can assume that I_i defines a pseudo EMV-subalgebra $\mathbf{I}_i = (I_i; \vee, \wedge, \oplus, 0)$ of N_i for $i = 1, 2$. We claim that there is a morphism $\phi \colon (N_1, I_1) \to (N_2, I_2)$ such that $\mathbf{\Phi}(\phi) = h$. To show that, we extend h to a homomorphism of pseudo MV-algebras ϕ from N_1 into N_2 for some objects (N_1, I_1) and (N_2, I_2) from $\mathcal{FPMV}$. By the proof of Theorem 8.7.2, $N_1 = I_1 \cup I_1^{\sim}$ and $N_2 = I_2 \cup I_2^{\sim}$. If $x \in I_1$, we put $\phi(x) = h(x)$, and if $x \in N_1 \setminus I_1$, there is a unique element $x_0 \in I_1$ such that $x = x_0^{\sim}$. Then we put $\phi(x) = h(x_0)^{\sim}$. We have $\phi(1) = 1$, $\phi(x^{\sim}) = (\phi(x))^{\sim}, x \in I_1$. Since for $x \in I_1, x^- = \varphi_\lambda(x)^{\sim}$, we have $\phi(x^-) = \phi(\varphi_\lambda(x)^{\sim}) = h(\varphi_\lambda(x))^{\sim} = (\varphi_\lambda(h(x)))^{\sim} = \phi(x)^-$.

If $x \in N_1 \setminus I_1$, then $x = x_0^{\sim}$ for a unique $x_0 \in I_1$. Whence,

$$\phi(x^{\sim}) = \phi(x_0^{\sim\sim}) = \phi(\varphi_\rho(x_0)) = h(\varphi_\rho(x_0)) = \varphi_\rho(h(x_0)) = h(x_0)^{\sim\sim} = \phi(x)^{\sim},$$

$$\phi(x^-) = \phi(x_0^{\sim-}) = \phi(x_0) = h(x_0) = h(x_0)^{\sim-} = \phi(x_0^{\sim})^- = \phi(x)^-.$$

Now let $x, y \in N_1$. By Theorem 8.7.2, there are four cases: (1) $x, y \in I_1$, then clearly $\phi(x \oplus y) = \phi(x) \oplus \phi(y)$. (2) $x = x_0^{\sim}$ and $y = y_0^{\sim}$ for some $x_0, y_0 \in I_1$. Then $\phi(x \oplus y) = \phi(x_0^{\sim} \oplus y_0^{\sim}) = \phi((y_0 \odot x_0)^{\sim}) = (\phi(y_0 \odot x_0))^{\sim} = (\phi(y_0) \odot \phi(x_0))^{\sim} = \phi(x_0)^{\sim} \oplus \phi(y_0)^{\sim} = \phi(x) \oplus \phi(y)$. (3) $x = x_0$ and $y = y_0^{\sim}$ for some $x_0, y_0 \in I_1$. Let $b \in \mathrm{Idm}(I_1)$ be an idempotent such that $x_0, y_0 \leq b$. Using Proposition 8.7.1, we have

$$x \oplus y = x_0 \oplus y_0^{\sim} = (y_0 \odot x_0^{-})^{\sim} = (y_0 \odot (x_0 \wedge y_0)^{-})^{\sim} = (y_0 \odot \lambda_b(x_0 \wedge y_0))^{\sim},$$

which entails

$$\begin{aligned}
\phi(x \oplus y) &= \phi((y_0 \odot \lambda_b(x_0 \wedge y_0))^{\sim}) = \big(\phi(y_0 \odot \lambda_b(x_0 \wedge y_0))\big)^{\sim} \\
&= \big(\phi(y_0) \odot \phi(\lambda_b(x_0 \wedge y_0))\big)^{\sim} = \big(\phi(y_0) \odot \lambda_{\phi(b)}(\phi(x_0 \wedge y_0))\big)^{\sim} \\
&= \big(\phi(y_0) \odot \lambda_{\phi(b)}(\phi(x_0) \wedge \phi(y_0))\big)^{\sim} \\
&= \big(\phi(y_0) \odot (\phi(x_0) \wedge \phi(y_0))^{-}\big)^{\sim} = (\phi(y_0) \odot \phi(x_0)^{-})^{\sim} \\
&= \phi(x_0) \oplus \phi(y_0)^{\sim} = \phi(x) \oplus \phi(y).
\end{aligned}$$

(4) $x = x_0^{\sim}$ and $y = y_0$ for some $x_0, y_0 \in I_1$. As in Case (iv) of the proof of Theorem 8.7.2, we have

$$x \oplus y = (y_0 \odot \lambda_b(x_0 \wedge y_0))^{\sim}.$$

Applying ϕ to this equality, we obtain as in (3), $\phi(x \oplus y) = \phi(x) \oplus \phi(y)$.

Therefore, ϕ is a homomorphism of pseudo MV-algebras, which is an extension of h. Whence, $\Phi(\phi) = h$, $\phi(I_1) \subseteq I_2$, and Φ is a full functor. $\qquad\square$

Proposition 8.7.5 *The functor Φ from the category $\mathcal{FPMV}$ into the category $\mathcal{PPEMV}$ has a left-adjoint.*

Proof We show that, for a proper pseudo EMV-algebra M, there is a universal arrow $((N, I), f)$ i.e., (N, I) is an object in $\mathcal{FPMV}$ and f is a morphism from M into $\Phi(N, I) = I$ such that if (N', I') is an object from $\mathcal{PPEMV}$ and f' is a morphism from M into $\Phi(N', I')$, then there exists a unique morphism $f^* : (N, I) \to (N', I')$ such that $\Phi(f^*) \circ f = f'$.

By Basic Representation Theorem 8.7.3, there are a unique (up to isomorphism) pseudo EMV-algebra N with top element and an injective EMV-homomorphism $f : M \to N$ such that $f(M)$ is a maximal and normal ideal of N. We assert that $((N, I), f)$ is a universal arrow for M. Let (N', I') be an object from $\mathcal{FPMV}$ and let f' be a morphism from M into $\Phi(N', I')$. We can define a mapping $f^* : N \to N'$ such that $f^*(f(x)) := f'(x)$ if $x \in M$ and if $y \in N \setminus f(M)$, there is $y_0 \in M$ such that $y = (f(y_0))^{\sim}$, and we set $f^*(y) = (f'(y_0))^{\sim}$. Then $f^* : N \to N'$ is a unique homomorphism of pseudo MV-algebras such that $\Phi(f^*) \circ f = f'$.

Define a mapping $\Psi : \mathcal{PPEMV} \to \mathcal{FPMV}$ by $\Psi(M) := (N, I)$ whenever $((N, I), f)$ is a universal arrow for M and if $f' : M \to M'$ is an EMV-homomorphism, there is a unique morphism $f^* : (N, I) \to (N', I')$, where $\Phi(N', I') = M'$, then we define $\Psi(f') := f^*$. Using Theorem 8.7.3, we have that Ψ is a left-adjoint functor of the functor Φ. $\qquad\square$

This gives:

Theorem 8.7.6 *The functor* Φ *defines a categorical equivalence between the category* $\mathcal{FPMV}$ *and the category of proper pseudo EMV-algebras* $\mathcal{PPEMV}$.

In what follows, we present another categorical equivalence for the category $\mathcal{FPPM}$ of proper pseudo EMV-algebras with fixed maximal and normal ideals using a special category of unital ℓ-groups. We only outline ideas; see [131, Sect. 7] for details.

We denote by $\mathcal{FULG}$ the category of unital ℓ-groups not necessarily commutative with a fixed maximal ℓ-ideal with a special property: The objects of $\mathcal{FULG}$ are triples (G, u, I) such that (G, u) is a unital ℓ-group and I is a fixed maximal ℓ-ideal of (G, u) such that $\mathbf{G}^+ = I^+ \cup I^u$, where

$$I^u = \{(-y + nu) + x : x \in I^+, n \geq 1, y \in I, 0 \leq y < nu\}, \tag{8.37}$$

and the ideal $I_u = I \cap [0, u]$ of $\Gamma(G, u)$ has enough idempotent elements but no top element. If (G_1, u_1, I_1) and (G_2, u_2, I_2) are two objects of $\mathcal{FULG}$, then a mapping $f : (G_1, u_1, I_1) \to (G_2, u_2, I_2)$ is a morphism if f is a homomorphism of unital ℓ-groups such that $f(I_1) \subseteq I_2$.

Let us define a functor $\Gamma_I : \mathcal{FULG} \to \mathcal{FPMV}$ as follows: If (G, u, I) is an object of $\mathcal{FULG}$, then

$$\Gamma_I(G, u, I) := (\Gamma(G, u), I \cap [0, u]),$$

and if f is a morphism from an object (G_1, u_1, I_1) into another one (G_2, u_2, I_2), then

$$\Gamma_I(f)(x) := f(x), \quad x \in \Gamma(G, u).$$

Finally, using [131, Theorem 7.7], we have another categorical equivalence:

Theorem 8.7.7 *The functor* Γ_I *defines a categorical equivalence of the category* $\mathcal{FULG}$ *and the category* $\mathcal{FPMV}$.

Finally, we have:

Corollary 8.7.8 *The categories* $\mathcal{FULG}$, $\mathcal{FPMV}$ *and* $\mathcal{PPEMV}$ *are mutually categorically equivalent.*

Bibliographical Remarks and Suggestions for Further Study

The late Nineties were "pregnant" with ideas of non-commutative logic algebras. They started with the paper by Georgescu and Iorgulescu [152] who introduced pseudo MV-algebras and Rachůnek [237] with generalized MV-algebras. They yielded a deeper study of non-commutative generalization of many-valued reasoning. Besides basic properties of pseudo MV-algebras, their article [152] also contains ideas about the Pierce representation of pseudo MV-algebras. The representation of pseudo MV-algebras by unital ℓ-groups, [109], allowed us to join two different areas of mathematics, pseudo MV-algebras and unital ℓ-groups. This fundamental result was also re-established in [149, 255] in completely different ways; the first one using residuated lattices and the second one showing the bricks and pseudo MV-algebras are equivalent. Pseudo MV-algebras in a residuated setup are also investigated in [215]. In addition, [264] shows that pseudo-MV algebras can also be treated as L-algebras. More fresh results on pseudo MV-algebras on the background of L-algebras can be found in the paper by W. Rump, see [105, 246, 250]. A relationship between some classes of pseudo MV-algebras and non-commutative Bézout domains is presented in [115].

Using this representation, in [185], it is possible to find results showing a near relationship between equational classes of unital ℓ-groups and subvarieties of pseudo-MV-algebras. So-called good and bad infinitesimals are studied in [94]. Convex chains and convex subalgebras in a pseudo MV-algebra are mentioned in [184, 186, 187] describes weak (m, n)-distributivity of lattice ordered groups and of generalized MV-algebras. To know more about top varieties of pseudo MV-algebras and unital ℓ-groups, we recommend consulting with [172], and for different covers of the Boolean variety of unital ℓ-groups see [90, 91]. Ideals and congruences were mentioned in [72], and De Finetti's approach to states on algebraic logic can be found in [201]. Hájek, [166], used pseudo-MV-algebras to study fuzzy logics with non-commutative conjunctions. To gain interesting information about states on MV-algebras and pseudo-MV-algebras, we recommend consulting [112, 199].

8.8   Exercises

8.8.1 Let $\mathbf{G} = (G; \vee, \wedge, \cdot, ^{-1}, e)$ be an ℓ-group written in a multiplicative way with a strong unit u. Let endow the interval $[u^{-1}, e]$ with the binary operations $x \circ y = xy \vee u^{-1}$ and unary operations $x^- = ex^{-1}$, $x^\sim = x^{-1}e$, $x, y \in [u^{-1}, e]$. Show that $([u^{-1}, e]; \circ, ^-, ^\sim, u^{-1}, e)$ is a prototypical example of a pseudo MV-algebra.

8.8.2 Show that a pseudo MV-algebra is a good BL-algebra $\mathbf{A} = (A; \cdot, \backslash, /, \wedge, \vee, 0, 1)$ satisfying $x^{-\sim} = x = x^{\sim -}$ for all $x \in A$.

8.8.3 Do all steps of the proof of Theorem 8.5.2.

8.8.4 Let $K_{I,J}^{\lambda, \rho}(\mathbf{G})$ be a kite algebra corresponding to an ℓ-group $\mathbf{G}$. On $A = (\mathbf{G}^-)^I$, restrict operations of divisions, $/$, and $\backslash$, and show that A with these restrictions and $1 := (e^{-1})^I$ is a pseudo hoop, and it is a hoop if and only if $\mathbf{G}$ is Abelian.

8.8.5 Show that in each pseudo EMV-algebra M, (8.28) holds, and that M is naturally ordered.

8.8.6 Let $(M; \vee, \wedge, \oplus, 0)$ be a pseudo EMV-algebra, $a, b \in \mathrm{Idm}I(M)$ such that $a \le b$. Show that for each $x \in [0, a]$, the following statements hold:

    (i) $\lambda_b(a) = \rho_b(a)$ is an idempotent, and $\lambda_a(a) = 0 = \rho_a(a)$.

    (ii) $\lambda_a(x) = \lambda_b(x) \wedge a$ and $\rho_a(x) = \rho_b(x) \wedge a$.

    (iii) $\lambda_b(x) = \lambda_a(x) \oplus \lambda_b(a) = \lambda_b(a) \oplus \lambda_a(x)$ and $\rho_b(x) = \rho_a(x) \oplus \rho_b(a) = \rho_b(a) \oplus \rho_a(x)$.

    (iv) $\rho_a(\lambda_a(x)) = x = \lambda_a(\rho_a(x))$.

    (v) $\lambda_a(x) \le \lambda_b(x)$ and $\rho_a(x) \le \rho_b(x)$.

8.8.7 Complete the proof of Proposition 8.7.1.

Finite Pocrims of Order ≤ 5

As demonstrated in Proposition 2.5.11 and Theorem 2.5.12, pocrims and BCK-algebras satisfying condition (P) are, in fact, equivalent. Therefore, to fully characterize pocrims of order ≤ 5, it suffices to examine BCK-algebras of order ≤ 5 satisfying condition (P).

Building upon the pioneering work of H. Jiang, who developed a computer program for classifying finite BCK-algebras, J. Meng and X. X. Niu conducted a comprehensive study of the commutativity, implicativity, and positive implicativity of all BCK-algebras with an order of 5 or less. This section draws upon the valuable insights presented in the work of J. Meng and Y. B. Jun (see [214, pp. 242–278] or [266, Appendix A]).

To simplify calculations and expedite the review process, we will implement the following tips:

(i) Every finite pocrim is bounded, and so has both a least and top element (Lemma 2.2.2). Therefore, if a finite BCK-algebra has no least element, then it is not a pocrim.

(ii) Existence of $\min\{z \in X : x \leq y \to z\}$ for all elements x and y of a BCK-algebra $(X; \to, 1)$. From $x \leq y \to x$, we get $\{z \in X : x \leq y \to z\} \neq \emptyset$, so every linearly ordered BCK-algebra is a pocrim.

Pocrims of Order One

Clearly, up to isomorphism, there exists a unique pocrim of order 1, i.e., the trivial pocrim $H_1 = \{1\}$, where $1 \odot 1 = 1 \to 1 = 1$. Then the pocrim $\mathbf{H}_1 = (H_1; \odot, \to, 1)$ is isomorphic to $\mathbf{1}$, clearly.

© The Author(s), under exclusive license to Springer Nature Switzerland AG 2026
A. Dvurečenskij et al., *Hoop Algebras*, Frontiers in Mathematics,
https://doi.org/10.1007/978-3-032-11736-6_9

Pocrims of Order Two

Up to isomorphism, there exists a unique pocrim $\mathbf{H}_2$ of order 2, where $H_2 = \{a, 1\}$ and the operations are defined as follows (Table 9.1).

Note that $\mathbf{H}_2 = (H_2; \odot, \rightarrow, 1)$ is isomorphic to $\mathbf{2}$.

Pocrims of Order Three

According to [266, Appendix A], there are 3 BCK-algebras with 3 elements, while 2 of them are pocrims. Let $H_3 = \{a, b, 1\}$.

(1) $\mathbf{H}_{3,1} = (H_3; \odot, \rightarrow, 1)$ is a pocrim where $\odot$ and $\rightarrow$ are defined in the following Table 9.2.

(2) $\mathbf{H}_{3,2} = (H_3; \odot, \rightarrow, 1)$ is a pocrim where $\odot$ and $\rightarrow$ are defined in the following Table 9.3.

Table 9.1 Operations of $\mathbf{H}_2$

$\odot$	a	1		$\rightarrow$	a	1		
a	a	a		a	1	1		$a \le 1.$
1	a	1		1	0	1		

Table 9.2 Operations of $\mathbf{H}_{3,1}$

$\odot$	a	b	1		$\rightarrow$	a	b	1		
a	a	a	a		a	1	1	1		$a \le b \le 1.$
b	a	a	b		b	b	1	1		
1	a	b	1		1	a	b	1		

Table 9.3 Operations of $\mathbf{H}_{3,2}$

$\odot$	a	b	1		$\rightarrow$	a	b	1		
a	a	a	a		a	1	1	1		$a \le b \le 1.$
b	a	b	b		b	a	1	1		
1	a	b	1		1	a	b	1		

Pocrims of Order Four

According to [266, Appendix A], there are 14 BCK-algebras with 4 elements, while 7 of them are pocrims. Let $H_4 = \{a, b, c, 1\}$.

(1) $\mathbf{H}_{4,1} = (H_4; \odot, \rightarrow, 1)$ is a pocrim where $\odot$ and $\rightarrow$ are defined in the following Table 9.4.

(2) $\mathbf{H}_{4,2} = (H_4; \odot, \rightarrow, 1)$ is a pocrim where $\odot$ and $\rightarrow$ are defined in the following Table 9.5.

(3) $\mathbf{H}_{4,3} = (H_4; \odot, \rightarrow, 1)$ is a pocrim where $\odot$ and $\rightarrow$ are defined in the following Table 9.6.

(4) $\mathbf{H}_{4,4} = (H_4; \odot, \rightarrow, 1)$ is a pocrim where $\odot$ and $\rightarrow$ are defined in the following Table 9.7.

Table 9.4 Operations of $\mathbf{H}_{4,1}$

$\odot$	a	b	c	1		$\rightarrow$	a	b	c	1	
a	a	a	a	a		a	1	1	1	1	
b	a	a	a	b		b	c	1	1	1	$a \leq b \leq c \leq 1.$
c	a	a	a	c		c	c	c	1	1	
1	a	b	c	1		1	a	b	c	1	

Table 9.5 Operations of $\mathbf{H}_{4,2}$

$\odot$	a	b	c	1		$\rightarrow$	a	b	c	1	
a	a	a	a	a		a	1	1	1	1	
b	a	a	a	b		b	c	1	1	1	$a \leq b \leq c \leq 1.$
c	a	a	b	c		c	b	c	1	1	
1	a	b	c	1		1	a	b	c	1	

Table 9.6 Operations of $\mathbf{H}_{4,3}$

$\odot$	a	b	c	1		$\rightarrow$	a	b	c	1	
a	a	a	a	a		a	1	1	1	1	
b	a	a	a	b		b	c	1	1	1	$a \leq b \leq c \leq 1.$
c	a	a	c	c		c	b	b	1	1	
1	a	b	c	1		1	a	b	c	1	

Table 9.7 Operations of $\mathbf{H}_{4,4}$

$\odot$	a	b	c	1
a	a	a	a	a
b	a	a	b	b
c	b	b	c	c
1	a	b	c	1

$\rightarrow$	a	b	c	1
a	1	1	1	1
b	b	1	1	1
c	a	b	1	1
1	a	b	c	1

$a \leq b \leq c \leq 1.$

(5) $\mathbf{H}_{4,5} = (H_4; \odot, \rightarrow, 1)$ is a pocrim where $\odot$ and $\rightarrow$ are defined in the following Table 9.8.

(6) $\mathbf{H}_{4,6} = (H_4; \odot, \rightarrow, 1)$ is a pocrim where $\odot$ and $\rightarrow$ are defined in the following Table 9.9.

(7) $\mathbf{H}_{4,7} = (H_4; \odot, \rightarrow, 1)$ is a pocrim where $\odot$ and $\rightarrow$ are defined in the following Table 9.10.

Table 9.8 Operations of $\mathbf{H}_{4,5}$

$\odot$	a	b	c	1
a	a	a	a	a
b	a	b	a	b
c	a	a	c	c
1	a	b	c	1

$\rightarrow$	a	b	c	1
a	1	1	1	1
b	c	1	c	1
c	b	b	1	1
1	a	b	c	1

$a \leq b, c \leq 1.$

Table 9.9 Operations of $\mathbf{H}_{4,6}$

$\odot$	a	b	c	1
a	a	a	a	a
b	a	b	b	b
c	a	b	b	c
1	a	b	c	1

$\rightarrow$	a	b	c	1
a	1	1	1	1
b	a	1	1	1
c	a	c	1	1
1	a	b	c	1

$a \leq b \leq c \leq 1.$

Table 9.10 Operations of $\mathbf{H}_{4,7}$

$\odot$	a	b	c	1
a	a	a	a	a
b	a	b	b	b
c	a	b	c	c
1	a	b	c	1

$\rightarrow$	a	b	c	1
a	1	1	1	1
b	a	1	1	1
c	a	b	1	1
1	a	b	c	1

$a \leq b \leq c \leq 1.$

Pocrims of Order Five

According to [266, Appendix A], there are 88 BCK-algebras with 5 elements while 50 of them do not have top elements, so they are not pocrims, clearly. In addition, 26 of them are pocrims. Let $H_5 = \{a, b, c, d, 1\}$.

(1) $\mathbf{H}_{5,1} = (H_5; \odot, \rightarrow, 1)$ is a pocrim where $\odot$ and $\rightarrow$ are defined in the following Table 9.11.

(2) $\mathbf{H}_{5,2} = (H_5; \odot, \rightarrow, 1)$ is a pocrim where $\odot$ and $\rightarrow$ are defined in the following Table 9.12.

Table 9.11 Operations of $\mathbf{H}_{5,1}$

$\odot$	a	b	c	d	1
a	a	a	a	a	a
b	a	a	a	a	b
c	a	a	a	a	c
d	a	a	a	a	d
1	a	b	c	d	1

$\rightarrow$	a	b	c	d	1
a	1	1	1	1	1
b	d	1	1	1	1
c	d	d	1	1	1
d	d	d	d	1	1
1	a	b	c	d	1

$a \leq b \leq c \leq d \leq 1.$

Table 9.12 Operations of $\mathbf{H}_{5,2}$

$\odot$	a	b	c	d	1
a	a	a	a	a	a
b	a	a	a	a	b
c	a	a	a	a	c
d	a	a	a	b	d
1	a	b	c	d	1

$\rightarrow$	a	b	c	d	1
a	1	1	1	1	1
b	d	1	1	1	1
c	d	d	1	1	1
d	c	d	d	1	1
1	a	b	c	d	1

$a \leq b \leq c \leq d \leq 1.$

(3) $\mathbf{H}_{5,3} = (H_5; \odot, \rightarrow, 1)$ is a pocrim where $\odot$ and $\rightarrow$ are defined in the following Table 9.13.

(4) $\mathbf{H}_{5,4} = (H_5; \odot, \rightarrow, 1)$ is a pocrim where $\odot$ and $\rightarrow$ are defined in the following Table 9.14.

(5) $\mathbf{H}_{5,5} = (H_5; \odot, \rightarrow, 1)$ is a pocrim where $\odot$ and $\rightarrow$ are defined in the following Table 9.15.

(6) $\mathbf{H}_{5,6} = (H_5; \odot, \rightarrow, 1)$ is a pocrim where $\odot$ and $\rightarrow$ are defined in the following Table 9.16.

Table 9.13 Operations of $\mathbf{H}_{5,3}$

$\odot$	a	b	c	d	1		$\rightarrow$	a	b	c	d	1
a	a	a	a	a	a		a	1	1	1	1	1
b	a	a	a	a	b		b	d	1	1	1	1
c	a	a	a	b	c		c	c	d	1	1	1
d	a	a	b	b	d		d	b	d	d	1	1
1	a	b	c	d	1		1	a	b	c	d	1

$a \leq b \leq c \leq d \leq 1.$

Table 9.14 Operations of $\mathbf{H}_{5,4}$

$\odot$	a	b	c	d	1		$\rightarrow$	a	b	c	d	1
a	a	a	a	a	a		a	1	1	1	1	1
b	a	a	a	a	b		b	d	1	1	1	1
c	a	a	b	b	c		c	b	d	1	1	1
d	a	a	b	b	d		d	b	d	d	1	1
1	a	b	c	d	1		1	a	b	c	d	1

$a \leq b \leq c \leq d \leq 1.$

Table 9.15 Operations of $\mathbf{H}_{5,5}$

$\odot$	a	b	c	d	1		$\rightarrow$	a	b	c	d	1
a	a	a	a	a	a		a	1	1	1	1	1
b	a	a	a	a	b		b	d	1	1	1	1
c	a	a	a	a	c		c	d	d	1	1	1
d	a	a	a	a	d		d	d	d	d	1	1
1	a	b	c	d	1		1	a	b	c	d	1

$a \leq b, c \leq d \leq 1.$

Table 9.16 Operations of $\mathbf{H}_{5,6}$

$\odot$	a	b	c	d	1
a	a	a	a	a	a
b	a	a	a	a	b
c	a	a	a	a	c
d	a	a	a	c	d
1	a	b	c	d	1

$\rightarrow$	a	b	c	d	1
a	1	1	1	1	1
b	d	1	1	1	1
c	d	d	1	1	1
d	c	c	d	1	1
1	a	b	c	d	1

$a \le b \le c \le d \le 1.$

(7) $\mathbf{H}_{5,7} = (H_5; \odot, \rightarrow, 1)$ is a pocrim where $\odot$ and $\rightarrow$ are defined in the following Table 9.17.

(8) $\mathbf{H}_{5,8} = (H_5; \odot, \rightarrow, 1)$ is a pocrim where $\odot$ and $\rightarrow$ are defined in the following Table 9.18.

(9) $\mathbf{H}_{5,9} = (H_5; \odot, \rightarrow, 1)$ is a pocrim where $\odot$ and $\rightarrow$ are defined in the following Table 9.19.

Table 9.17 Operations of $\mathbf{H}_{5,7}$

$\odot$	a	b	c	d	1
a	a	a	a	a	a
b	a	a	a	a	b
c	a	a	a	a	c
d	a	a	a	d	d
1	a	b	c	d	1

$\rightarrow$	a	b	c	d	1
a	1	1	1	1	1
b	d	1	1	1	1
c	d	d	1	1	1
d	c	c	c	1	1
1	a	b	c	d	1

$a \le b \le c \le d \le 1.$

Table 9.18 Operations of $\mathbf{H}_{5,8}$

$\odot$	a	b	c	d	1
a	a	a	a	a	a
b	a	a	a	b	b
c	a	a	a	b	c
d	a	b	b	d	d
1	a	b	c	d	1

$\rightarrow$	a	b	c	d	1
a	1	1	1	1	1
b	c	1	1	1	1
c	c	d	1	1	1
d	a	c	c	1	1
1	a	b	c	d	1

$a \le b \le c \le d \le 1.$

Table 9.19 Operations of $\mathbf{H}_{5,9}$

$\odot$	a	b	c	d	1
a	a	a	a	a	a
b	a	a	a	a	b
c	a	a	a	c	c
d	a	a	c	d	d
1	a	b	c	d	1

$\rightarrow$	a	b	c	d	1
a	1	1	1	1	1
b	d	1	1	1	1
c	c	c	1	1	1
d	b	b	c	1	1
1	a	b	c	d	1

$a \leq b \leq c \leq d \leq 1.$

(10) $\mathbf{H}_{5,10} = (H_5; \odot, \rightarrow, 1)$ is a pocrim where $\odot$ and $\rightarrow$ are defined in the following Table 9.20.

(11) $\mathbf{H}_{5,11} = (H_5; \odot, \rightarrow, 1)$ is a pocrim where $\odot$ and $\rightarrow$ are defined in the following Table 9.21.

(12) $\mathbf{H}_{5,12} = (H_5; \odot, \rightarrow, 1)$ is a pocrim where $\odot$ and $\rightarrow$ are defined in the following Table 9.22.

Table 9.20 Operations of $\mathbf{H}_{5,10}$

$\odot$	a	b	c	d	1
a	a	a	a	a	a
b	a	a	a	b	b
c	a	a	a	c	c
d	a	b	c	d	d
1	a	b	c	d	1

$\rightarrow$	a	b	c	d	1
a	1	1	1	1	1
b	c	1	1	1	1
c	c	c	1	1	1
d	a	b	c	1	1
1	a	b	c	d	1

$a \leq b \leq c \leq d \leq 1.$

Table 9.21 Operations of $\mathbf{H}_{5,11}$

$\odot$	a	b	c	d	1
a	a	a	a	a	a
b	a	a	a	b	b
c	a	a	b	c	c
d	a	b	c	d	d
1	a	b	c	d	1

$\rightarrow$	a	b	c	d	1
a	1	1	1	1	1
b	c	1	1	1	1
c	b	c	1	1	1
d	a	b	c	1	1
1	a	b	c	d	1

$a \leq b \leq c \leq d \leq 1.$

Table 9.22 Operations of $\mathbf{H}_{5,12}$

$\odot$	a	b	c	d	1
a	a	a	a	a	a
b	a	a	a	a	b
c	a	a	c	c	c
d	a	a	c	c	d
1	a	b	c	d	1

$\rightarrow$	a	b	c	d	1
a	1	1	1	1	1
b	d	1	1	1	1
c	b	b	1	1	1
d	b	b	d	1	1
1	a	b	c	d	1

$a \le b \le c \le d \le 1.$

(13) $\mathbf{H}_{5,13} = (H_5; \odot, \rightarrow, 1)$ is a pocrim where $\odot$ and $\rightarrow$ are defined in the following Table 9.23.

(14) $\mathbf{H}_{5,14} = (H_5; \odot, \rightarrow, 1)$ is a pocrim where $\odot$ and $\rightarrow$ are defined in the following Table 9.24.

(15) $\mathbf{H}_{5,15} = (H_5; \odot, \rightarrow, 1)$ is a pocrim where $\odot$ and $\rightarrow$ are defined in the following Table 9.25.

Table 9.23 Operations of $\mathbf{H}_{5,13}$

$\odot$	a	b	c	d	1
a	a	a	a	a	a
b	a	a	b	b	b
c	a	b	c	c	c
d	a	b	c	c	d
1	a	b	c	d	1

$\rightarrow$	a	b	c	d	1
a	1	1	1	1	1
b	b	1	1	1	1
c	a	b	1	1	1
d	a	b	d	1	1
1	a	b	c	d	1

$a \le b \le c \le d \le 1.$

Table 9.24 Operations of $\mathbf{H}_{5,14}$

$\odot$	a	b	c	d	1
a	a	a	a	a	a
b	a	a	a	a	b
c	a	a	c	c	c
d	a	a	c	c	d
1	a	b	c	d	1

$\rightarrow$	a	b	c	d	1
a	1	1	1	1	1
b	d	1	d	1	1
c	b	b	1	1	1
d	b	b	d	1	1
1	a	b	c	d	1

$a \le b, c \le d \le 1.$

Table 9.25 Operations of $\mathbf{H}_{5,15}$

$\odot$	a	b	c	d	1		$\to$	a	b	c	d	1	
a	a	a	a	a	a		a	1	1	1	1	1	
b	a	a	b	b	b		b	b	1	1	1	1	$a \leq b \leq c \leq d \leq 1.$
c	a	b	c	c	c		c	a	b	1	1	1	
d	a	b	c	d	d		d	a	b	c	1	1	
1	a	b	c	d	1		1	a	b	c	d	1	

(16) $\mathbf{H}_{5,16} = (H_5; \odot, \to, 1)$ is a pocrim where $\odot$ and $\to$ are defined in the following Table 9.26.

(17) $\mathbf{H}_{5,17} = (H_5; \odot, \to, 1)$ is a pocrim where $\odot$ and $\to$ are defined in the following Table 9.27.

(18) $\mathbf{H}_{5,18} = (H_5; \odot, \to, 1)$ is a pocrim where $\odot$ and $\to$ are defined in the following Table 9.28.

Table 9.26 Operations of $\mathbf{H}_{5,16}$

$\odot$	a	b	c	d	1		$\to$	a	b	c	d	1	
a	a	a	a	a	a		a	1	1	1	1	1	
b	a	a	a	a	b		b	d	1	1	1	1	$a \leq b \leq c \leq d \leq 1.$
c	a	a	c	c	c		c	b	b	1	1	1	
d	a	a	c	d	d		d	b	b	c	1	1	
1	a	b	c	d	1		1	a	b	c	d	1	

Table 9.27 Operations of $\mathbf{H}_{5,17}$

$\odot$	a	b	c	d	1		$\to$	a	b	c	d	1	
a	a	a	a	a	a		a	1	1	1	1	1	
b	a	a	a	b	b		b	c	1	1	1	1	$a \leq b \leq c \leq d \leq 1.$
c	a	a	c	c	c		c	b	b	1	1	1	
d	a	b	c	d	d		d	a	b	c	1	1	
1	a	b	c	d	1		1	a	b	c	d	1	

Table 9.28 Operations of $\mathbf{H}_{5,18}$

$\odot$	a	b	c	d	1
a	a	a	a	a	a
b	a	b	b	b	b
c	a	b	c	c	c
d	a	b	c	c	d
1	a	b	c	d	1

$\rightarrow$	a	b	c	d	1
a	1	1	1	1	1
b	a	1	1	1	1
c	a	b	1	1	1
d	a	b	d	1	1
1	a	b	c	d	1

$a \leq b \leq c \leq d \leq 1.$

Table 9.29 Operations of $\mathbf{H}_{5,19}$

$\odot$	a	b	c	d	1
a	a	a	a	a	a
b	a	b	b	b	b
c	a	b	c	c	c
d	a	b	c	d	d
1	a	b	c	d	1

$\rightarrow$	a	b	c	d	1
a	1	1	1	1	1
b	a	1	1	1	1
c	a	b	1	1	1
d	a	b	c	1	1
1	a	b	c	d	1

$a \leq b \leq c \leq d \leq 1.$

(19) $\mathbf{H}_{5,19} = (H_5; \odot, \rightarrow, 1)$ is a pocrim where $\odot$ and $\rightarrow$ are defined in the following Table 9.29.

(20) $\mathbf{H}_{5,20} = (H_5; \odot, \rightarrow, 1)$ is a pocrim where $\odot$ and $\rightarrow$ are defined in the following Table 9.30

(21) $\mathbf{H}_{5,21} = (H_5; \odot, \rightarrow, 1)$ is a pocrim where $\odot$ and $\rightarrow$ are defined in the following Table 9.31.

Table 9.30 Operations of $\mathbf{H}_{5,20}$

$\odot$	a	b	c	d	1
a	a	a	a	a	a
b	a	b	b	b	b
c	a	b	b	b	c
d	a	b	b	d	d
1	a	b	c	d	1

$\rightarrow$	a	b	c	d	1
a	1	1	1	1	1
b	a	1	1	1	1
c	a	d	1	1	1
d	a	c	c	1	1
1	a	b	c	d	1

$a \leq b \leq c \leq d \leq 1.$

Table 9.31 Operations of $\mathbf{H}_{5,21}$

$\odot$	a	b	c	d	1		$\rightarrow$	a	b	c	d	1
a	a	a	a	a	a		a	1	1	1	1	1
b	a	b	b	b	b		b	a	1	1	1	1
c	a	b	b	c	c		c	a	c	1	1	1
d	a	b	c	d	d		d	a	b	c	1	1
1	a	b	c	d	1		1	a	b	c	d	1

$a \le b \le c \le d \le 1.$

(22) $\mathbf{H}_{5,22} = (H_5; \odot, \rightarrow, 1)$ is a pocrim where $\odot$ and $\rightarrow$ are defined in the following Table 9.32.

(23) $\mathbf{H}_{5,23} = (H_5; \odot, \rightarrow, 1)$ is a pocrim where $\odot$ and $\rightarrow$ are defined in the following Table 9.33.

(24) $\mathbf{H}_{5,24} = (H_5; \odot, \rightarrow, 1)$ is a pocrim where $\odot$ and $\rightarrow$ are defined in the following Table 9.34.

Table 9.32 Operations of $\mathbf{H}_{5,22}$

$\odot$	a	b	c	d	1		$\rightarrow$	a	b	c	d	1
a	a	a	a	a	a		a	1	1	1	1	1
b	a	b	b	b	b		b	a	1	1	1	1
c	a	b	b	b	c		c	a	d	1	d	1
d	a	b	b	d	d		d	a	c	c	1	1
1	a	b	c	d	1		1	a	b	c	d	1

$a \le b \le c, d \le 1.$

Table 9.33 Operations of $\mathbf{H}_{5,23}$

$\odot$	a	b	c	d	1		$\rightarrow$	a	b	c	d	1
a	a	a	a	a	a		a	1	1	1	1	1
b	a	b	b	b	b		b	a	1	1	1	1
c	a	b	b	b	c		c	a	d	1	1	1
d	a	b	b	b	d		d	a	d	d	1	1
1	a	b	c	d	1		1	a	b	c	d	1

$a \le b \le c \le d \le 1.$

Table 9.34 Operations of $\mathbf{H}_{5,24}$

$\odot$	a	b	c	d	1
a	a	a	a	a	a
b	a	b	b	b	b
c	a	b	b	b	c
d	a	b	b	c	d
1	a	b	c	d	1

$\rightarrow$	a	b	c	d	1
a	1	1	1	1	1
b	a	1	1	1	1
c	a	d	1	1	1
d	a	c	d	1	1
1	a	b	c	d	1

$a \le b \le c \le d \le 1.$

Table 9.35 Operations of $\mathbf{H}_{5,25}$

$\odot$	a	b	c	d	1
a	a	a	a	a	a
b	a	b	b	b	b
c	a	b	b	b	c
d	a	b	b	c	d
1	a	b	c	d	1

$\rightarrow$	a	b	c	d	1
a	1	1	1	1	1
b	a	1	1	1	1
c	a	d	1	1	1
d	a	c	d	1	1
1	a	b	c	d	1

$a \le b \le c \le d \le 1.$

Table 9.36 Operations of $\mathbf{H}_{5,26}$

$\odot$	a	b	c	d	1
a	a	a	a	a	a
b	a	b	a	b	b
c	a	a	c	c	c
d	a	b	c	d	d
1	a	b	c	d	1

$\rightarrow$	a	b	c	d	1
a	1	1	1	1	1
b	c	1	c	1	1
c	b	b	1	1	1
d	a	b	c	1	1
1	a	b	c	d	1

$a \le b, c \le d \le 1.$

(25) $\mathbf{H}_{5,25} = (H_5; \odot, \rightarrow, 1)$ is a pocrim where $\odot$ and $\rightarrow$ are defined in the following Table 9.35.

(26) $\mathbf{H}_{5,26} = (H_5; \odot, \rightarrow, 1)$ is a pocrim where $\odot$ and $\rightarrow$ are defined in the following Table 9.36.

At the end of this section, it is a worth note that there are 129 pocrims of order 6. For more details see "Chapman University's Math Structures" in the following link: https://mathcs. chapman.edu.

References

1. Aaly Kologani, M., Borzooei, R.A.: On ideal theory of hoops. Math. Bohem. **145**(2), 141–162 (2020)
2. Aaly Kologani, M., Borzooei, R.A., Kim, H.S.: Graphs based on hoop algebras. Mathematics **7**(4), 362 (2019)
3. Aaly Kologani, M., Borzooei, R.A., Kouhestani, N.: On (semi) topological hoops. Quasigroups Related Syst. **27**(2), 161–174 (2019)
4. Aaly Kologani, M., Jun, Y.B., Borzooei, R.A.: Regular and Boolean elements in hoops and constructing Boolean algebras using regular filters. Analele ştiinţifice ale Universităţii" Ovidius" Constanţa. Seria Matematică 31(2), 5–22 (2023)
5. Aaly Kologani, M., Kouhestani, N., Borzooei, R.A.: On topological semi-hoops. Quasigroups Related Syst. **37**(2), 165–179 (2017)
6. Abad, M., Castaño, D.N., Varela, J.P.D.: MV-closures of Wajsberg hoops and applications. Algebra Univers. **64**, 213–230 (2010)
7. Aglianò, P.: Varieties of BL-algebras I, revisited. Soft. Comput. **21**, 153–163 (2017)
8. Aglianò, P.: Varieties of BL-algebras III: Splitting algebras. Stud. Logica. **107**(6), 1235–1259 (2019)
9. Aglianò, P.: Splittings in subreducts of hoops. Stud. Logica. **110**(5), 1155–1187 (2022)
10. Aglianò, P.: Quasivarieties of Wajsberg hoops. Fuzzy Sets Syst. **465**, 108514 (2023)
11. Aglianò, P., Ferreirim, I.M.A., Montagna, F.: Basic hoops: An algebraic study of continuous t-norms. Stud. Logica. **87**, 73–98 (2007)
12. Aglianò, P., Galatos, N., Marcos, M.A., et al.: Almost minimal varieties of commutative residuated lattices. Int. J. Algebra Comput. **34**(5), 807–836 (2024)
13. Aglianò, P., Montagna, F.: Varieties of BL-algebras I: general properties. J. Pure Appl. Algebra **181**(2–3), 105–129 (2003)
14. Aglianò, P., Montagna, F.: Varieties of BL-algebras II. Stud. Logica. **106**, 721–737 (2018)
15. Aglianò, P., Panti, G.: Geometrical methods in Wajsberg hoops. J. Algebra **256**(2), 352–374 (2002)
16. Aglianò, P., Ugolini, S.: Strictly join irreducible varieties of residuated lattices. J. Log. Comput. **32**(1), 32–64 (2022)

© The Editor(s) (if applicable) and The Author(s), under exclusive license to Springer Nature Switzerland AG 2026

A. Dvurečenskij et al., *Hoop Algebras*, Frontiers in Mathematics,
https://doi.org/10.1007/978-3-032-11736-6

17. Aglianò, P., Ugolini, S.: Projectivity in (bounded) commutative integral residuated lattices. Algebra Univers. **84**(1), 2 (2023)
18. Aglianò, P., Ugolini, S.: Structural and universal completeness in algebra and logic. Ann. Pure Appl. Logic **175**(3), 103391 (2024)
19. Aguzzoli, S., Bianchi, M.: On linear varieties of MTL-algebras. Soft. Comput. **23**(7), 2129–2146 (2019)
20. Amer, K.: Equationally complete classes of commutative monoids with monus. Algebra Univers. **18**, 129–131 (1984)
21. Anderson, M., Feil, T.: Lattice-Ordered Groups: An Introduction, vol. 4. Kluwer Academic Publishers, Springer, Dordrecht (1988)
22. Arthan, R., Oliva, P.: On pocrims and hoops (2014). arXiv:1404.0816
23. Bakhshi, M.: Prime $\mathcal{L}$-ideal spaces in hoop algebras. Soft. Comput. **27**(2), 629–644 (2023)
24. Belluce, L.P., Di Nola, A., Gerla, B.: Perfect MV-algebras and their logic. Appl. Categ. Struct. **15**, 135–151 (2007)
25. Belluce, L.P., Di Nola, A., Gerla, B.: Abelian ℓ-groups with strong unit and perfect MV-algebras. Order **25**, 387–401 (2008)
26. Berman, J., Blok, W.J.: Free Łukasiewicz and hoop residuation algebras. Stud. Logica. **77**, 153–180 (2004)
27. Bianchi, M.: The variety generated by all the ordinal sums of perfect MV-chains. Stud. Logica. **101**, 11–29 (2013)
28. Birkhoff, G.: Applications of lattice algebra. Math. Proc. Cambridge Philos. Soc. **30**(2), 115–122 (1934)
29. Birkhoff, G.: Lattice Theory, 3rd edn., vol. 25. American Mathematical Society Colloquium Publications (1973)
30. Blok, W.J., Ferreirim, I.M.A.: Hoops and their implicational reducts. Banach Center Publ. **28**(1), 219–230 (1993)
31. Blok, W.J., Ferreirim, I.M.A.: On the structure of hoops. Algebra Univers. **43**, 233–257 (2000)
32. Blok, W.J., Pigozzi, D.: On the structure of varieties with equationally definable principal congruences III. Algebra Univers. **32**, 545–608 (1994)
33. Blok, W.J., Raftery, J.G.: Varieties of commutative residuated integral pomonoids and their residuation subreducts. J. Algebra **190**(2), 280–328 (1997)
34. Blok, W.J., Van Alten, C.J.: The finite embeddability property for residuated lattices, pocrims and BCK-algebras. Algebra Univers. **48**(3), 253–271 (2002)
35. Blyth, T.S.: Lattices and Ordered Algebraic Structures. Springer, London (2005)
36. Boicescu, V., Filipoiu, A., Georgescu, G., Rudeanu, S.: Łukasiewicz-Moisil Algebras. Elsevier (1991)
37. Borzooei, R.A., Aaly Kologani, M.: Filter theory of hoop-algebras. J. Adv. Res. Pure Math. **6**(4), 72–86 (2014)
38. Borzooei, R.A., Aaly Kologani, M.: Stabilizer topology of hoops. Algebraic Struct. Appl. **1**(1), 35–48 (2014)
39. Borzooei, R.A., Aaly Kologani, M.: Local and perfect semihoops. J. Intell. Fuzzy Syst. **29**(1), 223–234 (2015)
40. Borzooei, R.A., Aaly Kologani, M., Ahn. S.: Product of ideals in hoops. Soft Comput. **27**(17), 11961–11971 (2023)
41. Borzooei, R.A., Aaly Kologani, M., Rezaei, G.: Radical of filters on hoops. Iran. J. Fuzzy Syst. **20**(7), 127–143 (2023)
42. Borzooei, R.A., Aaly Kologani, M., Xin, X.L., Jun, Y.B.: On annihilators in hoops. Soft. Comput. **26**(15), 6969–6980 (2022)
43. Borzooei, R.A., Aaly Kologani, M., Zahiri, O.: State hoops. Math. Slovaca **67**(1), 1–16 (2017)

44. Borzooei, R.A., Alavi, S., Aaly Kologani, M., Ahn, S.S.: Folding theory applied to pseudo-hoops. J. Intell. Fuzzy Syst. **39**(1), 1381–1390 (2020)

45. Borzooei, R.A., Dvurečenskij, A., Zahiri, O.: State BCK-algebras and state-morphism BCK-algebras. Fuzzy Sets Syst. **244**, 86–105 (2014)

46. Borzooei, R.A., Jun, Y., Aaly Kologani, M.: Ultra deductive systems and (nilpotent) Boolean elements in hoops. J. Algebraic Hyperstruct. Logical Algebras **3**(1), 77–93 (2022)

47. Borzooei, R.A., Rezaei, G.R., Aaly Kologani, M., Jun, Y.B.: Soju filters in hoop algebras. Bull. Sect. Logic **50**(1), 97–123 (2021)

48. Bosbach, B.: Komplementäre Halbgruppen. Axiomatik und Arithmetik. Fundamenta Mathematicae **64**, 257–287 (1969)

49. Bosbach, B.: Komplementäre Halbgruppen. Kongruenzen and Quotienten. Fundamenta Mathematicae **69**, 1–14 (1970)

50. Bosbach, B.: Concerning semiclans. Arch. Math. **37**, 316–324 (1981)

51. Bosbach, B.: Concerning cone algebras. Algebra Univers. **15**, 58–66 (1982)

52. Bosbach, B.: Residuation groupoids-again. RM **53**(1–2), 27–51 (2009)

53. Bosbach, B.: Divisibility groupoids-again. RM **57**(3), 257–285 (2010)

54. Botur, M., Chajda, I., Halaš, R., Kühr, J., Paseka, J., et al.: Algebraic Methods in Quantum Logic. Palacký University (2014)

55. Botur, M., Dvurečenskij, A.: State-morphism algebras–General approach. Fuzzy Sets Syst. **218**, 90–102 (2013)

56. Botur, M., Kowalski, T.: Kites and representations of pseudo MV-algebras. Fuzzy Sets Syst. **455**, 158–182 (2023)

57. Bou, F., Paoli, F., Ledda, A., Freytes, H.: On some properties of quasi-MV algebras and $\sqrt{\prime}$ quasi-MV algebras. Part II. Soft Comput. **12**(4), 341–352 (2008)

58. Bou, F., Paoli, F., Ledda, A., Spinks, M., Giuntini, R.: The logic of quasi-MV algebras. J. Log. Comput. **20**(2), 619–643 (2010)

59. Büchi, J.R., Owens, T.M.: Complemented Monoids and Hoops. Unpublished Manuscript

60. Buhagiar, D., Chetcuti, E., Dvurečenskij, A.: Loomis-Sikorski representation of monotone σ-complete effect algebras. Fuzzy Sets Syst. **157**(5), 683–690 (2006)

61. Burris, S., Sankappanavar, H.P.: A Course in Universal Algebra. Grad, vol. 78. Texts Math. Springer, New York, NY (1981)

62. Busaniche, M.: Decomposition of BL-chains. Algebra Univers. **52**(4), 519–525 (2004)

63. Busaniche, M., Montagna, F.: Hájek's logic BL and BL-algebras. In: Handbook of Mathematical Fuzzy Logic, vol. 1, pp. 355–447. College Publications, London (2011)

64. Buşneag, D.: A note on deductive systems of a Hilbert algebra. Kobe J. Math. **2**, 29–35 (1985)

65. Buşneag, D., (ed.).: Categories of Algebraic Logic. Academiei Romane (2006)

66. Buşneag, D., Ghiţă, M.: Some latticial properties of Hilbert algebras. Bull. mathématique de la Société des Sciences Mathématiques de Roumanie **53**(101)(2), 87–107 (2010)

67. Buşneag, D., Piciu, D.: Semi-g-filters, Stonean filters, MTL-filters, divisible filters, BL-filters and regular filters in residuated lattices. Iran. J. Fuzzy Syst. **13**(1), 145–160 (2016)

68. Bělohlávek, R.: Fuzzy Relational Systems: Foundations and Principles, vol. 20. Springer, New York, NY (2012)

69. Chajda, I., Halaš, R., Kühr, J.: Semilattice Structures, vol. 30. Heldermann Lemgo (2007)

70. Chang, C.C.: Algebraic analysis of many valued logics. Trans. Am. Math. Soc. **88**(2), 467–490 (1958)

71. Chang, C.C.: A new proof of the completeness of the Łukasiewicz axioms. Trans. Am. Math. Soc. **93**(1), 74–80 (1959)

72. Chen, W., Dudek, W.A.: Ideals and congruences in quasi-pseudo-MV algebras. Soft. Comput. **22**(12), 3879–3889 (2018)

73. Cignoli, R.L., d'Ottaviano, I.M., Mundici, D.: Algebraic Foundations of Many-Valued Reasoning, vol. 7. Kluwer Academic Publishers, Springer, Dordrecht (2000)

74. Cignoli, R.L., Torrens, A.: The poset of prime ℓ-ideals of an Abelian ℓ-group with a strong unit. J. Algebra **184**(2), 604–612 (1996)

75. Cignoli, R.L., Torrens, A.: Free cancellative hoops. Algebra Univers. **43**, 213–216 (2000)

76. Ciungu, L.C.: Algebras on subintervals of pseudo-hoops. Fuzzy Sets Syst. **160**(8), 1099–1113 (2009)

77. Ciungu, L.C.: On pseudo-BCK algebras with pseudo-double negation. Ann. Univ. Craiova-Math. Comput. Sci. Ser. **37**(1), 19–26 (2010)

78. Ciungu, L.C.: Bounded pseudo-hoops with internal states. Math. Slovaca **63**(5), 903–934 (2013)

79. Ciungu, L.C.: Non-Commutative Multiple-Valued Logic Algebras. Springer Monographs in Mathematics. Springer, Cham (2013)

80. Ciungu, L.C.: Involutive filters of pseudo-hoops. Soft. Comput. **23**, 9459–9476 (2019)

81. Ciungu, L.C.: Results in L-algebras. Algebra Univers. **82**(1), 7 (2021)

82. Ciungu, L.C., Dvurečenskij, A.: Measures, states and de Finetti maps on pseudo-BCK algebras. Fuzzy Sets Syst. **161**(22), 2870–2896 (2010)

83. Ciungu, L.C., Georgescu, G., Mureşan, C.: Generalized Bosbach states: Part I. Arch. Math. Logic **52**, 335–376 (2013)

84. Ciungu, L.C., Georgescu, G., Mureşan, C.: Generalized Bosbach states: Part II. Arch. Math. Logic **52**, 707–732 (2013)

85. Ciungu, L.C., Kuehr, J.: New probabilistic model for pseudo-BCK algebras and pseudo-hoops. J. Multiple-Valued Logic Soft Comput. 20 (2013)

86. Cohn, P.M.: Universal Algebra. Springer, Dordrecht (1981)

87. Conrad, P.F., Darnel, M.R.: Generalized Boolean algebras in lattice-ordered groups. Order **14**(4), 295–319 (1997)

88. Cornish, W.H.: BCK-algebras with a supremum. Math. Jpn. **27**, 53–73 (1982)

89. Darnel, M.R.: Theory of Lattice-Ordered Groups. Pure and Applied Mathematics. Marcel Dekker, New York (1995)

90. Darnel, M.R., Holland, W.C.: Solvable covers of the Boolean variety of unital ℓ-groups. Algebra Univers. **62**, 185–199 (2009)

91. Darnel, M.R., Holland, W.C.: More covers of the Boolean variety of unital ℓ-groups. Algebra Univers. **70**(2), 149–162 (2013)

92. Davey B.A., Priestley, H.A.: Introduction to Lattices and Order. Cambridge University Press (2002)

93. Di Nola, A., Dvurečenskij, A.: State-morphism MV-algebras. Ann. Pure Appl. Logic **161**(2), 161–173 (2009)

94. Di Nola, A., Dvurečenskij, A., Jakubík, J.: Good and bad infinitesimals, and states on pseudo MV-algebras. Order **21**, 293–314 (2004)

95. Di Nola, A., Dvurečenskij, A., Tsinakis, C.: Perfect GMV-algebras. Commun. Algebra **36**(4), 1221–1249 (2008)

96. Di Nola, A., Georgescu, G., Iorgulescu, A.: Pseudo-BL algebras: Part I. Multiple Valued Logic **8**(5/6), 673–716 (2002)

97. Di Nola, A., Georgescu, G., Iorgulescu, A.: Pseudo-BL algebras: Part II. Multiple Valued Logic **8**(5/6), 717–750 (2002)

98. Di Nola, A., Grigolia, R., Turunen, E.: Perfect MV-algebras. In: Fuzzy Logic of Quasi-Truth: An Algebraic Treatment, vol. 338, pp. 37–46. Springer, Cham (2016)

99. Di Nola, A., Grigolia, R., Turunen, E., et al.: Fuzzy Logic of Quasi-Truth: An Algebraic Treatment, Volume 338 of Studies in Fuzziness and Soft Computing. Springer (2016)

100. Di Nola, A., Lettieri, A.: Perfect MV-algebras are categorically equivalent to Abelian ℓ-groups. Stud. Logica. **53**, 417–432 (1994)

101. Di Nola, A., Lettieri, A.: Equational characterization of all varieties of MV-algebras. J. Algebra **221**(2), 463–474 (1999)
102. Di Nola, A., Lettieri, A.: Finite BL-algebras. Discret. Math. **269**(1–3), 93–112 (2003)
103. Di Nola, A., Leuştean, I.: Łukasiewicz logic and MV-algebras. In: Handbook of Mathematical Fuzzy Logic, vol. 2, pp. 469–583. College Publications, London (2011)
104. Diaconescu, D., Flaminio, T., Leuştean, I.: Lexicographic MV-algebras and lexicographic states. Fuzzy Sets Syst. **244**, 63–85 (2014)
105. Dietzel, C., Rump, W.: The structure group of a non-degenerate effect algebra. Algebra Univers. **81**, 1–30 (2020)
106. Dvurečenskij, A.: Loomis-Sikorski theorem for σ-complete MV-algebras and ℓ-groups. J. Aust. Math. Soc. **68**(2), 261–277 (2000)
107. Dvurečenskij, A.: On pseudo MV-algebras. Soft. Comput. **5**, 347–354 (2001)
108. Dvurečenskij, A.: States on pseudo MV-algebras. Stud. Logica. **68**, 301–327 (2001)
109. Dvurečenskij, A.: Pseudo MV-algebras are intervals in ℓ-groups. J. Aust. Math. Soc. **72**(3), 427–446 (2002)
110. Dvurečenskij, A.: Loomis-Sikorski theorem for monotone σ-complete effect algebras. J. Aust. Math. Soc. **79**(3), 305–318 (2005)
111. Dvurečenskij, A.: Aglianò-Montagna type decomposition of linear pseudo hoops and its applications. J. Pure Appl. Algebra **211**(3), 851–861 (2007)
112. Dvurečenskij, A.: States on pseudo effect algebras and integrals. Found. Phys. **41**(7), 1143–1162 (2011)
113. Dvurečenskij, A.: Riesz decomposition properties and the lexicographic product of po-groups. Soft. Comput. **20**(6), 2103–2117 (2016)
114. Dvurečenskij, A.: Representation of perfect and n-perfect pseudo effect algebras. Fuzzy Sets Syst. **455**, 19–34 (2023)
115. Dvurečenskij, A., Fuchs, L., Zahiri, O.: Non-commutative bézout domains and pseudo MV-algebras. J. Math. Anal. Appl. **549**(2), 129545 (2025)
116. Dvurečenskij, A., Giuntini, R., Kowalski, T.: On the structure of pseudo BL-algebras and pseudo hoops in quantum logics. Found. Phys. **40**, 1519–1542 (2010)
117. Dvurečenskij, A., Holland, W.C.: Top varieties of generalized MV-algebras and unital lattice-ordered groups. Commun. Algebra **35**(11), 3370–3390 (2007)
118. Dvurečenskij, A., Holland, W.C.: Covers of the Abelian variety of generalized MV-algebras. Commun. Algebra **37**(11), 3991–4011 (2009)
119. Dvurečenskij, A., Holland, W.C.: Komori's characterization and top varieties of GMV-algebras. Algebra Univers. **60**(1), 37–62 (2009)
120. Dvurečenskij, A., Kowalski, T.: On decomposition of pseudo BL-algebras. Math. Slovaca **61**(3), 307–326 (2011)
121. Dvurečenskij, A., Kowalski, T.: Kites and pseudo BL-algebras. Algebra Univers. **71**(3), 235–260 (2014)
122. Dvurečenskij, A., Kowalski, T., Montagna, F.: State morphism MV-algebras. Int. J. Approximate Reasoning **52**(8), 1215–1228 (2011)
123. Dvurečenskij, A., Pulmannová, S.: New Trends in Quantum Structures, vol. 516. Kluwer Academid Publishers, Dordrecht, and Ister Science, Bratislava, Springer, Dordrecht (2000)
124. Dvurečenskij, A., Rachůnek, J.: Probabilistic averaging in bounded $R\ell$-monoids. Semigroup Forum **72**, 191–206 (2006)
125. Dvurečenskij, A., Vetterlein, T.: Pseudoeffect algebras. I. Basic properties. Int. J. Theor. Phys. **40**, 685–701 (2001)
126. Dvurečenskij, A., Vetterlein, T.: Pseudoeffect algebras. II. Group representations. Int. J. Theor. Phys. **40**, 703–726 (2001)

127. Dvurečenskij, A., Zahiri, O.: When the lexicographic product of two po-groups has the Riesz decomposition property. Algebra Univers. **78**, 67–91 (2017)
128. Dvurečenskij, A., Zahiri, O.: The Loomis-Sikorski theorem for EMV-algebras. J. Aust. Math. Soc. **106**(2), 200–234 (2019)
129. Dvurečenskij, A., Zahiri, O.: On EMV-algebras. Fuzzy Sets Syst. **373**, 116–148 (2019)
130. Dvurečenskij, A., Zahiri, O.: Pseudo EMV-algebras. I. Basic properties. J. Appl. Logic–IfCoLog J. Logics Appl. 6, 1285–1327 (2019)
131. Dvurečenskij, A., Zahiri, O.: Pseudo EMV-algebras. II. Representation and states. J. Appl. Logic–IfCoLog J. Logics Appl. 6, 1329–1372 (2019)
132. Dvurečenskij, A., Zahiri, O.: MV-algebras and their corresponding Bézout domains. Commun. Algebra **52**(12), 5165–5179 (2024)
133. Dvurečenskij, A., Zahiri, O.: Hoops and domains. Fuzzy Sets Syst. **549**(2), 109404 (2025)
134. Dvurečenskij, A., Zahiri, O.: A representation of symmetric Wajsberg pseudo hoops. Fuzzy Sets Syst. **519**, 109541 (2025)
135. Dvurečenskij, A., Zahiri, O.: Some results on quasi MV-algebras and perfect quasi MV-algebras. In: Studia Logica, pp. 1–37 (2025)
136. Esteva, F., Godo, L.: Monoidal t-norm based logic: Towards a logic for left-continuous t-norms. Fuzzy Sets Syst. **124**(3), 271–288 (2001)
137. Esteva, F., Godo, L., Hájek, P., Montagna, F.: Hoops and fuzzy logic. J. Log. Comput. **13**(4), 532–555 (2003)
138. Etingof, P., Schedler, T., Soloviev, A.: Set-theoretical solutions to the quantum Yang-Baxter equation. Duke Math. J. **100**(2), 169–209 (1999)
139. Evans, T.: Some connections between residual finiteness, finite embeddability and the word problem. J. Lond. Math. Soc. **2**(1), 399–403 (1969)
140. Farahani, H., Zahiri, O.: Algebraic view of MTL-filters. Ann. Univ. Craiova-Math. Comput. Sci. Ser. **40**(1), 34–44 (2013)
141. Ferreirim, I.M.A.: On Varieties and Quasivarieties of Hoops and Their Reducts. Ph.D. thesis, University of Illinois at Chicago (1992)
142. Flaminio, T., Montagna, F.: An algebraic approach to states on MV-algebras. In: Proceedings of EUSFLAT'07, vol. 2, pp. 201–206 (2007)
143. Flaminio, T., Montagna, F.: MV-algebras with internal states and probabilistic fuzzy logics. Int. J. Approximate Reasoning **50**(1), 138–152 (2009)
144. Fleischer, I.: Every BCK-algebra is a set of residuables in an integral pomonoid. J. Algebra **119**(2), 360–365 (1988)
145. Font, J.M., Rodríguez, A.J., Torrens, A.: Wajsberg algebras. Stochastica **8**(1), 5–31 (1984)
146. Fu, Y.L., Xin, X.L., Wang, J.T.: State maps on semihoops. Open Math. **16**(1), 1061–1076 (2018)
147. Fuchs, L.: Partially Ordered Algebraic Systems. Pergamon Press, Oxford, London, New York (1963)
148. Galatos, N., Jipsen, P., Kowalski, T., Ono, H.: Residuated Lattices: An Algebraic Glimpse at Substructural Logics. Studies in Logic and the Foundations of Mathematics. Elsevier, Amsterdam (2007)
149. Galatos, N., Tsinakis, C.: Generalized MV-algebras. J. Algebra **283**(1), 254–291 (2005)
150. Georgescu, G.: Bosbach states on fuzzy structures. Soft. Comput. **8**, 217–230 (2004)
151. Georgescu, G., Iorgulescu, A.: Pseudo-MV algebras: A noncommutative extension of MV algebras. In: The Proceedings of the Fourth International Symposium on Economic Informatics, vol. 11(1), pp. 961–968. Bucharest, Romania (1999)
152. Georgescu, G., Iorgulescu, A.: Pseudo MV-algebras. Multiple Valued Logic **6**, 95–135 (2001)
153. Georgescu, G., Leuştean, L., Preoteasa, V.: Pseudo-hoops. J. Multiple-Valued Logic. Soft. Comput. **11**(1), 153–184 (2005)

154. Gerla, B., Russo, C., Spada, L.: Representation of perfect and local MV-algebras. Math. Slovaca **61**(3), 327–340 (2011)
155. Ghorbani, S.: Logic for abstract hoop twist-structures. Ann. Pure Appl. Logic **169**(10), 981–996 (2018)
156. Giuntini, R., Ledda, A., Paoli, F.: Expanding quasi-MV algebras by a quantum operator. Stud. Logica. **87**(1), 99–128 (2007)
157. Giustarini, V., Manfucci, F., Ugolini, S.: Free constructions in hoops via ℓ-groups. Studia Logica 1–49 (2024)
158. Glass, A.M., Holland, W.C.: Lattice-Ordered Groups: Advances and Techniques, vol. 48. Kluwer Academic Publishers, Springer, Dordrecht (2012)
159. Glass, A.M.W.: Partially Ordered Groups, vol. 7. World Scientific (1999)
160. Goodearl, K.R.: Partially Ordered Abelian Groups with Interpolation, vol. 20. American Mathematical Society (1986)
161. Grätzer, G.: General Lattice Theory, 2nd edn., vol. 52. Birkhäuser Basel (2003)
162. Grätzer, G.: Universal Algebra. Springer, New York, NY (2008)
163. Grätzer, G.: Lattice Theory: First Concepts and Distributive Lattices. A Series of Books in Mathematics. W. H. Freeman, San Francisco (2009)
164. Grätzer, G.: Lattice Theory: Foundation. Birkhäuser Basel (2011)
165. Grishin, V.N.: Impossibility of defining the class of L_0-algebras by means of identities. Matematicheskie Zametki **38**(5), 641–651 (1985)
166. Hájek, P.: Fuzzy logics with noncommutative conjuctions. J. Log. Comput. **13**(4), 469–479 (2003)
167. Hájek, P.: Metamathematics of Fuzzy Logic, vol. 4. Kluwer Academic Publishers, Springer, Dordrecht (2013)
168. Halaš, R., Botur, M.: On very true operators on pocrims. Soft. Comput. **13**(11), 1063–1072 (2009)
169. Haniková, Z.: Varieties generated by standard BL-algebras. Order **31**(1), 15–33 (2014)
170. Harrop, R.: On the existence of finite models and decision procedures for propositional calculi. Math. Proc. Cambridge Philos. Soc. **54**(1), 1–13 (1958)
171. Harrop, R.: Measures on clans and on MV-algebras. In: Pap, E. (ed.) Handbook of Measure Theory, vol. II, pp. 911–945. Elsevier Science, Amsterdam (2002)
172. Harrop, R.: Small varieties of lattice-ordered groups and MV-algebras. In: Chajda, I., (ed.) Proceedings of the Dresden Conference 2004 (AAA 68) and the Summer School on General Algebra 2004. Verlag Johannes Heyn, Klagenfurt, pp. 107–114 (2005)
173. Higgs, D.: Dually residuated commutative monoids with identity element as least element do not form an equational class. Math. Jpn. **29**, 69–75 (1984)
174. Hungerford, T.W.: Algebra, vol. 73. Springer, New York, NY (1974)
175. Imai, Y., Iséki, K.: On axiom systems of propositional calculi. I. Proc. Jpn. Acad. **41**(6), 436–439 (1965)
176. Imai, Y., Iséki, K.: On axiom systems of propositional calculi. XIV. Proc. Jpn. Acad. **42**(1), 19–22 (1966)
177. Iorgulescu, A.: Some direct ascendents of Wajsberg and MV algebras. Scientiae Mathematicae Japonicae **57**(3), 583–647 (2003)
178. Iorgulescu, A.: On BCK algebras-part Ia: An attempt to treat unitarily the algebras of logic. New algebras. J. Univ. Comput. Sci. **13**(11), 1628–1654 (2007)
179. Iorgulescu, A.: Algebras of Logic as BCK-algebras. Colecţia Informatică, Editura ASE (2008)
180. Iorgulescu, A.: On BCK algebras-part Ib: An attempt to treat unitarily the algebras of logic. New algebras. J. Univ. Comput. Sci. **14**(22), 3686–3715 (2008)
181. Iorgulescu, A.: Non-commutative Algebras: Pseudo-BCK Algebras versus m-pseudo-BCK Algebras, vol. 107. College Publications, Studies in Logic (2024)

182. Iséki, K.: An algebra related with a propositional calculus. Proc. Jpn. Acad. **42**(1), 26–29 (1966)
183. Iséki, K.: An introduction to the theory of BCK-algebras. Math. Jpn. **23**, 1–26 (1978)
184. Jakubík, J.: Convex chains in a pseudo MV-algebra. Czechoslov. Math. J. **53**(1), 113–125 (2003)
185. Jakubík, J.: On varieties of pseudo MV-algebras. Czechoslov. Math. J. **53**(4), 1031–1040 (2003)
186. Jakubík, J.: Weak (m, n)-distributivity of lattice ordered groups and of generalized MV-algebras. Soft. Comput. **10**, 119–124 (2006)
187. Jakubík, J.: Convex subalgebras of upper bounded GMV-algebras. Math. Slovaca **66**(2), 379–386 (2016)
188. Jipsen, P.: An overview of generalized basic logic algebras. Neural Netw. World **13**(5), 491–500 (2003)
189. Jipsen, P., Ledda, A., Paoli, F.: On some properties of quasi-MV algebras and $\sqrt{'}$ quasi-MV algebras. Part IV. Rep. Math. Logic **48**, 3–36 (2013)
190. Jipsen, P., Montagna, F.: On the structure of generalized BL-algebras. Algebra Univers. **55**(2), 227–238 (2006)
191. Jipsen, P., Tsinakis, C.: A survey of residuated lattices. In: Ordered Algebraic Structures: Proceedings of the Gainesville Conference. Sponsored by the University of Florida 28th February–3rd March, 2001, pp. 19–56. Springer (2002)
192. Kaarli, K., Pixley, A.F.: Polynomial Completeness in Algebraic Systems. Chapman and Hall/CRC (2000)
193. Khorami, R.T., Borumand Saeid, A.: Some unitary operators on hoop-algebras. Fuzzy Inf. Eng. **9**(2), 205–223 (2017)
194. Kish, M.S., Borzooei, R.A., Jabbari, S.H., Aaly Kologani, M.: A note on Noetherian and Artinian hoops. Discussiones Mathematicae: General Algebra Appl. **44**(1), 177–198 (2024)
195. Komori, Y.: Super- Łukasiewicz propositional logics. Nagoya Math. J. **84**, 119–133 (1981)
196. Kondo, M.: Some types of filters in hoops. In: 2011 41st IEEE International Symposium on Multiple-Valued Logic, pp. 50–53. IEEE (2011)
197. Kopytov, V.M., Medvedev, N.Y.: The Theory of Lattice-Ordered Groups, vol. 307. Kluwer Academic Publishers, Springer, Dordrecht (2013)
198. Kowalski, T., Paoli, F.: On some properties of quasi-MV algebras and $\sqrt{'}$ quasi-MV algebras. Part III. Rep. Math. Logic **45**, 161–199 (2010)
199. Kroupa, T.: Every state on semisimple MV-algebra is integral. Fuzzy Sets Syst. **157**(20), 2771–2782 (2006)
200. Kühr, J.: Pseudo-BCK Algebras and Related Structures. Ph.D. thesis, Univerzita Palackého v Olomouci (2007)
201. Kühr, J., Mundici, D.: De finetti theorem and borel states in [0, 1]-valued algebraic logic. Int. J. Approximate Reasoning **46**(3), 605–616 (2007)
202. Kuratowski, K.: Topology I, vol. 7. Mir, Moskva (In Russian) (1996)
203. Ledda, A., Konig, M., Paoli, F., Giuntini, R.: MV-algebras and quantum computation. Stud. Logica. **82**, 245–270 (2006)
204. Leuştean, L.: Representations of Many-Valued Algebras. Ph.D. thesis, University of Bucharest (2003)
205. Liu, M., Liu, H.: Pseudo-Ehoops. IAENG Int. J. Appl. Math. **55**(1), 90–103 (2025)
206. Lubomirsky, N., San Martín, H.J., Zuluaga Botero, W.J.: Relatively compatible operations in BCK-algebras and some related algebras. Logic J. IGPL **25**(3), 348–364 (2017)
207. Mac Lane, S.: Categories for the Working Mathematician, vol. 5. Springer Science & Business Media, New York, Heidelberg, Berlin (1971)
208. Mac Lane, S.: Categories for the Working Mathematician, 2nd edn., volume 5 of Graduate Texts in Mathematics. Springer New York, NY (2013)
209. Mal'tsev, A.I.: Multiplication of classes of algebraic systems. Sib. Math. J. **8**(2), 254–267 (1967)

210. Manfucci, F., Ugolini, S.: Free product hoops. In: Conference of the European Society for Fuzzy Logic and Technology, pp. 518–529. Springer (2023)

211. Marra, V., Mundici, D.: The Lebesgue state of a unital Abelian lattice-ordered group. J. Group Theory **10**, 655–684 (2007)

212. McNaughton, R.: A theorem about infinite-valued sentential logic. J. Symbolic Logic **16**(1), 1–13 (1951)

213. Meng, J., Jun, Y.B.: BCK-Algebras. Kyung Moon Sa Company, Seoul (1994)

214. Metcalfe, G., Montagna, F., Tsinakis, C.: Amalgamation and interpolation in ordered algebras. J. Algebra **402**, 21–82 (2014)

215. Metcalfe, G., Paoli, F., Tsinakis, C.: Residuated Structures in Algebra and Logic, volume 277 of Mathematical Surveys and Monographs. American Mathematical Society (2023)

216. Molkhasi, A., Shum, K.P.: Algebraic geometry over complete lattices and involutive pocrims. Discussiones Mathematicae-General Algebra Appl. **42**(2), 339–347 (2022)

217. Montagna, F., Noguera, C., Horčík, R.: On weakly cancellative fuzzy logics. J. Log. Comput. **16**(4), 423–450 (2006)

218. Montagna, F., Ugolini, S.: A categorical equivalence for product algebras. Stud. Logica. **103**(2), 345–373 (2015)

219. Motamed, S., Torkzadeh, L., Borumand Saeid, A., Mohtashamnia, N.: Radical of filters in BL-algebras. Math. Logic Q. **57**(2), 166–179 (2011)

220. Mundici, D.: Interpretation of AF C^*-algebras in Łukasiewicz sentential calculus. J. Funct. Anal. **65**(1), 15–63 (1986)

221. Mundici, D.: MV-algebras are categorically equivalent to bounded commutative BCK-algebras. Math. Jpn. **31**, 889–894 (1986)

222. Mundici, D.: A constructive proof of McNaughton's theorem in infinite-valued logic. J. Symbolic Logic **59**(2), 596–602 (1994)

223. Mundici, D.: Averaging the truth-value in Łukasiewicz logic. Stud. Logica. **55**, 113–127 (1995)

224. Mundici, D.: Tensor products and the Loomis-Sikorski theorem for MV-algebras. Adv. Appl. Math. **22**(2), 227–248 (1999)

225. Mundici, D.: Advanced Łukasiewicz Calculus and MV-Algebras, vol. 35. Springer, Dordrecht (2011)

226. Namdar, A., Borzooei, R.A.: Nodal filters in hoop algebras. Soft. Comput. **22**(21), 7119–7128 (2018)

227. Namdar, A., Borzooei, R.A.: Special hoop algebras. Italian J. Pure Appl. Math. **39**, 334–349 (2018)

228. Namdar, A., Borzooei, R.A., Borumand Saeid, A., Aaly Kologani, M.: Some results in hoop algebras. J. Intell. Fuzzy Syst. **32**(3), 1805–1813 (2017)

229. Nganteu Tchikapa, C.: A coalgebraic study of BL-algebras. Ph.D. thesis, Université de Yaoundé (2022)

230. Ono, H., Komori, Y.: Logics without the contraction rule. J. Symbolic Logic **50**(1), 169–201 (1985)

231. Pałasiński, M.: An embedding theorem for BCK-algebras. Math. Semin. Notes, Kobe Univ. **10**, 749–751 (1982)

232. Panti, G.: Invariant measures in free MV-algebras. Commun. Algebra **36**(8), 2849–2861 (2008)

233. Paoli, F., Ledda, A., Giuntini, R., Freytes, H.: On some properties of quasi-MV algebras and $\sqrt{\prime}$ quasi-MV algebras. Rep. Math. Logic **44**, 31–63 (2009)

234. Paoli, F., Ledda, A., Spinks, M., Freytes, H., Giuntini, R.: Logics from $\sqrt{\prime}$ quasi-MV algebras. Int. J. Theor. Phys. **50**(12), 3882–3902 (2011)

235. Pei, D.: A unified normal residuated based logic system and its completeness. Southeast Asian Bull. Math. **28**(6), 1089–1098 (2004)

236. Porta, H.: Sur un théorème de Skolem. Comptes Rendus Académie des Scince Paris **256**, 5262–5264 (1963)
237. Rachůnek, J.: A non-commutative generalization of MV-algebras. Czechoslov. Math. J. **52**(2), 255–273 (2002)
238. Raftery, J.G.: On the variety generated by involutive pocrims. Rep. Math. Logic **42**, 71–86 (2007)
239. Ravindran, K.: On a Structure Theory of Effect Algebras. Ph.D. thesis, Kansas State University (1996)
240. Roman, S.: Lattices and Ordered Sets. Springer, New York, NY (2008)
241. Ruan, X., Liu, X.: On the cancellation problem for L-algebras. J. Algebra Appl. **23**(14), 2550006 (2024)
242. Rudeanu, S.: Sets and Ordered Structures. Bentham Science Publishers (2012)
243. Rump, W.: A decomposition theorem for square-free unitary solutions of the quantum Yang-Baxter equation. Adv. Math. **193**(1), 40–55 (2005)
244. Rump, W.: L-algebras, self-similarity, and ℓ-groups. J. Algebra **320**(6), 2328–2348 (2008)
245. Rump, W.: A general glivenko theorem. Algebra Univers. **61**, 455–473 (2009)
246. Rump, W.: The category of L-algebras. Theory Appl. Categories **39**(21), 598–624 (2023)
247. Rump, W.: The geometry of discrete L-algebras. Adv. Geom. **23**(4), 543–565 (2023)
248. Rump, W.: L-algebras and topology. J. Algebra Appl. **22**(2), 2350034 (2023)
249. Rump, W.: Prime L-algebras and right-angled Artin groups. Semigroup Forum **106**(2), 481–503 (2023)
250. Rump, W.: Non-commutative effect algebras, L-algebras, and local duality. Math. Slovaca **74**(2), 451–468 (2024)
251. Rump, W.: A complete invariant system for Noetherian BL-algebras and more general L-algebras. Ann. Pure Appl. Logic **176**, 103580 (2025)
252. Rump, W., Vendramin, L.: The prime spectrum of an L-algebra. Proc. Am. Math. Soc. **152**, 3197–3207 (2024)
253. Rump, W., Yang, Y.C.: A note on Bosbach's cone algebras. Stud. Logica. **98**, 375–386 (2011)
254. Sikorski, R.: Boolean Algebras, vol. 2. Springer, Berlin, Heidelberg, New York (1964)
255. Subrahmanyam, N.V.: Bricks and pseudo MV-algebras are equivalent. Math. Slovaca **58**(2), 131–142 (2008)
256. Ugolini, S.: The polyhedral geometry of Wajsberg hoops. J. Log. Comput. **34**(3), 557–589 (2024)
257. Wang, G.J.: Non-Classical Mathematical Logic and Approximate Reasoning. Science in China Press, Beijing (2000)
258. Ward, M., Dilworth, R.P.: Residuated lattices. In: The Dilworth Theorems: Selected Papers of Robert P. Dilworth, pp. 317–336. Springer Birkhäuser Boston (1939)
259. Wroński, A.: BCK-algebras do not form a variety. Math. Jpn. **28**, 211–213 (1983)
260. Xie, F., Liu, H.: Ehoops. J. Multiple-Valued Logic Soft. Comput. **37**, 77–106 (2021)
261. Xu, Y., Ruan, D., Qin, K., Liu, J.: Lattice-valued logic. Stud. Fuzziness Soft Comput. **132**, 207–257 (2003)
262. Yang, Y.C.: ℓ-groups and Bézout domains. Ph.D. thesis, Universitat Stuttgart (2006)
263. Yang, Y.C., Rump, W.: Bézout domains with nonzero unit radical. Commun. Algebra **38**(3), 1084–1092 (2010)
264. Yang, Y.C., Rump, W.: Pseudo-MV algebras as L-algebras. J. Multiple-Valued Logic Soft Comput. **19**(5), 621–632 (2012)
265. Yisheng, H.: BCI-Algebra, vol. 16. Science Press Beijing (2006)
266. Yutani, H.: The class of commutative BCK-algebras is equationally definable. Math. Semin. Notes **5**(2), 207–210 (1977)
267. Zahiri, O., Farahani, H.: n-fold filters of MTL-algebras. Afr. Mat. **25**(4), 1165–1178 (2014)

Index

© The Editor(s) (if applicable) and The Author(s), under exclusive license to Springer Nature Switzerland AG 2026
A. Dvurečenskij et al., *Hoop Algebras*, Frontiers in Mathematics,
https://doi.org/10.1007/978-3-032-11736-6

MIX
Papier aus verantwortungsvollen Quellen
Paper from responsible sources
FSC® C105338

FSC
www.fsc.org

If you have any concerns about our products,
you can contact us on
ProductSafety@springernature.com

In case Publisher is established outside the EU,
the EU authorized representative is:
Springer Nature Customer Service Center GmbH
Europaplatz 3, 69115 Heidelberg, Germany

Printed by Libri Plureos GmbH
in Hamburg, Germany